Eclipse on the Coral Sea: Cycle 24 Ascending

(GONG 2012, LWS/SDO-5, and SOHO 27)

Journal of Physics: Conference Series Volume 440

Palm Cove, Queensland, Australia
12-16 November 2012

ISBN: 978-1-62748-672-9
ISSN: 1742-6588

Printed from e-media with permission by:

Curran Associates, Inc.
57 Morehouse Lane
Red Hook, NY 12571

Some format issues inherent in the e-media version may also appear in this print version.

Printed by Curran Associates, Inc. (2013)

For permission requests, please contact the Institute of Physics
at the address below.

Institute of Physics
Dirac House, Temple Back
Bristol BS1 6BE UK

Phone: 44 1 17 929 7481
Fax: 44 1 17 920 0979

techtracking@iop.org

Additional copies of this publication are available from:

Curran Associates, Inc.
57 Morehouse Lane
Red Hook, NY 12571 USA
Phone: 845-758-0400
Fax: 845-758-2634
Email: curran@proceedings.com
Web: www.proceedings.com

TABLE OF CONTENTS

ACTIVITY IN THE SOLAR INTERIOR

SEISMOLOGY

SMALL SCALE TRANSIENTS AND SPACE WEATHER

LARGE SCALE TRANSIENTS AND SPACE WEATHER

MAGNETIC FIELD MEASUREMENT AND EXTRAPOLATION

SOLAR CYCLE EFFECTS ON ACTIVITY

CONVECTION AND WAVES

WAVES THROUGH THE SOLAR ATMOSPHERE

ATMOSPHERIC WAVES DRIVEN BY P-MODES

SUNSPOTS

Eclipse on the Coral Sea: Cycle 24 Ascending Preface

Paul Cally[1], **Robert Erdélyi**[2] **and Aimee Norton**[3]

[1] Monash Centre for Astrophysics and School of Mathematical Sciences, Monash University, Clayton 3800, Victoria, Australia
[2] Solar Physics and Space Plasma Research Centre (SP^2RC), University of Sheffield, Hicks Building, Hounsfield Road, Sheffield S3 7RH, UK
[3] HEPL Solar Physics, Stanford University, CA 94305-4085 USA

E-mail: paul.cally@monash.edu, robertus@sheffield.ac.uk, aanorton@stanford.edu

A total solar eclipse is the most spectacular and awe-inspiring astronomical phenomenon most people will ever see in their lifetimes. Even hardened solar scientists draw inspiration from it. The eclipse with 2 minutes totality in the early morning of 14 November 2012 (local time) drew over 120 solar researchers (and untold thousands of the general public) to the small and picturesque resort town of Palm Cove just north of Cairns in tropical north Queensland, Australia, and they were rewarded when the clouds parted just before totality to reveal a stunning solar display.

The eclipse was also the catalyst for an unusually broad and exciting conference held in Palm Cove over the week 12–16 November. *Eclipse on the Coral Sea: Cycle 24 Ascending* served

Journal of Physics: Conference Series **440** (2013) 011001 doi:10.1088/1742-6596/440/1/011001

as GONG 2012, LWS/SDO-5, and SOHO 27, indicating how widely it drew on the various sub-communities within solar physics. Indeed, as we neared the end of the ascending phase of the peculiar Solar Cycle 24, it was the perfect time to bring the whole community together to discuss our Sun's errant recent behaviour, especially as Cycle 24 is the first to be fully observed by the *Solar Dynamics Observatory* (SDO). The whole-Sun perspective was a driving theme of the conference, with the cycle probed from interior (helioseismology), to atmosphere (the various lines observed by the Atmospheric Imaging Assemble (AIA) aboard SDO, the several instruments on *Hinode*, and other modern observatories), and beyond (CMEs etc.). The quality of the presentations was exceptional, and the many speakers are to be commended for pitching their talks to the broad community present.

These proceedings draw from the invited and contributed oral presentations and the posters exhibited in Palm Cove. They give an (incomplete) snapshot of the meeting, illustrating its broad vistas. The published contributions are organized along the lines of the conference sessions, as set out in the Contents, leading off with a provocative view of Cycle 24 thus far from Sarbani Basu. Other invited papers presented here include an appreciation of *Hinode's* view of solar activity as the cycle rises by Toshifumi Shimizu; a first taxonomy of magnetic tornadoes and chromospheric swirls by Sven Wedemeyer *et al.*; an analysis of *Hinode*/EIS observations of transient heating events; a timely re-examination of solar dynamo theory by Paul Charbonneau; an exciting teaser for the solar potential of the Murchison Widefield Array now operating in Western Australia by Steven Tingay *et al.*; an overview and critique of the state of nonlinear force-free magnetic field extrapolation theory and practice by Mike Wheatland and Stuart Gilchrist; and a masterful review of atmospheric MHD wave coupling to the Sun's internal *p*-mode oscillations by Elena Khomenko and Irantzu Calvo Santamaria. The many contributed papers published here are no less exciting.

All papers have been refereed to a high standard. The editors thank all the referees, drawn both from conference attendees and the wider community, who have taken their tasks very seriously and provided very detailed and helpful reports. Nearly all contributions have been substantially improved by the process.

We must also thank our financial sponsors. Both the Global Oscillations Network Group (GONG) and LWS/SDO were generous in their support, as were the School of Mathematical Sciences and the Monash Centre for Astrophysics (MoCA) at Monash University, Melbourne, and the Centre for Astronomy at James Cook University, Townsville. The Local Organizing Committee and the many students who assisted before and during the conference also deserve high praise for facilitating such a memorable meeting.

Peer review statement

All papers published in this volume of *Journal of Physics: Conference Series* have been peer reviewed through processes administered by the proceedings Editors. Reviews were conducted by expert referees to the professional and scientific standards expected of a proceedings journal published by IOP Publishing.

Eclipse on the Coral Sea: Cycle 24 Ascending (GONG 2012, LWS/SDO-5, and SOHO 27) IOP Publishing
Journal of Physics: Conference Series **440** (2013) 012001 doi:10.1088/1742-6596/440/1/012001

The peculiar solar cycle 24 — where do we stand?

Sarbani Basu

Department of Astronomy, Yale University, P. O. Box 20801, New Haven CT 06520-8101, U. S. A.

E-mail: `sarbani.basu@yale.edu`

Abstract. The minimum that preceded solar cycle 24 was unusual in its depth and duration. It was the quietest minimum recorded in the era of detailed data. Cycle 24 started off extremely slow and has continued to be weak. We review the conditions of the minimum that preceded cycle 24. We discuss ignored or missed signs that cycle 24 would not be normal, and finally comment on the behaviour of the cycle thus far.

1. Introduction

Solar cycle 24 has been very weak so far. It was preceded by an extremely quiet and long solar minimum. Data from the solar interior, the solar surface and the heliosphere all show that cycle 24 began from an unusual minimum and is unlike the cycles that preceded it.

We begin this review of where solar cycle 24 stands today with a look at the antecedents of this cycle, and examine why the minimum preceding the cycle is considered peculiar (§ 2). We then examine in § 3 whether we missed early signs that the cycle could be unusual. § 4 describes where cycle 24 is at today.

2. The peculiar minimum

The minimum that preceded solar cycle 24 was unusual in its depth — the 10.7 cm flux was the lowest recorded, while there were sunspot free days than have been recorded among recent cycles. The minimum was also of a much longer duration than any other minimum in recent history.

The minimum preceding the cycle showed other unusual characteristics. For instance, the polar fields were lower than those of previous cycles. In Fig. 1 we show the polar fields as observed by the Wilcox Solar Observatory. It is very clear that the fields were much lower than those at the minimum before cycle 22 and also smaller than the fields during the minimum before cycle 23. Unfortunately, the data do not cover a period much before cycle 21 maximum so we cannot compare the polar fields during the last minimum with those of even earlier minima. Other, more recent data sets, such as the Kitt Peak and MDI magnetograms, and they too also show that the polar fields were weak during the cycle 24 minimum compared with the cycle 23 minimum (de Toma 2011; Gopalswamy et al. 2012).

The structure of the solar corona was also quite different from what is expected during a normal minimum. As can be seen from the LASCO images shown in Fig. 2 the solar corona has the canonical solar-minimum structure during the cycle 23 minimum, but the coronal did not have a simple configuration of streamers in an equatorial belt as it was during the previous minimum in 1996.

Journal of Physics: Conference Series **440** (2013) 012001 doi:10.1088/1742-6596/440/1/012001

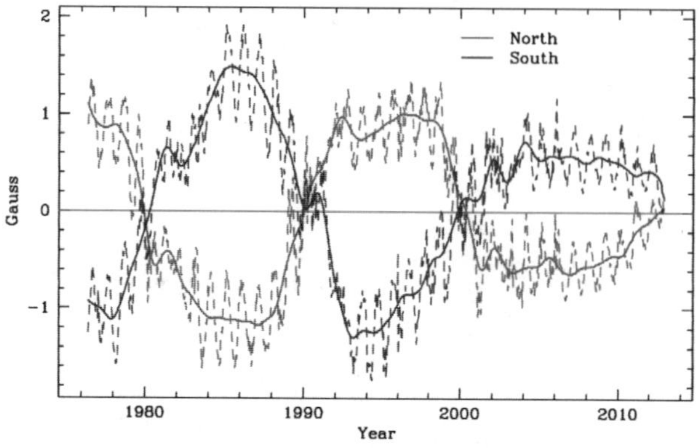

Figure 1. The variation of the polar magnetic fields as observed by the Wilcox Solar Observatory. The thick lines are the smoothed values

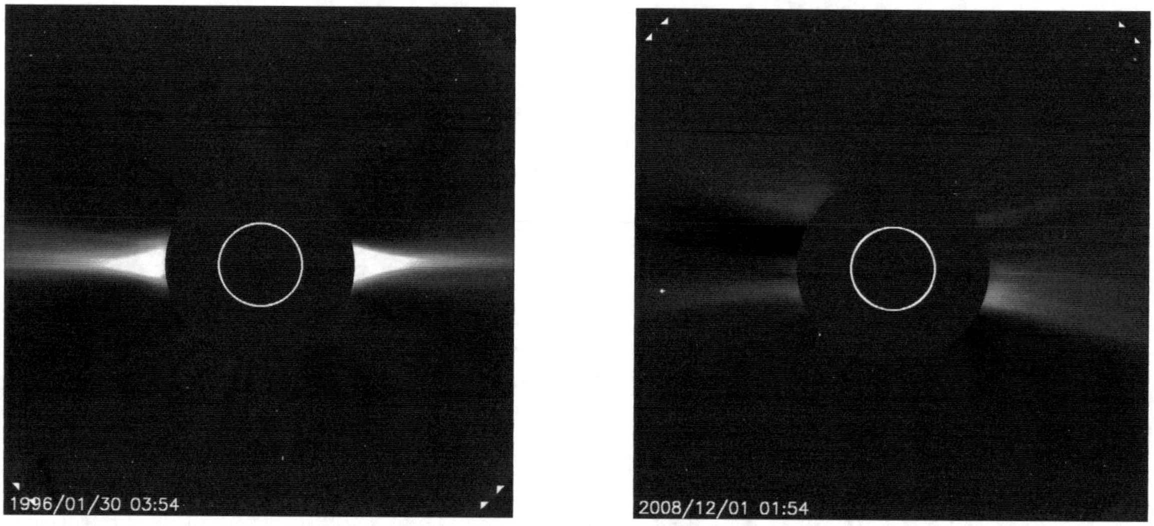

Figure 2. LASCO C2 images showing the solar corona in January 1996 (the minimum before cycle 23; left) and December 2009 (minimum before cycle 24; right). Note that the coronal structures are quite different, with the corona being confined to the equatorial plane during the cycle 23 minimum but with considerable latitudinal structure during the cycle 24 minimum. (Images courtesy of NASA.)

The above examples are just some of the unusual characteristics of the minimum that have been documented in the proceedings if the SOHO 23 workshop entitled "Understanding a Peculiar Solar Minimum" (ASPCS Vol, 428). Other noteworthy differences include the fact that the fast solar winds were confined to higher latitudes — using interplanetary scintillation measurements Manoharan (2012) showed that at 125 $R_\odot$ fast (> 600 km s^{-1}) winds were seen at latitudes $> 30°$ at the cycle 23 minimum, but these were confined to much higher latitudes during the cycle 24 minimum (see Fig. 3). De Toma (2011) showed that the speed distribution of solar winds at earth was also different. While during the cycle 23 minimum in 1996, the speed-distribution had a peak at 370 km/s, in 2009 the distribution peaked at 340 km/s. In 2007 and 2008 the distribution was bimodal with peaks at 350 and 600 km/s in 2007 and 330 and 600 km/s in 2008.

The differences between the cycle 24 minimum and the previous ones were not confined to phenomena exterior to the Sun, dynamics of the solar interior showed differences too. For instance, Basu & Antia (2010) showed that the nature of the meridional flow during the cycle 24

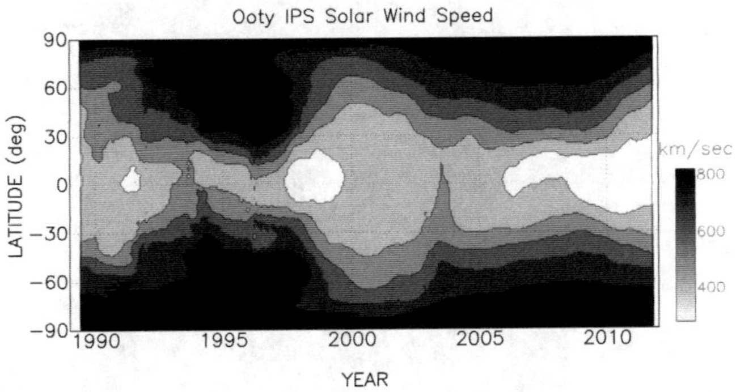

Figure 3. Helio-latitudinal distributions of the solar wind speed obtained from Ooty interplanetary scintillation data. (Image courtesy of P. K. Manoharan).

minimum was quite different from that during cycle 23. This is significant because meridional flows are believed to play an important role in solar dynamo models (see e.g., Dikpati et al. 2010, Nandy et al. 2011, etc.). The main difference was that the meridional flow in the immediate sub-surface layers at higher latitudes was faster during the cycle 23 minimum that during the cycle 24 minimum. The difference can be seen in Fig. 3 of Basu & Antia (2010). Since the solar cycle is almost certainly driven by a dynamo, the differences in meridional flow between the last two minima, and between cycle 23 and the first part of cycle 24, may be important factors in creating the cycle differences, which extend into the corona and even cosmic rays (Gibson et al. 2009). Differences were also seen in the solar zonal flows (Howe et al. 2009; Antia & Basu 2010 etc.), and it was found that the equator-ward migration of the prograde mid-latitude flow was slower during the cycle 24 minimum compared with that of cycle 23.

The peculiar solar minimum has been followed by a rather weak activity cycle, and of course the question is what will happen next. Predictions of the behaviour of cycle 24 spanned across a wide range of cycle-strengths (see Pesnell, 2012 for a compilations of cycle 24 predictions), but only a handful predicted the weak cycle that we are tending towards.

While there is enough evidence to show that the cycle 24 minimum was very different when compared with that of cycle 23, it is not completely clear that the minimum was peculiar compared with minima that occurred in the early part of the 20th century or late in the 19th century. The biggest problem in making such a comparison is that very little data exist before the 1950s. The only data available are the sunspot numbers, and some observations of the solar corona during eclipses.

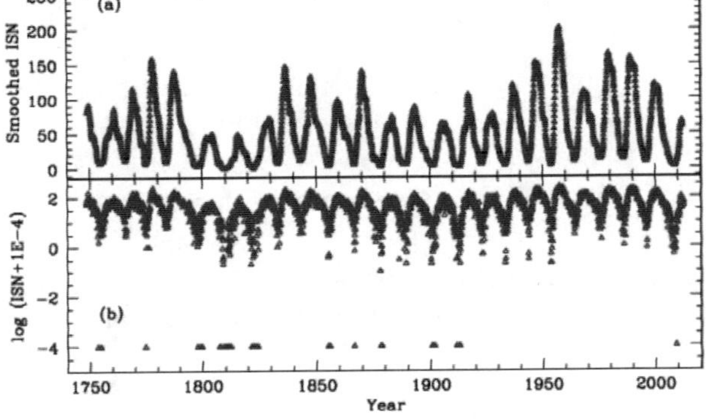

Figure 4. (a) The monthly smoothed sunspot number, and (b) the log of the monthly sunspot number plotted as a function of time. A constant was added before taking the logarithm of the ISM to handle the sunspot-free periods. The data were downloaded from the National Geophysical Data Center (ngdc.noaa.gov).

As discussed by Sheeley (2010), looking at the historic record of sunspot numbers we find that there were sun-spot free solar minima in the early part of the 20th century (the minima

Figure 5. The solar corona as seen during solar eclipses. *Left:* A photograph of the solar eclipse over Sumatra in 1901 (the minimum before cycle 14) processed with modern image-processing techniques, and *Right:* Solar eclipse in 2009 over the Marshall Islands. Note that the configuration of the solar corona was quite similar during the two minima. (Images courtesy of Miloslav Druckmüller and the HAO historical eclipse database [Judge et al. 2010].)

of 1900 and 1910), going even further back, we see a lot of other minima that were as deep at the cycle 24 minimum, at least as far as sunspot numbers are concerned. This can be seen in Fig. 4. Note that plotting the logarithm of the sunspot numbers accentuates the deep minima and shows that the last minimum was not unique in its depth.

Another issue is that of the configuration of the solar corona during the last minimum. As Judge et al. (2010) pointed out, there are eclipse records that show that the solar corona during the solar minimum circa 1901 was very similar to that of 2009 as seen in eclipse photographs (see Fig. 5). Thus it is quite possible that the 'anomalous' and 'peculiar' solar minimum was not so anomalous after all, and the seeming anomaly is simply the result of not having enough data from earlier cycles for a proper comparison.

3. Did we miss signs of an unusual cycle?

There is evidence coming to light that the Sun had started behaving unusually (at least unusually compared to other epochs for which we have good data) during the cycle 23 maximum and perhaps even earlier. For instance Janardhan et al. (2011) published studies of inter-planetary scintillation that showed that solar wind turbulence has been decreasing steadily from about the time of the minimum before cycle 23 (Fig. 6).

There were signs from the solar interior too that something unusual was going on inside the Sun. Data on solar oscillation frequencies that can be used to deduce conditions inside the Sun (see e.g. Christensen-Dalsgaard 2002). Solar oscillation frequencies are known to change with solar activity (Woodard & Noyes 1985; Elsworth et al. 1990; Libbrecht & Woodard 1990; and more recently Basu 2002; Howe et al. 1999). These solar-cycle related changes are also correlated with solar activity indices (Jain & Bhatnagar 2003). The solar cycle-related change in the frequency of a given mode is known to be a function of the frequency of that mode. High-frequency modes have larger shifts than lower-frequency modes. and this implies that solar-cycle related changes in the solar structure occur predominantly in a thin sub-surface layer of the Sun (Nishizawa & Shibahashi 1995). The Birmingham Solar Oscillation Network (BiSON) has been

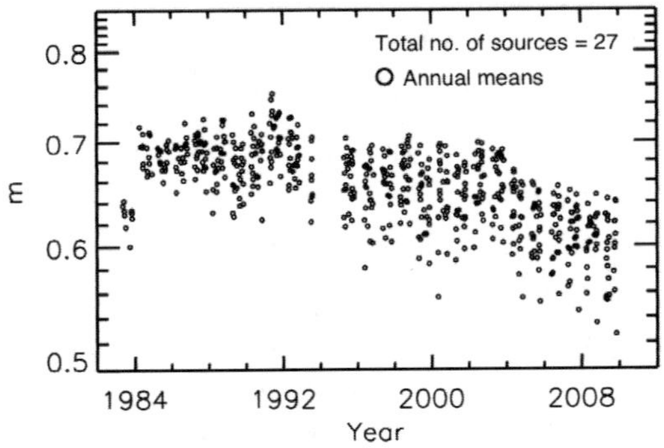

Figure 6. The annual mean interplanetary scintillation index plotted as a function of time. The spread in the normalized scintillation index (m) is due to the spread in the angular sizes of the extragalactic radio sources. This ranges in the sample from about 10 mas to about 200 mas. (Figure courtesy of P. Janardhan).

collecting solar oscillation data for over thirty years. Its observations cover cycles 22 and 23 in their entirety and of all cycle 24 to date. The data also cover a few epochs of cycle 21. Basu et al. (2012) showed that the frequency-dependence of solar-cycle related shifts was different during cycle 23 compared to cycle 22. The correlation of the different solar activity indices with the frequency shifts was affected, and in particular, they showed that while the International Sunspot Number (ISN) and the 10.7 cm flux was correlated with the change frequencies during cycle 22, this correlation was lost after the maximum of cycle 23 (Figures 2 and 3 of Basu et al. 2012). Had this change in behaviour been detected in a timely fashion, it would have alerted us to the fact that the Sun was going through an unusual phase. The correlation between frequency-change and activity indices was regained for the high-frequency modes ($\nu > 2400\ \mu$Hz) in cycle 24, but not for modes with lower frequencies.

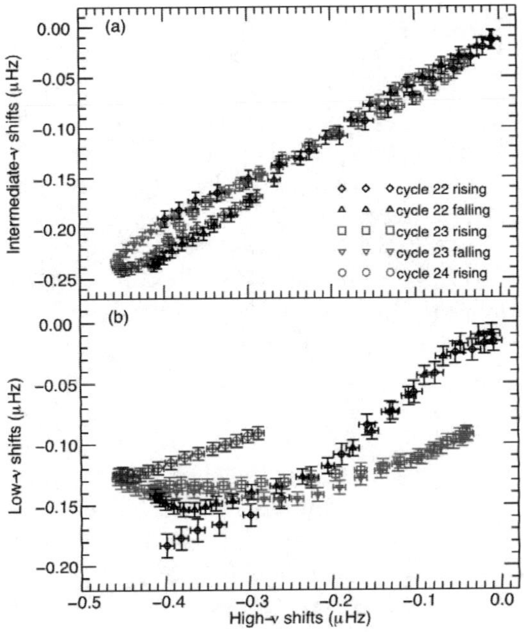

Figure 7. Frequency shifts in the mid-frequency range ($2400 < \nu \leq 2920\ \mu$Hz; panel a) and low-frequency range ($1860 < \nu \leq 2400\ \mu$Hz; panel b) plotted as a function of the frequency shifts in the high-frequency range ($2920 < \nu \leq 3450\ \mu$Hz for cycles 22, 23, and 24. The frequency shifts have been smoothed to remove the signature of the two-year cycle (see Fletcher et al. 2010).

Basu et al. (2012) also found that the change of modes an intermediate frequency range ($2400 < \nu \leq 2920\ \mu$Hz) as a function of the change in higher frequencies ($2920 < \nu \leq 3450\ \mu$Hz) was the same during cycles 22 and 23, the change in the low frequency modes ($1860 < \nu \leq 2400\ \mu$Hz) was very different during cycle 23 compared with cycle 22, in fact there was very little change in the low frequency modes during cycle 23 (see Fig. 7). This can be interpreted as the layer of magnetic field had effectively become thinner after cycle 22. Thus there seem have been early indications that the Sun was behaving in a manner that was different from its earlier behaviour.

4. Cycle 24 thus far

As can be seen from Fig. 8, solar cycle 24 started slowly compared with the previous three cycles. Also slow in rising was the number of flares. Biesecker (2011) showed that while in the first 27 months of the cycle the number of M class flares per sunspot was comparable with those of cycle 21-23, the number if X class flares was much smaller, with only 1 X-class flare compared with 18, 20 and 17 for cycles 21, 22 and 23 respectively. The behaviour of the polar fields (Fig. 1 appears to show that we are very close to the solar maximum, if not there already, making this one of the weakest cycles in recent history.

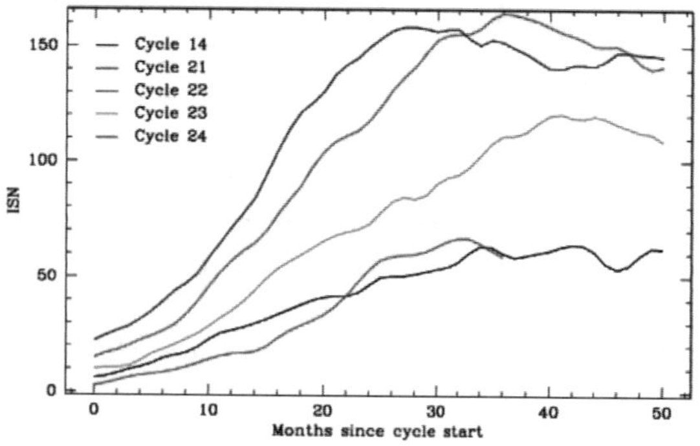

Figure 8. The rise of the monthly averaged smoothed sunspot number for different solar cycles. Note that the rise of cycle 24 is the slowest among all recent cycles.

This solar cycle is showing a number of other differences compared to previous cycles. For instance the poleward migration of magnetic fields is weaker (Hathaway 2013). This poleward movement is believe to be a precursor of the next cycle, and could indicate that the next cycle will be weak too. Altrock (2011, 2012) also commented the absence of this poleward movement (which he dubbed "the rush to the poles") in coronal Fe XIV emission. Flows on the solar surface are also different. Hathaway & Upton (private communication) find that the meridional flows as obtained by tracking of magnetic elements observed by the MDI and HMI instruments are faster now compared with cycle 23 maximum. Doppler measurements of the solar surface also show large variations in the meridional flows (Ulrich 2010 and private communication).

The dynamics of the solar interior is quite different from what has been seen before. Of course one has to keep in mind that we only have 16 years of data on solar interior dynamics, and this allows us to compare this cycle only with cycle 23. As can be seen in Antia & Basu (this volume), the latitude independent part of the solar rotation rate is slower during this cycle than it was in cycle 23. The zonal flows are different too, in particular, the poleward flow is very weak. Plots of the zonal flow seen at different latitudes plotted at a function of radius and time show other differences between the flows during cycles 23 and 24, as can be seen in Fig. 3 of Antia & Basu (*this volume*). In fact the poleward flow is so weak, it was at first believed to be absent and led to the speculation that cycle 25 may not occur at all and that we may be delayed or that we may be headed towards a Maunder-type minimum (Hill et al. 2011). What Antia & Basu (*this volume*) also find is that while the zonal flows at all latitudes in cycle 23 can be fitted with three harmonics of an 11.7 year period, one cannot simultaneously fit the mid-latitudes of cycle 24.

It is too early to be able to say whether or not solar-cycle dependent change in solar structure is the same during this cycle as it was in cycle 23. Structure changes during cycle 23 were so

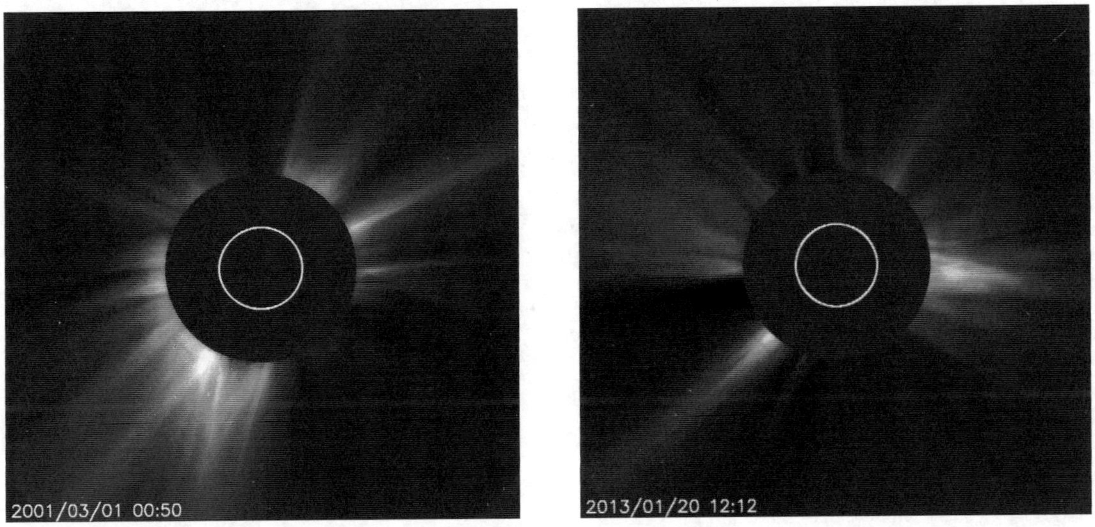

Figure 9. LASCO C2 images showing the solar corona in March 2001 (cycle 23 maximum; left) and January 2013 (almost cycle 24 maximum; right). Note that the coronal structures are quite similar in terms of the latitudinal extent; the current coronal emission however, seem weaker. (Images courtesy of NASA.)

small that it required an analysis of the entire cycle 23 dataset before any statistically significant change could be determined, particularly at the convection-zone base (see Baldner & Basu 2008). It is a bit too to do a similar analysis for cycle 24. The changes in the immediate sub-surface layers of the Sun also remain unclear. As can be seen from Fig. 7, in relation to shifts in the high-frequency solar modes, the frequency-shifts in the low-frequency modes are different not only compared with cycle 23, they seems to be following a different path compared with cycle 22 too. Data over a few more epochs are needed to examine and interpret the trend properly.

Not all observations show differences with respect to the previous cycle. For instance, the solar corona seems to be what a typical solar-maximum corona is expected to be (Fig. 9). The solar winds seem to have the same speed distribution as a function of latitude (Fig. 3). However, interplanetary turbulence still appears low compared with that during the cycle 23 maximum and seems to be getting lower (Fig. 6).

5. What comes next?

Cycle 24 has raised a number of questions and the observations gathered over the next few years will be crucial in answering these. Among the important questions is what cycle 25 will be like. E.g., will the rush to the poles remain weak, or will that strengthen? This could affect what the beginning of the next cycle will be like. While the weakness of the rush to the poles and the poleward branch of the zonal flows appear to indicate a delayed cycle 25, the higher speeds of the meridional flows appear to indicate otherwise.

Another issue is the strength of the sunspot magnetic fields. Penn & Livingston (2011) had claimed that the maximum field strength in sunspots shows a gradual decrease over the last several years and that sunspots may completely disappear in the not-so-distant future. This led Pevtsov et al. (2011) to look into historic data and claim that the Penn & Livingston conclusions are based on limited data collected over just the descending part of cycle 23. Looking into data over several solar cycles they found that the maximum field strength of sunspots is dependent on the phase of the solar cycle. However, in a follow-up paper, Nagovitsyn et al. (2012) claimed that the magnetic field strength data show a solar cycle variation when only sunspots with the strongest magnetic fields are included; they also found a negative correlation between the numbers of small and large sunspots and that during the the period of 1998-2011, the number of large sunspots gradually decreased, while the number of small sunspots steadily increased. Of course, since it is believed that sunspots need a minimum strength to emerge, the question will be whether the trend of decreasing numbers of large sunspots will continue over the current cycle, and hence make a spotless cycle possible. The long-term trends in the strength and size of sunspots will perhaps be a better indicator than solar flows as to whether the Sun is moving towards a magnetically quiet phase. How the Sun behaves in 2013 will be exciting to observe.

Acknowledgements

Grateful thanks go to Miloslav Druckmüller, P. K. Manoharan and P. Janardhan for their figures. This work was supported in part by NASA grant NXX10AE60G.

References

[1] Altrock R C 2012 arXiv:1209.2963
[2] Altrock R C 2011 *Solar Phys.*, **274**, 251
[3] Antia H M and Basu S 2010 *ApJ* **720**, 494
[4] Baldner C S and Basu S 2008 *ApJ*, **686**, 1349
[5] Basu S 2002 *ESA SP* bf 508, 7
[6] Basu S and Antia H M 2010 *ApJ* **717**, 488
[7] Basu S, Brromhall A-M, Chaplin W J and Elsworth Y 2012 *ApJ* **758**, 43
[8] Biesecker D., Space Weather Workshop 2011 (http://www.swpc.noaa.gov/sww/sww11/Schedule_w_presentations_2011.html)
[9] Christensen-Dalsgaard J 2001 *Rev. Mod. Phys.* **74**, 1073
[10] de Toma G 2011 *Solar Phys.* **274**, 195
[11] Dikpati M, Gilman P A, de Toma G and Ulrich R K 2010 *GRL* **37**, L14107
[12] Elsworth Y, Howe R, Isaak G et al. 1990 *nature* **345**, 322
[13] Fletcher S T, Broomhall A-M, and Salabert D et al. 2010 *ApJ* **718**, L19
[14] Gibson S E, Kozyra J. U., de Toma G et al. 2009 *JGRA* **114**, A09105
[15] Gopalswamy N, Yashiro Y, Mäkela P et al. 2012 *ApJ* **750**, L42
[16] Hathaway D H 2013 http://solarscience.msfc.nasa.gov/images/magbfly.jpg
[17] Hill, F. et al. 2011 *BAAS* **43**, 16.10
[18] Howe R, Komm R, Hill F 1999 *ApJ* **524** 1084
[19] Howe R, Christensen-Dalsgaard J, Hill F et al. 2009 *ApJ* **701**, L87
[20] Jain K and Bhatnagar A 2003 *Solar Phys.* **213**, 257
[21] Janardhan P, Bisoi S K, Ananthakrishnan S et al. 2011 *Geophys. Res. Lett.* **38**, L20108
[22] Judge P G, Burkepile J, de Toma G and Druckmüller M 2010 *ASPCS* **428** 171
[23] Libbrecht K G and Woodard M F 1990 *Nature* **345** 779
[24] Manoharan P K 2012 *ApJ* **751**, 128
[25] Nagovitsyn Y A, Pevtsov A A and Livingston W C 2012 *ApJ* **758**, L20
[26] Nandy D, Muñoz-Jaramillo A, Martens P C H 2011 *Nature* **471**, 80
[27] Nishizawa Y and Shibahashi H 1995 *ASPCS* **76**, 280
[28] Penn M J and Livingston W C 2011, *Proc. IAU Symp. 273*, p126
[29] Pesnell W D 2012 *Solar Phys.* **281**, 507
[30] Pevtsov A A, Nagovitsyn Y. A., Tlatov A G and Rybak A L 2011 *ApJ* **742**, L36
[31] Sheeley N R 2010 *ASPCS* **428**, 3
[32] Ulrich R K 2010 *ApJ* **725**, 658
[33] Woodard M and Noyes R W 1985 *Nature* **318** 449

Hinode observations of flares and active region emergence

Toshifumi Shimizu

Institute of Space and Astronautical Science, Japan Aerospace Exploration Agency, 3-1-1 Yoshinodai, Chuo-ku, Sagamihara, Kanagawa 252-5210, Japan

E-mail: shimizu@solar.isas.jaxa.jp

Abstract. After observing the quiet Sun during a long and deep minimum, *Hinode*'s observing priority is now the active Sun, i.e., flares and active regions. Because of its small field-of-view instruments and telemetry restrictions, capturing major flares with good observing coverage needs challenging efforts on operation plannings for *Hinode* and we have been using the latest available AIA and HMI full-Sun images for targeting. Also, capturing the initial phase of large-scale emerging flux activities is not easy for *Hinode* without having the forecast of flux emergence before they are visible on the surface. Nevertheless, *Hinode* has some good examples of observations of such events. In this paper, we briefly discuss two specific observations; X-class flares in March 2012 and a large-scale flux emergence in December 2009. The Stokes polarimetry data for the flares presented in this paper reveals significant gas flows along the neutral line of the sheared magnetic distributions and the change of magnetic field inclination at the areas where penumbral development and disappearance was observed with the occurrence of the flares. The flux emergence and the sunspot formation presented were well monitored by the *Hinode* three telescopes from the beginning through the end of the emergence, Of particular interest is the finding of a precursory signature in the chromosphere preceding the appearance of sunspot penumbra.

1. Towards better Hinode observations of emerging active regions

Hinode [1], launched on September 2006, has been observing the Sun to understand the processes of magnetic field generation, the processes responsible for energy transfer from the photosphere to the corona, and the mechanisms responsible for heating, dynamics and eruptive phenomena. The *Solar Optical Telescope (SOT)* [2, 3, 4, 5] provides observations at 0.2-0.3 arcsec resolution of magnetic and velocity fields at the solar surface, while the *EUV Imaging Spectrometer (EIS)* [6] and the *X-Ray Telescope (XRT)* [7, 8] provide EUV spectroscopic data and soft X-ray images, respectively, for diagnosing thermal properties and dynamics in the corona. Their observing field of views are limited to the target of interest, in order to obtain high spatial resolution and high precision, and thus they are complementary to the AIA and HMI full-Sun images from the *Solar Dynamics Observatory.*

The snapshot successfully acquired during the transit of Venus became one of the *Hinode*'s famous pictures in 2012. The impressive image (Figure 1) called attention from the public and Japanese major newspapers published this colorful image on the top page. This image was quickly distributed through the internet media. A honorable note was attached to the picture in one media site (gizmode): "Not only this image wins the internet for the most amazing image

Figure 1. Transit of Venus captured by Solar Optical Telescope with Ca II H filter. 22:27 UT on 5 June 2012

of the Venus transit, but to me it's also one of the most impressive images in the history of astronomy and space exploration. The scale and the feeling left me in awe."

With the *Hinode* telescopes, we observed *quiet* regions including coronal holes and ephemeral regions without sunspots during a long and deep minimum (2007-2009), but we were surprised that *quiet* regions are full of dynamical natures of magnetic structures. From 2009, the Sun started to show the appearance of new-cycle active regions on the surface and produce flares (Table 1). In order to investigate the active Sun with *Hinode*, active regions, especially regions with the potential to create flares, are given the highest observing priority. Capturing major flares with good observing coverage, however, is operationally challenging because *Hinode* has narrow field-of-view instruments and strong limitation on telemetry volume due to onboard X-band modulator malfunction [9]. Telemetry resources should be also shared with synoptic observations performed repeated on a regular basis and non-flare observations proposed by the international user community (called HOP, *Hinode* Operation Plan). The selection of active regions for the observations is made by checking AIA and HMI full-Sun images available at the time of selections. We may also consider the Maximum Millennium Flare Watch target designation. As soon as possible after the issuance of a Maximum Millennium Flare Watch, we attempt to follow that target and use flare watch observing sequences. In addition, the *Hinode* team may declare "*Hinode* flare watch" to shift all resources to observations of flare-potential regions.

2. Flare observations

Hinode data have been extensively analyzed for three X-class flares during December 2006, producing more than 30 papers in refereed journals. After four years of solar quietness, the *Hinode* telescopes started to capture major flares from 2011. Table 1 shows the statistics for all flares recorded in the SolarSoft flare list (SSW list), in which not only X-, M-, and C-class flares but also B- and A-class flares are included. The number of flares acquired with the XRT, SOT

Journal of Physics: Conference Series **440** (2013) 012002 doi:10.1088/1742-6596/440/1/012002

filtergraph (FG) and EIS is based on the *Hinode* flare catalog; flares are counted as successful flares in this table, if more than one sample (image for XRT and SOT or map for EIS) is acquired during the time of the event and the flare location is within the instrument field of view [10]. More than half of the flares were captured with the large field-of-view XRT, but only a limited number of flares were captured with the narrow field-of-view SOT and spectroscopic instrument EIS.

Table 1. Number of flares in *Hinode* data (courtesy of K. Watanabe)

All classes	*XRT*	*SOT(FG)*	*EIS*	*SSWlist*
2006 (25 Oct-)	160	171	68	432
2007	443	267	209	713
2008	24	8	7	134
2009	125	83	65	320
2010	690	264	125	1248
2011	1269	464	276	2389
2011(-31 Oct)	1010	390	245	1958
total	3721	1647	995	7194
percent of SSW	51.7%	22.9%	13.8%	

Table 2. Number of major flares included in Table 1 (courtesy of K. Watanabe)

GOES class	*XRT*	*SOT(FG)*	*EIS*	*SSWlist*
X	14	10	7	19
M	143	70	53	260

The number of successful X- and M-class flare detections are given in Table 2 and details of the successful X-class flares are listed in Table 3. The table gives GOES class, location on the solar disk, and number of images (for XRT and SOT) or raster-scanning maps (for EIS) during the flare period. "Flare mode" in remarks means that the telescopes might perform observations with high-cadence flare watch program. The onboard mission data processor analyzes XRT flare patrol images for large intensity increases indicative of a flare [8]. If a flare is found a flare flag is issued that allows the telescopes to terminate their pre-scheduled sequence and switch to a flare watch program.

3. 7 March 2012 X5.4/X1.3 flares

These flares occurred in the complex active region NOAA 11429. XRT images show a bright sheared feature along the neutral line, suggesting a sheared structure in the coronal magnetic fields. The X5.4 peak was observed as strong heating in a local loop structure existing along the neutral line. This compact brightening evolved into the formation of a large flare arcade. This morphological change was triggered by the X1.3 flare.

For accurate determination of magnetic fields and dynamics at the photosphere, SOT acquired Stokes-polarimter (SP) [11] maps every a few hours. The SP yields the full-polarization states

Table 3. Hinode's X-class flares in 2006-2012 (courtesy of K. Watanabe)

Flare date/time	*GOES*	*Location*	*XRT*	*SOT(FG)*	*EIS*	*Remarks*
2006/12/06 18:29	X6.5	S05E57	0	30	0	
2006/12/13 02:14	X3.4	S07W22	44	44	1	
2006/12/14 21:07	X1.5	S06W46	84	80	1	
2011/02/15 01:44	X2.2	S20W10	116	102	0	Flare mode
2011/03/09 23:13	X1.5	N08W11	54	0	0	Flare mode
2011/08/09 07:48	X6.9	N14W69	59	0	0	
2011/09/06 22:12	X2.1	N14W18	82	0	0	Flare mode
2011/09/24 09:21	X1.9	N13E61	0	0	6	
2011/11/03 20:16	X1.9	N21E64	122	118	0	Flare mode
2012/01/27 17:37	X1.7	N33W85	447	381	7	Flare mode
2012/03/05 02:30	X1.1	N19E58	220	275	0	Flare mode
2012/03/07 00:02	X5.4	N18E31	296	6	0	Flare mode
2012/03/07 01:05	X1.3	N15E26	63	0	0	
2012/07/06 23:01	X1.1	S13W59	161	0	1	Flare mode
2012/07/12 15:37	X1.4	S13W03	278	43	1	Flare mode
2012/10/23 03:13	X1.8	S13E58	60	101	3	Flare mode

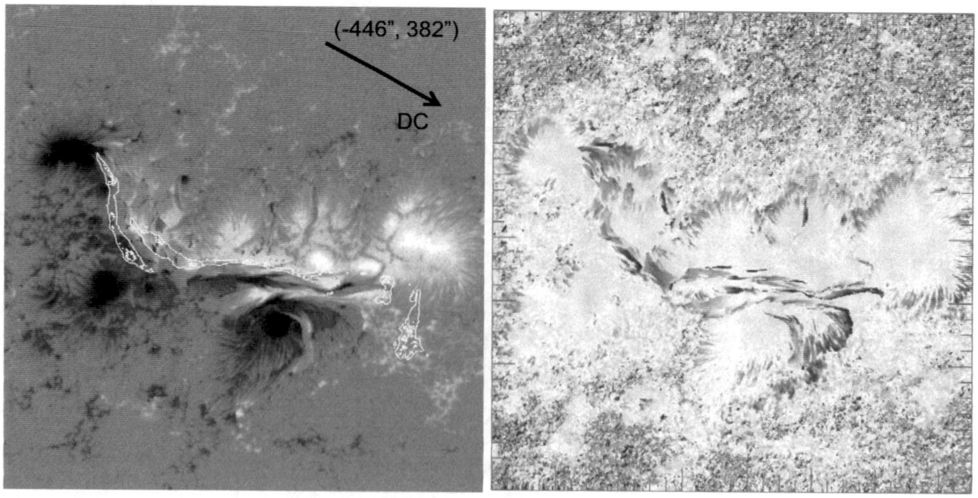

Figure 2. Two physical parameters derived from SOT Stokes-polarimeter data. Left: B_z, the magnetic flux component vertical to the solar surface. Contours give the position of flare ribbons seen in Ca II H images at initial phase (7 March 2012 00:07:25 UT). Right: Doppler shift of the Fe I 630.15 nm line. Red is away from the observer and blue is toward the observer. The arrow gives the direction to the solar disk center.

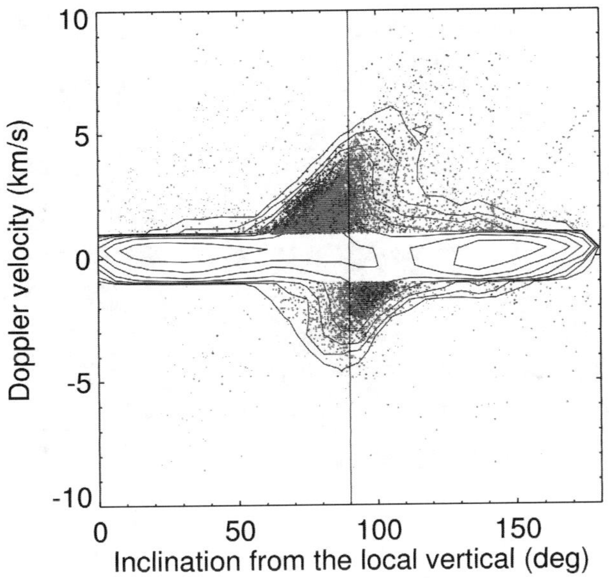

Figure 3. Measured Doppler velocity compared with the inclination of magnetic flux. The inclination is the angle on the local coordinate frame; 0 and 180 deg is vertical to the solar surface and 90 deg is horizontal orientation. Red points are Doppler features (higher than 1 km/s) observed along the neutral line. Blue points are from the penumbral areas where Evershed flow signatures are observed. Contours represent the density of data points in the entire field of view.

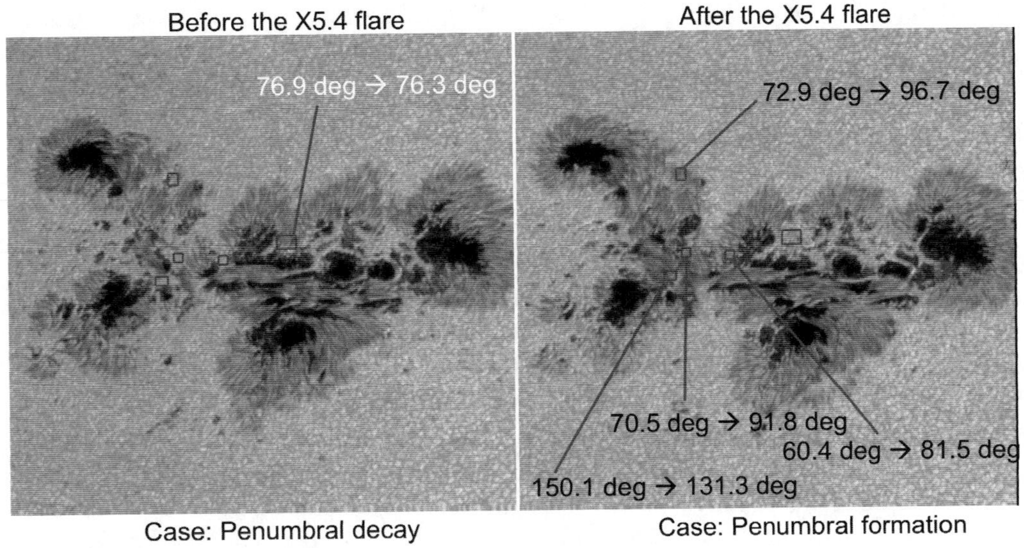

Figure 4. Continuum images recorded before and after the X5.4 flare. The change of magnetic field inclination is given for one location where penumbra disappeared (values given in the left image) and four locations where penumbra developed (in the right image).

of line profiles of the two magnetically sensitive Fe lines at 630.15 and 630.25 nm and Stokes inversions allow us to deduce magnetic field vectors and thermo-dynamics parameters at the photosphere. Figure 2 shows the spatial distribution of the magnetic flux component vertical to the solar surface and Doppler shift signals observed in this active region. The flare ribbons were located at the both sides of the neutral line, where highly sheared elongated features (both

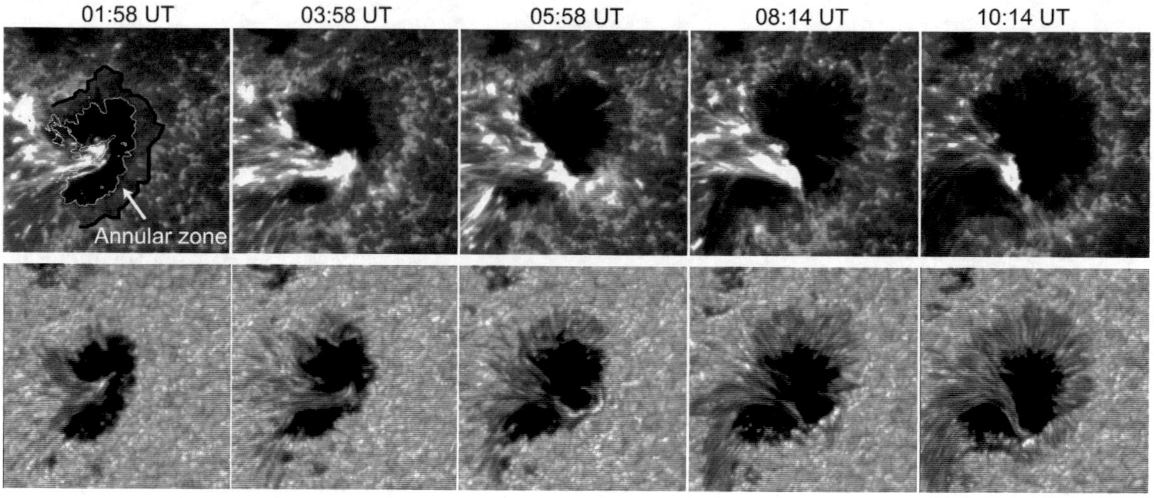

Figure 5. Development of a sunspot newly formed with magnetic emergence. Upper panels: observed with Ca II H filter, lower panels: with G-band filter. The contours in the first panel indicate the annular zone newly discovered with *Hinode* SOT. 31 December 2009.

negative and positive polarities) exist. Remarkable is strong Doppler shift signals (the order of ± 5 km/s, which corresponds to the sound speed at the photosphere) distributed along the neutral line. The strong Doppler shift signals are dominantly observed in horizontally or highly inclined magnetic fields (Figure 3), indicating that gas flows are excited in horizontal flux tubes. Investigating the roles of these gas dynamics may be important for understanding dynamics and triggering mechanisms of flares.

Continuum images from the SP maps show both the decay and formation of sunspot penumbra in several areas at the time of these flares (Figure 4). Such rapid penumbral changes have been observed in other flares [12, 13]. The changes in field inclination at each penumbral change can be accurately evaluated with the SP data. Both cases shows that the field is changed toward the horizontal orientation in the region where penumbral changes are observed. This would be due to magnetic field restructuring associated with the flares.

4. Initial phase of large-scale emerging flux activities

Emerging processes of magnetic flux are another important target for observations during the time when the solar activity is high. Flux emergence is a key process for energy storage and the triggering of flares and microflares (transient brightenings) [14, 15]. We lack knowledge about magnetic field evolution and dynamics during the large-scale emerging flux activities, because of little number of high resolution observations available for the initial phase. Forecasting is not possible before the emerging flux elements appear on the photosphere, and only chance encounters enable narrow-field-of-view telescopes to properly point to the emerging site from before the start of emergence. In the period from 29 December 2009 through 1 January 2010, the SOT has continuously monitored the active region NOAA 11039. Fortunately, a large-scale emerging flux appeared in the SOT field of view on 30-31 December.

Sunspots were formed in a sequence of successive magnetic flux emergence to the solar surface. Pores, which are essentially small sunspots without penumbra (naked umbra), first formed at both edges of emerging flux regions, and then they developed to large sunspots with penumbra (see Figure 6). In this evolution, details of the penumbral formation are hidden in a veil of

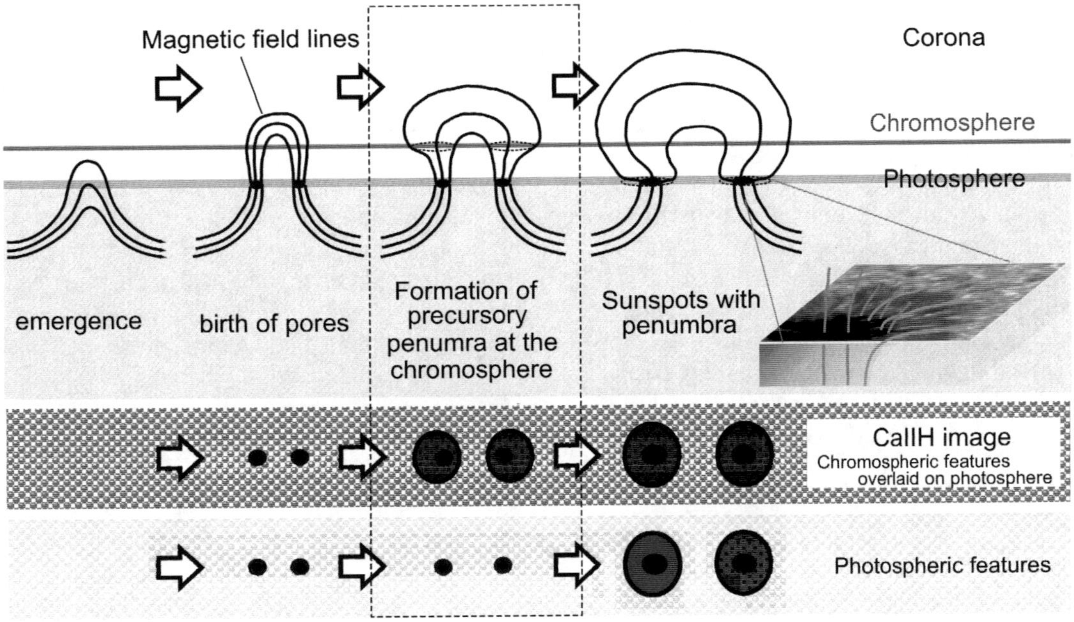

Figure 6. The history of magnetic field lines from the emergence to the appearance of sunspots with penumbra, with illustrating sunspot features seen in Ca II H and G-band images.

mystery. Figure 5 shows the morphological evolution of a pore (sunspot) formed at the leading edge of the emerging flux region in chromospheric Ca II H and photospheric G-band images taken every about 2 hours. Emerging flux activities happened at the east side of the spot, and remarkable chromospheric brightenings are associated with the elongated structure formed in the umbra. The penumbra formed in sectors; the penumbra was developed at the north side, and then it formed at the west and south.

In the Ca II H images, we noticed a zone between the umbra and ambient bright features before the penumbral formation (Figure 5) [16]. This zone appeared soon after the pore formation and was seen for about 10 hrs until the penumbral formation. The zone is annular with the width of 3"-5" and it exists at the outer side of the umbral edge (indicated by the dotted contour) on almost all the locations excepting the east side of the spot. The outer boundary of this zone is remarkably sharp. The penumbra was developed as if to fill the annular zone.

The observations indicate that the annular zone is different from sunspot moat flow region and that it likely represents a structure in the chromosphere. A possible interpretation is that the annular zone may reflect the formation of a magnetic canopy at the chromospheric level overlying the region surrounding the umbra, much before the formation of the penumbra at the photospheric level (Figure 6). The magnetic structure observed as the annular zone goes down to the photospheric level, resulting in appearance of sunspot penumbra. This observation clearly showed the importance of magnetic structures at the chromospheric level in the development of photospheric sunspot penumbrae.

5. Final remark

As shown in the flare statistics tables, there are many unique *Hinode* datasets for X- and M-class flares. The field of view may be restricted in the flare region (a part of active region) but it is with high spatial resolution and precision. These data can be enhanced by combining with complementary AIA and HMI full-Sun images and thus they can contribute to solar physics and space weather researches. We hope many researches use *Hinode* data more extensively in their researches.

Acknowledgments

The author is grateful that Paul Cally and the SOC members invited him to attend this meeting held with a unique event. He also thanks Kyoko Watanabe for providing flare statistics and Shinsuke Imada for a flare data (not included in the proceedings). *Hinode* is a Japanese mission developed and launched by ISAS/JAXA, collaborating with NAOJ as a domestic partner, NASA and STFC (UK) as international partners. It is operated by these agencies in co-operation with ESA and NSC (Norway). This work was supported by JSPS KAKENHI Grant Number 23540278 and JSPS Core-to-Core Program 22001.

References

[1] Kosugi T et al. 2007 *Solar Phys.* **243** 3
[2] Tsuneta S et al. 2008 *Solar Phys.* **249** 167
[3] Suematsu Y et al. 2008 *Solar Phys.* **249** 197
[4] Shimizu T et al. 2008 *Solar Phys.* **249** 221
[5] Ichimoto K et al. 2008 *Solar Phys.* **249** 233
[6] Culhane J L 2007 *Solar Phys.* **243** 19
[7] Golub L et al. 2007 *Solar Phys.* **243** 63
[8] Kano R et al. 2008 *Solar Phys.* **249** 263
[9] Shimizu T 2009 *The Second Hinode Science Meeting: Beyond Discovery-Toward Understanding* ed. Lites B, Cheung M, Magara T, Mariska J. and Reeves K, Astron. Soc. Pac. **CS-415** 1
[10] Watanabe K, Masuda S, and Segawa T 2012 *Solar Phys.* **279** 317
[11] Lites B W et al. 2013 *Solar Phys.* in press
[12] Wang H et al. 2004 *Astrophys. J.* **601** L195
[13] Deng N et al. 2005 *Astrophys. J.* **623** 1195
[14] Shimizu T, Shine R A, Title A M, Tarbell T D, and Frank Z 2002 *Astrophys. J.* **574** 1074
[15] Kano R, Shimizu T and Tarbell T D 2010 *Astrophys. J.* **720** 1136
[16] Shimizu T, Ichimoto K, and Suematsu Y 2012 *Astrophys. J.* **747** L18

PICARD SOL, a new ground-based facility for long-term solar radius measurements: first results

M Meftah[1], T Corbard[2], A Irbah[1], F Morand[2], R Ikhlef[2,3], C Renaud[2], A Hauchecorne[1], P Assus[2], B Chauvineau[2], M Crepel[1], F Dalaudier[1], D Djafer[4], M Fodil[3], F Laclare[2], P Lesueur[1], M Lin[1], and G Poiet[1]

[1] Laboratoire Atmosphères, Milieux et Observations Spatiales, UMR 8190, CNRS, Université de Versailles Saint-Quentin-en-Yvelines, Paris VI, IPSL, 78280 Guyancourt, France
[2] Laboratoire Lagrange, UMR 7293, Université de Nice Sophia-Antipolis, CNRS, Observatoire de la Côte d'Azur, Bd. de l'Observatoire, 06304 Nice, France
[3] CRAAG, Observatoire d'Alger, Alger, Algérie
[4] Unité de Recherche Appliquée en Energies Renouvelables, CDER, B.P. 88, Ghardaïa, Algérie

E-mail: `Mustapha.Meftah@latmos.ipsl.fr`

Abstract. PICARD SOL is the ground component of the PICARD mission and is operational since March 2011. A set of instruments including the replica of the space instrument and several atmospheric monitors was set up at Calern observatory in order to compare solar radius measured in space and on ground and to better understand and calibrate atmospheric effects on ground based measurements. SODISM II provides full disk images of the chromosphere and photosphere of the Sun in five narrow pass bands ranging from the near ultraviolet to the near infrared. Our preliminary results show a very good instrumental stability. After plate scale calibration using star doublet observations and corrections for atmospheric refraction, first estimates of the mean solar radius at the five wavelengths (393.37, 535.7, 607.1, 782.2, and 1025.0 nm) are deduced from measurements recorded between May 2011 and December 2012.

1. Introduction

The PICARD program [1] owes its name to Jean Picard (1620-1682), considered as a pioneer of precise modern astrometry who measured the solar radius. The project includes not only a space mission but also a ground component set up at the Calern observatory where a solar radius survey was initiated in 1975 with the visual Solar Astrolabe [2]. This was later completed by DORAYSOL (Définition et Observation du RAYon SOLaire), an automatized astrolabe using charge-coupled device records and a variable prism in order to increase the number of daily measurements [3]. A mean solar radius of 959.48" with a standard deviation (σ) of 0.32" was deduced from 19,169 measurements using DORAYSOL between 2000 and 2006. During the same period, 371 measurements made with the solar astrolabe led to a mean value of 959.55" with σ=0.26". Ground measurements made over two solar cycles at Calern observatory, show a significant variation of about 0.2" peak-to-peak anti-correlated with solar activity [2]. Space measurements have not confirmed this observation but they are covering relatively short periods and were not dedicated to solar metrology. A debate therefore exists to know if this observed variation from ground is of solar origin, purely atmospheric or a combination of both

Journal of Physics: Conference Series **440** (2013) 012003 doi:10.1088/1742-6596/440/1/012003

effects. Atmospheric turbulence has an effect on these measurements and numerical simulations were developed in order to better understand it. The uncertainties and bias introduced by optical turbulence decreases with the seeing but it is also strongly conditioned by turbulence coherence times. It is therefore important to record all the spatio-temporal parameters of optical turbulence at the time of the recorded image. A generalized daytime seeing monitor was built for this goal. Furthermore, general atmospheric conditions (sky brightness, scattering by aerosols, clouds, ...) can affect image quality and impact astrometric measurements. This is why a set of complementary monitors was also installed on the site. A continuation of ground-based series represents our best chance to detect possible long-term variations of solar radius. Short-term space missions dedicated to solar metrology can be used for cross calibrating measurements during their lifetime. In the next sections, we present the different instruments of PICARD ground-based facility, and give some preliminary results obtained from the first twenty months of continuous operations.

2. PICARD SOL instruments

2.1. SOlar Diameter Imager and Surface Mapper (SODISM II)

SODISM II is an 11-cm diameter telescope with a charge-coupled device (CCD) at its focal plane [4]. It is a Ritchey-Chrétien telescope with a focal length of 2,626 mm, an aperture of F/23 with a central obscuration of 40% in diameter (F/30 and 50% for the main channel). The field of view is $\sim 36' \times 36'$ and one image is recorded every minute. The CCD array has 2048×2048 square pixels of $13.5\,\mu$m side ($\sim$ 1.06" per pixel). The whole SODISM II instrument is in primary vacuum ($\sim$10 mbar), and thermally regulated ($\sim$20°C). The Sun image is stabilized on the detector by the equatorial mount (better than $\pm$0.5" during the exposure time). SODISM II main optical path consists essentially of a front window, a primary mirror (M1), a secondary mirror (M2), two filter wheels (FW) with interchangeable interference filters, and a CCD regulated at -10°C $\pm$0.2^{o}C. SODISM II provides full disk-images of the Sun at five wavelengths: 393.37, 535.7, 607.1, 782.2, and 1025.0 nm with narrow bandwidth of 0.7, 0.5, 0.7, 1.6, and 6.4 nm respectively. Several campaigns of star doublet observations have provided the image plate scale: $\sim$ 1.0611" per pixel with a relative uncertainty of 10^{-4}. This introduces an uncertainty of 0.09" in the determination of the solar radius.

2.2. Moniteur d'Images SOLaires Franco-Algérien (MISOLFA)

MISOLFA is a solar seeing monitor associated with a CCD detector at its focal plane [5]. It is a Cassegrain-coudé telescope with a focal length of 10,000 mm, a main entrance pupil of 252 mm, and an aperture of F/40. The field of view is $\sim 2.1' \times 1.6'$. The CCD array has 640 by 480 square pixels of $9.9\,\mu$m side ($\sim$ 0.20" per pixel). An Alt-Azimuth mount supports and rotates MISOLFA and a prismatic entrance window allows to record simultaneously two opposite solar limbs. Thirty-two pairs of opposite solar limbs are recorded every second. The MISOLFA design is based on the statistical analysis of Angle of Arrivals (AA-fluctuations) defined as the slope in each point of the wavefront. Two measurement channels are provided:

- A direct channel in which the Sun's limb images are formed on a CCD camera with suitable magnification. This channel enables the evaluation of the spatial coherence parameters of the wavefront (Fried parameter r_0, outer scale, and iso-planatism domain).
- A second channel (pupil channel) that forms the image of the pupil through a diaphragm placed on the solar limb is used to evaluate the wavefront temporal parameters using high cadence photo-electric detectors (photodiodes).

2.3. Additional instrumentation

In addition to a weather station we have set up three instruments in order to provide continuous quality index for atmospheric conditions. The instruments listed below are used:

- A Sun tracking photometer which measures Sun and sky radiance using a combination of seven spectral filters in order to derive total column water vapor, ozone and aerosols properties. The chosen wavelengths are nearly the same as for the SODISM II instrument.
- A pyranometer is used to measure broadband solar irradiance (from 200 to 3600 nm), and is designed to measure the solar radiation flux density from a field of view of 180 degrees.
- A visible wide-field camera monitors weather conditions and cloud coverage. The field of view is slightly greater than 180 degrees in the horizontal direction and the path of the solar disk across the sky is masked.

3. Results

SODISM II images were obtained between May 2011 and December 2012. For this preliminary work, we simply defined the solar radius by the position of the inflection point of solar limb profiles taken at different position angles. The inflection point position is obtained by computing the location of the maximum of the derivative squared of the limb profiles. After computing all contour inflection points, we can find the best circle fit using a least squares method. An evolution of the estimated solar radius by this technique, and from SODISM II images, is shown on Figure 1 as a function of time for different wavelengths. The corresponding mean solar radii and standard deviations are provided in Table 1. Solar radius measurements at Calern since 1978 are shown on Figure 2. The first important point to notice is that the standard deviation obtain with this new series (around 0.23″) is of the same order or slightly lower than the one obtained with previous astrolabe measurements. We thus believe that the instrument has been fairly stable during its first twenty months of operations and we do not observe any degradation or systematic effects with time.

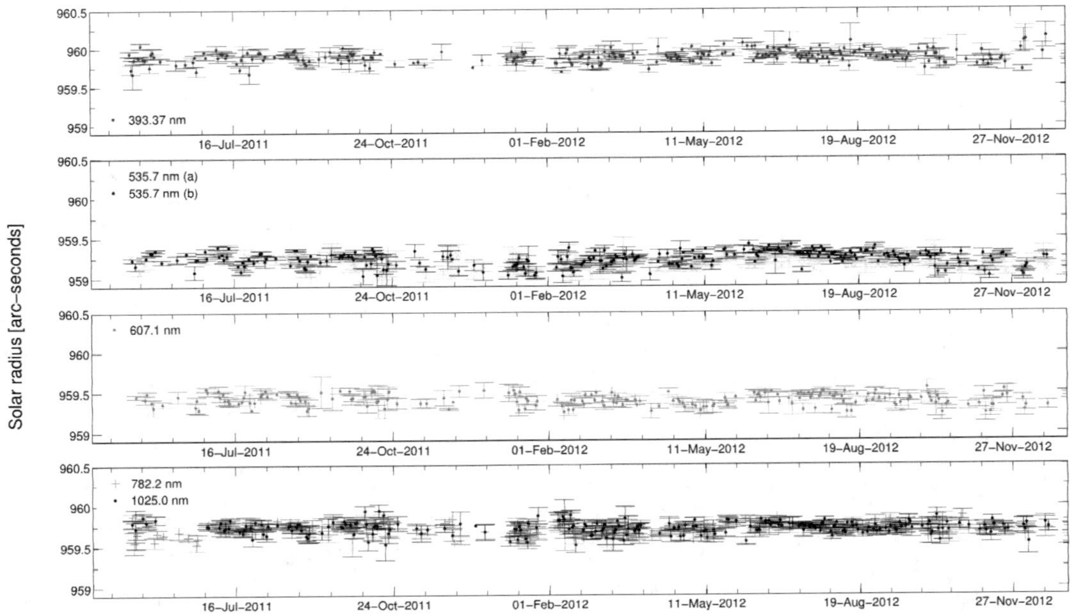

Figure 1. SODISM II solar radius measurements (daily mean) with refraction correction.

Table 1. SODISM II mean solar radii and standard deviation σ for N measurements (corrected for refraction only) - from May 2011 to December 2012.

Wavelength λ	Exposure	N	Solar radius	σ	Comparison with instrument(s)
393.37 nm	1.70 s	5,120	959.884"	0.21"	–
535.7 nm (a)	1.30 s	11,813	959.198"	0.25"	Astrolabes at 548 nm (959.51")
535.7 nm (b)	8.90 s	6,032	959.250"	0.18"	Astrolabes at 548 nm (959.51")
607.1 nm	1.28 s	4,904	959.428"	0.26"	SDS [6] at 620 nm (959.53")
782.2 nm	1.43 s	5,657	959.714"	0.24"	SDM [7] at 800 nm (959.68")
1025.0 nm	1.70 s	5,700	959.715"	0.25"	–

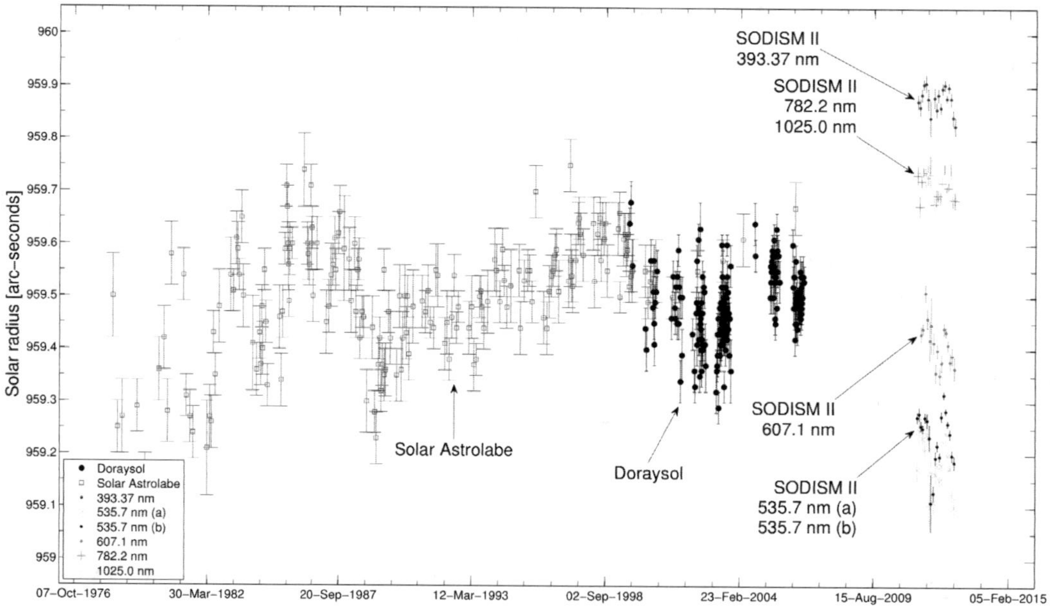

Figure 2. Solar radius at Calern site of the Observatoire de la Côte d'Azur since 1978. SODISM II solar radius measurements were corrected for refraction only.

Another important result is that there is a clear difference between the mean radius values observed at different wavelengths. Images were not de-convolved for diffraction which varies with wavelength. This effect however cannot explain the observed variation with wavelength. The broadening due to larger point spread function (PSF) for larger wavelengths will effectively affect (but by lowering) the inflection point position, but this is compensated by the fact that the continuum limb darkening functions (LDF) themselves become steeper for increasing wavelengths. The overall effect is less than 0.1". The Fried parameter on the other hand varies as $\lambda^{6/5}$, and therefore the bias induced by turbulence on radius estimates will tend to decrease as wavelength increase. This bias being always negative, we therefore expect radius estimates not corrected from turbulence to effectively increase with wavelength as reported here in Table 1. Corrections from the bias induced by turbulence will then decrease this observed wavelength dependence in the continuum but this will be discussed in more details in a future paper. Finally, we note that the CaII line at 393.37 nm is formed higher in the lower chromosphere and the results cannot be directly compared to other observations made in the photospheric continuum.

Journal of Physics: Conference Series **440** (2013) 012003 doi:10.1088/1742-6596/440/1/012003

A more detailed study is needed to confront these mean values and their separation as a function of wavelength to theoretical predictions. The mean values obtained at 607.1 nm and 782.2 nm are consistent with the value found at similar wavelengths by Solar Disk Sextant (SDS) and Solar Diameter Monitor (SDM) respectively (see Table 1). The mean radius value (959.22") obtained at 535.7 nm is however significantly lower than the mean value (959.51") of previous astrolabe series (using a broad bandwidth). Different effects can introduce a bias in solar radius measurements. The most important ones are atmospheric refraction and turbulence. Scattering of the solar light by aerosols plays a non-dominant role. At first order, astrolabe measurements are not affected by refraction as we observe always at the same zenith distance. Solar radius measurements obtained from SODISM II images (Figure 1) were corrected for refraction but not for atmospheric turbulence. We can estimate from simulations the mean bias introduced by turbulence for a given Fried parameter r_0 [8]. Monthly Fried parameters obtained with MISOLFA between 2010 and 2012 are shown in Table 2.

Table 2. Fried parameter r_0 measurement performed between June 2010 and May 2012 [8] (monthly median values for this period and standard deviation σ).

Month	r_0 [cm]	σ [cm]	Month	r_0 [cm]	σ [cm]
Jan.	2.82 cm	0.94 cm	Jul.	3.88 cm	1.76 cm
Feb.	3.42 cm	1.26 cm	Aug.	2.50 cm	1.16 cm
Mar.	2.92 cm	1.11 cm	Sep.	3.70 cm	1.24 cm
Apr.	3.71 cm	1.28 cm	Oct.	3.45 cm	1.19 cm
May	4.10 cm	1.32 cm	Nov.	3.50 cm	0.91 cm
Jun.	4.20 cm	1.32 cm	Dec.	2.75 cm	0.50 cm

The broadening due to seeing affect the inflection point position and tend to bias radius estimates toward lower values as the seeing increase. Estimates of this bias with its one sigma uncertainty are given as a function of the Fried parameter by Ikhlef et al. [8]. The average diurnal Fried parameter r_0 for Calern is 3.41 cm from which we can infer that the mean bias on solar radius measurement with SODISM II at 535.7 nm is around 0.55", and that this mean value of the bias can fluctuate within $\pm$0.2" at one σ [8]. This would lead to a fairly good agreement between SODISM II solar radii at 535.7 nm ($\sim$959.75") and the value of Arthur Auwers in 1891 (959.63" with an heliometer) currently used in all ephemeris calculations.
At 607.1 nm, the convolution of the LDF (with SODISM II) and the Earth's atmospheric turbulence cause a shift of the inflection point position. This effect is at the order of 0.4"$\pm$0.16" (at one σ) for Fried parameter r_0 equal to $\sim$4.0 cm. With this bias, the mean solar radius at 607.1 nm is found to be equal to 959.83", whereas the estimated uncertainties of the measures, random plus systematic, are typically less than 0.2". For MDI, at a wavelength of 676.78 nm, the solar radius was equal to 960.12" [9] during Mercury transits (May 7, 2003 and November 8, 2006). SODISM II solar radius (with appropriate corrections) is consistent with MDI result.
At 782.2 nm, the estimated Fried parameter r_0 is $\sim$5.4 cm from which we can infer that the mean bias on solar radius measurement with SODISM II is 0.16"$\pm$0.09" (at one σ). In this case, the mean solar radius at 782.2 nm is found to be equal to 959.87".

Journal of Physics: Conference Series **440** (2013) 012003 doi:10.1088/1742-6596/440/1/012003

4. CONCLUSION

PICARD SOL is a new ground-based facility for long-term solar radius measurements, which includes a multi-wavelength full disk solar imager stabilized in temperature and a set of atmospheric monitors working simultaneously. It has been operated for 20 months since May 2011 and shows good instrumental stability. First estimates of the mean solar radius observed at five different wavelengths were obtained after plate scale calibration and atmospheric refraction correction. Using the mean value of the Fried parameter given by the turbulence monitor MISOLFA, rough estimates have been made of the bias introduced by optical turbulence on our solar radius measurements. These preliminary results are consistent with previous measurements (from ground and space) and give us confidence on our ability to detect possible long-term variation of the solar radius by continuing this series. The complete set of atmospheric monitors has been only very partially used for this preliminary work. The generalized seeing monitor gives us the ability to truly monitor the full spatio-temporal parameters of atmospheric turbulence at the time of each image record and this will be used to correct each measurement individually. Other monitors give us general quality index that will also be included in our future analysis. Finally we note that cross calibrations with space missions such as PICARD, Solar Dynamics Observatory (SDO) or any future dedicated mission are highly desirable in order to validate the corrections that can be made for compensating the bias introduced by optical turbulence.

Acknowledgments

This project is supported by the CNRS and the French space agency (CNES), as part of the PICARD mission. We thank Laboratoire d'Optique Atmosphérique (University of Lille 1, France), and Laboratoire d'Etudes Spatiales et d'Instrumentation en Astrophysique (LESIA, France) for their support. Numerous individuals have been involved in this project. They are all gratefully acknowledged for their respective contributions although they cannot be listed here as many of them joined the development for only few months.

References

[1] Thuillier G, Dewitte S and Schmutz W 2006 *Advances in Space Research* **38** 1792 – 1806
[2] Laclare F, Delmas C, Coin J P and Irbah A 1996 *Solar Phys.* **166** 211–229
[3] Morand F, Delmas C, Corbard T, Chauvineau B, Irbah A, Fodil M and Laclare F 2010 *Comptes Rendus Physique* **11** 660–673
[4] Meftah M, Irbah A, Corbard T, Morand F, Thuillier G, Hauchecorne A, Ikhlef R, Rouze M, Renaud C, Djafer D, Abbaki S, Assus P, Chauvineau B, Cissé E M, Dalaudier F, D'Almeida E, Fodil M, Laclare F, Lesueur P, Lin M, Marcovici J P and Poiet G 2012 *SPIE* (*Society of Photo-Optical Instrumentation Engineers Conference Series* vol 8446)
[5] Irbah A, Corbard T, Assus P, Borgnino J, Dufour C, Ikhlef R, Martin F, Meftah M, Morand F, Renaud C and Simon E 2010 *SPIE* (*Society of Photo-Optical Instrumentation Engineers Conference Series* vol 7735)
[6] Sofia S, Heaps W and Twigg L W 1994 *Astrophys. J.* **427** 1048–1052
[7] Brown T M and Christensen-Dalsgaard J 1998 *Astrophys. J. Lett.* **500** L195
[8] Ikhlef R, Corbard T, Irbah A, Morand F, Fodil M, Chauvineau B, Assus P, Renaud C, Meftah M, Abbaki S, Borgnino J, Cissé E M, D'Almeida E, Hauchecorne A, Laclare F, Lesueur P, Lin M, Martin F, Poiet G, Rouzé M, Thuillier G and Ziad A 2012 *SPIE* (*Society of Photo-Optical Instrumentation Engineers Conference Series* vol 8444)
[9] Emilio M, Kuhn J R, Bush R I and Scholl I F 2012 *Astrophys. J.* **750** 135 (*Preprint* 1203.4898)

Journal of Physics: Conference Series **440** (2013) 012004 doi:10.1088/1742-6596/440/1/012004

On the co-alignment of solar telescopes. A new approach to solar pointing.

J Staiger
Kiepenheuer Institut, 79104 Freiburg, Schöneckstrasse 6, Germany

E-mail: joe@kis.uni-freiburg.de

Abstract. Helioseismological measurements require long observing times and thus may be adversely affected by lateral image drifts as caused by pointing instabilities. At the Vacuum Tower Telescope VTT, Tenerife we have recorded drift values of up to 5" per hour under unstable thermal conditions (dome opening, strong day-to-day thermal gradients). Typically drifts of 0.5" - 1.0" per hour may be encountered under more favorable conditions.
Past experience has shown that most high-resolution solar telescopes may be affected by this problem to some degree. This inherent shortcoming of solar pointing is caused by the fact that the guiding loop can be closed only within the guiding beam but not within the telescope's main beam.
We have developed a new approach to this problem. We correlate continuum brightness patterns observed from within the telescope main beam with patterns originating from a full disk telescope. We show that brightness patterns of sufficient size are unique with respect to solar location at any instant of time and may serve as a location identifier. We make use of the fact that averaged location information of solar structures is invariant with respect to telescope resolution. We have carried out tests at the VTT together with SDO. We have used SDO as a full disk reference. We were able to reduce lateral image drifts by an order of magnitude.

1. Pointing Systems

Solar pointing systems have to provide for easy and convenient access to solar targets. The pointing precision should match the telescope's resolution. Existing systems fall into two categories which may be distinguished by their feedback properties: Closed-loop and open-loop systems.

1.1. Closed Loop Systems

Closed loop systems require the presence of a full disk telescope (guiding telescope) which has to be physically connected to the primary telescope. The disk center as derived from limb sensors or PSD's will define the coordinates system's origin. Opto-mechanical feedback will hold the disk image in place (Fig. 1: VTT Limb Guider).
Examples: Dick Dunn Telescope, NSO Sacramento Peak; Vacuum Tower Telescope (VTT), KIS Tenerife

Journal of Physics: Conference Series **440** (2013) 012004 doi:10.1088/1742-6596/440/1/012004

1.2. Open Loop Systems

Open loop systems do not require specific opto-mechanical pointing components. Solar coordinates are derived from ephemerisis tables and geographical data. Atmospheric refraction effects together with mechanical and thermal telescope properties have to be taken into account by model algorithms. Telescope drive encoders will follow the model data to access a desired target [1]
Examples: GREGOR, KIS Tenerife (Future); ATST, NSO Hawaii (Future)

Fig. 1: VTT Limb Guider

Fig. 2: FOV Drift with time

2. The stability- and co-alignment problem.

Spurious image drifts are a common problem with solar telescopes during observations of extended duration. Fig. 2 displays the relative migration of a 50"-by-50" FOV within the primary focal plane of the VTT, Tenerife during 5 hours of observation. The lateral shifts sum up to approx. 10".
The problem will cause a loss of position and will reduce the effective size of the observing area with time. It may complicate the spatial co-alignment of multiple telescopes as the contributing systems may suffer from the same problem in a non-synchronized way. This will especially be the case when slit-spectrograph observations with narrow footprints are to be synchronized.

If the structures to be observed can be uniquely identified by their instantaneous shape (sunspots, pores etc.), correlation based tools may allow to freeze a region of interest into the viewing area. A most elaborate example for this type of main-beam closed-loop feedback has been developed for SOT telescope onboard space-based HINODE. This unit allowed for image stabilization at a level of below 0.03 arcsec (3 σ) [2]. These unique stability values have been a primary cornerstone for the huge success of HINODE/SOT [3];

Yet the basic problem of establishing a stable and absolute heliographic reference frame within the primary focal plan cannot be resolved by any of the classical guiding means. This is owed to the fact that the only existing quasi shape-invariant fiducial on the sun is the solar limb. Spatial feedback in absolute heliographic terms may thus only be established from full-disk guiding telescopes where the limb is permanently present but not from primary telescopes where this is not the case.
Tight opto-mechanical coupling between the guiding beam and the main beam will partially compensate for these inherent shortcomings. But any shape changes of telescope structure as caused by unbalanced thermal or mechanical loads which are beyond the reach of the guiding beam will not be detected. These undetected loads are a primary source of pointing instabilities and image drifts.

Any project to address this longstanding issue will thus have to find a solution for the following tasks:

- Find main-beam fiducial
- Establish main-beam absolute heliographic feedback.

3. The new approach

3.1. Hypothesis

The instantaneous continuum brightness pattern surrounding a given observational position as viewed by the primary telescope is a unique and unambiguous identifier in determining the absolute heliographic coordinates of this position.

3.2. Stochastic View

The autocorrelation analysis of quiet sun continuum fluctuations lets suggest that adjacent pixels have only a low degree of spatial and temporal coherence when viewed at image resolutions of lower than 1". This goes back to the irregular and non-predictable distribution of bright granular and dark inter-granular areas across the solar surface.

A probability measure for the occurrence of a specific intensity pattern may thus be derived when assuming a random formation process. It will depend on the number of observed pixels and the range of intensity variations within the digital resolution of the receiver. In the case of a 10-by-10 pixel field and the minimum 2-bit dynamic sub-range for the brightness fluctuations (binary representation) the result would be of the order of

$$P \approx 7.8 * 10^{-29}$$

3.3. Pointing Relevancy

At such low probability levels a pattern may be regarded as to be unique with respect to all other concurrently existing patterns across the solar disk. This should allow for unambiguously locating the primary beam pattern within a simultaneously taken full-disk image. Absolute heliographic coordinate information may then be derived. This in turn may be fed back into a main-beam closed-loop circuit to compensate for any pattern displacements.

4. Technical requirements

Table 1. The following table will give an overview over the opt-mechanical and computational equipment which has to be provided for the new pointing tools:

.

Component	*Requirement*
Main Beam Imaging:	A slit-jaw image (1D or 2D) of the observation field must be provided
Full-disk Imaging	A full-disk image must be provided. Telescope aperture requirements of 10 – 15 cm are sufficient. No mechanical coupling to primary telescope required. No coordinate information from other guiding-telescope required. Telescope may be stand-alone or space-based.

Detectors	CCD camera for slit-jaw image. CCD camera for full-disk image (2K-by-2K or higher).
Networking	Primary telescope and full-disk telescope must be linked for imaging synchronization.
Computing	Real-time pattern processing must be available.

5. Procedural Steps

Several steps have to be taken in sequential order to receive a correctional signal and feed it back into the control loop.

- Take slit-jaw image.
- Take full-disk image.
- Search full-disk for match with slit-jaw pattern.
- Cross-correlate both patterns for closest fit
- Get solar center from full-disk.
- Get offset of full-disk match pattern with respect to solar center.
- Get offset of slit-jaw pattern with respect to full-disk match pattern.
- Add both offsets to retrieve main beam absolute heliographic coordinates .
- Feed back (optional) correctional displacement signal into main beam controls.

6. Test Results with SDO and VTT

A proof-of-concept experiment was carried out at the Izana Observatory, Tenerife in July and October 2012. Continuum data were recorded with the new HELLRIDE instrument [4]. SDO was used as remote full-disk guiding telescope for the new pointing model. SDO web-server time delays required post-facto processing. Performance of the VTT‘s standard limb guider was compared with new system.

Figures 3 and 4 below show stability performance during 5 hours duration. Green circles mark 1“, 5“ and 10“ deviation range with respect to start position. White dot marks instantaneous correlation peak of cross-correlation between full-disk pattern and slit-jaw pattern during final measurement step.

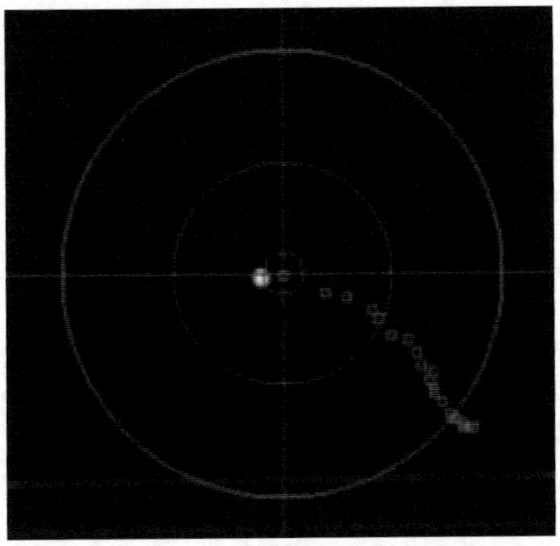

Fig. 3: Stability of "Classical" Limb Guider

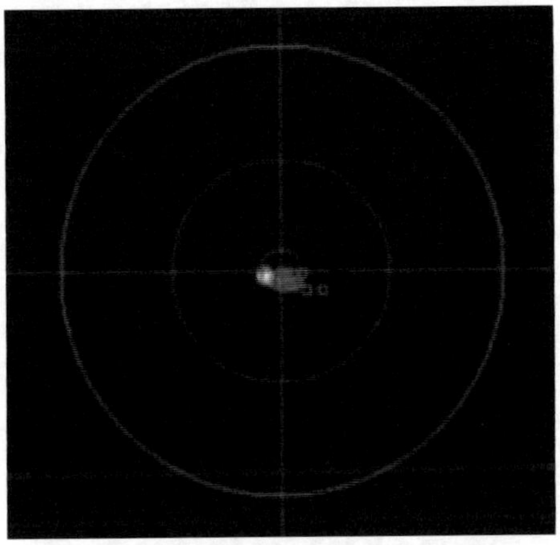

Fig. 4: Stability of New System

7. Benefits

We expect the new pointing model to offer various advantages over existing systems within different fields as listed in the following table

Field	*Possible Benefits*
Engineering	Minimum technical development efforts required. (no mechanical coupling to primary telescope) Frequently used software tools may be applied. Moderate maintenance requirements due to simplicity of system.
Cost	Off the shelf-components may be used (computing, cameras, networking etc.). No customized guiding optics required. Standard astronomical telescope of moderate aperture with standard drives may be used as full-disk guiding telescope. Multiple telescopes may share single guiding telescope
Operational	Precise co-alignment of multiple telescopes possible over extended times Thermal image drifts reduced by appr. order of magnitude Re-adjustment process by more than two orders of magnitude faster than with existing procedures (no main beam limb access required). New approach can be applied to any solar telescope. System may provide input for 'absolute' correlation tracking.

Science	High cadence 'pinpoint' spectrograph observations possible ('sit-and-stare'). Determination of precise horizontal flow velocities possible due to 'absolute' heliographic pointing. No loss of observational area during long-duration measurements. Precise and fast access to dedicated locations (e.g. flare events etc.).

8. Operation Modes

The new pointing method will allow for a standard drift-compensation mode (slow) and a combined drift/jitter compensation mode (fast). In standard operational mode the cadence for updating the reference full-disk image and the main-beam image will be of the same order and will define the cadence of compensational feedback. We consider an update cadence of the order of 10 to 20 sec sufficient to keep the image motion during this time of below 0.05 arcsec as caused by solar rotation. The tools may also be used as an add-on to classical guiding.
In fast mode the reference imaging cadence and the main-beam cadence are differing. The main-beam image is now updated at high cadence. The reference image may then be used to detect any rapid image motions to serve as feedback-input to piezo-optics. In this mode the new system would operate as a correlation tracking device freezing the object of interest for the first time into an absolute heliographic framework.

9. Conclusions

We have developed a new approach to solar pointing. It should allow the precise co-alignment of multiple telescopes. New forms of determining lateral velocities in absolute heliographic terms should be possible eg. the recording of the proper motions of magnetic elements within their field environment without the necessity to lock to these elements.
We used SDO and VTT as a proof-of-concept system for the new tools. During an observation campaign in July 2012 we were able to reduce the thermally related image drifts of the VTT by an order of magnitude. Due to a 20-to-40 min. delay in availability of SDO continuum images we have not yet been able to test the system under real-time conditions. We plan to carry out such tests in 2013 at the VTT.
We think that the closed-loop procedures described here will enable us for the first time to continuously span up an absolute heliographic coordinate grid within a solar telescope's main beam which remains in place with time.

References

[1] Wallace P.T., 2006 *ATST Pointing and Tracking PTW/001*

[2] Shimizu T, Nagata S, Tsuneta S. et al., 2008 Solar Physics, **249** 221

[3] Tsuneta S., Itchimoto K, Katsukawa Y et al., 2008 Solar Physics, **249** 167

[4] Staiger J, 2011 *Astron.&Astroph.* **835** 83

Observations of the radial velocity of the Sun as measured with the novel SONG spectrograph: results from a 1-week campaign

P.L. Pallé [1,2], F. Grundahl [3], A. Triviño Hage[1,2], J. Christensen-Dalsgaard[3], S. Frandsen[3], R. A. García[4], K. Uytterhoeven [1,2], M. F. Andersen[3], P. K. Rasmussen[5], A. N. Sørensen[5], H. Kjeldsen[3], P. Spano[8] , H. Nilsson[7], H. Hartman[7] , U.G. Jørgensen[5,6], J. Skottfelt[5,6], K. Harpsøe [5,6], M. I. Andersen[5]

[1] Instituto de Astrofísica de Canarias, E-38200 La Laguna, Tenerife, Spain
[2] Dept. de Astrofísica, Universidad de La Laguna, E-38206 La Laguna, Tenerife, Spain
[3] Stellar Astrophysics Centre, Department of Physics and Astronomy. Aarhus University, Denmark
[4] Laboratoire AIM, CEA/DSM-CNRS-Université Paris Diderot; CEA, IRFU, SAp, F-91191, Gif-sur-Yvette, France
[5] Niels Bohr Institute, University of Copenhagen, Juliane Maries Vej 30, 2100, Kbenhavn Ø, Denmark
[6] Centre for Star and Planet Formation, Natural History Museum of Denmark, University of Copenhagen, Øster Voldgade 5-7, 1350, København K, Denmark
[7] Lund Observatory, Lund University, Box 43, SE-221 00 Lund, Sweden
[8] INAF, Osservatorio Astronomico di Brera, Italy

E-mail: `pere.l.palle@iac.es`

Abstract. Deployment of the prototype node of the *SONG* project took place in April 2012 at Observatorio del Teide (Canary Islands). Its key instrument (echelle spectrograph) was installed and operational a few weeks later while its 1 m feeding telescope suffered a considerable delay to meet the required specifications. Using a fibre-feed, solar light could be fed to the spectrograph and we carried out a 1-week observing campaign in June 2012 to evaluate its performance for measuring precision radial velocities. In this work we present the first results of this campaign by comparing the sensitivity of the SONG spectrograph with other helioseismology reference instruments (*Mark-I* and *GOLF*) when simultaneous data are considered.

1. Introduction

The Stellar Observations Network Group (SONG) is an international initiative, led by the Stellar Astrophysics Centre (Aarhus University) and Niels Bohr Institute (University of Copenhagen), to design and build a global network of small telescopes (Grundahl et al. [5]). The goal of SONG is to become a key facility in both asteroseismology and planet-search research programs, a facility that provides state-of-the-art data of a quality that could not be achieved by use of any other space-based or ground-based facility. SONG is proposed to have a total of eight nodes in the northern and southern hemisphere. By placing the telescopes at roughly equally spaced longitudes, long-term nearly continuous observations can be obtained. Each telescope has an aperture of 1 m and will be equipped with a high-resolution spectrograph for measuring very precise stellar radial velocities and dual-colour lucky-imaging cameras for photometry of faint stars in crowded fields. The first-prototype node was completely deployed in April 2012 at Observatorio del Teide, a well-reputed observing site already hosting nodes of the major helioseismology networks (BiSON and GONG). In this work, we present the results obtained with the novel SONG spectrograph when observing a reference star, the Sun, and further to compare the achieved sensitivity to velocity fluctuations with simultaneous data obtained with specific helioseismic instruments such as Mark-I (located few hundred meters away from SONG) and GOLF (aboard the SoHO satellite).

2. Observations and data analysis

2.1. SONG spectrograph

As part of the original design, the system included a fibre feed which allows us to direct sunlight into the spectrograph (see Fig. 1). The first test results were very encouraging and we subsequently mounted the fibre on a normal-incidence pyrheliometer tracking the Sun and carried out a short campaign of a week (11 to 16 June 2012) in which we observed continuously during a 10 h period each day. The SONG echelle spectrograph is equipped with an Iodine cell for wavelength reference, covering the interval 4800 to 6700 Å and a spectral resolution up to 110000 and CCD focal plane of 2K x 2K pixels. Images were taken continuously every 2.8 s and basic reduction was performed using an ad-hoc pipeline written by one of us (FG) and still under development (Triviño et al. [9]). The spectra are extracted using the REDUCE package developed by Piskunov and Valenti [8], while the precise line-of-sight velocity determination (`iSONG` package) is based on the technique developed by Butler et al. [2]. In Fig. 2 the obtained solar radial velocity, residuals and their Power Spectrum Density (PSD) are shown.

Figure 1. SONG telescopic installation at Observatorio del Teide (telescope & laboratory) and a zoom at the normal incidence pyrheliometer with the attached fibre optic to feed the SONG echelle-spectrograph (bottom right) with solar light.

2.2. Mark-I spectrophotometer

Mark-I is a sun-as-a-star resonant scattering spectrophotometer (Brookes et al. [1]) operating at the blue and red wings of the potassium KI 7699 Å line. It is a reference instrument in helioseismology and the seed node of the BiSON Network that has been continuously operated since 1976 at Observatorio del Teide. Daily data consists of a 2 s interval measurement of

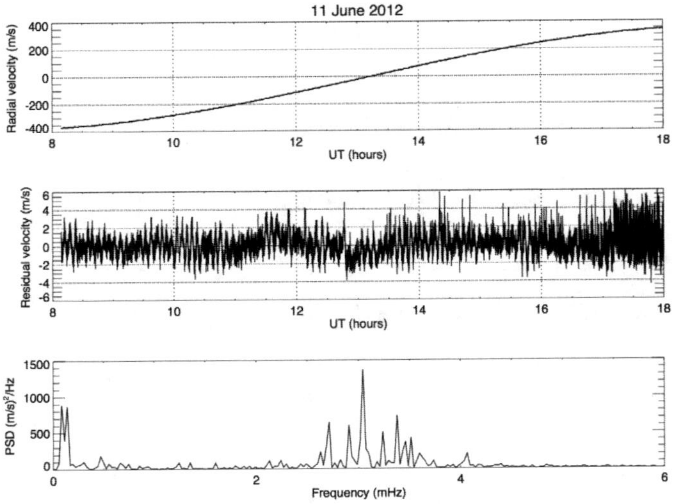

Figure 2. Reduced raw data output from a daily solar run (11 June 2012) with SONG spectrograph. The relative line-of-sight velocity (top), residuals after a polynomial fitting (center) and its PSD (bottom).

the instrumental line-of sight velocity which is further calibrated and detrended to obtain the residual velocity (see the whole procedure in Pallé et al. [7]).

2.3. GOLF/SoHO spectrophotometer

This sun-as-a-star resonant scattering spectrophotometer (Gabriel et al. [3]) provides continuous measurements (since 1996) at a single wing of the Na D1 - D2 solar spectral lines (5890 and 5895 Å). The calibrated data (García et al. [4]) consists on the solar residual velocities every 20 s in which barycentric velocity corrections were performed and low frequency trends (T > 3 days) removed. At present, the GOLF photon noise is more than 20 times higher than at the beginning of the mission.

3. Comparison between simultaneous ground-based and space observations

In order to assess and to compare the SONG spectrograph sensitivity, a well known small solar velocity feature has been chosen: the acoustic 5-min solar eigenmode signal. The characteristics of this signal as measured simultaneously by a neighboring instrument (Mark-I) and with another located in space, constitute a robust indicator of its performance. To undertake this comparison, the following procedures were applied: a) synchronization of data acquisition times; b) selection of a common sampling time for all data sets, long enough not to loose the 5-minute p-mode signal. A sampling time of 60 s was selected and averages were performed over 30 Mark-I consecutive points, 21 for SONG and 3 for GOLF. A prior filtering of high frequencies ($\nu >$ 8 mHz) was performed to avoid aliasing, with a FIR digital filter with zero-phase distortion (Oppenheim, A.V. and R.W. Schafer [6]); c) Removal of low frequencies ($\nu <$ 0. 8 mHz) to enhance and to isolate the 5-min band. The same filtering technique as in b) has been used, with a high-pass ad hoc filter design. In Fig. 3, the expanded residuals for some of the observing days and for SONG and Mark-I data, are shown.

A step further in this comparison was undertaken by computing the PSD for each daily run from 11 to 16 June, for each of the data sets and all of them having the same observational window (length). In Fig. 4 the daily averaged power spectra of the three data sets are shown.

4. Conclusions and perspectives

The stellar-designed SONG spectrograph has proved to perform extremely well, in terms of sensitivity to small velocity fluctuations and mid-term stability, along its *first-light* measurements

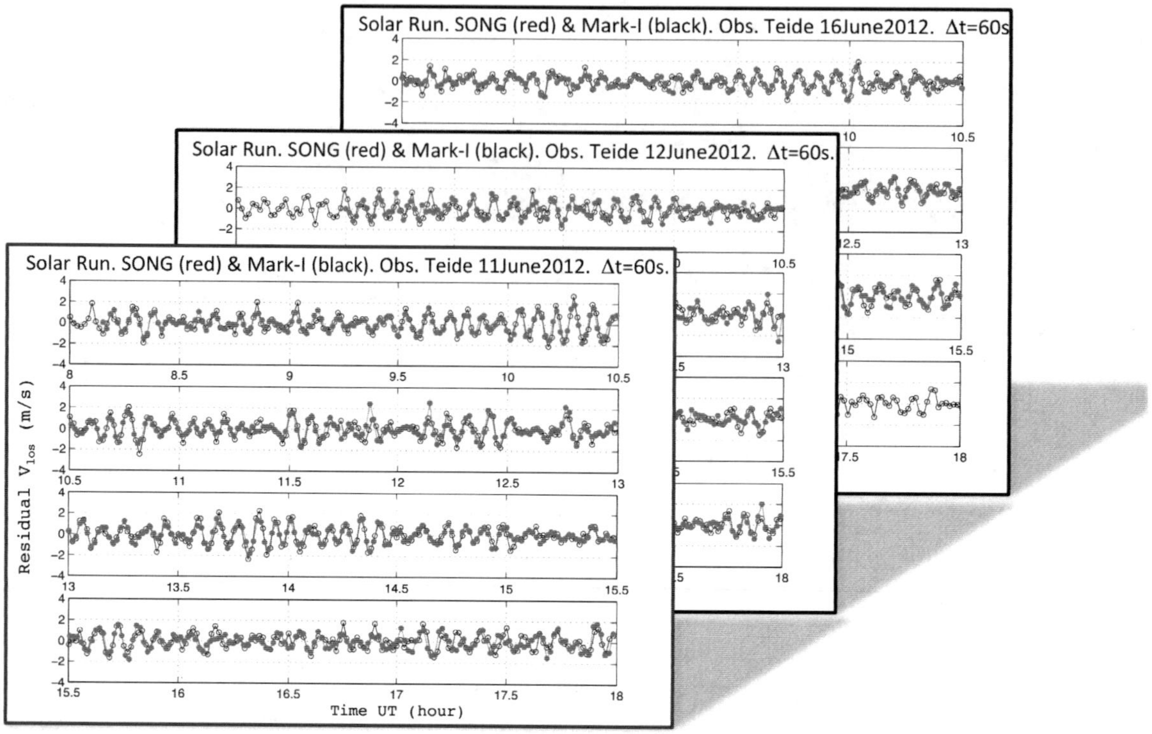

Figure 3. SONG and Mark-I residual velocity signals for different observing days of the run. Each daily signal is expanded in 2.5 hours raw to clearly distinguish the coincidence of the signals and the 5-min feature.

of a particular stellar object: the Sun. Comparative analysis with simultaneous ground and space data gathered with other instruments (Mark-I and GOLF/SoHO) has been performed and confirmed the expected performance of SONG spectrograph. A more extensive analysis based of a six days time series is currently ongoing and it will allow (although limited resolution of 2 μHz) to compare some properties of individual solar p-modes as seen with the different instruments (wavelengths). These obtained results are interesting enough to consider the possibility to perform future routine solar observations with SONG.

Acknowledgments
Based on observations made with the SONG telescopic installation operated on the island of Tenerife in the Spanish Observatorio del Teide of the Instituto de Astrofísica de Canarias. Funding for the Stellar Astrophysics Centre is provided by The Danish National Research Foundation (Grant DNRF106). The research is supported by the ASTERISK project (ASTERoseismic Investigations with SONG and Kepler) funded by the European Research Council (Grant agreement n. 267864) and by the Spanish Ministry of Economy Competitiveness (MINECO) grant AYA2010-17803. The Danish SONG team gratefully acknowledges financial support from the Villum Foundation, the Carlsberg Foundation and the Danish Natural Sciences Research Council. SoHO is a space mission of international cooperation between ESA and NASA.

Journal of Physics: Conference Series **440** (2013) 012051 doi:10.1088/1742-6596/440/1/012051

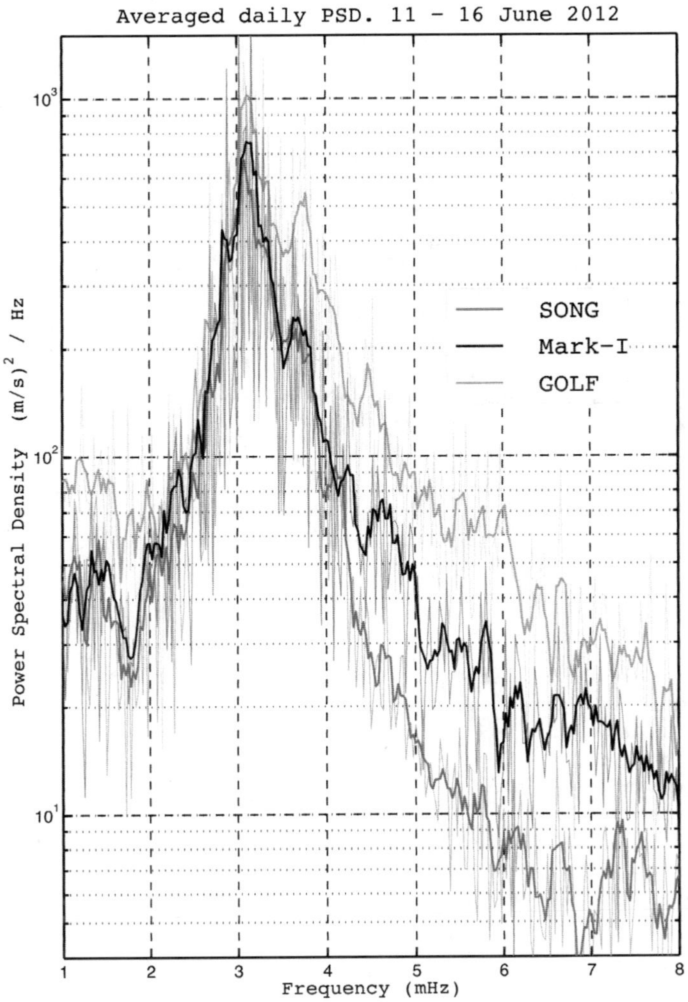

Figure 4. Average of the 6 daily power spectrum density (PSD) computed for each of the three simultaneous data sets. Thin lines correspond to the raw spectra and thick lines are the smoothed ones (5 points). The frequency resolution of the spectra is of $\sim$ 27 μHz. The power of the peaks around 3 mHz is slightly different for the various data sets (40 %) and most likely related with the different sensitivity to the sounded layers at the solar atmosphere. At high frequencies (dominated by photon noise), SONG data shows a 2.5 and 4.4 lower values than Mark-I and GOLF respectively.

References

[1] Brookes J R, Isaak G R and van der Raay H B 1978 *MNRAS* **185** 1
[2] Butler R P, Marcy G W, Williams E, McCarthy C, Dosanih P and Vogt S S 1996 *Astrophys. J.* **108** 500
[3] Gabriel A H, Grec G, Charra J, Robillot J-M, Roca Cortés T, Turck-Chièze S, Bocchia R, Boumier P, Cantin M, Cespedes E, Cougrand B, Crétolle J, Damé L, Decaudin M, Delache P, Denis N, Duc R, Dzitko H, Fossat E, Fourmond J-J, García R et al. 1995 *Solar Phys.* **162** 61
[4] García R A, Turck-Chièze S, Boumier P, Robillot J-M, Bertello L, Charra J, Dzitko H, Gabriel A H, Jiménez-Reyes S J, Pallé P L, Renaud C, Roca Cortés T and Ulrich R K 2005 *Astron. & Astroph.* **442** 385
[5] Grundahl F, Christensen-Dalsgaard J, Græ Jørgensen U, Frandsen S, Kjeldsen H and Rasmussen P K 2011 *J. Phys.: Conf. Series* **271** 012083
[6] Oppenheim A V and Schafer R W 1989 *Discrete-Time Signal Processing*, (New Jersey: Prentice-Hall) pp 284-285
[7] Pallé P L, Fossat E, Régulo C, Loudagh S, Schmider F X, Ehgamberdiev S, Gelly B, Grec G, Khalikov S and Lazrek M 1993 *Astron. & Astroph.* **280** 324
[8] Piskunov N E and Valenti J A 2002 *Astron. & Astroph.* **385** 1095
[9] Triviño Hage A, Uytterhoeven K, Grundahl F, Pallé P.L. and the SONG Team 2012 *Astron. N.* **333** 1107

Journal of Physics: Conference Series **440** (2013) 012005 doi:10.1088/1742-6596/440/1/012005

Magnetic tornadoes and chromospheric swirls – Definition and classification.

Sven Wedemeyer[1], Eamon Scullion[1], Oskar Steiner[2], Jaime de la Cruz Rodriguez[3] and Luc Rouppe van der Voort[1]

[1] Institute of Theoretical Astrophysics, University of Oslo, P.O. Box 1029 Blindern, N-0315 Oslo, Norway

[2] Kiepenheuer Institute for Solar Physics, Schöneckstr. 6-7 D-79104 Freiburg, Germany

[3] Department of Physics and Astronomy, Uppsala University, Box 516, SE-75120 Uppsala, Sweden

E-mail: sven.wedemeyer@astro.uio.no

Abstract. Chromospheric swirls are the observational signatures of rotating magnetic field structures in the solar atmosphere, also known as magnetic tornadoes. Swirls appear as dark rotating features in the core of the spectral line of singly ionized calcium at a wavelength of 854.2 nm. This signature can be very subtle and difficult to detect given the dynamic changes in the solar chromosphere. Important steps towards a systematic and objective detection method are the compilation and characterization of a statistically significant sample of observed and simulated chromospheric swirls. Here, we provide a more exact definition of the chromospheric swirl phenomenon and also present a first morphological classification of swirls with three types: (I) Ring, (II) Split, (III) Spiral. We also discuss the nature of the magnetic field structures connected to tornadoes and the influence of limited spatial resolution on the appearance of their photospheric footpoints.

1. Introduction

The combination of photospheric vortex flows and magnetic fields produces rotating magnetic field structures in the solar atmosphere, which are both seen in high-resolution observations and numerical simulations (Wedemeyer-Böhm et al. 2012, hereafter Paper I). The streamlines, which trace the simulated velocity field of the plasma in the rotating field structures, have a narrow footpoint in the photosphere, which broadens into a wide funnel in the chromosphere above (see Fig. 1). This appearance is reminiscent of tornadoes on Earth, which led to the name 'magnetic tornadoes' for this solar phenomenon. The term 'tornado' has already been used before for vertically aligned spiral structures in connection with prominences (see Chap. 10 by Tandberg-Hansen in Bruzek & Durrant 1977). Such 'solar tornadoes'[1] were observed with the Coronal Diagnostic Spectrometer (CDS, Harrison et al. 1995) onboard the Solar and Heliospheric Observatory (SOHO, Domingo et al. 1995), implying that rotation may play an important role for the dynamics of the solar transition region (Pike & Harrison 1997; Pike & Mason 1998; Banerjee et al. 2000). More recent observations with the Solar Dynamics Observatory (SDO, Lemen et al. 2012) revealed more details of this phenomenon (Li et al. 2012). There is some

[1] See the ESA press release at http://www.esa.int/esaCP/Pr_15_1998_i_EN.html

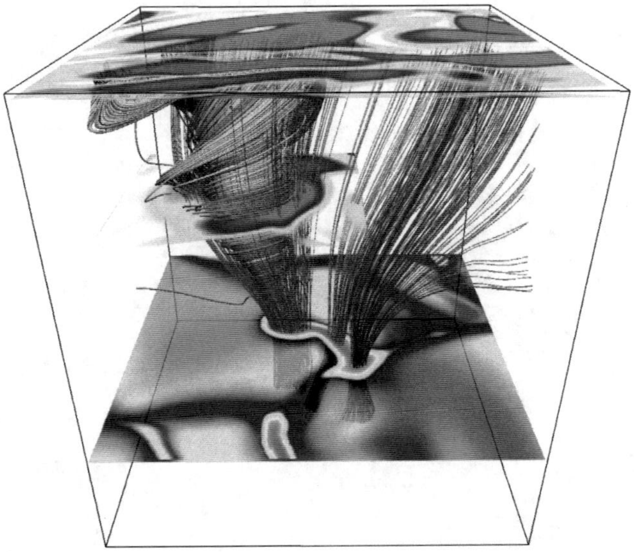

Figure 1. Visualization of a close-up region from the numerical simulation with a magnetic tornado. The red (mostly vertical) lines represent magnetic field lines, whereas the red/orange/yellow wound lines trace the velocity field in the tornado in this snapshot. The lower surface shows the granulation superposed with the absolute magnetic field strength at the bottom of the photosphere. The other layers show the horizontal velocity at heights of 1000 km and 2000 km with pink marking the highest speeds. Created with VAPOR (Clyne et al. 2007).

indication that these giant tornadoes can be explained as rotating magnetic structures driven by underlying photospheric vortex flows just like the magnetic tornadoes described in Paper I (Su et al. 2012; Orozco Suárez et al. 2012). The connection between the prominence-related giant tornadoes and the smaller magnetic tornadoes is not clear yet. However, it is established that photospheric vortex flows, which are an essential ingredient of tornadoes, are a common phenomenon on the Sun. Vortex flows are seen both in observations (e.g. Brandt et al. 1988; Bonet et al. 2008, 2010; Steiner et al. 2010; Vargas Domínguez et al. 2011) and in numerical simulations (e.g. Nordlund 1985; Stein & Nordlund 1998; Vögler et al. 2005; Steiner et al. 2010; Kitiashvili et al. 2011; Shelyag et al. 2011; Moll et al. 2012, Paper I).

Here, we present a first definition and classification of chromospheric swirls (Wedemeyer-Böhm & Rouppe van der Voort 2009, hereafter Paper II), which is the essential observational signature of magnetic tornadoes, and address the appearance of the magnetic footpoints, which are another essential indicator of a magnetic tornado.

2. Numerical simulations

The analysis in Paper I was primarily based on numerical simulations with the 3-D radiation magnetohydrodynamics code CO^{5}BOLD (Freytag et al. 2012; Wedemeyer et al. 2004). The computational box of these models has a horizontal size of 8 Mm × 8 Mm and extends vertically from 2.4 Mm below the optical depth level $\tau_c = 1$ to 2.0 Mm above it, i.e., to the top of the chromosphere. The initial model was derived from a non-magnetic simulation, which was supplemented with an initially vertical, homogeneous magnetic field with a field strength of $B_0 = 50$ G. Periodic lateral boundaries and an open lower boundary were used, whereas the top boundary is transmitting for hydrodynamics and outward radiation. The tangential magnetic field component vanishes at the top boundary, i.e., there, the magnetic field is vertical.

Journal of Physics: Conference Series **440** (2013) 012005 doi:10.1088/1742-6596/440/1/012005

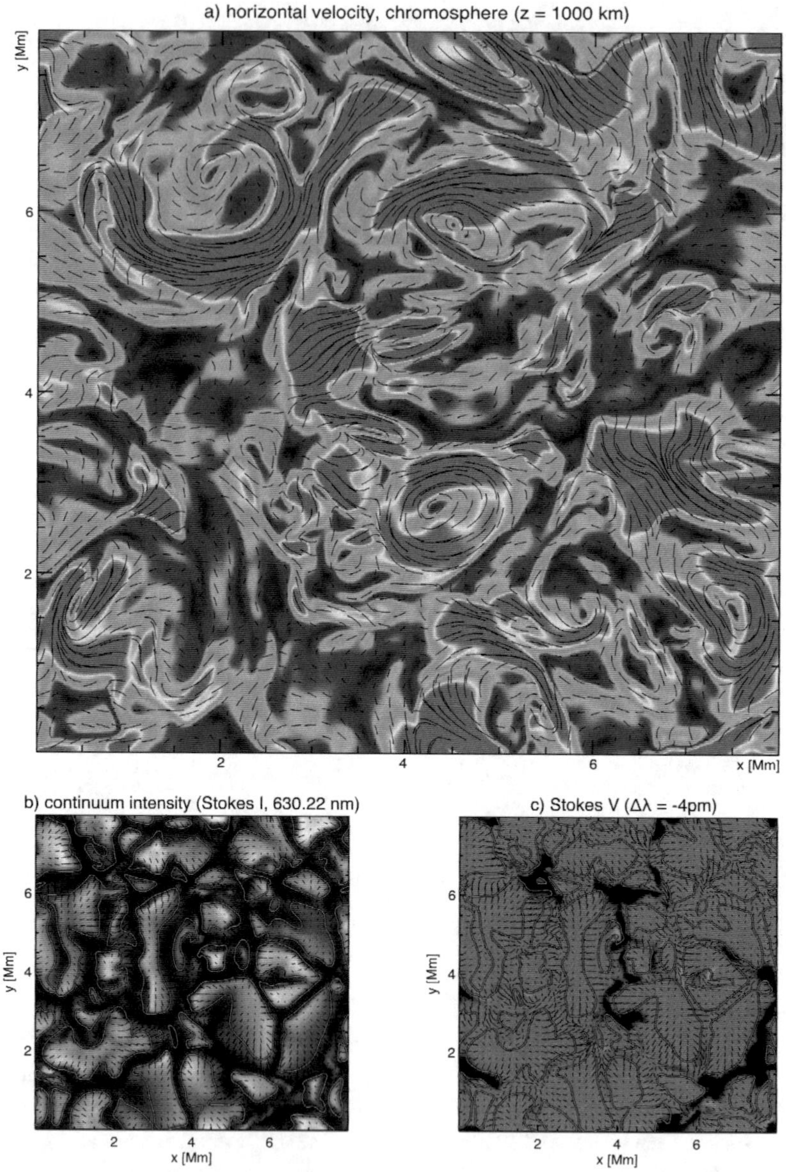

Figure 2. **(a)** Horizontal cross-section through a simulation snapshot at $z = 1000\,\mathrm{km}$ (chromosphere) displaying in colors the absolute value of the horizontal velocity and the projected streamlines. **(b)** The corresponding continuum intensity (Stokes I at $\lambda = 630.22\,\mathrm{nm}$) and **(c)** Stokes V at $\Delta\lambda = -4\,\mathrm{pm}$ from line center. The contours for $v_z = 0\,\mathrm{km/s}$ in the lower panels outline the granule boundaries, while the streamlines trace the horizontal velocity field.

No artificial limiting of the Alfvén speed has been imposed. The analysis of the magnetic tornadoes in Paper I was done on a sequence with 1 s cadence. In Fig. 2, cross-sections through one of these snapshots are shown. The upper panel displays the (horizontal) flow pattern in the chromosphere. There are about a dozen chromospheric swirls of different types and sizes (some cases more obvious than others), indicating that swirl events are very common in the simulations. Typically, every second magnetic footpoint in the photosphere is connected to a swirl in the chromosphere above.

3. Magnetic tornadoes and chromospheric swirls – Definition and classification

3.1. Conditions for the formation of tornadoes

In the photosphere, the plasma flows away from the granule interiors and towards the intergranular lanes, where the cooled plasma shoots back down into the convection zone (see Fig. 2b). The flow carries a net angular momentum. Due to the conservation of angular momentum, the downflows in the lanes can create vortex flows ('bathtub effect', see, e.g., Nordlund 1985). Vortex flows are formed most easily at the vertices of lanes, where plasma meets from the neighbouring granules. The flows also advect magnetic field into the intergranular lanes, where the field is concentrated and thus amplified. Consequently, vortex flows and magnetic field concentrations often exist at the same locations. A stationary magnetic tornado is generated when the magnetic field concentration is perfectly co-located with the vortex flow, resulting in the rotation of the entire magnetic structure. However, exact co-location is not always perfectly fulfilled. A magnetic field structure can be partially pushed into a vortex and out again, which results in a partial rotation of the magnetic field structure but not in a stationary tornado. We therefore expect tornadoes to exhibit a spectrum of rotational behavior, from partial to stationary rotation.

3.2. Chromospheric swirls as observable signature of tornadoes

Swirls have been discovered as dark rings in the line core of the calcium infrared triplet line at a wavelength of 854.2 nm during an observation campaign at the Swedish 1-m Solar Telescope (SST) (Scharmer et al. 2003) in 2008 (Paper II). We refer to this observational imprint as the Ca line core signature (CLCS) or Ca 854 swirl hereafter. It is the only direct indicator of chromospheric swirls and magnetic tornadoes known so far. This is due to the fact that detecting swirls is a challenging task, which requires stable, high-quality image series with high spatial and temporal resolution for a purely chromospheric diagnostic. In this respect, a narrow wavelength transmission range is crucial because the subtle swirl signature is only visible if the intensity signal is not contaminated by large contributions from the line wings, which are formed lower in the atmosphere. The CRISP instrument (Scharmer et al. 2008) at the SST offers this possibility, but even with this instrument, the observation of chromospheric swirls is still challenging.

3.3. Definition – What is a chromospheric swirl?

In accordance with Paper II, we note that the swirl pattern can be seen as darkening in the Ca II 854.2 nm line core at wavelengths down to $\Delta\lambda \approx 70\,\mathrm{pm}$. The intensity at wavelengths further away from the line core originate from deeper down in the atmosphere, where the rotating magnetic structure has a smaller extent and is thus less clearly visible. There is a corresponding brightening with mostly the same pattern in the red line wing close to the line core, produced by the Doppler shift within the swirl. Based on these observations and supporting numerical simulations, we formulate the following conditions that need to be fulfilled for a chromospheric swirl:

(i) Ca II 854 nm line core images exhibit a dark ring, a ring fragment or a spiral (see Fig. 3).

(ii) The Ca 854 swirl is rotating.

(iii) The Ca 854 swirl is visible for several minutes.

(iv) Vertical velocities (Doppler-shifts at disk centre) on the order of at least 2 km/s and more can be measured at locations coinciding with the Ca 854 swirl.

(v) A photospheric magnetic field concentration is visible at the same location, e.g., in the form of a magnetic bright point and/or in the magnetogram.

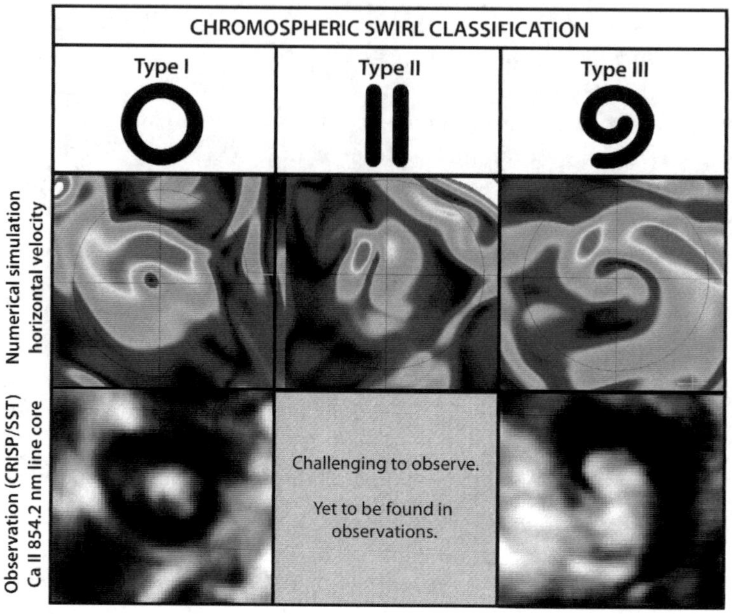

Figure 3. The three major types of chromospheric swirls: Type I (Ring), type II (Split), and type III (Spiral). The middle row shows the color-coded horizontal velocity in horizontal cross-sections at $z = 1000\,\mathrm{km}$ for swirls in the numerical simulation (Paper I). Observed examples of type I and III swirls are presented in the bottom row. The images are taken in the line core of the Ca II infrared triplet line at 854.2 nm (Paper II).

3.4. Classification

Swirls can have different shapes in the Ca line core images as already stated in Paper II. The shapes are also seen in horizontal cross-sections of the horizontal velocity at chromospheric heights in the numerical models. The three most prominent types are shown in Fig. 3:

Type I:	Ring.	Example in Fig. 2a: $[x = 4.3\,\mathrm{Mm},\ y = 2.7\,\mathrm{Mm}]$
Type II:	Split.	Example in Fig. 2a: $[x = 4.0\,\mathrm{Mm},\ y = 4.6\,\mathrm{Mm}]$
Type III:	Spiral.	Example in Fig. 2a: $[x = 1.9\,\mathrm{Mm},\ y = 6.4\,\mathrm{Mm}]$

All three types are seen in numerical simulations and types I and III are observed. A type I swirl is not necessarily a closed ring but can consist of ring fragments, too. Some simulation examples suggest that type I swirls can evolve into type II as the circular flow pattern becomes increasingly elongated until the swirl signature fades away. A type III swirl can also have multiple spiral arms. Based on a preliminary analysis of the simulations, half of the cases are classified as type I, and one fourth as type II and III, respectively. Type I swirls are possibly connected to open field structures, which can rotate more freely than the legs of closed loops. The latter build up more twist, which might result in spiral-like type III swirls.

4. Photospheric footpoints of magnetic tornadoes

The existence of a photospheric bright point as magnetic field indicator below a chromospheric swirl is a necessary condition for the detection of a magnetic tornado. However, the sizes of magnetic bright points (MBP) can be very small so that their identification and tracking over time is affected by the spatial resolution of the employed instrument. In the following, we discuss implications for the detection of tornado-related MBPs.

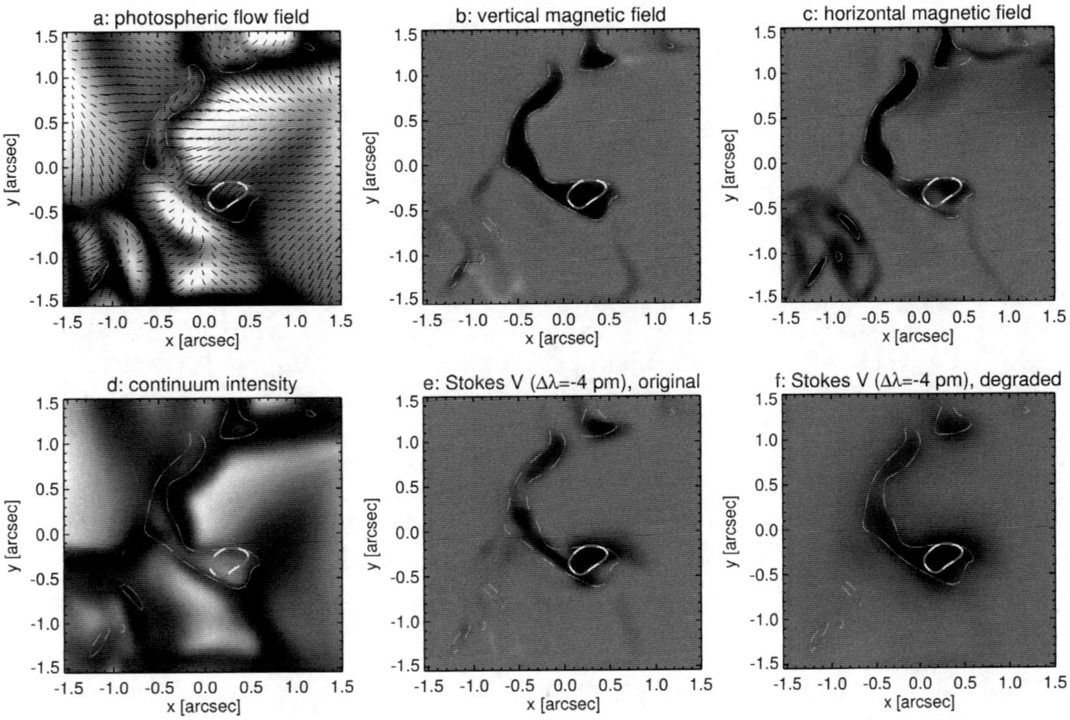

Figure 4. Effect of limited spatial resolution on the appearance of the magnetic flux topology shown for a small close-up region from a magnetohydrodynamical simulation (Fig. 2b-c, Paper I). **a)** Flow field at the bottom of the photosphere. The vertical velocity component is plotted in grey-scale, whereas the horizontal components in the plane are represented by arrows. **b)** The vertical magnetic field component and **c)** the horizontal magnetic field ($(B_x^2 + B_y^2)^{1/2}$) in the same plane. **d)** The continuum intensity (Stokes I at a wavelength of $\lambda = 630.22$ nm). **e)** Original Stokes V map at $\Delta\lambda = -4$ pm from line center and **f)** the Stokes V map after a PSF has been applied. The blue-yellow dashed lines are contours of the absolute magnetic field strength at $|B| = 400$ G, whereas the red-white dashed contour outlines the magnetic concentration that is visible in the degraded image (f).

The magnetized plasma of the solar atmosphere – A continuous medium. Starting from the initial model, the magnetic field quickly concentrates in the vertices of intergranular lanes and a complex and entangled magnetic field structure evolves. It is possible to identify groups of magnetic field lines that rotate and/or sway together. These groups usually do not persist for long because the magnetic field is continuously rearranged on short time scales connected to the lifetime of granules and the resulting changes in the photospheric flow field. Like other authors before (e.g., Stein & Nordlund 2006), we argue that the appearance of flux tubes is (at least partially) the product of the limited spatial resolution of the observations, resulting in apparently separated MBPs. It would be more appropriate to talk of a continuous magneto-fluid in which the field lines at neighbouring locations exhibit similar dynamics, rather than to describe the magnetic field as individual, isolated flux tubes.

The effect of limited spatial resolution is illustrated in Fig. 4, which displays a close-up region in a numerical simulation snapshot (see Sect. 2). As result of the convective flows (Fig. 4a), the magnetic field is arranged in the form of continuous sheets and knots (see B_z in panel b and

$B_h = (B_x^2 + B_y^2)^{1/2}$ in panel c). The strongest field concentrations appear as bright features in the continuum intensity (panel d), although the correlation between magnetic field strength and intensity is not strict. The corresponding magnetogram for the Fe I line at a wavelength of 630.2 nm is shown in panel e. This magnetogram, i.e., the Stokes V signal, is calculated with the full-Stokes radiative transfer code NICOLE (Socas-Navarro 2011; de la Cruz Rodríguez et al. 2012) along vertical lines of sight. The Stokes signal is formed over a height range in the atmosphere, whereas the magnetic field in panels b and c refers to a single horizontal cross-section at the bottom of the photosphere only. The resulting magnetogram appears therefore more intermittent and less continuous than the magnetic field in the horizontal plane. Next, the magnetogram is convolved with a point spread function (PSF, see Wedemeyer-Böhm 2008) that simulates the effect of a telescope with an aperture of 50 cm, which is equivalent to the Solar Optical Telescope (SOT, Tsuneta et al. 2008; Suematsu et al. 2008; Ichimoto et al. 2008; Shimizu et al. 2008) onboard the Hinode spacecraft (Kosugi et al. 2007). The resulting degraded magnetogram is shown in Fig. 4f. It should be noted that this degradation is optimistic. In practice, the derivation of the Stokes components from measurements with limited spatial, temporal, and spectral resolution would result in magnetograms that are even more diffuse and less reliable than the synthetic example presented here. Nevertheless, the effect of the limited spatial resolution is obvious. Only individual blobs of enhanced Stokes V signal remain instead of the continuous magnetic flux sheet in the original model atmosphere. When observed, these blobs would be identified as MBPs. The most prominent MBP is outlined with a red-white dashed contour in Fig. 4. Most of the remaining magnetic field (blue-yellow contours) would remain undetected. In particular, the thin magnetic flux sheets in the intergranular lanes are barely visible. In view of this effect, the questions arises how uniquely a MBP can be identified. During an observation run, a MBP can be tracked from its first appearance until its disappearance as long as it remains sufficiently separated from other MBPs. In that case a lifetime can be determined. However, the numerical simulations suggest that the magnetic field is continously reorganised. Some of the magnetic field, which would be observed as part of the MBP, may be advected into a neighbouring sheet and, in return, additional field may be advected into the MBP. These changes would remain undetected, whereas the same MBP appears to persist. The lifetimes of observed MBPs are consistent with the timescales on which the granulation pattern evolves (e.g., Jafarzadeh et al. 2013, and references therein) but the magnetic field (incl. individual MBPs) can nevertheless last beyond the lifetime of the surrounding granules.

5. Discussion and Conclusions

The primary way to detect a chromospheric swirl and thus a magnetic tornado is the Ca II 854.2 nm line core signature (CLCS) so far, which requires a narrow filter transmission range. These requirements make it challenging to observe chromospheric swirls and, to our knowledge, only CRISP/SST observations have repeatedly produced successful swirl detections so far. Ultimately, more observations, preferably with a variety of instruments, are needed in order to gather a statistically significant sample that allows for a detailed characterization of magnetic tornadoes, including their occurrence, distribution of sizes and lifetimes and the related energy transport rates. These small-scale events represent a challenging test for the next generation of solar telescopes like ATST (Keil et al. 2011) and EST (Collados et al. 2010). High spatial resolution is not only important for the observations but also for the numerical simulations. The successful modelling of magnetic tornadoes requires the sufficient resolution of the narrow intergranular lanes, where photospheric vortex flows and magnetic fields coalesce.

Acknowledgments

SW thanks the organizers of the conference "Eclipse on the Coral Sea: Cycle 24 Ascending" GONG 2012/LWS/SDO-5/SOHO 27, which has been held in Palm Cove, Australia, in 2012.

References

Banerjee, D., O'Shea, E., & Doyle, J. G. 2000, A&A, 355, 1152

Bonet, J. A., Márquez, I., Sánchez Almeida, J., Cabello, I., & Domingo, V. 2008, ApJ, 687, L131

Bonet, J. A., Márquez, I., Sánchez Almeida, J., et al. 2010, ApJ, 723, L139

Brandt, P. N., Scharmer, G. B., Ferguson, S., Shine, R. A., & Tarbell, T. D. 1988, Nature, 335, 238

Bruzek, A. & Durrant, C. J., eds. 1977, Astrophysics and Space Science Library, Vol. 69, Illustrated glossary for solar and solar-terrestrial physics

Clyne, J., Mininni, P., Norton, A., & Rast, M. 2007, New J. Phys, 9, 1

Collados, M., Bettonvil, F., Cavaller, L., et al. 2010, Astronomische Nachrichten, 331, 615

de la Cruz Rodríguez, J., Socas-Navarro, H., Carlsson, M., & Leenaarts, J. 2012, A&A, 543, A34

Domingo, V., Fleck, B., & Poland, A. I. 1995, Sol. Phys., 162, 1

Freytag, B., Steffen, M., Ludwig, H.-G., et al. 2012, Journal of Computational Physics, 231, 919

Harrison, R. A., Sawyer, E. C., Carter, M. K., et al. 1995, Sol. Phys., 162, 233

Ichimoto, K., Katsukawa, Y., Tarbell, T., et al. 2008, in Astronomical Society of the Pacific Conference Series, Vol. 397, First Results From Hinode, ed. S. A. Matthews, J. M. Davis, & L. K. Harra, 5

Jafarzadeh, S., Solanki, S. K., Feller, A., et al. 2013, A&A, 549, A116

Keil, S. L., Rimmele, T. R., Wagner, J., Elmore, D., & ATST Team. 2011, in Astronomical Society of the Pacific Conference Series, Vol. 437, Solar Polarization 6, ed. J. R. Kuhn, D. M. Harrington, H. Lin, S. V. Berdyugina, J. Trujillo-Bueno, S. L. Keil, & T. Rimmele, 319

Kitiashvili, I. N., Kosovichev, A. G., Mansour, N. N., & Wray, A. A. 2011, ApJ, 727, L50

Kosugi, T., Matsuzaki, K., Sakao, T., et al. 2007, Sol. Phys., 243, 3

Lemen, J. R., Title, A. M., Akin, D. J., et al. 2012, Sol. Phys., 275, 17

Li, X., Morgan, H., Leonard, D., & Jeska, L. 2012, ApJ, 752, L22

Moll, R., Cameron, R. H., & Schüssler, M. 2012, A&A, 541, A68

Nordlund, A. 1985, Sol. Phys., 100, 209

Orozco Suárez, D., Asensio Ramos, A., & Trujillo Bueno, J. 2012, ApJ, 761, L25

Pike, C. D. & Harrison, R. A. 1997, Sol. Phys., 175, 457

Pike, C. D. & Mason, H. E. 1998, Sol. Phys., 182, 333

Scharmer, G. B., Bjelksjo, K., Korhonen, T. K., Lindberg, B., & Petterson, B. 2003, in Society of Photo-Optical Instrumentation Engineers (SPIE) Conference Series, ed. S. L. Keil & S. V. Avakyan, Vol. 4853, 341

Scharmer, G. B., Narayan, G., Hillberg, T., et al. 2008, ApJ, 689, L69

Shelyag, S., Keys, P., Mathioudakis, M., & Keenan, F. P. 2011, A&A, 526, A5

Shimizu, T., Nagata, S., Tsuneta, S., et al. 2008, Sol. Phys., 249, 221

Socas-Navarro, H. 2011, A&A, 529, A37

Stein, R. F. & Nordlund, A. 1998, ApJ, 499, 914

Stein, R. F. & Nordlund, Å. 2006, ApJ, 642, 1246

Steiner, O., Franz, M., Bello González, N., et al. 2010, ApJ, 723, L180

Su, Y., Wang, T., Veronig, A., Temmer, M., & Gan, W. 2012, ApJ, 756, L41

Suematsu, Y., Tsuneta, S., Ichimoto, K., et al. 2008, Sol. Phys., 249, 197

Tsuneta, S., Ichimoto, K., Katsukawa, Y., et al. 2008, Sol. Phys., 249, 167

Vargas Domínguez, S., Palacios, J., Balmaceda, L., Cabello, I., & Domingo, V. 2011, MNRAS, 416, 148

Vögler, A., Shelyag, S., Schüssler, M., et al. 2005, A&A, 429, 335

Wedemeyer, S., Freytag, B., Steffen, M., Ludwig, H.-G., & Holweger, H. 2004, A&A, 414, 1121

Wedemeyer-Böhm, S. 2008, A&A, 487, 399

Wedemeyer-Böhm, S. & Rouppe van der Voort, L. 2009, A&A, 507, L9 (Paper II)

Wedemeyer-Böhm, S., Scullion, E., Steiner, O., et al. 2012, Nature, 486, 505 (Paper I)

Journal of Physics: Conference Series **440** (2013) 012006 doi:10.1088/1742-6596/440/1/012006

GONG Inter-site Hα Flare Comparison

Owen Giersch[1,2]
[1]Australian Bureau of Meteorology, IPS Radio and Space Service, Learmonth Solar Observatory, Exmouth, Western Australia, Australia
[2]Curtin Universty of Technology, Bentley, Western Australia

Email: owengiersch@curtin.edu.au , owen@ips.gov.au

Abstract. A challenge of the past few decades for the Solar Observing Optical Network (SOON), operated by the United States Air Force Weather Agency (AFWA), has been to obtain consistent flare brightness reporting for the same flare from different sites. Flare area is usually considered to be a more reliable measure, but significant variation of values between sites still occurs. The Global Oscillation Network Group (GONG) deployed a Hα patrol system in 2010. This provides a modern system with near identical equipment to compare flares from six different sites. The classification of flares and techniques of flare measurement will be briefly discussed. The results presented here suggest that even though different GONG sites report different flare areas and brightnesses, for some sites they vary in a consistent way allowing correction factors to be applied.

1. Flare Classification

Hα flares are classified by their area (importance) and intensity (brightness). Area is measured in millionths of the solar hemisphere (usually referred to as mils). Intensity is compared with the background Sun, with the background given a value of 100%. For a flare to be placed in a particular brightness category, it must have at least 10 mils of area above the category threshold. This classification scheme was introduced after the International Astronomical Union meeting in 1964 [1]. Tables 1 and 2 show the area and intensity categories. Figures 1 to 4 show sample results from a flare observed by GONG Hα at Big Bear. This flare reached the 1N category.

Table 1. Flare Area Categories.

Area Range (mils)	Importance
$10 \leq A < 100$	0
$100 \leq A < 250$	1
$250 \leq A < 600$	2
$600 \leq A < 1200$	3
$A > 1200$	4

Table 2. Flare Brightness Categories.

Intensity Range (%)	Brightness
$150 < I \leq 250$	F (Faint)
$250 < I \leq 350$	N (Normal)
$I > 350$	B (Brilliant)

Most observatories report only those flares that have been continuously above the minimum flare thresholds for 2 minutes or more. Plage regions that have flaring areas that do not have an area greater than 10 mils are sometimes referred to as point brightenings. Flares that have long duration but low

Journal of Physics: Conference Series **440** (2013) 012006 doi:10.1088/1742-6596/440/1/012006

brightness, often briefly dropping below threshold levels may be referred to as plage fluctuations [2].

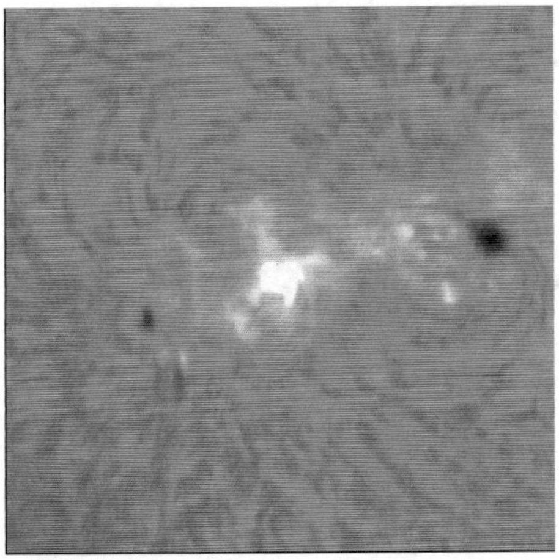

Figure 1. Sample flare image from GONG Big Bear obtained on 2012/07/30 at 15:45:54 UT (location S21.6E28.1). Image is 535 × 535 arcseconds.

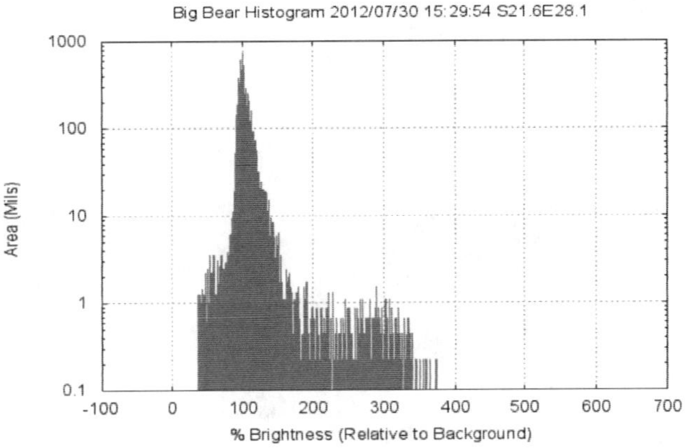

Figure 2. Sample flare histogram from GONG Big Bear.

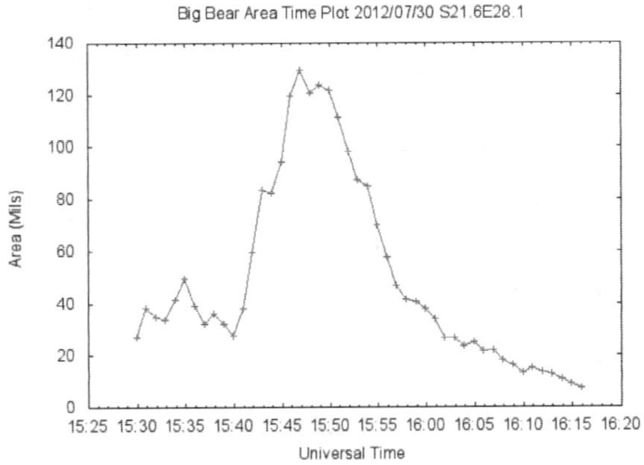

Figure 3. Sample Area vs Time Plot from GONG Big Bear.

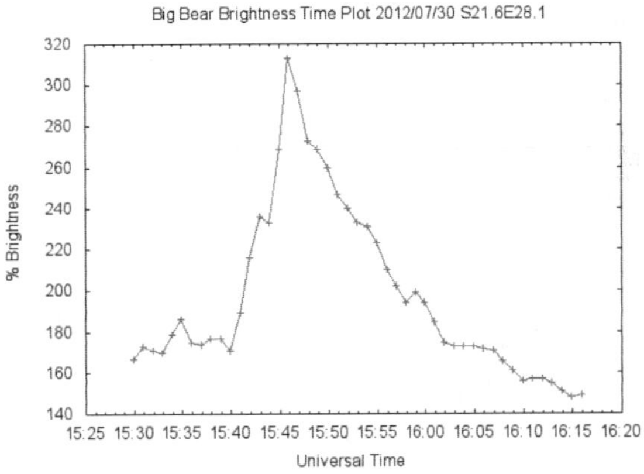

Figure 4. Sample Brightness vs Time Plot from GONG Big Bear.

2. AFWA Site Comparison

AFWA has been performing flare patrol since 1979. The SOON sites that have been operational for all or part of that time are Learmonth, San Vito, Holloman and Ramey. Flare importance and brightness were compared for the same flares from the different sites. The flare reports were obtained from the National Geophysical Data Center (NGDC) [3]. Figures 5 and 6 show results for the Holloman and Ramey comparison. These sites were chosen as they are close together (29 degrees apart in longitude). Ramey was closed in 2002.

A line can be fitted to the area, but the scatter in the points is quite high. Holloman recorded higher flare areas than Ramey. An analysis of flare brightness showed that Faint flares were in agreement

Journal of Physics: Conference Series **440** (2013) 012006 doi:10.1088/1742-6596/440/1/012006

80% of the time. However when restricting analysis to Normal and Brilliant flares (as shown in figure 6), there was agreement between sites less than 50% of the time.

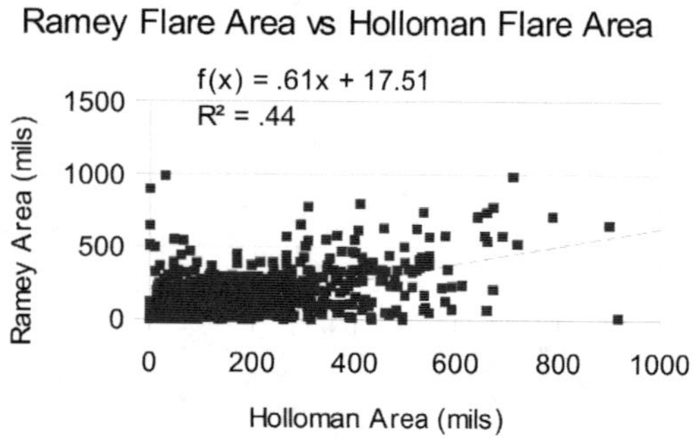

Figure 5. Ramey vs Holloman Flare Area.

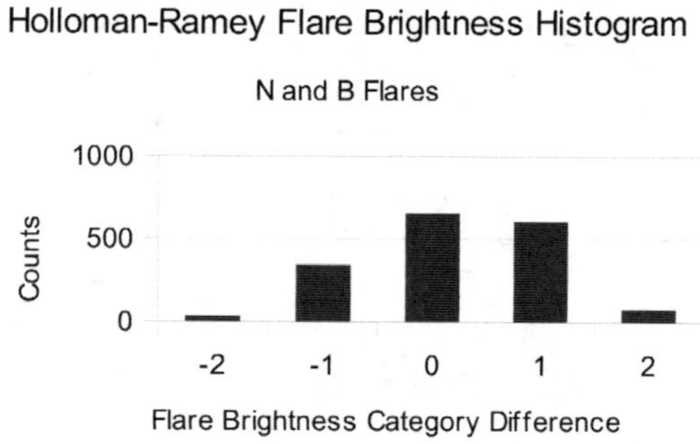

Figure 6. Ramey vs Holloman Flare Brightness. Note that individual bin values are not available in archived data, thus only flare categories were used.

3. GONG Flare Analysis

Each image is normalised by finding the central disk brightness and dark sky brightness. The images are limb darkening corrected by fitting a 5th order polynomial to the brightness at various radii of the disk. The images are checked to ensure they are cloud free. Using the limb corrected image, pixels which are 1.5 times brighter than the background Sun are flaring. Flaring pixels within 5 degrees of each other are grouped to create a flaring region. This region is analysed to produce a histogram (as shown in Figure 2). If the region meets flare criteria then the flare parameters are stored and updated as necessary. The entire process is automated in order to simulate a real time flare analysis system.

4. GONG Site Comparison

GONG Hα images were obtained from the GONG website [4]. The images selected were for July 2012 and were chosen to be in the range 1 hour either side of flares found in the Solar Geophysical Activity Summaries obtained from the Space Weather Prediction Center (SWPC) [5]. July 2012 had a particularly high number of flares (for cycle 24). Once the images were processed, the same flares were compared between the different sites. In the case of flare brightness, we have the actual bin values for the flares, and not just the categories as was the case with the USAF data. Figures 7 and 8 show a sample GONG site comparison for flare area and brightness.

Table 3 shows the slope (m), intercept (b) and correlation coefficient (r) for all of the GONG sites. Mauna Loa shows the best correlation. Few flares were detected at Udaipur due to rain and cloud and thus Udaipur as been omitted from the analysis.

Since Mauna Loa has the best correlation with all of the sites, Mauna Loa is used as a standard. Simply inverting the lines of regression from Table 3 for site comparisons involving Mauna Loa produces the correction factors. These results are summarised in Table 4.

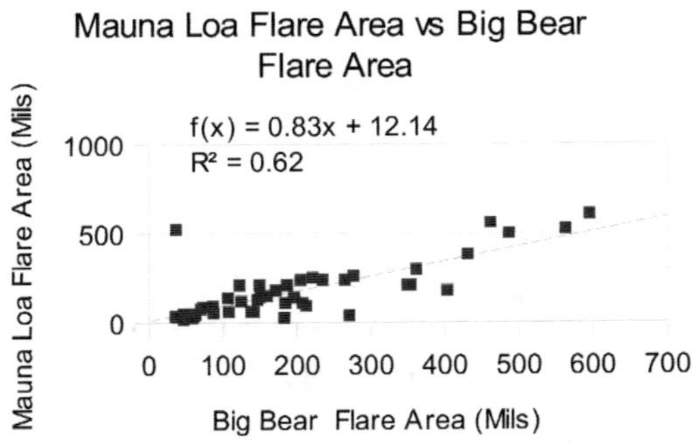

Figure 7. Mauna Loa Flare Area vs Big Bear Flare Area.

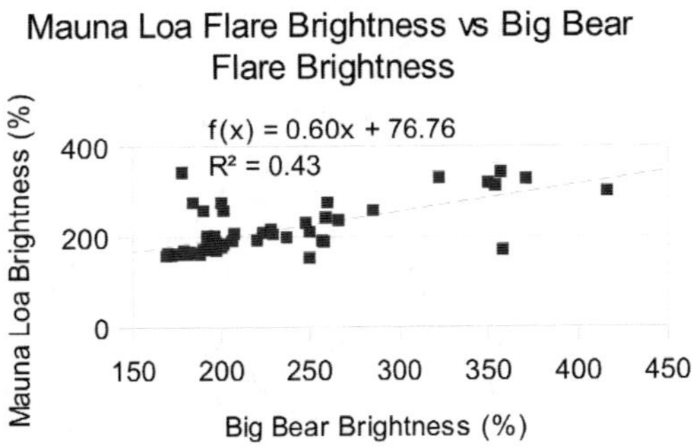

Figure 8. Mauna Loa Flare Brightness vs Big Flare Brightness

Table 3. Lines of Regression for GONG site Flare Comparison.

Sites Compared	Area			Brightness		
	m	b	r	m	b	r
Big Bear vs Cerra Tololo	0.84	88.97	0.51	1.00	38.51	0.64
Big Bear vs Learmonth	0.60	43.00	0.55	0.38	123.71	0.50
Big Bear vs Teide	0.70	150.06	0.44	1.25	6.67	0.66
Cerra Tololo vs Teide	1.19	28.21	0.93	0.89	41.46	0.70
Mauna Loa vs Big Bear	0.83	12.14	0.79	0.60	76.76	0.66
Mauna Loa vs Cerra Tololo	0.86	78.99	0.81	0.90	38.29	0.86
Mauna Loa vs Learmonth	1.17	-39.25	0.72	1.05	-19.50	0.97
Mauna Loa vs Teide	1.21	81.65	0.92	1.42	-47.24	0.91
Teide vs Learmonth	0.06	38.45	0.21	0.00	173.16	0.00

Table 4. Correction Factors for GONG Hα Flares.

Site	Area Correction		Brightness Correction	
	m	b	m	b
Big Bear	1.20	-14.63	1.67	-127.93
Cerra Tololo	1.16	-91.85	1.11	-42.54
Learmonth	0.85	33.55	0.95	18.57
Teide	0.83	-67.48	0.70	33.27

5. Summary

GONG Hα flares have been successfully corrected for all sites except for Udaipur. Mauna Loa had the best correlation with the other sites, and subsequently is being used as a standard. The corrections are a simple linear expression for each site for both flare brightness and area. This is an improvement over the USAF SOON sites, where none of the sites are able to regularly agree on flare brightness.

The disagreement between some sites could be due to cloud coverage which interrupted flares or variations in the Hα filters. A longer term study needs to be conducted to determine which, if any of these effects is responsible for these discrepancies.

With further analysis as to the causes of the GONG sites that disagree, it can be expected that the correlation between sites will improve.

6. References

[1] M. C. Ballario, A. Bruzek, E. E. Dubov, M. A. Ellison, E. R. Hedeman, V. Lincoln, R. Michard, Y. Ohman, H. J. Smith, C. S. Warwick, H. D. Prince, "Hα Flare Classification: Report of the Working Committee on the Impovement in Assignment of Hα Flare Importance", *Report to Commission on Solar Activity at the XII General Assembly of the International Astronomical Union*, Hamburg, 1964

[2] AFWAMAN 15-2: http://static.e-publishing.af.mil/production/1/afwa/publication/afwaman15-1/afwaman15-1.pdf

[3] National Geophysical Data Centre Hα flare reports: ftp://ftp.ngdc.noaa.gov/STP/space-weather/solar-data/solar-features/solar-flares/halpha/events/

[4] Global Oscillation Network Group Hα images: ftp://gong2.nso.edu/HA/haf/

[5] Space Weather Prediction Centre Solar Geophysical Activity Summary: ftp://ftp.swpc.noaa.gov/pub/warehouse/2012/SGAS/

Journal of Physics: Conference Series **440** (2013) 012007 doi:10.1088/1742-6596/440/1/012007

Ellerman bombs: fallacies, fads, usage

Robert J. Rutten[1,2], Gregal J. M. Vissers[2], Luc H. M. Rouppe van der Voort[2], Peter Sütterlin[3] and Nikola Vitas[4]

[1] Lingezicht Astrophysics, 't Oosteneind 9, 4158 CA Deil, The Netherlands
[2] Institute of Theoretical Astrophysics, University of Oslo, P.O. Box 1029, Blindern, N–0315 Oslo, Norway
[3] Institute for Solar Physics, Albanova University Center, SE–106 91 Stockholm, Sweden
[4] Instituto de Astrofísica de Canarias, C/ Via Lactea S/N, E–38200 La Laguna, Tenerife, Spain

E-mail: R.J.Rutten@uu.nl

Abstract. Ellerman bombs are short-lived brightenings of the outer wings of Hα that occur in active regions with much flux emergence. We point out fads and fallacies in the extensive Ellerman bomb literature, discuss their appearance in various spectral diagnostics, and advocate their use as indicators of field reconfiguration in active-region topography using AIA 1700 Å images.

1. Introduction

Ellerman [1] described "solar hydrogen bombs" in 1917 as intense brightenings of the extended wings of Hα, Hβ and Hγ, not visible in other lines and with the line cores unaffected. They last a few minutes and occur repetitively in active regions with much flux emergence, preferentially near and especially between penumbrae. These properties define the Ellerman bomb (EB) phenomenon. The subsequent EB literature cannot be fully reviewed here but we point out some fads and fallacies below. Bray and Loughhead [2] concluded in 1974 that *"extensive modern observations have added little to Ellerman's original description"*. This lack of progress changed with high-resolution observing, first in the Flare Genesis flight [3] [4] [5] [6] [7] [8] and more recently with the Swedish 1-m Solar Telescope (SST) [9] [10].

2. Magnetic concentrations as pseudo Ellerman bombs

Figure 1 shows snapshots from 3-hour multi-wavelength Dutch Open Telescope (DOT) movies[1]. The Hα wing image in the lower-left panel shows many bright grains that in the movie adhere to Ellerman's description: *"they seem to follow one another like the balls of a Roman candle"*. About half exceed the mean intensity by over 30%, a few are brighter than 54% ($4\,\sigma$) excess. Traditional EB threshold criteria might classify these as EBs. However, movie inspection including comparison with the parallel Ca II H and G-band movies shows that all the bright grains are simply "network bright points" marking strong-field magnetic concentrations (MCs). They are not EBs but "moving magnetic features" described by [12] [13] [14] [15] [16] [17]. Nor was there flux emergence in this region evident in SOHO/MDI magnetograms.

[1] Links: four-panel movie, wing-only movie.

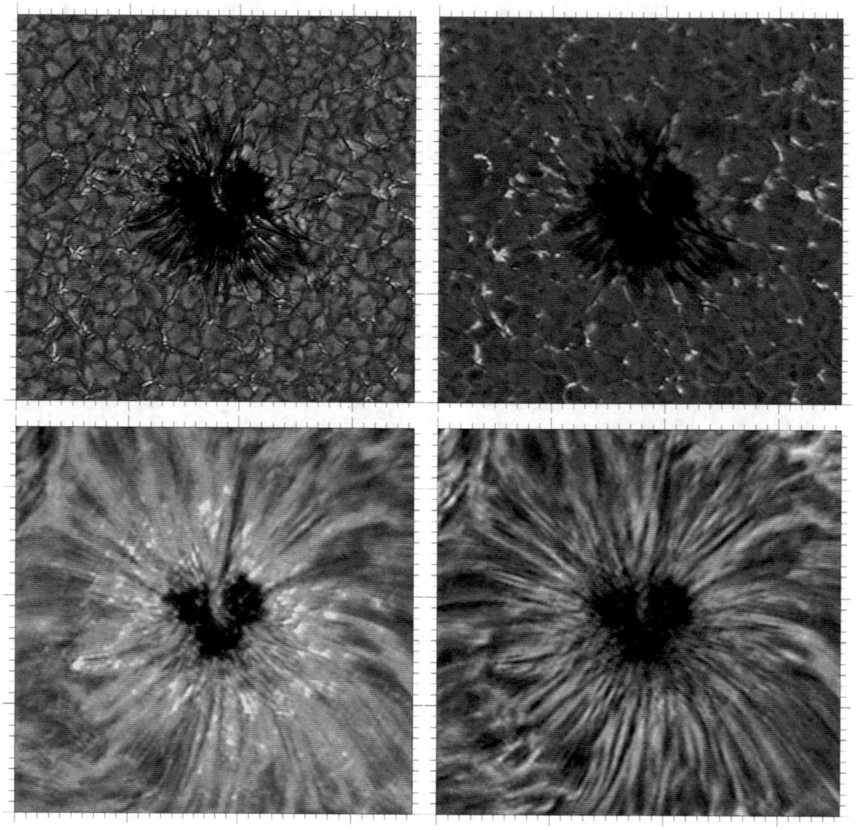

Figure 1. Simultaneous DOT images of the main spot of decaying AR 10789 at view angle $\mu = 0.86$. Clockwise: G band; blue Ca II H wing at $\Delta\lambda = -2.35$ Å; sum of the Hα wings at $\Delta\lambda = \pm 0.5$ Å; Hα core. Tick marks: arcseconds.

MCs are bright in the continuum from hot-wall radiation [18]. The classic magnetostatic thin fluxtube/sheet paradigm [19] [20] [21] [22] [23] [24] [25] [26] explains their physics well and is well applicable [27], although actual MCs show complex morphology with rapid changes [28] [29]. Since MCs are small and reside in dark intergranular lanes, their continuum brightening is only seen at sub-arcsecond resolution [30]. MCs also appear as bright "line gaps" in the cores of neutral-metal lines due to ionization from evacuation, particularly in Mn I lines [31] [32] for which the contrast is not weakened by Dopplershifts in the surrounding granulation [11]. They also appear markedly bright in the outer wings of Hα [33], at larger contrast than in other diagnostics including the G band [34]. For the G band the contrast enhancement comes from dissociation of the CH molecules producing this feature, for the Hα wings from smaller collisional damping. Both result also from MC evacuation.

The Hα core image in Fig. 1 shows the chromospheric fibril canopy that overlies and shields the deep photosphere imaged in the far Hα wings. Hα does not sample the intermediate layers (seen in Ca II H as shock-ridden clapotisphere [35]) due to its low-temperature opacity gap [36] [37] [38], so that at any wavelength its formation jumps between chromosphere and deep photosphere depending on the chromospheric fibril opacity. The $\Delta\lambda = \pm 0.5$ Å sampling wavelength of Fig. 1 mixes deep-photosphere and upper-chromosphere brightness contributions, in this case MCs and superpenumbral fibrils. The fibrils look like the line-core ones but are thinner. The MCs show about the same morphology as in the Ca II-wing and G-band images

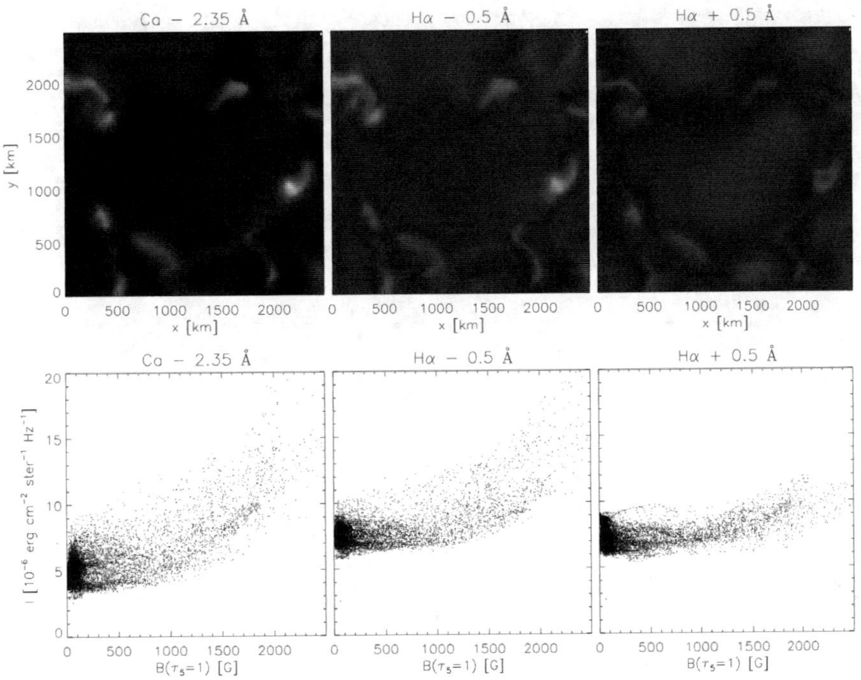

Figure 2. Synthesis of Ca II H wing and Hα wing images for the numerical MHD simulation snapshot of [11]. The scatter plots show brightness against magnetic field strength per pixel. Downdrafts make some MCs, including the brightest, brighter in the blue Hα wing.

because their brightness originates at about the same depth. The overlying fibrils block some MCs. Where they are sufficiently thin that the photospheric MC brightness shines through, they degrade the MC image sharpness. This defocus, compared to the sharp LTE-formed G-band and Ca II H-wing brightness features, results from ray spreading across the opacity gap and scattering through the effectively thin fibrils.

Note in the Hα wing image that the shielding by overlying fibrils appears to be less for the MCs in the moat than for MCs further away. Larger transparency of superpenumbral fibrils may result from repetitive shock heating by running penumbral waves [39] [40], ionizing hydrogen too frequently to let it reach population equilibrium since hydrogen ionization/recombination balancing is exceptionally slow in cooling shock aftermaths due to the large $n = 1$–2 Lyα transition [41] [42].

In the DOT movie some of the larger MCs do brighten momentarily, especially in the blue Hα wing. Blue-wing brightening may result from MC downdrafts, as in the numerical simulation shown in Fig. 2. Downdrafts often occur in moat MCs that have opposite polarity to the sunspot [43]. MC downdrafts tend to produce shocks higher up [44] that are best seen in Na I D Dopplergrams [45]. Inspection of simultaneous Hα and Na I D active-region Dopplergram sequences from the SST shows that such MC shock occurrence is often accompanied by Hα blue-wing brightening.

MC brightening may also result from field concentration by bathtub vorticity in granular swirls [46]. Such brightening also reflects increasing hot-wall radiation, and must not be misinterpreted as MHD heating or reconnection [47]. However, swirl brightening seems a rare phenomenon. In the DOT data of Fig. 1 Hα wing brightening shows no correlation with vorticity in the granular flows measured from the G-band movie. Rather, it happens in concert with

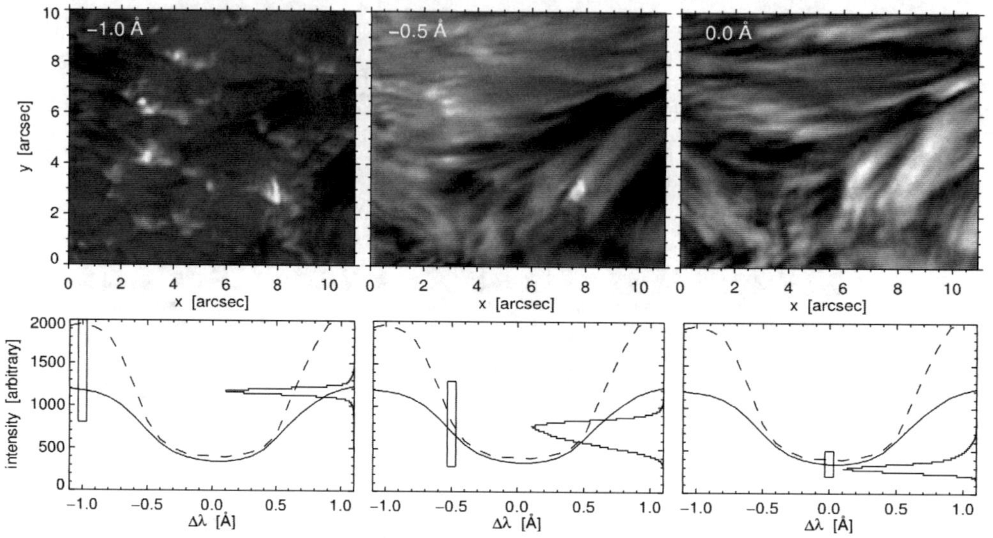

Figure 3. Upper row: SST EB images in Hα at the specified wavelength separations from line center, taken from Fig. 1 of [9]. View angle $\mu = 0.67$ with the limb direction to the top. Lower row: Hα profile (solid) averaged over the full 67×67 arcsec^2 field of view of the observations and Hα profile (dashed) for a pixel in the EB at $(x, y) = (8.0, 2.5)$. The histograms show the normalized intensity occurrence distribution over the full observed field of view at each wavelength. The rectangle widths show the passband FWHM, the vertical extents the greyscale clipping of the corresponding images (each bytescaled for maximum contrast).

Ca II H-wing brightenings that seem mostly due to episodes of magnetic patch [48] compression in the outward moat flow.

Upshot regarding EB fallacies: part of the EB literature (and the moustache literature, Severny's [49] term for far-extended bright line wings) concerns MCs, not EBs. This is the case when dense network was not resolved, moustache brightening of the continuum and neutral-metal lines was observed, or when low Hα-wing contrast thresholds were applied. In our opinion this warning should be heeded when reading [49] [50] [51] [52] [53] [54] [55] [56] [57] [58] [59] [60].

3. Ellerman bombs are not chromospheric

Figure 3 after Fig. 1 of [9] shows EBs at the unprecedented resolution of the SST. In this slanted limbward high-resolution view EBs appear in the outer Hα wings as tiny bright upright flames that flicker rapidly while their footpoints travel along MC-filled network strands. The ensemble may be bright for many minutes but the subflames last only seconds. Inspection of the corresponding SST movies[2] demonstrates this behavior irrevocably. The elongated shape was noted before in lower-resolution data [61] [4] [62], as was their intermittent substructure [63].

Figure 3 also demonstrates that EBs are purely photospheric phenomena. They are not seen at Hα line center because the flames, even when tall (a few hundred to a thousand km [64] [61] [65]), do not break through the overlying dense canopy of chromospheric Hα fibrils that always covers a growing active region. Comparison of the greyscale ranges (narrow rectangles in the lower panels of Fig. 3) shows that the EB wing emission is much brighter than the brightest line-core features. Some diffuse line-center brightening might again result from photospheric EB emission that passes through the opacity gap and scatters through the fibrils, but our SST

[2] `http://iopscience.iop.org/0004-637X/736/1/71/fulltext`; one example.

Journal of Physics: Conference Series **440** (2013) 012007 doi:10.1088/1742-6596/440/1/012007

movies suggest that such line-center brightening above EBs is uncommon. At $\Delta\lambda = -0.5$ Å the fibrils show the same morphology as at line center but they are less opaque and therefore brighter [38], except for blueshifted dark ones seen also at $\Delta\lambda = -1.0$ Å. At $\Delta\lambda = -0.5$ Å the EBs do shine through, again with loss of sharpness from scattering. Some MCs also shine through, but less brightly than in Fig. 1 because the bytescale now includes much brighter EBs and the Hα fibril canopy is thicker in this flux-emergence region than around the little decaying spot of Fig. 1.

Thus, an EB top may reach higher than the 400–500 km nominal height of the temperature minimum in standard one-dimensional static-equilibrium models of the solar atmosphere [66] [67] [68] [69]. The onset of the actual solar outward temperature rise likely varies between 200 km in MCs and 2000 km in internetwork, varying temporally as well [42] [70], but in any case, the jet-like EB flame protrusions originate from the deep photosphere, in the network at the EB footpoint, and do not affect or poke through the overlying chromosphere defined [35] by the Hα fibrils.

Apparent blue-brighter-than-red EB wing asymmetry [71] usually results from inverse Evershed flows [72] [73] [74] along the fibrils, the dark chromospheric line core shifting into the red-wing emission. Thus, asymmetries and wavelengths of EB emission peaks do not define Dopplershifts of the EB emitting material [59] [6] but are set by the absorbing overlying fibrils [75].

It has been suggested that EBs are accompanied by Hα surges [76] [77] [78] [79] [80] [81]. However, the high-resolution SST data of [9] contained only two tentative cases. Our newer SST data sampled in Fig. 4 contain at most one questionable case [10]. An EB–surge connection is certainly not ubiquitous [4].

Upshot regarding EB fallacies: most EB papers err in describing the EB phenomenon as chromospheric. It is not. EBs have no systematic counterpart in the overlying chromosphere, transition region or corona. Dark chromospheric Hα cores are not EB ingredients, nor are their Dopplershifts.

Comment regarding EB fads: EBs may combine strong downflows in the low photosphere [82] with outward flows higher up [81] [9]. Such bi-directional flows are reminiscent of "chromospheric anemone jets" seen in Hinode Ca II H data [83] [84] [85]. If such Ca II H jets actually are EBs, they are also confined to the photosphere. Just as for Hα, not every bright Ca II H feature is necessarily chromospheric.

4. Ellerman bomb visibility

Even though they are photospheric, proper EBs (those that adhere to Ellerman's description) do not show up in neutral-metal lines nor in the continuum. At sufficient angular resolution they do show continuum moustaches, G-band brightening [86], and narrow neutral-line gaps at their footpoint, sampling the MC from which they arise. Ellerman wrote: *"they frequently appear in the faculae so that their spectra are superposed on those of faculae, thus giving the appearance of great extension to the bright "bomb" band"* which wasn't heeded in most pseudo-EB literature.

EB flames are also observable in Ca II 8542 Å, Ca II H & K, and in mid-UV continua [87] [4] [6] [88] [89] [62] [90] [91] [86] [10]. They are not identically the same in these different diagnostics, but they often show up at the same space-time locations. We illustrate this for the 1700 Å continuum in Fig. 4 by combining image cutouts from the new SST data of [10] with co-aligned image cutouts from SDO. Clearly, AIA's 1700 Å shows the larger EBs also, at much lower resolution but with less interference by chromospheric fibrils. Since this continuum originates in the upper photosphere [67] [92], the bottom part of an Hα EB is hidden in slanted 1700 Å viewing but this is not noticeable at AIA's angular resolution.

AIA's 1600 Å images (not shown) show the EBs too, at yet larger contrast, but these also show a few more extended and yet brighter transition-region transients for this area and period.

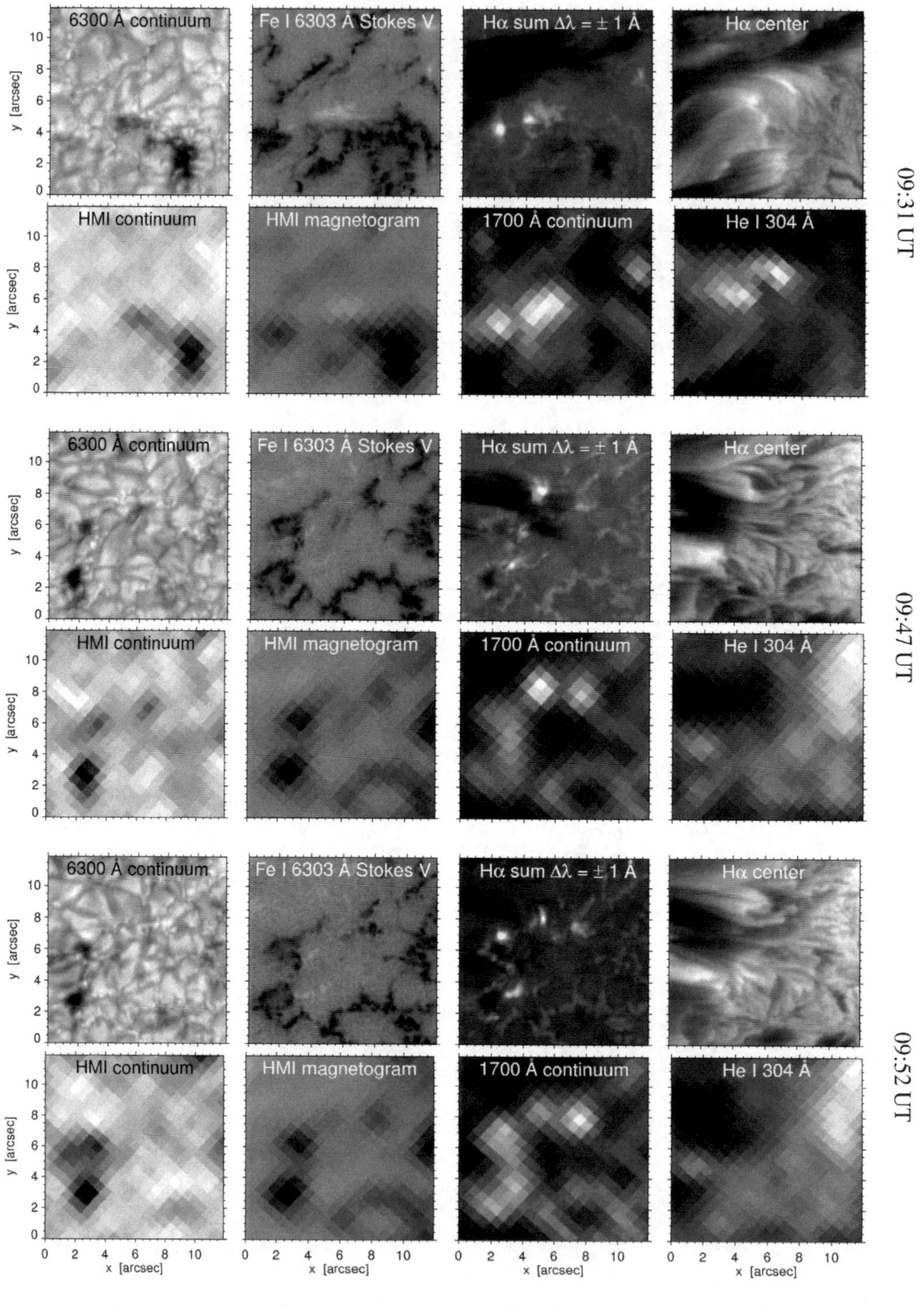

Figure 4. Two small-field cutouts of corresponding SST images (upper row of each set) and SDO images (lower rows) from the data of [10]. Magnification per pdf viewer is recommended for the SST images. View angle $\mu = 0.89$; the panel tops are in the limb direction. The third set shows the same cutout as the second, but five minutes later. Each panel is bytescaled independently. The arrows overlaid on the SST magnetograms show surface flows measured from the continuum sequence by correlation tracking.

They have short-loop morphology and correspond to bright patches in Hα line-center, are also seen in He I 304 Å, and sometimes in Fe IX-X 171 Å. These loop-like bright transients appear unrelated to underlying EBs. We attribute their visibility in the 1600 Å images to the C IV lines in this passband and tentatively identify them as upper-atmosphere transients such as small flaring arch filaments and microflares visible in transition-region diagnostics and Hα line center. The He I 304 Å images in Fig. 4 indeed mimic the Hα line-center morphology which appears indifferent to EBs underneath.

It is illustrative and recommended to play and blink SDO 1700 Å, 1600 Å, and He I 304 Å movies of a very active region with much flux emergence, for example AR 11654 on January 10, 2013 and the following days when it was crackling with short-lived brightenings in these diagnostics. The plethora of transient brightness features in 1600 Å are the sum of different sorts, respectively seen in 1700 Å and in He I 304 Å. Very short-lived, pointlike 1700 Å brightenings were likely photospheric Hα-wing EBs, whereas more spread-out loop-like longer-duration brightenings were likely flaring arch filaments and small flares, bright in the transition-region diagnostics and probably bright at Hα line center. Inspection of such flux-emergence AIA movies and of our high-resolution SST data suggests that distinction should be made between point-like (at SDO resolution) EBs and loop-like flaring arch filaments as different entities, respectively photospheric and upper-atmosphere. AIA's 1700 Å shows fewer of the latter type and is therefore the best SDO diagnostic for EB studies. Figure 4 demonstrates that SDO's resolution is poor compared with the SST's, but the obvious AIA advantage is that it samples the stronger EBs in any Earth-side active region at any time.

Comment regarding EB fallacies: the low resolution of older data has led to claims of apparent cospatiality of higher-atmosphere brightenings with EBs. Examples: coincidence with EBs was claimed for 6 out of 16 microflares by [93] but the example pair in their Figs. 4 and 5 has 8 arcsecond separation. The Hα line-center brightening observed and modeled by [94] was not an EB. The H$\alpha \pm 0.35$ Å feature claimed by [95] to be an EB at 1 MK doesn't look like an EB to us. The long-lived EB of [96] at the foot of an arch filament looks like a large MC. Similarly for the Ca II H bright points bordering the flaring arch filament in [97], much like the bright points of [98]. [6] reported one coincidence of X-ray brightening at an exceptionally bright EB but doubted ubiquitous co-spatiality. We have aligned our new SST data with SDO/AIA image sequences and searched for but did not find any significant EB impact on the overlying transition region (He I 304 Å) and corona (Fe IX 171 Å) [10].

Comment regarding EB fads: most EB studies are based on observations in the Hα wings and typically describe a single, a few, or at most some dozen EBs in a single active region during a short period. This type of study should exploit IRIS in the near future since EBs will be well visible in Mg II h & k, as in Ca II H & K. On the other hand, it seems time to exploit the AIA 1700 Å database to study very many more with respect to occurrence patterns.

5. Ellerman bomb detection

In [9] EBs were identified and selected manually on the basis of their flame morphology in the SST Hα-wing movies, meaning bright, narrow, tall, upright, flickering appearance in the limbward view at this unprecedented resolution. More formal detection criteria are formulated for two SST data sets in [10]. One of the requirements defining an EB kernel is 55% excess intensity in the Hα wings over the spatial average of the active region, while setting a threshold of 5 σ or higher above the mean seems a good counterpart to select EBs in AIA 1700 Å images. Both thresholds are passed by the EBs in Fig. 4, whereas almost none of the bright Hα-wing features in the movie sampled in Fig. 1 passed a 55% Hα-wing excess threshold, nor a 5 σ excess threshold in simultaneous 1600 Å images from TRACE[3].

[3] Recently [60] claimed to detect over 3000 EBs during 90 minutes around a small spot rather like the one in Fig. 1. Our application of the above threshold to the simultaneous SDO 1700 Å images suggests that they mostly

Journal of Physics: Conference Series **440** (2013) 012007 doi:10.1088/1742-6596/440/1/012007

In addition, small spatial extent is a discriminator, as already suggested by Ellerman: *"On rarer occasions they [EBs] are superposed on bright reversals of Hα over eruptive regions, but this is an uncommon occurrence, and the distinction is easily made between the two phenomena by the flickering of the "bomb" band compared to the Hα reversal, due to the effect of seeing."* – as using scintillation to distinguish stars from planets. Small size, fast variation, large brightness, and appearing in a region with much flux emergence together become the recipe to locate EBs in AIA 1700 Å image sequences. We are presently refining such automated detection for multiple viewing angles [99].

6. Ellerman bomb radiation mechanism

In traditional one-dimensional stratification modeling along a vertical line of sight the emergent Hα profile maps the variation of the source function with depth, with the outermost wings sampling the deepest layers [100]. Postulating a suited temperature perturbation with appropriate structure and motion to a standard model atmosphere can then explain any Hα excess emission profile [101] [88], but not unambiguously; different models may produce similar Hα profiles [91].

In the slanted perspective of Figs. 3 and 4 the EB flames appear as extended, dense, hot slabs that stick up from the network MCs into the otherwise Hα-transparent upper photosphere, making cloud modeling more appropriate than one-dimensional plane-parallel modeling. The EB excess emission and its profile including its moustache extent are properties of such a slab, while the core absorption and Dopplershift patterns are properties of the overlying fibrils. Further away from line center the slab is optically thinner but not "deeper".

We attribute the non-visibilty of the EB slabs in neutral-metal lines to neutral-metal ionization in the hot EB flames, and similarly their transparency in the continuum to H^- ionization. Considerable ionization of neutral hydrogen is also likely, making cascade recombination the main producer of the EB Hα photons. Even if the ionization occurs only briefly during a momentary reconnection event, Hα will so shine in a longer-duration afterglow because hydrogen recombination balancing is slow in the aftermath [41] [42].

Following the suggestion of [102] we attribute the extended bright moustaches to subsequent thermal Thomson scattering of these photons, notwithstanding the small process crosssection, because the flames combine high temperature with very high (mid-photosphere) density. While the line-center peak of the EB brightness profile remains hidden by overlying fibrils, the wings may gain brightness at large thermal electron Dopplershift whenever a line-core photon meets an electron and is scattered our way. Since EBs are very optically thick at Hα line center, being non-transparent already in the outer wings, we suggest that Hα resonance scattering confines line-core photons in a random walk within the EB until they are Dopplershifted into a wing and escape. This scattering mechanism will produce similar moustaches for other chromospheric lines (principally Mg II h & k, Ca II H & K, and Ca II 8542 Å) if an EB similarly confines scattering line photons, which is likely unless these ions ionize too much. Obviously, numerical simulation with detailed spectral synthesis including non-monochromatic Thomson scattering may vindicate this moustache mechanism.

In this view large moustache extent happens *because* EBs are photospheric and makes them visible beyond the overlying fibrils in the spectrum. For Hα such scattering occurs only within the flame, there being neither free electrons nor hydrogen atoms in the $n = 2$ level in the surrounding atmosphere, and so it remains local. Flame images taken in the outer Hα wings therefore remain sharp, as demonstrated in Figs. 3 and 4. In Ca II H & K EBs get a diffuse halo from additional resonance scattering in the surrounding upper photosphere [90].

Comment regarding EB fallacies: Severny [49] [50], recognizing that the moustache extent

detected pseudo-EBs of which the Hα-wing brightness has nothing to do with MHD heating.

cannot be explained by Stark broadening, invoked an EB scenario of nuclear explosion and relativistic particle beams. This way of thinking became a school of thought in which flares and moustaches are regarded as similar phenomena, often mentioned together, in which the photosphere is hit from above by particle beams due to reconnection in the upper atmosphere. Such beams would then explain the lack of Hα line-core brightening by passing through and not affecting the Hα chromosphere. Linear line-core polarization attributed to the beam impact became the diagnostic for this notion [103] [104] [105] [106] [107] [108] [109] [110] [88] [111]. In our opinion these authors failed to recognize that EBs are photospheric and occur without anything happening overhead. We note that in the example of [103] the polarization is not cospatial with the moustache, and that the excess emission profiles of [104] have no dip at line center. We suspect that the measured line core polarizations stem from flaring arch filaments, possibly through the mechanism of [112].

7. Ellerman bomb occurrence

The SST samples in Fig. 4 include magnetograms that show examples of what we find to be characteristic behavior in these data. Small white opposite-polarity Stokes-V patches produce EB flames in the Hα wing when colliding at relatively high speed with larger black patches, and then vanish. This is good evidence, similar to [90] but at higher resolution and with better statistics, that EBs mark strong-field cancelation and supports the notion that the bright EB flames are caused by photospheric reconnection [113] [114] [4] [80] [81] [115] [63] [9]. The gasdynamical instability proposed by [116] seems a less likely explanation since it does not have field emergence as necessary condition.

Strong opposite-polarity field emergence may happen with small-scale $\cap$ loop shape in classical Ω emergence [117] or with additional $\cup$ loop shape in serpentine emergence from the Parker instability [118] [119], with connecting arch filaments [120] accompanied by dipped fields and bald patches [3] [7] [8] [65] [121] [122] [115] [123] (see cartoons in [65] [123] [124]). In this picture chromospheric convergence of $\cup$ loop sides is thought to give reconnection in the line-tied regime [80] [124], but the EBs in our SST data occur at colliding photospheric flows in the frozen-in regime (Fig. 4).

Comment regarding EB fads: serpentine patterning into opposite-polarity pairs of the MCs in the moat flow around decaying spots was already suggested by [14], but since EBs are emerging-flux phenomena it is tempting to assume serpentine emergence and $\cup$ loop patterning. This scenario may well be valid, but convincing proof requires long-duration wide-field imaging spectroscopy with spectral, spatial, and temporal resolution at least comparable to our SST sequences. Obviously an IRIS topic. Photospheric feature tracking following [125] would be a good technique to analyze such data. Likewise, the notion of bi-directional anemone-jet flows with reconnection a few hundred km up is attractive but needs further verification.

8. Conclusion: Ellerman bomb usage

EBs are the most spectacular solar photosphere phenomenon. They seem especially informative as space-time markers of strong-field reconnection events in the photosphere, better than searching for bipolar cancelation in magnetograms. Comparison of the SST and HMI magnetograms in Fig. 4 illustrates the need for superhigh resolution and sensitivity in the latter approach and shows that even at the high SST quality one would not be able to identify reconnection events from magnetogram sequences alone. Larger Stokes sensitivity is desirable but longer integration would degrade the temporal and spatial resolution fatally. Thus, EB detection seems a better way to locate small-scale reconnection events in emerging flux regions. Since the larger and brighter EBs are also visible in 1700 Å images even at the AIA resolution, the AIA database permits monitoring active-region field re-configuration through EB identification in large data volumes [99]. EBs may so become useful photospheric telltales in studying chromospheric active-region field topology evolution and energy loading.

Acknowledgments

We thank R. Rezaei for informative discussions and the referee for valuable suggestions. RJR acknowledges that the late C. Zwaan often suggested EBs as research topic, and thanks the Leids Kerkhoven-Bosscha Fonds for travel support. We made much use of NASA's Astrophysics Data System Bibliographic Services. The macro to make the citations above link to ADS (at least in the arXiv preprint) was contributed by EDP Sciences.

References

[1] Ellerman F 1917 *Astrophys. J.* **46** 298
[2] Bray R J and Loughhead R E 1974 *The solar chromosphere* Chapman & Hall, London
[3] Bernasconi P N, Rust D M, Georgoulis M K and Labonte B J 2002 *Solar Phys.* **209** 119
[4] Georgoulis M K, Rust D M, Bernasconi P N and Schmieder B 2002 *Astrophys. J.* **575** 506
[5] Schmieder B, Pariat E, Aulanier G, Georgoulis M K, Rust D M and Bernasconi P N 2002 in *Solar Variability: From Core to Outer Frontiers* (*ESA Special Pub.* 506) ed Wilson A 911
[6] Schmieder B, Rust D M, Georgoulis M K, Démoulin P and Bernasconi P N 2004 *Astrophys. J.* **601** 530
[7] Pariat E, Aulanier G, Schmieder B, Georgoulis M K, Rust D M and Bernasconi P N 2004 *Astrophys. J.* **614** 1099
[8] Pariat E, Aulanier G, Schmieder B, Georgoulis M K, Rust D M and Bernasconi P N 2006 *Adv. Space Research* **38** 902
[9] Watanabe H, Vissers G, Kitai R, Rouppe van der Voort L and Rutten R J 2011 *Astrophys. J.* **736** 71
[10] Vissers G, Rouppe van der Voort L H M and Rutten R J 2013 *ApJ* submitted
[11] Vitas N, Viticchiè B, Rutten R J and Vögler A 2009 *Astron. Astrophys.* **499** 301
[12] Sheeley Jr N R 1969 *Solar Phys.* **9** 347
[13] Vrabec D 1971 in *Solar Magnetic Fields* (*IAU Symposium* vol 43) ed Howard R 329
[14] Harvey K and Harvey J 1973 *Solar Phys.* **28** 61
[15] Vrabec D 1974 in *Chromospheric Fine Structure* (*IAU Symposium* vol 56) ed Athay R G 201
[16] Muller R and Mena B 1987 *Solar Phys.* **112** 295
[17] Hagenaar H J and Shine R A 2005 *Astrophys. J.* **635** 659
[18] Spruit H C 1976 *Solar Phys.* **50** 269
[19] Spruit H C 1977 *Magnetic flux tubes and transport of heat in the convection zone of the Sun.* Ph.D. Thesis Utrecht University
[20] Spruit H C 1981 in *The Sun as a Star* ed Jordan S *NASA Special Pub.* **450** 385
[21] Deinzer W, Hensler G, Schüssler M and Weisshaar E 1984 *Astron. Astrophys.* **139** 426
[22] Knölker M, Schüssler M and Weisshaar E 1988 *Astron. Astrophys.* **194** 257
[23] Spruit H C, Schüssler M and Solanki S K 1991 in *Solar Interior and Atmosphere* Univ Arizona Press 890
[24] Solanki S K and Brigljevic V 1992 *Astron. Astrophys.* **262** L29
[25] Bünte M, Solanki S K and Steiner O 1993 *Astron. Astrophys.* **268** 736
[26] Solanki S K 1993 *Space Science Rev.* **63** 1

[27] Yelles Chaouche L, Solanki S K and Schüssler M 2009 *Astron. Astrophys.* **504** 595

[28] Berger T E, Rouppe van der Voort L H M, Löfdahl M G, Carlsson M, Fossum A, Hansteen V H, Marthinussen E, Title A and Scharmer G 2004 *Astron. Astrophys.* **428** 613

[29] Rouppe van der Voort L H M, Hansteen V H, Carlsson M, Fossum A, Marthinussen E, van Noort M J and Berger T E 2005 *Astron. Astrophys.* **435** 327

[30] Title A M and Berger T E 1996 *Astrophys. J.* **463** 797

[31] Livingston W and Wallace L 1987 *Astrophys. J.* **314** 808

[32] Livingston W, Wallace L, White O R and Giampapa M S 2007 *Astrophys. J.* **657** 1137

[33] Leenaarts J, Rutten R J, Sütterlin P, Carlsson M and Uitenbroek H 2006 *Astron. Astrophys.* **449** 1209

[34] Leenaarts J, Rutten R J, Carlsson M and Uitenbroek H 2006 *Astron. Astrophys.* **452** L15

[35] Rutten R J 2012 *Roy. Soc. London Phil. Trans. Series A* **370** 3129

[36] Schoolman S A 1972 *Solar Phys.* **22** 344

[37] Rutten R J and Uitenbroek H 2012 *Astron. Astrophys.* **540** A86

[38] Leenaarts J, Carlsson M and Rouppe van der Voort L 2012 *Astrophys. J.* **749** 136

[39] Zirin H and Stein A 1972 *Astrophys. J. Lett.* **178** L85

[40] Christopoulou E B, Georgakilas A A and Koutchmy S 2001 *Astron. Astrophys.* **375** 617

[41] Carlsson M and Stein R F 2002 *Astrophys. J.* **572** 626

[42] Leenaarts J, Carlsson M, Hansteen V and Rutten R J 2007 *Astron. Astrophys.* **473** 625

[43] Zhang J, Solanki S K, Woch J and Wang J 2007 *Astron. Astrophys.* **471** 1035

[44] Kato Y, Steiner O, Steffen M and Suematsu Y 2011 *Astrophys. J. Lett.* **730** L24

[45] Rutten R J, Leenaarts J, Rouppe van der Voort L H M, De Wijn A G, Carlsson M and Hansteen V 2011 *Astron. Astrophys.* **531** A17

[46] Bonet J A, Márquez I, Sánchez Almeida J, Cabello I and Domingo V 2008 *Astrophys. J. Lett.* **687** L131

[47] Jess D B, McAteer R T J, Mathioudakis M, Keenan F P, Andic A and Bloomfield D S 2007 *Astron. Astrophys.* **476** 971

[48] De Wijn A G, Rutten R J, Haverkamp E M W P and Sütterlin P 2005 *Astron. Astrophys.* **441** 1183

[49] Severny A B 1956 *The Observatory* **76** 241

[50] Severnyi A B 1957 *Soviet Astronomy* **1** 668

[51] Bruzek A 1968 in *Structure and Development of Solar Active Regions* (*IAU Symposium* vol 35) ed Kiepenheuer K O 293

[52] Turon P J and Léna P J 1970 *Solar Phys.* **14** 112

[53] Bruzek A 1972 *Solar Phys.* **26** 94

[54] Zachariadis T G, Alissandrakis C E and Banos G 1987 *Solar Phys.* **108** 227

[55] Stellmacher G and Wiehr E 1991 *Astron. Astrophys.* **251** 675

[56] Denker C, de Boer C R, Volkmer R and Kneer F 1995 *Astron. Astrophys.* **296** 567

[57] Denker C 1997 *Astron. Astrophys.* **323** 599

[58] Nindos A and Zirin H 1998 *Solar Phys.* **182** 381

[59] Georgakilas A A, Christopoulou E B and Koutchmy S 1999 in *Magnetic Fields and Solar Processes* (*ESA Special Pub.* 448) ed Wilson A and et al 279

[60] Nelson C J, Doyle J G, Erdélyi R, Huang Z, Madjarska M S, Mathioudakis M, Mumford S J and Reardon K 2013 *Solar Phys.* **283** 307

[61] Kurokawa H, Kawaguchi I, Funakoshi Y and Nakai Y 1982 *Solar Phys.* **79** 77

[62] Pariat E, Schmieder B, Berlicki A, Deng Y, Mein N, López Ariste A and Wang S 2007 *Astron. Astrophys.* **473** 279

[63] Hashimoto Y, Kitai R, Ichimoto K, Ueno S, Nagata S, Ishii T T, Hagino M, Komori H, Nishida K, Matsumoto T, Otsuji K, Nakamura T, Kawate T, Watanabe H and Shibata K 2010 *Pub. Astron. Soc. Japan* **62** 879

[64] Harvey J W 1963 *The Observatory* **83** 37

[65] Watanabe H, Kitai R, Okamoto K, Nishida K, Kiyohara J, Ueno S, Hagino M, Ishii T T and Shibata K 2008 *Astrophys. J.* **684** 736

[66] Gingerich O, Noyes R W, Kalkofen W and Cuny Y 1971 *Solar Phys.* **18** 347

[67] Vernazza J E, Avrett E H and Loeser R 1981 *Astrophys. J. Suppl.* **45** 635

[68] Fontenla J M, Avrett E H and Loeser R 1993 *Astrophys. J.* **406** 319

[69] Avrett E H and Loeser R 2008 *Astrophys. J. Suppl.* **175** 229

[70] Leenaarts J, Rutten R J, Reardon K, Carlsson M and Hansteen V 2010 *Astrophys. J.* **709** 1362

[71] Koval A N and Severny A B 1970 *Solar Phys.* **11** 276

[72] Beckers J M 1962 *Austr. J. Physics* **15** 327

[73] Dere K P, Schmieder B and Alissandrakis C E 1990 *Astron. Astrophys.* **233** 207

[74] Teriaca L, Curdt W and Solanki S K 2008 *Astron. Astrophys.* **491** L5

[75] Dara H C, Alissandrakis C E, Zachariadis T G and Georgakilas A A 1997 *Astron. Astrophys.* **322** 653

[76] Roy J R 1973 *Solar Phys.* **28** 95

[77] Roy J R and Leparskas H 1973 *Solar Phys.* **30** 449

[78] Carlqvist P 1979 *Solar Phys.* **63** 353

[79] Shibata K, Nishikawa T, Kitai R and Suematsu Y 1982 *Solar Phys.* **77** 121

[80] Isobe H, Tripathi D and Archontis V 2007 *Astrophys. J. Lett.* **657** L53

[81] Matsumoto T, Kitai R, Shibata K, Otsuji K, Naruse T, Shiota D and Takasaki H 2008 *Pub. Astron. Soc. Japan* **60** 95

[82] Shimizu T, Martínez-Pillet V, Collados M, Ruiz-Cobo B, Centeno R, Beck C and Katsukawa Y 2007 in *New Solar Physics with Solar-B Mission* (*Astron. Soc. Pacific Conf. Series* vol 369) ed Shibata K, Nagata S and Sakurai T 113

[83] Shibata K, Nakamura T, Matsumoto T, Otsuji K, Okamoto T J, Nishizuka N, Kawate T, Watanabe H, Nagata S, UeNo S, Kitai R, Nozawa S, Tsuneta S, Suematsu Y, Ichimoto K, Shimizu T, Katsukawa Y, Tarbell T D, Berger T E, Lites B W, Shine R A and Title A M 2007 *Science* **318** 1591

[84] Morita S, Shibata K, Ueno S, Ichimoto K, Kitai R and Otsuji K I 2010 *Pub. Astron. Soc. Japan* **62** 901

[85] Nishizuka N, Nakamura T, Kawate T, Singh K A P and Shibata K 2011 *Astrophys. J.* **731** 43

[86] Herlender M and Berlicki A 2011 *Central European Astrophys. Bull.* **35** 181

[87] Qiu J, Ding M D, Wang H, Denker C and Goode P R 2000 *Astrophys. J. Lett.* **544** L157

[88] Fang C, Tang Y H, Xu Z, Ding M D and Chen P F 2006 *Astrophys. J.* **643** 1325

[89] Socas-Navarro H, Martínez Pillet V, Elmore D, Pietarila A, Lites B W and Manso Sainz R 2006 *Solar Phys.* **235** 75

[90] Matsumoto T, Kitai R, Shibata K, Nagata S, Otsuji K, Nakamura T, Watanabe H, Tsuneta S, Suematsu Y, Ichimoto K, Shimizu T, Katsukawa Y, Tarbell T D, Lites B W, Shine R A and Title A M 2008 *Pub. Astron. Soc. Japan* **60** 577

[91] Berlicki A, Heinzel P and Avrett E H 2010 *Mem. Soc. Astron. Italiana* **81** 646

[92] Fossum A and Carlsson M 2005 *Astrophys. J.* **625** 556

[93] Shimizu T, Shine R A, Title A M, Tarbell T D and Frank Z 2002 *Astrophys. J.* **574** 1074

[94] Guglielmino S L, Bellot Rubio L R, Zuccarello F, Aulanier G, Vargas Domínguez S and Kamio S 2010 *Astrophys. J.* **724** 1083

[95] Madjarska M S, Doyle J G and De Pontieu B 2009 *Astrophys. J.* **701** 253

[96] Herlender M and Berlicki A 2010 *Central European Astrophys. Bull.* **34** 65

[97] Jess D B, Mathioudakis M, Browning P K, Crockett P J and Keenan F P 2010 *Astrophys. J. Lett.* **712** L111

[98] Kitai R and Muller R 1996 *Solar Phys.* **165** 155

[99] Vissers G, Rouppe van der Voort L H M and Rutten R J 2013 in preparation

[100] Rutten R J 2003 *Radiative Transfer in Stellar Atmospheres* Lecture Notes Utrecht University

[101] Kitai R 1983 *Solar Phys.* **87** 135

[102] Engvold O and Maltby P 1968 in *Mass Motions in Solar Flares and Related Phenomena* (Procs 9th Nobel Symposium) ed Oehman Y 109

[103] Babin A N and Koval A N 1985 *Solar Phys.* **98** 159

[104] Firstova N M 1986 *Solar Phys.* **103** 11

[105] Ding M D, Henoux J C and Fang C 1998 *Astron. Astrophys.* **332** 761

[106] Henoux J C, Fang C and Ding M D 1998 *Astron. Astrophys.* **337** 294

[107] Kazantsev S A, Firstova N M, Kashapova L K, Bulatov A V and Petrashen' A G 1998 *Russian Phys. J.* **41** 1258

[108] Kosovichev A G and Zharkova V V 2001 *Astrophys. J. Lett.* **550** L105

[109] Kashapova L K 2002 *Astron. Reports* **46** 918

[110] Zharkova V V and Kashapova L K 2005 *Astron. Astrophys.* **431** 1075

[111] Babin A N and Koval A N 2011 *Bulletin Crimean Astrophys. Obs.* **107** 36

[112] Henoux J C, Chambe G, Heristchi D, Semel M, Woodgate B, Shine D and Beckers J 1983 *Solar Phys.* **86** 115

[113] Yokoyama T and Shibata K 1995 *Nature* **375** 42

[114] Litvinenko Y E 1999 *Astrophys. J.* **515** 435

[115] Archontis V and Hood A W 2009 *Astron. Astrophys.* **508** 1469

[116] Diver D A, Brown J C and Rust D M 1996 *Solar Phys.* **168** 105

[117] Zwaan C 1985 *Solar Phys.* **100** 397

[118] Parker E N 1966 *Astrophys. J.* **145** 811

[119] Nozawa S, Shibata K, Matsumoto R, Sterling A C, Tajima T, Uchida Y, Ferrari A and Rosner R 1992 *Astrophys. J. Suppl.* **78** 267

[120] Strous L H and Zwaan C 1999 *Astrophys. J.* **527** 435

[121] Cheung M C M, Schüssler M, Tarbell T D and Title A M 2008 *Astrophys. J.* **687** 1373

[122] Pariat E, Masson S and Aulanier G 2009 *Astrophys. J.* **701** 1911

[123] Pariat E, Masson S and Aulanier G 2012 in *4th Hinode Science Meeting: Unsolved Problems and Recent Insights* (*Astron. Soc. Pacific Conf. Series* 455) ed Bellot Rubio L, Reale F and Carlsson M 177

[124] Pariat E, Schmieder B, Masson S and Aulanier G 2012 in *EAS Publications Series* 55) ed Faurobert M, Fang C and Corbard T 115

[125] Strous L H, Scharmer G, Tarbell T D, Title A M and Zwaan C 1996 *Astron. Astrophys.* **306** 947

Perturbations in the wave parameters near active regions

M. C. Rabello-Soares[1], R. S. Bogart[2], P. H. Scherrer[2]

[1]Physics Department, Universidade Federal de Minas Gerais, Belo Horizonte, MG 30380, Brazil,
[2]W. W. Hansen Experimental Physics Laboratory, Stanford University, Stanford, CA 94305, USA

E-mail: cristina@fisica.ufmg.br

Abstract. The wave characteristics derived from ring-diagram analysis of HMI Doppler data in magnetically quiet regions near active regions are compared with those with no nearby active regions using 5° patches during 2.5 years of cycle 24 ascending phase. We search for perturbations that may be associated with propagation of the acoustic oscillations through the nearby sunspot. We observe significant variations in the mode parameters and flows. We analyse their dependence on the direction of the wave propagation. The observed mode dependence of the variations in mode amplitude, line width and frequency does not have the same functional form as that observed for the differences between quiet and active regions.

1. Introduction

The characteristics of solar acoustic waves are modified as they propagate through a sunspot. Braun, Duvall & LaBonte ([1]) were the first to observe that sunspots absorb up to half of the incident p-mode power. Since then the variation of mode characteristics and flows in active regions has been extensively studied using different local helioseismology techniques (see [2] and references therein). Although much progress has been done, the physical origin of the changes in the mode parameters is not well understood. Here, instead of studying the wave characteristics in active regions, we look at quiet regions close to an active region using ring-diagram analysis of HMI/SDO data. We search for perturbations in the wave characteristics as the solar acoustic oscillations are affected as they propagate inside the nearby sunspot. Understanding the variations due to the nearby active region will help us understand the interaction of the acoustic waves with the active region; they may also shed some light on the determination of wave parameters made in the presence of strong magnetic field. Next, we describe the data and method used (section 2), present our results (section 3) and summarize our conclusions (section 4).

2. Data and Method

We use the wave characteristics obtained by the HMI ring-analysis pipeline ([3]) during Carrington rotations 2097 to 2130 (from May 2010 to November 2012). The data consist of full-disk Dopplergrams obtained at a 45-second cadence by HMI/SDO. Small patches of the solar disk with size 5° × 5° were tracked at the Carrington rotation rate for 9.6 hours.

The wave parameters were estimated using two different methods for parametric fitting of the ridges in the 3-D power spectrum. Module 'rdfitc' fits a 13-parameter model including an asymmetry term ([4]). While module 'rdfitf', fits only 6 parameters ([5]). Both methods use a Lorentzian profile. In the case of rdfitc, there is the addition of an asymmetry term (S):

$$P(k_x, k_y, \nu) = \frac{A}{R_\nu^2 + 1}\left[S^2 + (1 + R_\nu S)^2\right] + B(k) \tag{1}$$

where $R_\nu = (\nu - \nu_0 + k_x U_x + k_y U_y)/\gamma$. The value at the maximum is A, γ is the half width at half maximum, Ux and Uy are zonal and meridional flows respectively and B(k) is the background term. In rdfitf, S=0, $\gamma = \exp(w)$ and $A = \exp(a)/\exp(w)$; the 6 fitted parameters are: a, w, ν_0, U_x, U_y and one background term. In rdfitc, $A = \exp[A_0 + A_1(k - k_0) + A_2(k_x/k)^2 + A_3(k_x k_y/k^2)]$, $\gamma = w_0 + w_1(k - k_0)$ and the mean frequency $\nu_0 = ck^p$. The 13 fitted parameters are: $A_0, A_1, A_2, A_3, c, p, U_x, U_y, w_0, w_1, S$ and two background terms. In this work, we re-write A in the form $A = \exp[a_0 + a_2\cos(2\theta) + a_3\sin(2\theta) + A_1(k - k_0)]$ where $a_0 = A_0 + A_2/2$, $a_2 = A_2/2$ and $a_3 = A_3/2$.

The level of magnetic activity in each patch is given by the Magnetic Activity Index (MAI), which is the unsigned average of all HMI magnetogram pixels with a flux greater than 50G, ($|B_z| \geq 50$G) taken every 48 minutes (i.e. each 64th HMI magnetogram) and using the same temporal and spatial apodizations as the ones used in ring-diagram power spectra calculation. The MAI for each patch has the contribution from about 12 images. For a detailed description, using MDI magnetograms, see [6]. We select all quiet patches (MAI $<$ 5G) in our data that have at least one large magnetic active patch (MAI $>$ 100G) in its vicinity and it is not more than 8° away. We will call these quiet tiles with a nearby active region "target" tiles from now on and the angle of the nearby active region (AR) in relation to the target tile θ_{AR}. Since there are other tiles with MAI $>$ 5G in the vicinity of a target tile besides the large active region, θ_{AR} is an average of the angles of all tiles with MAI $>$ 5G in the vicinity of the target tile weighted by their MAI and distance from the target.

When mapping patches of the solar disc from their surface spherical coordinates to the plane, those located farther away from disk center will have a larger geometric foreshortening and a lower effective spatial resolution, affecting the power spectrum that will be fitted. To avoid theses geometric effects, we will analyse the wave characteristics differences between the target tile and a set of quiet tiles (ranging from 2 to 6 tiles) with no nearby active region, but at the same latitude and Stonyhurst longitude as the target patch (from now on, 'comparison' tiles). For a given target patch, the fitted ring parameters obtained for its comparison patches are averaged for each mode and subtracted from the fitted parameters obtained for the target patch. For control purposes, we arbitrarily pick up one of the comparisons for each target patch and used it as the "target". We repeat the calculation above, by calculating the mean (excluding the randomly chosen patch) and subtracted from the chosen one (from now on, 'control' set).

Each relative parameter (p) difference, (p[target] - p[comparison]) / p[comparison], is averaged over intervals of 10° in θ_{AR} for each mode, the results are discussed in section 3.2. Outliers larger than 3σ are removed from the average. We next average over all θ_{AR} for each mode, the results are in section 3.1.

As mentioned in the introduction, the variation of the mode parameters in active regions in relation to quiet regions has been well studied. For comparison with the results obtained for the target tiles (from now on, 'nearby' variation), we also calculate the difference in the fitted parameters between patches with MAI $>$ 100G and a set of quiet tiles (2 to 6 tiles) at the same latitude and Stonyhurst longitude. Each parameter difference for each mode is fitted as a function of its MAI (see Fig. 1 in [7] where a similar fit was done for 15° tiles). We use the results of our fitting for MAI = 100G, and call them from now on, 'AR' variation.

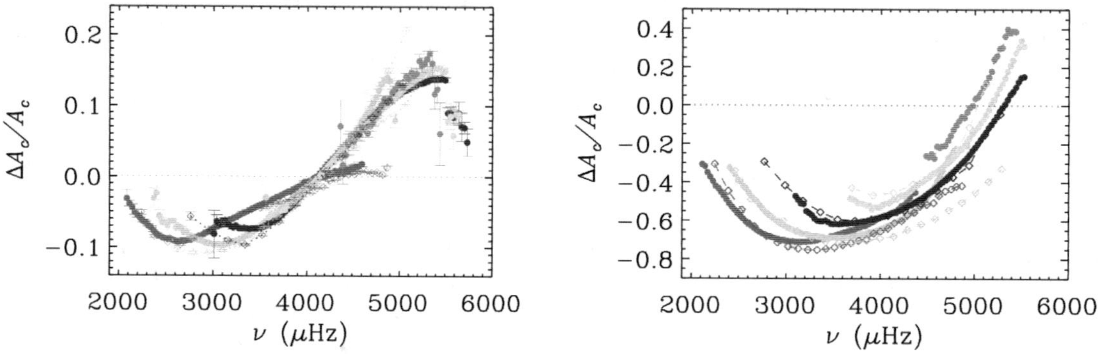

Figure 1. Relative 'nearby' (left) and 'AR' (right) amplitude variation given by the constant term in A (equation 1) obtained using rdfitc (circles) and rdfitf (diamonds) for f (red), p_1 (green), p_2 (blue), p_3 (cyan), p_4 (salmon) modes.

3. Observed perturbations in the wave parameters

3.1. Mode dependence

Amplitude. Fig. 1 (left panel) shows the 'nearby' variation of mean power in the ring given by the constant term in A defined in equation (1). There is a good agreement between rdfitc and rdfitf results. The relative amplitude variation is mainly a function of frequency. Comparing with the 'AR' variation (Fig. 1 right panel), the maximum of attenuation around 3 mHz is displaced towards lower frequencies (specially for the f mode) and it is, as expected, much smaller (almost an order of magnitude smaller). At frequencies higher than ≈5.2 mHz, the 'AR' variation is positive. This is known as the "acoustic halo" effect. Enhanced power surrounding active regions has been observed at frequencies above the photospheric acoustic cut-off by several authors (see [8] and references therein) and it is often thought to be the result of acoustic to magnetoacoustic mode conversion ([9]). The 'nearby' variation, on the other hand, becomes positive at a much lower frequency, around 4.1 mHz.

Linewidth. There is also a good agreement between rdfitc and rdfitf in the 'nearby' variation of mean half-width of the ring given by the constant term in γ (equation 1). Like the amplitude, the relative 'nearby' linewidth variation is approximately a function of frequency alone. As the amplitude, the 'nearby' width variation changes sign around 4 mHz, from 2% around 3 mHz to -7% around the cut-off frequency.

Frequency. Fig. 2 (left panel) shows the 'nearby' variation of mode's central frequency given by ν_0 in equation (1). Although they are not completely dissimilar, there is a clear disagreement between rdfitc and rdfitf. In the presence of strong magnetic fields, it is well observed an increase of the mode frequency which is larger at larger frequencies until near the cut-off frequency where it sharply decreases (Fig. 2 right panel). The 'AR' frequency variation is an order of magnitude larger than the 'nearby' variation.

3.2. Anisotropy

Only the variation in the flow (given by U_x and U_y) and the non-constant terms of the amplitude in rdfitc (given by a_2 and a_3) are anisotropic in relation to nearby active region (θ_{AR}).

Amplitude. The anisotropic terms in A defined in equation (1) is given by a_2 and a_3 (rdfitc): $A_\theta = \exp[a_2 \cos(2\theta') + a_3 \sin(2\theta')]$, where $\theta' = \theta + \theta_{AR}$. Fig. 3 (top left panel) shows A_θ variation for a given mode as a function of θ' where the different colors are for different θ_{AR}. As expected, the amplitude of the waves propagating in the direction of the nearby active region are

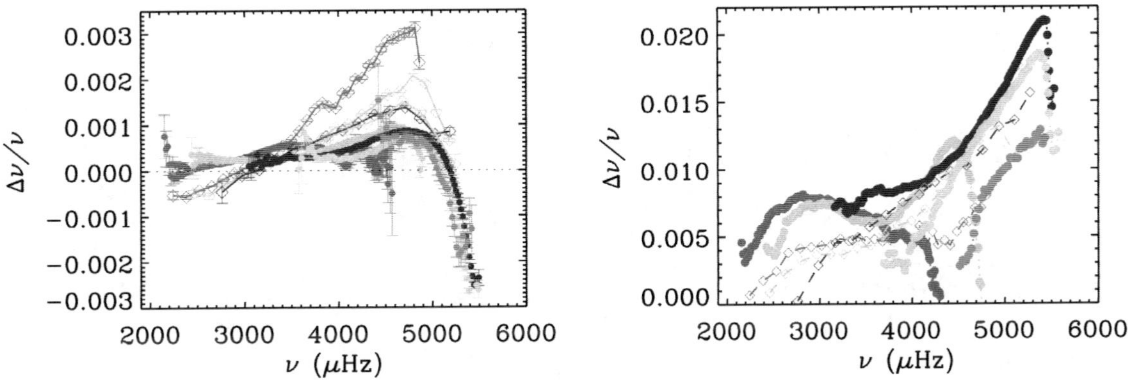

Figure 2. Relative 'nearby' (left panel) and 'AR' (right panel) frequency variation obtained using rdfitc (circles) and rdfitf (diamonds) for f (red), p_1 (green), p_2 (blue), p_3 (cyan), p_4 (salmon) modes.

attenuated in relation to ones propagating perpendicular to it. The variation given by the control set (black crosses) is much smaller than the target variation. Fig. 3 (top right panel) shows A_θ variation in the direction of the nearby active region for all modes. For frequencies smaller than ≈4.2 mHz, the anisotropic variation is of the same order of magnitude as the constant term in A (Fig. 1 left panel). For larger frequencies, the anisotropic variation is small (about 20%) compared with the variation in the constant term in A, with a slight amplification in the waves propagating in the direction of the nearby active region and a slight attenuation in the direction perpendicular to it (Fig. 3 top right). As a result, the modes are amplified (acoustic halo) if there is an active region nearby with very little dependence on their propagation direction. Fig. 3 (bottom panels) shows A_θ plus the variation of the constant term (given by Fig. 1 left panel) in the direction of the nearby AR (left panel) and perpendicular to it (right panel).

Flow. The fitted zonal and meridional flows (Ux and Uy) are expressed in relation to the direction of the nearby active region as $U_{\|}$ and $U_{\perp}$. We do not observe any variation in $U_{\perp}$. The variation in $U_{\|}$ varies from -20 to 5 m/s and it is much larger than the control set. The two different fitting methods give different results. The results for both methods show no clear dependence on frequency, degree or ν/l.

4. Conclusions

We compare the wave characteristics obtained by the HMI ring-analysis pipeline for magnetically quiet 5° patches, that have at least one large magnetic active region (MAI > 100G) in their vicinity, and it is not more than 8° away, with those with no nearby active regions, but at the same latitude and Stonyhurst longitude. The parameter variations were averaged over all visible sunspots observed during 2.5 years (or 34 Carrington rotations) of the ascending phase of cycle 24. We found significant variations in the mode parameters.

Their amplitude varies from -20% to +20% depending on the frequency and direction of the wave propagation in relation to the nearby active region. Waves propagating in the direction of the nearby active region and with frequencies around 3 mHz present the largest attenuation. While modes with frequencies around 5 mHz present the largest amplification, with those propagating in the direction of the nearby active region ≈20% more than those propagating perpendicular to it. The linewidth varies a few percent and it is mainly a function of frequency. The mode lifetime increases with frequency for $\nu \lesssim 4.5$ mHz. The frequency varies within ±0.2%

Journal of Physics: Conference Series **440** (2013) 012008 doi:10.1088/1742-6596/440/1/012008

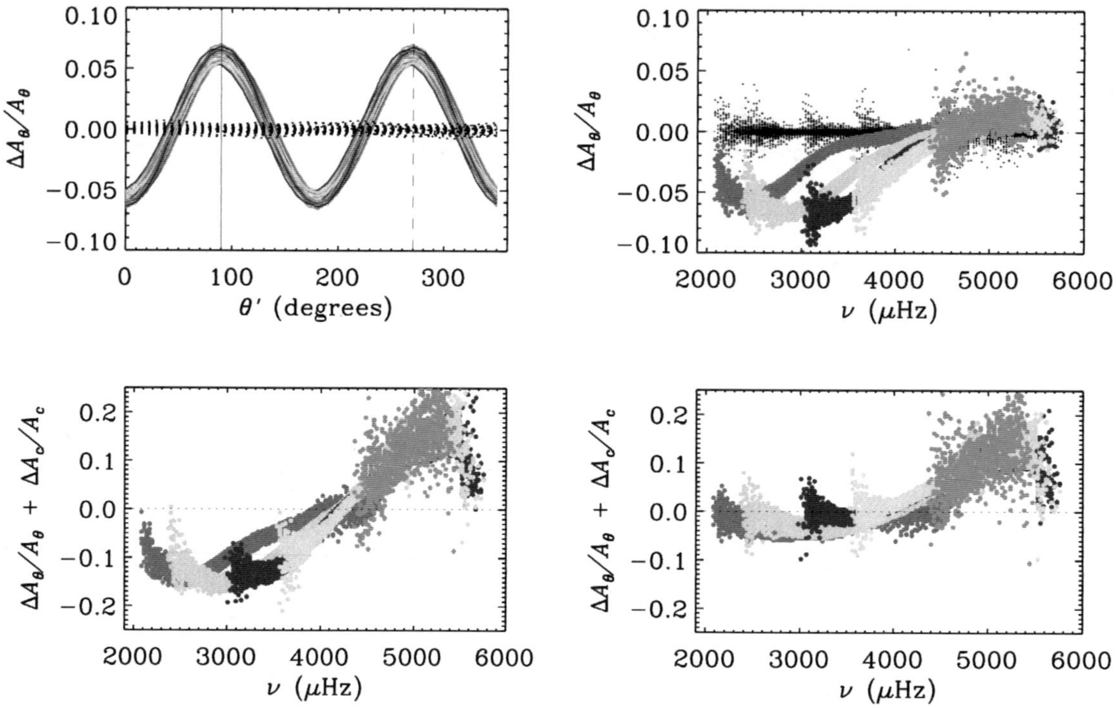

Figure 3. Top panels: Relative A_θ variation for a given mode (n=0, l=615 and ν=2488 μHz) on the left where the different colors are for different θ_{AR} and, on the right, in the direction of the nearby active region for all modes where red, green, blue, cyan and salmon correspond to n = 0, 1, 2, 3 and 4 respectively. The black crosses are the results for the control set. Bottom panels: Variation of A_θ plus the constant term in the amplitude (given by Fig. 1 left panel) in the direction of the nearby active region (left panel) and perpendicular to it (right panel). The colour scheme is same as for the top right panel.

(or ± 10 μHz). The variation of the flow in the direction of the nearby active region is less than 20 m/s and has no clear dependence with the lower turning point depth. There is no flow variation in the perpendicular direction.

Acknowledgments

M. C. Rabello-Soares is thankful to CAPES-Brazil for travel support.

References

[1] Braun D C, Duvall T L Jr and Labonte B J 1987 *ApJ* **319** L27
[2] Baldner C S, Basu S, Bogart R S, et al. 2012 *Solar Phys.* (in press)
[3] Bogart R S, Baldner C, Basu S, Haber D A and Rabello-Soares M C 2011 *J. Conf. Ser* 271 012008
Bogart R S, Baldner C, Basu S, Haber D A and Rabello-Soares M C 2011 *J. Conf. Ser* 271 012009
[4] Basu S and Antia H M 1999 *ApJ* **525** 517
[5] Haber D A, Hindman B W, Toomre J, Bogart R S, Thompson M J and Hill F 2000 *Solar Phys.* **192** 335
[6] Basu S, Antia H M, Bogart R S 2004 *ApJ* **610** 1157
[7] Rabello-Soares M C, Bogart R S and Basu S 2008 *J. Conf. Ser.* **118** 012084
[8] Rajaguru S P, Couvidat S, Sun X, Hayashi K, Schunker H 2012 *Solar Phys.* **in press**
[9] Khomenko E and Collados M 2009 *A&A* **506** L5

Journal of Physics: Conference Series **440** (2013) 012009 doi:10.1088/1742-6596/440/1/012009

Millimeter emission of solar flares

V. G. Nagnibeda[1], **V. V. Smirnova**[1,2], **V. S. Ryzhov**[3] **and A. V. Zhiltsov**[3]

[1] Sobolev Astronomical Institute, Saint Petersburg State University, Universitetsky pr.28, St. Petergof, 198504, Russia

[2] Central (Pulkovo) Astronomical Observatory, Russian Academy of Sciences, Pulkovskoe shausse 65, St.Petersburg, 196140, Russia

[3] Bauman Moscow State Technical University, 2 Bauman street 5, Moscow, 105005, Russia

E-mail: vvsvid.smirnova@yandex.ru

Abstract. We analyzed two unique radio observations of millimeter solar flares at 93 and 140 GHz with the THz component in these spectra. Data were obtained from ground-based radio-telescope RT-7.5 operated by Bauman Moscow State Technical University (BMSTU) with the spacial resolution of 2.5 and 1.5 arc-minutes. We analyzed temporal structure of observed radio-bursts and their dynamics in comparison with soft and hard X-ray light-curves obtained from GOES and RHESSI space-based observations. It was found the 140 GHz emission enhancement at the spectra that is new independent confirmation of the THz component existance. Also, we analyzed data of microwave solar flares with the emission enhanced at 35 GHz obtained from Nobeyama radioheliograph (NoRH) and radiopolarimeter (NoRP). It was found that the maximum of the flux density spectra was shifted toward high frequencies that didn't agree with the model spectrum obtained from microwave observations. We assumed, that such kind of spectra are associated with the gyro-synchrotron radiation of the significant number of high-energetic electrons emission with energies of about 500 keV. The emission mechanism at millimeter waves is the gyro-synchrotron radiation. However, if the energetic electrons are not enough the thermal emission mechanism could be dominant factor at millimeter radiation that could also explain the spectral maximum shift to high frequencies.

1. Introduction

Millimeter and sub-millimeter radio-burst observations allow us to obtain the important information about flare plasma parameters, the spatial distribution of energetic electrons, and flare source location in the solar atmosphere. But short millimeter solar observations are very limited by the instrument requirements and weather conditions. It is well known that observations obtained from the Solar Submillimeter Telescope (SST) at 212 and 405 GHz [5] and observations from the Köln Observatory for Submillimeter and Millimeter Astronomy (KOSMA) at 230 and 345 GHz [9] confirm THz emission in solar flares. But these observations are not enough to provide the interpretation of the THz component appearing in flares because they consist only of a few points and not of a complete microwave-millimeter spectrum.

It is generally accepted that non-thermal microwave emission during large solar flares is produced by the gyro-synchrotron mechanism which involves coronal magnetic fields of at least a few hundreds of Gauss and electrons of hundreds of keV [1],[3],[16].

In the typical microwave spectrum (Figure 1, left panel) the maximum of the flux density corresponds to frequency in the range of 5–10 GHz [7]. But some flares show peculiarities in

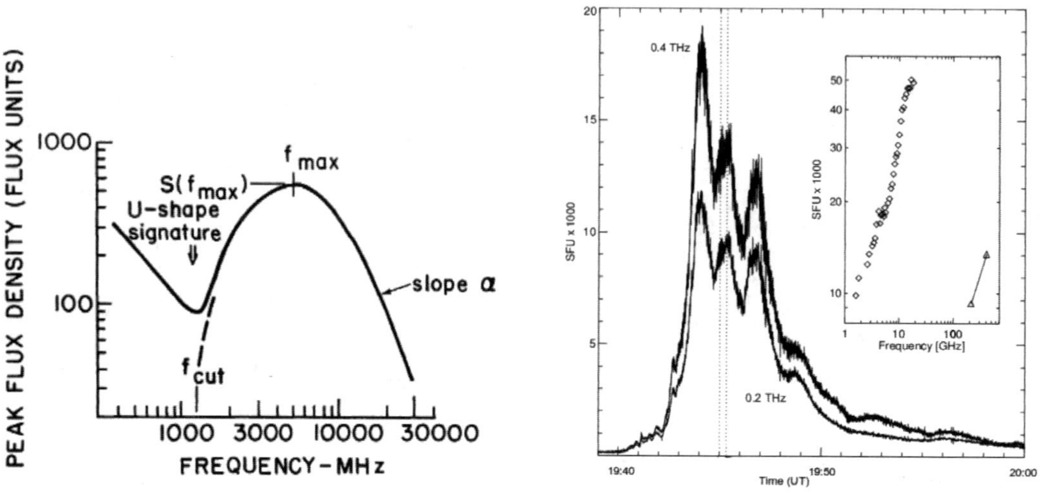

Figure 1. Left panel: typical spectrum of microwave radio bursts [7]. Right panel: the largest sub-mm solar burst observed by the SST telescope that occurred on November 4, 2003 showing the time profiles at 0.2 and 0.4 THz. The inner panel shows the double spectral features, with the well known microwaves and the new THz component [10].

their spectra at high frequencies [2]. These peculiarities, such as a spectral flattening or even an excess of the high-frequency emission, are interpreted by authors as the superposition of multiple sources of gyro-synchrotron radiation or optically thin bremsstrahlung of the evaporated flare plasma.

In some flares the short millimeter part of the spectrum is significantly 'hard', in disagreement with the model spectrum (see Figure 1, right panel) [10]. In these kinds of events the flux density of millimeter emission is high and sometimes even exceeds the microwave flux. This behavior has been observed more often in X-class flares (by GOES classification) and interpreted by several authors as synchrotron radiation of relativistic electrons with energies more than 500 keV [8]. But it is difficult to give a clear interpretation about what mechanism dominates at short millimeter wavelengths during the flare process, because there are not enough multi-frequency millimeter data with high spatial resolution.

In this contribution we present new unique observations of solar flares at 93 and 140 GHz obtained from the RT-7.5 radio-telescope operated by Bauman Moscow State Technical University (BMSTU). Relatively few data at 93 GHz have been obtained with any instruments. But at 140 GHz radio-frequency such observations have never been provided. Thus, data obtained from RT-7.5 are unique and very useful for the analysis of short millimeter emission in flares. We also present an analysis of microwave radio-burst data obtained from the Nobeyama radioheliograph (NoRH) and the Nobeyama radiopolarimeter (NoRP).

2. RT-7.5 BMSTU: observations at 93 and 140 GHz

The RT-7.5 radio-telescope is a Cassegrain-type of antenna with a diameter of 7.75 m. It located in Dmitrov (Moscow region, Russia). It has two receivers for 93 GHz and 140 GHz (3.2 mm and 2.2 mm). The beam size of the telescope is 2.5 arc-minutes at 93 GHz and 1.5 arc-minutes at 140 GHz. The antenna can provide solar disk mapping, active region mapping, and tracking of a selected point. The maximum temporal resolution is 0.125 s [12], [13], [14].

We analysed two radio-bursts that occurred in active region NOAA 11515. Dates of

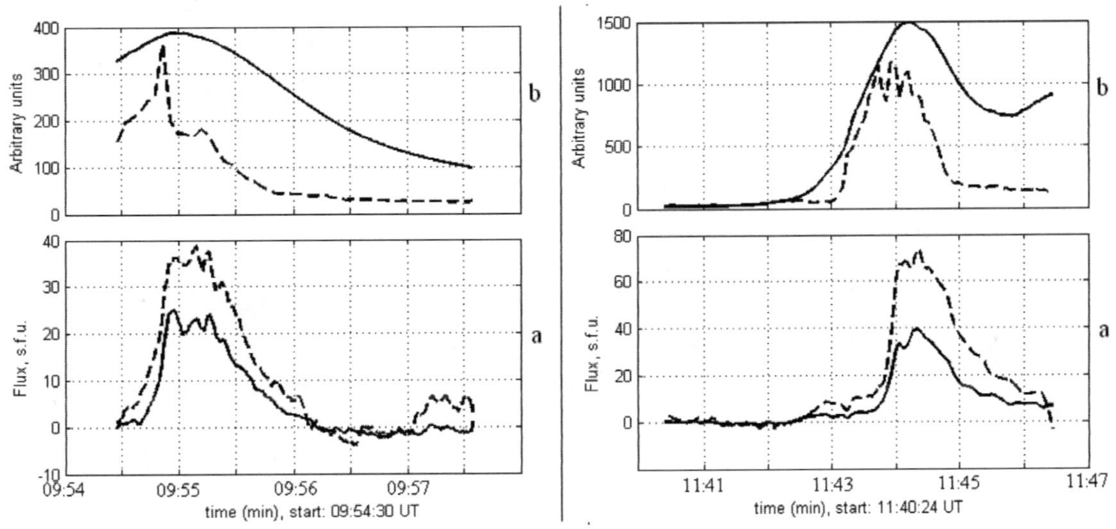

Figure 2. Left panel: a) radio flux density temporal profiles obtained from the observations at 93 GHz (solid line) and 140 GHz (dashed line) for SOL2012-07-04. b) GOES 1-8 Å SXR light-curve (solid line) and the RHESSI 25-50 keV HXR light-curve (dashed line). Right panel: the same, but for SOL2012-07-05.

observations: 2012/07/04 and 2012/07/05. GOES class of flares: M5.3 and M6.1. The weather conditions were good during the radio-observations.

Figure 2 (panels (a) solid and dashed lines) shows the temporal structure of the bursts at 93 and 140 GHz. We also examined soft and hard X-ray data simultaneously with radio emission (see Figure 2, panels (b)). Soft X-ray data were obtained from GOES-15 observations at 1-8 Å. The temporal evolution of the SXR emission during the impulsive phase of flares was smooth and longer than radio flux of the bursts but the maximum value was reached at the same time. Hard X-ray data were obtained from the RHESSI observations at the energetic channel 25-50 keV. The maximum of the HXR emission at 25-50 keV appeared earlier than the radio maximum in both cases.

The radio flux density spectra for the moments of the flare maxima are shown in Figure 3. The errors in the flux density were estimated using the beam pointing accuracy and sensitivity of the receivers.

We found the 11.7 GHz data obtained from the Metsähovi radio-observatory (Finland) for the SOL2012-07-04 event (see Figure 3, left panel). Unfortunately there were no microwave data for the SOL2012-07-05 flare. Observations at 11.7 GHz were additionally obtained for the event of 2012/07/04. But in both cases the flux density at 140 GHz is higher than at 93 GHz. This situation is similar to the observations in KOSMA (Figure 1, right panel).

3. NoRH and NoRP: observations, data analysis and results

We analysed several radio bursts using Nobeyama radioheliograph (NoRH) data at 17 and 34 GHz and Nobeyama radiopolarimeter (NoRP) data at 1, 2, 3.75, 9.4, 17, 35 and 80 GHz [11]. Temporal structures of the chosen bursts were complex with a number of impulses on the extended rise-and-fall background.

We present the flare event analysis for 2005/09/13 in NOAA 10808. The GOES class of the flare is X 1.7. The temporal structure of the radio-burst at 35 GHz is presented in Figure 4

Journal of Physics: Conference Series **440** (2013) 012009 doi:10.1088/1742-6596/440/1/012009

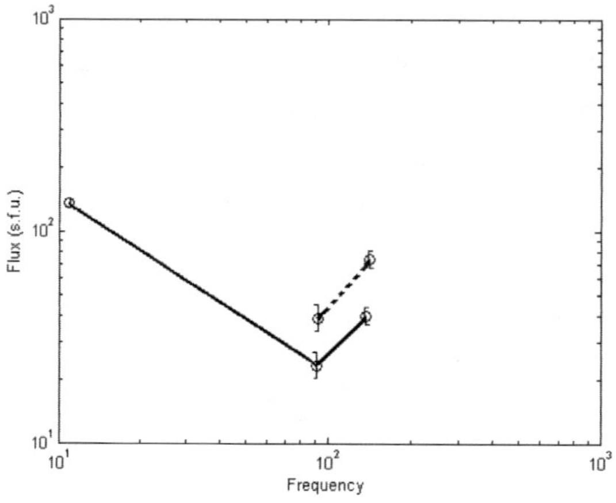

Figure 3. Flux density spectra. Left panel: spectrum obtained at 11, 93 and 140 GHz. Date of observations: 2012/07/04, time: 09:55:00 UT. Right panel: spectrum obtained at 93 and 140 GHz. Date of observations: 2012/07/05, time: 11:44:30 UT.

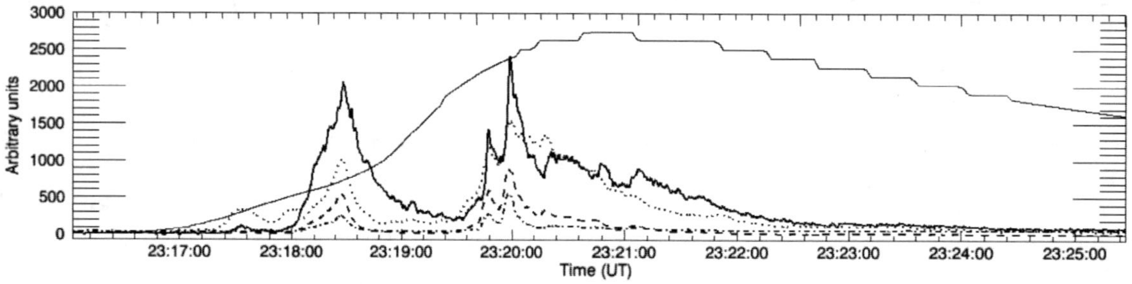

Figure 4. 35 GHz light-curve (solid bold), GOES plot (——) and HXR plots: 25-50 keV (······), 50-100 keV (- - - -), 100-300 keV (— · —).

(solid-bold line). Here one can see the GOES 1-8 Å light-curve (solid line) and HXR light-curves obtained from RHESSI observations. It is known that the soft X-ray emission is generated by the thermal emission of hot plasma. So, time-variations of SXR emission reflects the thermal development of the flare source. Figure 4 shows the temporal profile of the GOES soft X-ray flux that coincides well with the microwave and millimeter burst background. Thus we can use the GOES profile to divide the microwave bursts into an impulsive component, which we could interpret as a non-thermal part of a burst, and the thermal background. Hard X-ray emission was associated with the impulsive phase of the flare. In Figure 4 it is clear that the burst's impulsive component corresponded well to the impulses in HXR emission.

In Figure 5 the radio-burst spectra at the moments of the first and second impulse are shown. The spectral maximum is shifted toward the millimeter range. That is different from the classic microwave spectrum (see Figure 1, left panel) where the maximum is located at 5 GHz.

4. Discussion

There is no simple interpretation of the impulsive non-thermal component of the spectrum with the millimeter emission enhancement. However, in connection with a rather good correlation of the millimeter burst impulsive part and the hard X-ray emission, we suggest in this case that the

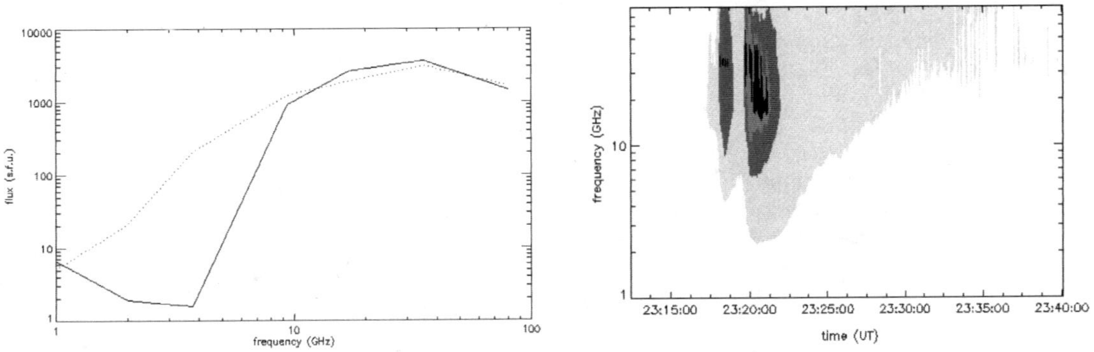

Figure 5. Left panel: radio-intensity spectra: first maximum (——), second maximum (······). Right panel: frequency dynamic spectrum.

maximum shift in the radio-spectra is associated with the injection of high energetic electrons in the flare source. During the powerful flares a greater density of energetic electrons could affect the microwave spectrum, and be responsible for the flux density increasing at high frequencies.

The gyrosynchrotron emission depends not only on the particle energy, but also on the magnetic field strength. It is important to know the position of the radio source in the foot-points or in the top of the flaring loop, to estimate the value of the magnetic field strength, using existing models. However, insufficient angular resolution in radio observations does not allow us to localize the burst radio source in this case.

At the present time interest in the investigation of millimeter and sub-millimeter solar flares is significantly increased. But data available for this range are very limited and do not provide the full spectrum, because of inadequate frequency coverage. Obviously, large systems, such as major radiointerferometers could not provide monitoring observations of the Sun. These instruments have a lot of observational tasks that require time and special calibration methods. For this reason we have started our observations on RT-7.5 at 93 and 140 GHz frequencies that could give the possibility of obtaining new very important points at short millimeter range.

5. Conclusions

1. We have obtained new results that confirmed the THz emission existence in solar flares at short radio-frequencies. New unique flare observations at 93 and 140 GHz were analyzed. The spectra showed the flux density enhancement at 140 GHz. The maximum of the HXR emission did not exceed 25–50 keV in both cases. So, we assumed that these spectra could be interpreted as thermal emission enhancement in the flare sources.
2. We analyzed NoRP and NoRH microwave data, obtained at 1, 2, 3.75, 9, 17, 35 and 80 GHz radio-frequencies. One event on 2005/09/13, appearing in NOAA 10808, was taken as an example. We found, using the dynamic spectra analysis, that the spectral maximum was shifting during the flare toward high frequencies. The HXR emission of the flare reached 300 KeV and was well-correlated with the radio-emission (see Figure 4).

References

[1] Benz A.O., Living Reviews in Solar Physics, vol. 5, no. 1
[2] Chertok, I.M.; Fomichev,V.V.; Gorgutsa, R.V.; Hildebrandt,J.; Krüger, A.; Magun, A.; Zaitsev,V.V., Sol. Phys., Vol. 160, Iss. 1, pp.181-198
[3] Hudson H., 2011, Space Sci. Rev., 158, pp. 541

[4] Kaufmann P.; Marcon R.; Giménez de Castro G.; White S., and 7 co-authors, ApJ, Vol. 742, Iss. 2, article id. 106, 2011
[5] Kaufmann, P., Trottet, G., Giménez de Castro, C. G., Raulin, J.-P., Krucker, S., Shih, A. Y., Levato, H., Sol.Phys. Vol.255, pp. 131-142, 2009
[6] Kaufmann P.; Raulin J-P.; Melo A.M.; Correia E.; Costa J.E.R.; Giménez de Castro C.G.; and 9 co-authrs, ApJ, 574:10591065, 2002
[7] Kundu M.R., Vlahos L., Space Science Reviews 32, 405-462, 1982.
[8] Kundu M.R. et al., The Astrophysical Journal, 547: 1090-1099, 2001.
[9] Lüthi, T., Lüdi, A. and Magun, A., A& A, 420, pp. 361-370, 2004
[10] Marcon R, Kaufmann P, Fernandes L.O.T., and 6 co-authors, 2012, Journal of Infrared, Millimeter, and Terahertz Waves, Volume 33, Issue 2, pp.192-205
[11] Nakajima, H.; Nishio, M.; Enome, S.; Shibasaki, K.; and 13 co-authors, 1994, IEEE Proceedings Vol. 82, pp. 705-713
[12] Rozanov, B. A., 1981, Radioelektronika, vol. 24, p. 3-8.
[13] Solov'ev G.N., Rozanov, B. A., Ivanov V.N., Nagnibeda V.G., Piotrovich V.V., 1990, Radioelectronic devises
[14] Smirnova, V., Riehokainen, A., Ryzhov, V., Zhiltsov, A., Kallunki, J. A& A, 2011, Volume 534, id.A137
[15] Trottet G., Vilmer N., Barat C., Benz A., Magun A., Kuznetsov A., Sunyaev A., and Terekhov O., 1998, A& A, 334, 10991111
[16] White S.M., Krucker S., Shibasaki K., Yokoyama T., Shimojo M., and Kundu M.R., 2003, ApJ, pp. 111-114

Journal of Physics: Conference Series **440** (2013) 012010 doi:10.1088/1742-6596/440/1/012010

Spectral Signatures of Transient Heating in the Solar Corona

Youra Taroyan and Keiran Burrows

Institute of Mathematics and Physics, Aberystwyth University, Aberystwyth, SY23 3BZ, UK

E-mail: yot@aber.ac.uk

Abstract. Hinode/EIS observations of an active region are presented. Intensity and Doppler velocity maps for different emission lines are constructed. Consistent with previous results, the Doppler velocities near the footpoints of coronal loops appear to be blue shifted for emission lines with temperatures above 1 MK. The gradual transition from blue shifts at temperatures above 1 MK to red shifts at temperatures below 1 MK is addressed through numerical modeling of loop dynamics. The simulation results are converted into synthetic EIS observations and compared with the actual measurements. Persistent blue shifts, blue wing enhancements, and red shifts observed in EUV lines are interpreted as a signature of repetitive heating occurring near the loop footpoints on a time scale of several minutes.

Hinode/EIS observations have revealed persistent blue-shifts in active regions (Hara et al. 2008; Doschek et al. 2008; Tian et al. 2011). The blue-shifts are usually seen near the footpoints of coronal magnetic loops and may last for days or weeks. Blue shifts of up to 20 km s^{-1} were derived using single Gaussian fits to approximate coronal emission line profiles. Significant deviations from a single Gaussian profile were found in the blue wing of the line profiles. These deviations were studied using multiple Gaussian fitting (Peter 2010) and the blue-red asymmetry technique (Tian et al. 2011). Depending on the applied method, blue shifts between 30 and 100 km s^{-1} were derived.

A study carried out by Patsourakos & Klimchuk (2006) treats a single loop as a bundle of thin threads which are continually heated by nanoflares. The nanoflares result in upflows and broad line widths in the corona. The model has been critically discussed by Peter (2010). Hansteen et al. (2010) studied heating caused by the braiding of the photospheric magnetic field which produces a bi-directional flow of material from the heating site: there are downflows in transition region lines and slight upflows of a few km s^{-1} in low coronal lines. Murray et al. (2010) proposed an outflow scenario based on active region expansion, whereas Bradshaw et al. (2011) interpreted the observations by Hara et al. (2008) in terms of interchange reconnection at the boundaries between open and closed magnetic field regions.

Recently Warren et al. (2011), McIntosh et al. (2012) have detected upflows accompanied by enhanced blue wings in high temperature lines and downflows in low temperature lines. SDO/AIA observations by Kamio et al. (2011) have revealed high speed hot upflows and cool downflows in active regions. A link between short lived ($\sim$ 100 s) type II spicules and upflows in coronal loops was suggested by De Pontieu et al. (2011).

The present Letter will address the nature of the EUV Doppler shifts through a combination of data analysis and numerical forward modeling.

Journal of Physics: Conference Series **440** (2013) 012010 doi:10.1088/1742-6596/440/1/012010

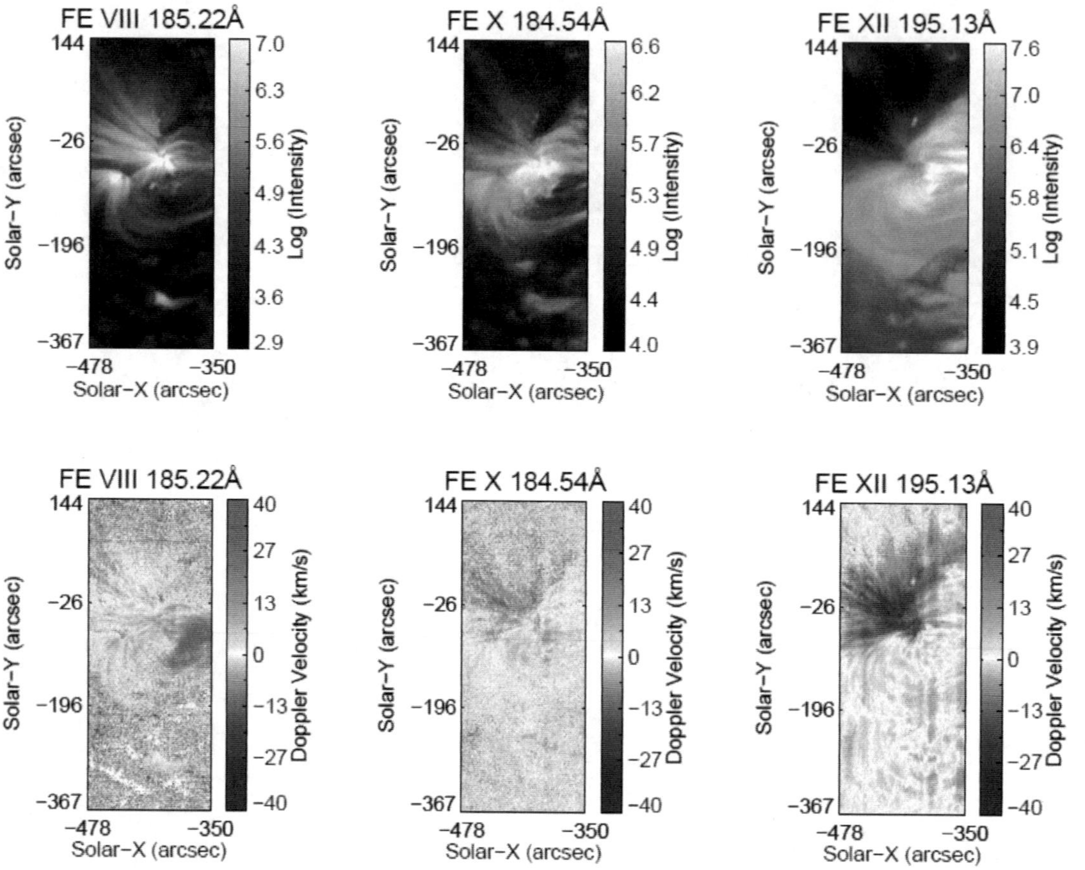

Figure 1. Intensity and Doppler velocity maps of an active region in three selected lines: Fe VIII 185 Å, Fe X 184.5 Å and Fe XII 195 Å.

1. Observations

The EUV imaging-spectrometer (EIS) aboard Hinode has two CCDs covering the wavelength ranges 171 – 212 Å and 245 – 291 Å. EIS has both narrow (1″ and 2″ wide) slits, and wider (40″ and 266″) imaging slots, all with up to 512″ in the solar Y direction. Spectroscopic observations can be carried out either in sit-and-stare or in scanning modes. Further details of Hinode/EIS characteristics are given by Culhane et al. (2007), Kosugi et al. (2007).

For this study, we concentrated on an active region visible on 20th February 2007. Hinode/EIS observations of the active region were carried out using the 1″ wide slit in scanning mode. Spectral data for a number of wavelengths were produced. The data have been studied by Warren et al. (2011), Martinez-Sykora et al. (2011).

In the following analysis, we select the Fe VIII 185.21 Å, Fe X 184.54 Å and Fe XII 195.12 Å spectral lines with formation temperatures of $\log T \approx 5.8$, 6.0 and 6.2. EIS level zero data files are processed with the eis_prep software, using the default options. The eis_wave_corr procedure and the level one data are used to calculate the wavelength corrections for each spectral line. We fit a single Gaussian profile to each line spectrum, using the eis_auto_fit software. The reference wavelength of each line profile has been updated to match the rest wavelength found in a laboratory, and is taken from Brown et al. (2008). EIS Doppler velocities need calibrating against a reference wavelength, where the quiet Sun velocity averages to zero. For this work the Fe VIII 185 Å line is chosen as a reference. The other lines are calculated relative to the final wavelength of this line. Doppler velocity measurements require the slit tilt and thermal drift of the instrument to be calculated accurately. This is achieved using the eis_update_fitdata procedure, which updates these parameters, based on the fitted line profile and the first approximations from eis_wave_corr. Finally we output intensity and Doppler velocity maps for the spectral lines.

The intensity maps in Figure 1 show loops that become bright in the higher temperature line of Fe XII 195 Å. The loops are connected to the footpoints which can be seen near the center of the intensity image in Fe VIII 185 Å. The Doppler velocity maps in the lower panel reveal the presence of blue shifts in Fe XII near the footpoints. The blue shifts fade out in Fe X and turn into red shifts in the lower temperature Fe VIII line. The blue shifts at high temperatures are a persistent feature of EIS observations and may last for prolonged periods of time (e.g., Hara et al. 2008). The footpoints areas of the cool loops in Fe VIII are red shifted whereas the footpoint areas of the hot loops in Fe XII are blue shifted. The velocities for the intermediate Fe X line are close to zero. This gradual transition from red to blue will be addressed in the forthcoming analysis which is based on numerical forward modeling.

In addition to loop like structures, Figure 1 shows fan like structures at the top left corner of the image. These structures mainly appear bright in the lower temperature line of Fe VIII 185 Å. The question whether the blue shifts seen in the Fe XII velocity map are related to the fan like structures has been addressed by Warren et al. (2011).

2. Numerical Forward Modeling

In order to explain the features seen in Figure 2, in particular, the gradual transition from red in Fe VIII to blue in Fe XII near the loop footpoints, we carry out numerical forward modeling. A single loop with a typical length of 40 Mm is initially in a cooling state due to the combined action of thermal conduction and radiation. The loop is gravitationally stratified and remains in the vertical plain. Cooling leads to downflows along the loop legs. Only longitudinal motions are considered. Motions in other directions are ignored. This is a valid approximation as long as the ratio of the thermal pressure to magnetic pressure is small. The initial peak temperature at the apex is about 0.6 MK and the dense chromospheric footpoints are kept at a constant temperature of 20 000 K. The governing equations of continuity, momentum and energy are integrated numerically to investigate the loop response to transient heating. The one dimensional treatment and the adaptive mesh refinement technique allows us to accurately model the dynamics in the steep transition region and the emission in EUV lines. The details of the HYDRAD code used in our modeling are described by Bradshaw & Mason (2003).

A phenomenological heating term is included in the energy equation. The heating is located 3 Mm inside the chromospheric boundary and it has an exponential scale height of 4 Mm. Localizing the heating near the footpoints has the effect of driving strong upflows. For simplicity, a symmetric profile about the apex is chosen. The heating occurs on a time scale of a few minutes. It smoothly increases from $t = 0$ s, reaches its peak rate at $t = 150$ s and vanishes at $t = 300$ s. The subsequent dynamics is determined by thermal conduction and radiation. The amount of energy input is proportional to the cross sectional area of the loop. For a cross sectional diameter of 1 Mm the total energy input near each footpoint is about 10^{26} erg. The selected heating parameters are best suited for reproducing the observed temporal evolution.

Figure 2 displays the evolution of the loop. The vertical axes denote distance along the loop. Temperature starts to rise after heating catches up with losses. The loop reaches a peak temperature of 3 MK at about $t = 200$ s. The second panel of Figure 2 shows strong upflows along both legs during the heating stage which leads to high temperatures. During this stage a maximum upflow speed of 110 km s^{-1} is reached. The subsequent evolution of the loop is dominated by downflows as the loop cools to lower temperatures. The initial state is approximately restored at the end of the simulation.

Journal of Physics: Conference Series **440** (2013) 012010 doi:10.1088/1742-6596/440/1/012010

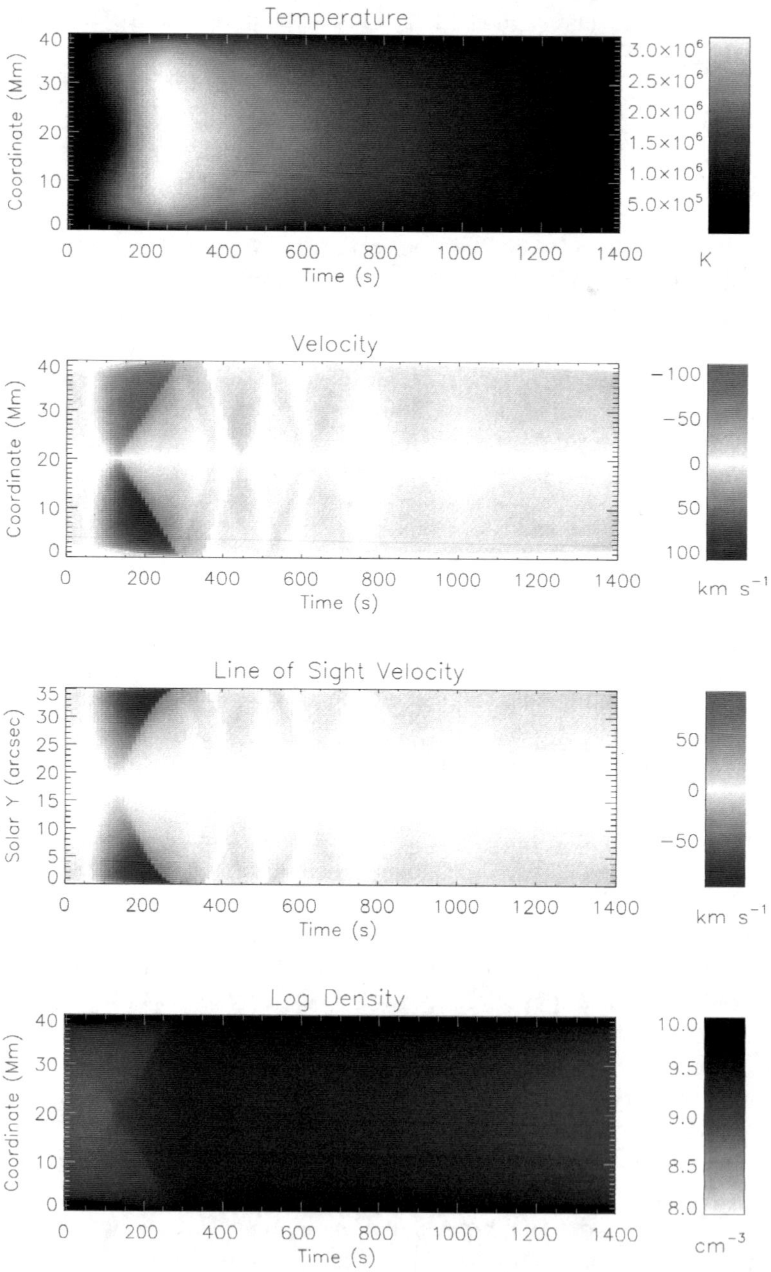

Figure 2. The evolution of the loop during the first 1400 s. Temperature, velocity along the loop and density are plotted as functions of time and loop coordinate. The third panel displays the line of sight velocity.

The velocities projected onto the line of sight are shown in the third panel of Figure 2. For simplicity, it is assumed that the loop is located in the center of the solar disc. Note the shift of the peak velocities towards the footpoints due to the semicircular geometry of the loop.

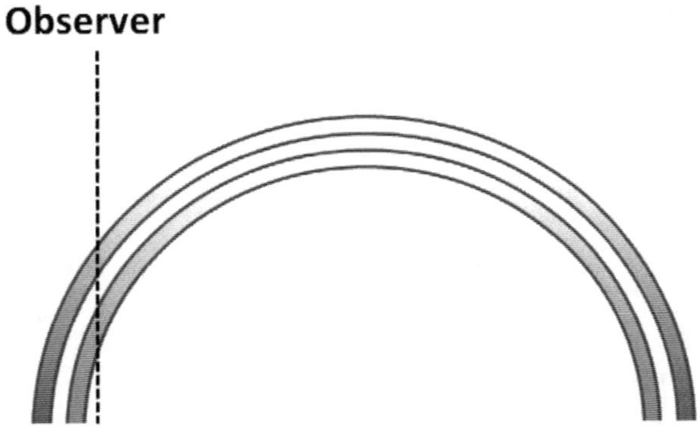

Figure 3. Three loops intersecting with the line of sight. The colours indicate different states: upflows (blue), downflows, (red), no flows (white).

Downflows correspond to positive red-shifts, and upflows correspond to negative blue shifts. The heating pulse leads to blue shifts of up to 97 km s^{-1}. The bottom panel of Figure 2 displays the evolution of density: heating and upflows are followed by enhanced density whereas cooling and downflows reduce the amount of material in the loop.

3. Discussion

Figure 2 suggests that the evolution of the loop proceeds in three main stages: there are blue shifts near the footpoints at high temperatures and red shifts at low temperatures. The velocities are small at intermediate temperatures when there is a transition from heating to cooling or vice-versa. However, the blue shifts observed in high temperature lines are usually quite persistent and last much longer than the flows in our simulations (Hara et al. 2008). Another important difference is that Figure 1 shows an entire active region, whereas our simulations are carried out for a single loop. It is therefore not possible to directly reproduce the observed behavior from the simulations.

According to the observations presented in Figure 1, there might be several loops along the line of sight. In general, the loops have different densities, temperatures, lengths, widths, inclinations, etc, and are in different stages of evolution as shown in Figure 3. All these loops converege at the footpoints where the blue shifts are mainly seen. In the absence of detailed knowledge of the involved parameters, we assume that there are four loops along the line of sight which are almost identical if their length is large compared to the cross sectional radius. Each loop follows the cycle shown in Figure 2 which is repeated upon completion and there is a lag time of 250 s between the loops.

The simulation results are converted into synthetic spectral observations using the EIS response function. The full details of this procedure are described by Taroyan et al. (2006). The emission from the three loops shown in Figure 3 is integrated along a segment near the footpoints. The length of the segment corresponds to a $1''$ EIS slit. The instrumental width of the EIS instrument is added to the thermal width of the line profile.

Figure 4 shows synthetic line profiles constructed for Fe VIII, Fe X Fe XII at different moments of time. Each plotted curve represents a superposition of three spectral line profiles representing different loops.

A comparison between the observations and the synthesized spectral line profiles reveals similarities. In both cases, the low temperature Fe VIII lines are predominantly red shifted. These are cooling loops dominated by downflows. The Fe X Doppler shifts are small. This type of emission mainly corresponds to loops at intermediate temperatures that are in a transition

Journal of Physics: Conference Series **440** (2013) 012010 doi:10.1088/1742-6596/440/1/012010

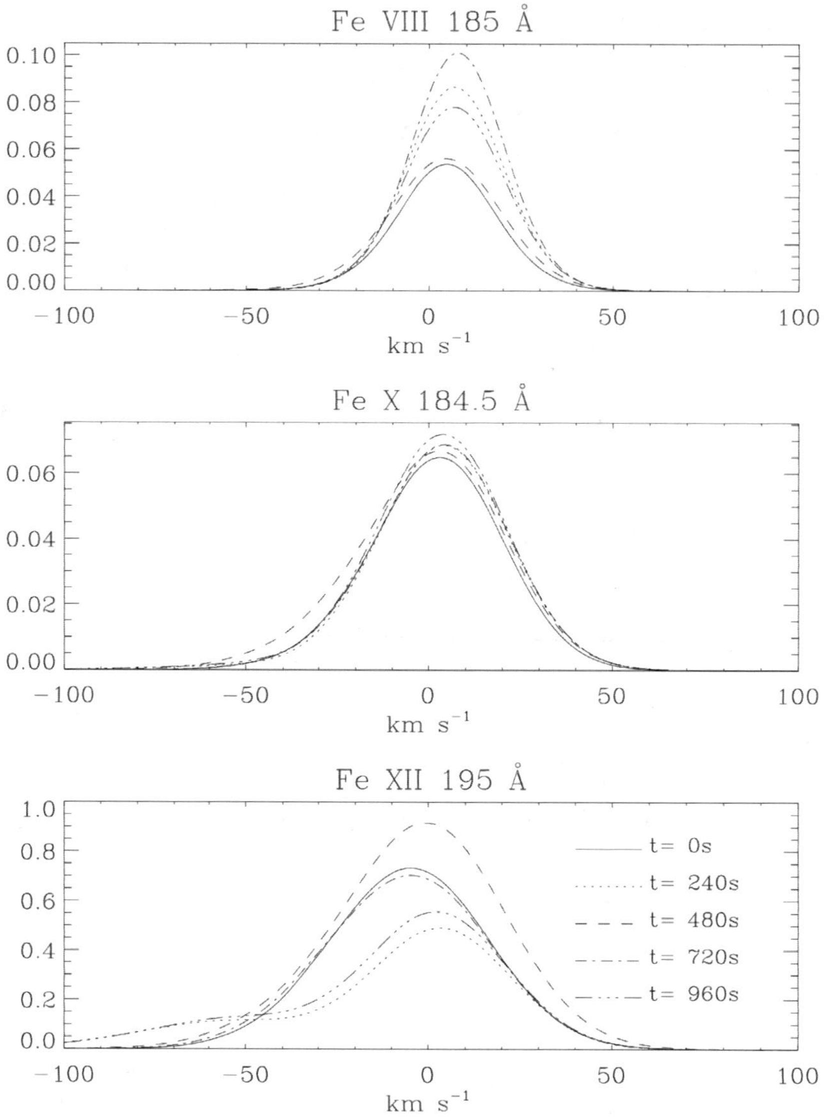

Figure 4. Synthesized spectral line profiles at different moments of time. Each curve represents a superposition of four different line profiles which correspond to identical loops in different states.

from heating to cooling or vice-versa. Blue shifts and enhanced blue wings appear with increasing temperature in Fe XII and Fe XV. The enhanced blue wings representing high speed flows become most prominent in the high temperature Fe XV line. These types of line profiles are associated with loops that are in the heating stage.

The relatively higher intensities in Fe VIII and Fe X can be explained by the presence of underlying cool structures that are not taken into account by the model. Despite the mentioned differences that are mainly due to the simplicity of our model, the agreement with the observations is encouraging.

A comparison of spectral observations with forward modeling supports the idea of heating on a time scale of a few minutes. Blue shifts and blue wing enhancements are associated with heating, whereas the red shifts are a signature of cooling. No such clear association exists when the heating time scale is significantly reduced. We have performed a sample simulation of heating with a similar total energy input but with a duration of 60 s: strong red shifts appear during the heating stage and blue shifts appear during the cooling stage as the rapid pressure pulse bounces back and forth along the loop. It is therefore no longer possible to reproduce the observations shown in Figures 1. A significant increase in the total heat input leads to higher maximum temperatures but results in red shifts at temperatures above 1 MK. We have therefore found important constraints on models of loop heating.

Below we explain the absence of significant net shifts in the case of rapid heating. The initial heating event drives evaporation from the chromosphere which lasts for a short period of time. The emission is expected to peak during the cooling phase when the densities are high. However, these brief upflows propagate away from the footpoint segment of the loop by the time it cools to the Fe XII passband and no significant upflows are seen. Subsequently the pulse is propagating and bouncing back and forth along the loop as the loop cools to Fe X and Fe VIII passbands. This may result in blue, red or no shifts. Therefore a superposition of rapidly heated loops leads to a situation whereby no significant net flows are detected. Therefore, no significant net flows will be detected if the heating time is short compared with the pulse travel time. The travel time is roughly determined as the ratio of the loop length and the sound speed. When the loops are longer than about 20 Mm, the reflection effects become small and therefore blue shifts and red shifts are seen at high and low temperatures, respectively.

4. Conclusions

The coexistence of blue shifts in coronal lines and red shifts in transition region lines has been known for many years. These persistent features have mainly been interpreted as a manifestation of a bi-directional flow of material from the reconnection site or nanoflares (Teriaca et al. 1999; Hansteen et al. 2010; Patsourakos & Klimchuk 2006). Hinode/EIS observations have revealed new features near the footpoints of active region loops. Figures 1 and 4 show red shifts at low temperatures (Fe VIII line), blue shifts and enhanced blue wings at high temperatures (Fe XII and Fe XV lines). All these features are reproduced from numerical simulations of transient loop heating if the convergence of several loops at the footpoints is taken into account. The heating occurs on a time scale of a few minutes. The nature of the heating process remains unknown. However, the obtained results provide important constraints on models of loop heating.

[1] Bradshaw S. J. & Mason H. E., 2003, A&A, 407, 1127
[2] Bradshaw, S. J., Aulanier, G., Del Zanna, G., 2011, ApJ, 743, 66
[3] Brown, C. M., Feldman, U., Seely, J. F., et al., 2008, ApJS, 176, 511
[4] Culhane, J. L., et al., 2007 Sol. Phys., 243, 19
[5] De Pontieu, B., McIntosh, S. W., Carlsson, M., et al., 2011, Science, 331, 55
[6] Doschek, G. A., Warren, H. P., Mariska, J. T., et al., 2008, ApJ, 686, 1362
[7] Hansteen, V. H., Hara, H., De Pontieu, B., et al., 2010, ApJ, 718, 1070
[8] Hara, H., Watanabe, T., Harra, L. K., et al., 2008, ApJ, 678, L67
[9] Kamio, S., Peter, H., Curdt, W., et al., 2011, A&A, 532, A96
[10] Kosugi, T., et al., 2007, Sol. Phys., 243, 3
[11] Martinez-Sykora, J., De Pontieu, B., Hansteen, V., et al. 2011, ApJ, 732, 84
[12] McIntosh, S. W., Tian, H., Sechler, M., et al., 2012, ApJ, 749, 60
[13] Murray, M. J., Baker, D., van Driel-Gesztelyi, L., et al. 2010, Sol. Phys., 261, 253
[14] Patsourakos, S. Klimchuk, J. A., 2006, ApJ, 647, 1452
[15] Peter, H., 2010, A&A, 521, A51

[16] Peter, H. & Judge, P. H., 1999, ApJ, 52
[17] Taroyan, Y., Bradshaw, S. J., & Doyle, J. G., 2006, A&A, 446, 315
[18] Teriaca, L., Doyle, J. G., Erdlyi, R. et al., 1999, A&A, 352, L99
[19] Tian, H., McIntosh, S. W., De Pontieu, B., et al., 2011, ApJ, 738, 18
[20] Warren, H. P., Ugarte-Urra, I., Young, P. R., et al., 2011, ApJ, 727, 58

Low-degree multi-spectral *p*-mode fitting

R Howe[1], **A-M Broomhall**[1,2,3], **W J Chaplin**[1], **Y Elsworth**[1] **and K Jain**[4]

[1]School of Physics and Astronomy, University of Birmingham, UK
[2]Centre for Fusion, Space and Astrophysics, Department of Physics, University of Warwick, Coventry, CV4 7AL
[3]Institute of Advanced Study, University of Warwick, Coventry, CV4 7AL
[4]National Solar Observatory, 950 N Cherry Ave., Tucson, AZ 85719, USA

E-mail: rhowe@nso.edu

Abstract. We combine unresolved-Sun velocity and intensity observations at multiple wavelengths from the Helioseismic and Magnetic Imager and Atmospheric Imaging Array onboard the Solar Dynamics Observatory to investigate the possibility of multi-spectral mode-frequency estimation at low spherical harmonic degree. We test a simple multi-spectral algorithm using a common line width and frequency for each mode and a separate amplitude, background and asymmetry parameter, and compare the results with those from fits to the individual spectra. The preliminary results suggest that this approach may provide a more stable fit than using the observables separately.

1. Introduction

The 1700 and 1600 Å bands of the Atmospheric Imaging Assembly (AIA) onboard the Solar Dynamics Observatory (SDO) are formed higher in the solar atmosphere than the lines typically used for Doppler velocity measurements – around 360 km for the 1700 Å and 480 km for the 1600 Å band [1], extending from the upper photosphere into the chromosphere. They neverthless show a clear *p*-mode spectrum for integrated-Sun measurements [2]. Although the velocity and intensity spectra formed at different heights differ in many details and reflect different mode physics, the underlying mode frequency and lifetime should be independent of the height of formation or the details of the observation, allowing us to use several spectra together to obtain an improved estimate of the underlying mode parameters. In this article, we attempt to improve our estimates of the parameters of low-degree *p*-modes, using data from AIA and from the Helioseismic and Magnetic Imager (HMI) also onboard SDO.

2. Data

The data used cover 691 days starting with 13 May 2010, giving a frequency resolution of 17 nHz. Time series were formed from the AIA 1600 Å and 1700 Å bands and the HMI Doppler Velocity [V], Continuum Intensity [Ic], and Line Core Intensity [Lc, formed by subtracting the Line Depth observable from Ic], using the DATAMEAN keyword supplied with each HMI and AIA image in the JSOC database[1]. The heights of formation of the HMI observables are around

[1] http:/jsoc.stanford.edu/

25 km for HMI Ic, 100 km for HMI V, and 270 km for HMI Lc [3; 4]. All time series were interpolated to a uniform 45-second cadence; long-term and daily trends were also removed.

3. Analysis

In this preliminary work we took a simple approach to multi-spectral fitting, in which all five spectra were fitted simultaneously for each $\ell = 0, 2$ and $\ell = 1, 3$ pair of modes. Each mode had only one frequency, linewidth and rotational splitting parameter, but the amplitude, asymmetry and background parameters were independent for each spectrum. The peak profile was that of [5]:

$$P(x) = A\frac{(1+Sx)^2 + S^2}{(1+x^2)} + B, \tag{1}$$

where A is the amplitude, S is the asymmetry parameter, B is the background parameter and $x = 2(\nu - \nu_0)/\gamma$, γ being the FWHM and ν_0 the central frequency. For the sake of a more stable fit we fixed the rotational splitting at 0.41 μHz and also fixed the amplitude ratios of the rotationally split components, but the relative amplitudes of $\ell = 1, 3$ and $\ell = 0, 2$ mode pairs were not constrained.

We use an AMOEBA-based [6] maximum likelihood estimation fitting algorithm, with a first pass on a lightly (15-bin) smoothed power spectrum to improve the first guesses for the final fit. We fit each $\ell = 0, 2$ and $\ell = 1, 3$ pair separately in a 50-μHz frequency window. For each mode pair, the observables were scaled so that all of the smoothed spectra had the same maximum value and roughly the same weight in the fit. Formal uncertainties on the estimated parameters were approximated using the inverse of a numerically computed Hessian matrix, neglecting off-diagonal terms to save computing time.

In Figure 1 we show some sample multi-spectral fits with the first-guess and final models superimposed on the spectrum. This figure also serves to illustrate the differing signal-to-noise ratios of the various observables, with HMI Ic having the lowest signal-to-noise and HMI V the highest. After fitting, we rejected results where the fit did not converge properly (non-finite error), as well as those with unreasonably small line widths or negative amplitude parameters. This weeding does not remove all of the obviously "out of line" fits, particularly when signal-to-noise is poor, and some further criteria will be needed in future work. To test the efficacy of the simultaneous fitting, we fitted the five spectra both together and individually using the same model and first-guess parameters.

4. Results and discussion

In Figure 2 we show the linewidth and frequency parameters from the multi-spectral and single-spectral fits, with frequency being represented by the difference between the input first-guess and final estimated frequency so as to show the scatter. By definition all of the frequency and width estimates fall on top of one another for the multi-spectral case, so HMI V dominates the left-hand panels because it was plotted last and has the largest number of successful fits. There are a couple of cases where the multi-spectral fit failed for HMI V but succeeded for HMI Lc, which show as green points. Clearly the scatter in the frequency and width estimates is larger in the single-fit case for the noiser observables, with many of the width estimates falling so far off the curve as to indicate that the fit has effectively failed in these cases.

Figures 3, 4 and 5 show the amplitude, background and asymmetry for each observable. For every observable except HMI V (which has much the highest signal-to-noise ratio), the simultaneous fit results in a larger number of successful parameter determinations than the individual fits, where success is defined as having all of the parameters for a given mode and observable pass the simple checks outlined above. The amplitude and background results also seem to follow a smoother curve for the multi-spectral fits, particularly for the weak $\ell = 3$

Journal of Physics: Conference Series **440** (2013) 012011 doi:10.1088/1742-6596/440/1/012011

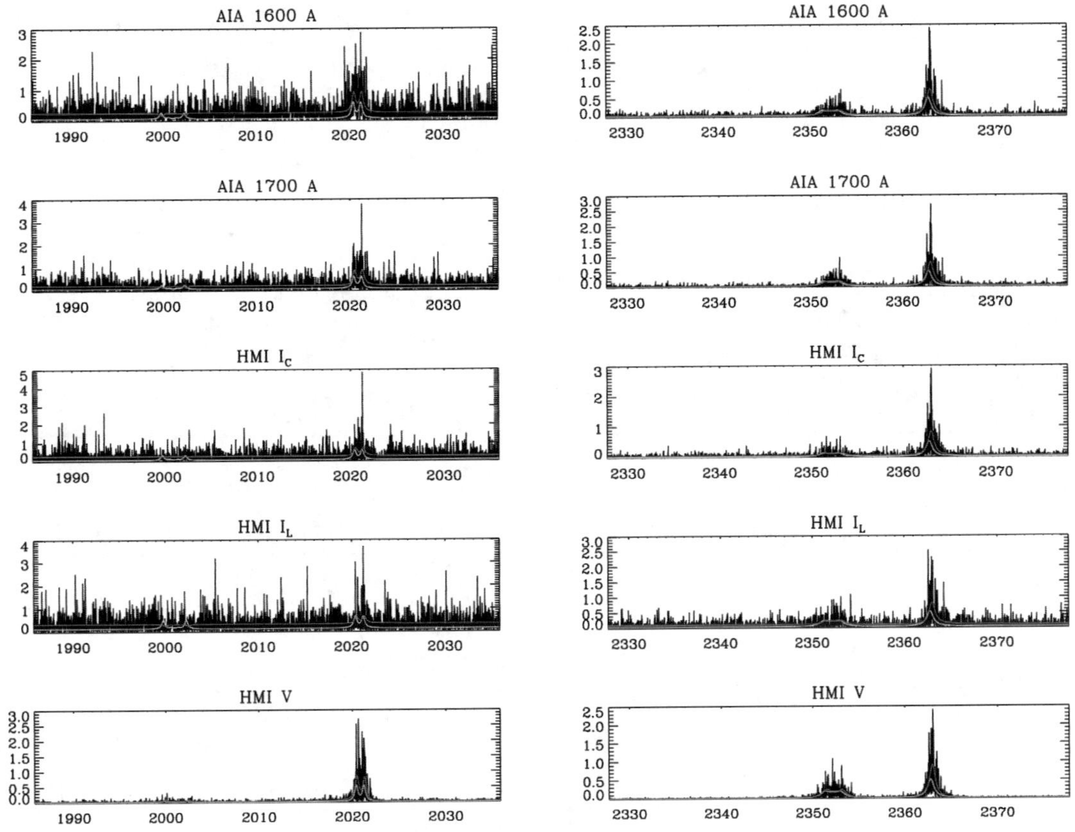

Figure 1. Sample segments of spectra for the various observables. The light-blue curve shows the first-guess model and the red curve the final fit.

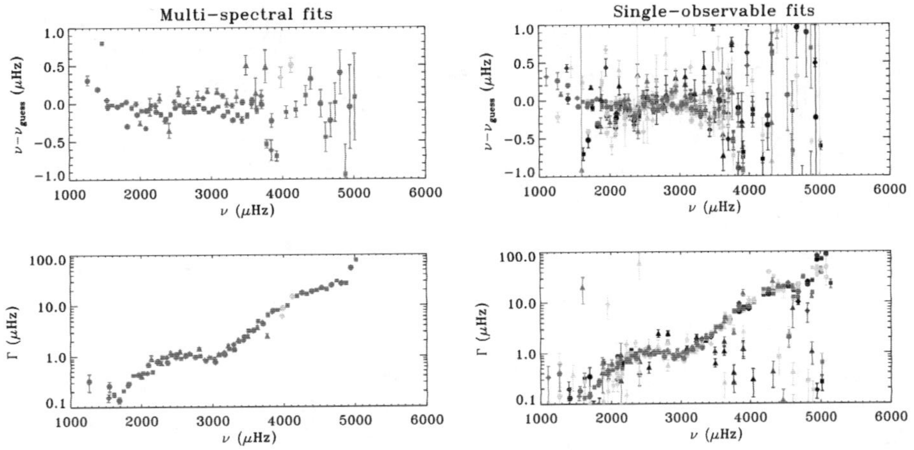

Figure 2. Fitted frequency relative to first-guess frequency (top) and fitted line width (bottom) for the multi-spectral fit (left) and for the individual spectral fits (right). Black represents AIA 1600 A, blue AIA 1700 A, light blue HMI Ic, green HMI Lc, and red HMI V; the different degrees are shown as circles ($\ell = 0$), squares ($\ell = 1$), diamonds ($\ell = 2$) and triangles ($\ell = 3$).

Journal of Physics: Conference Series **440** (2013) 012011 doi:10.1088/1742-6596/440/1/012011

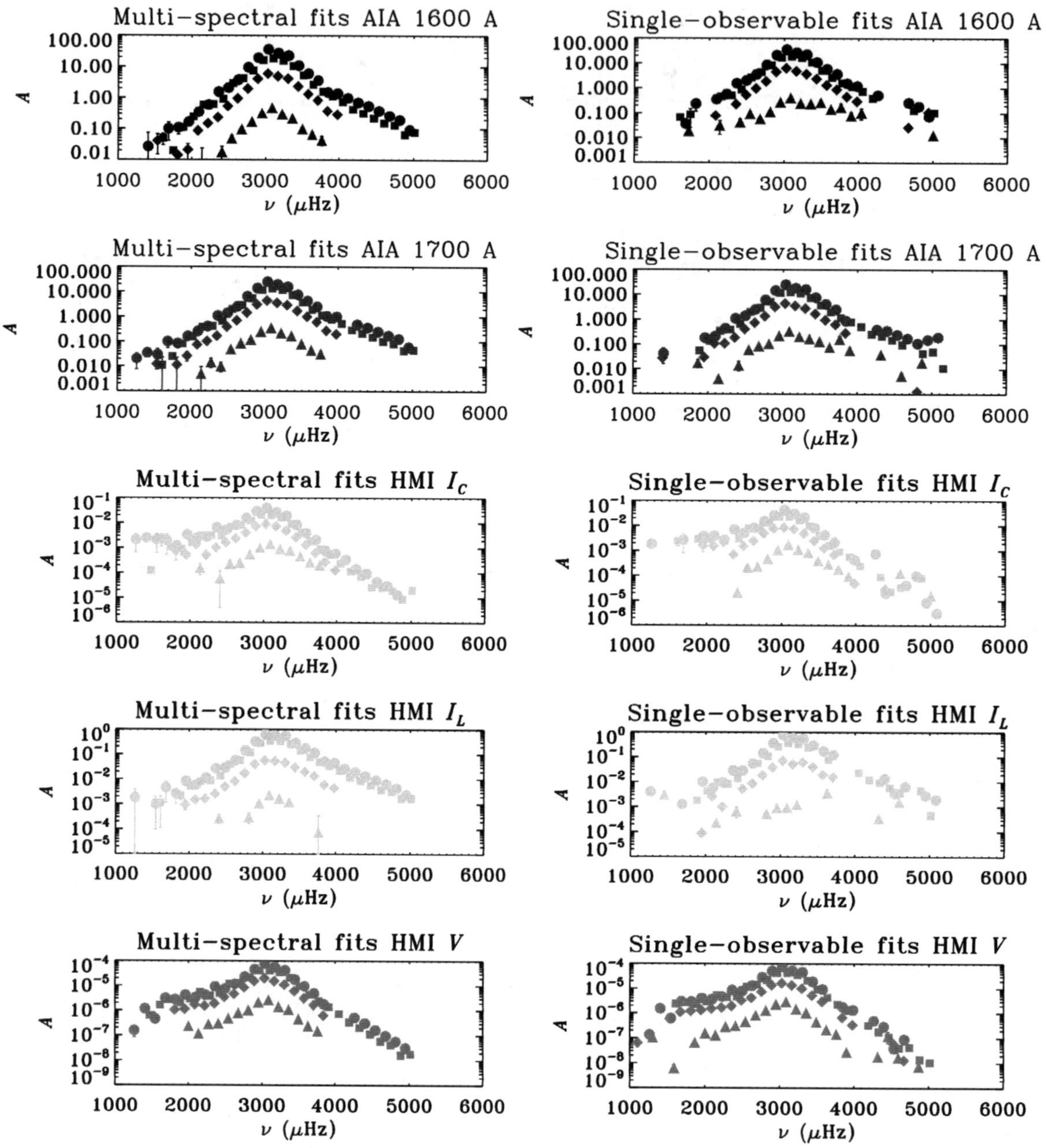

Figure 3. Comparison of fitting results from multi-spectra (left) and single (right) fits showing amplitude for each observable. The different degrees are shown as circles ($\ell = 0$), squares ($\ell = 1$), diamonds ($\ell = 2$) and triangles ($\ell = 3$).

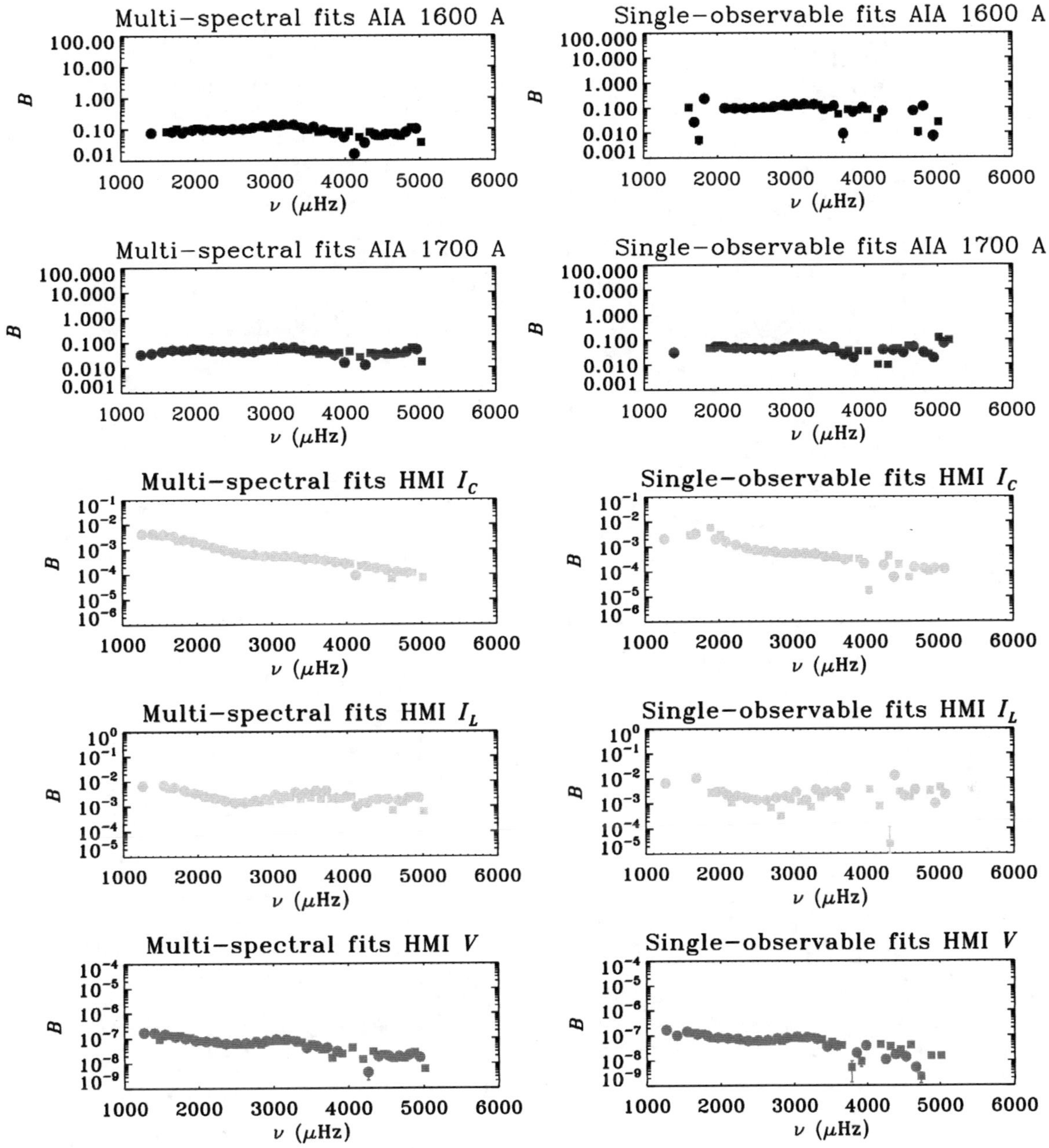

Figure 4. Comparison of fitting results from multi-spectral (left) and single-spectrum (right) fits showing background parameter for each observable. Circles indicate $\ell = 0, 2$ pairs and squares $\ell = 1, 3$.

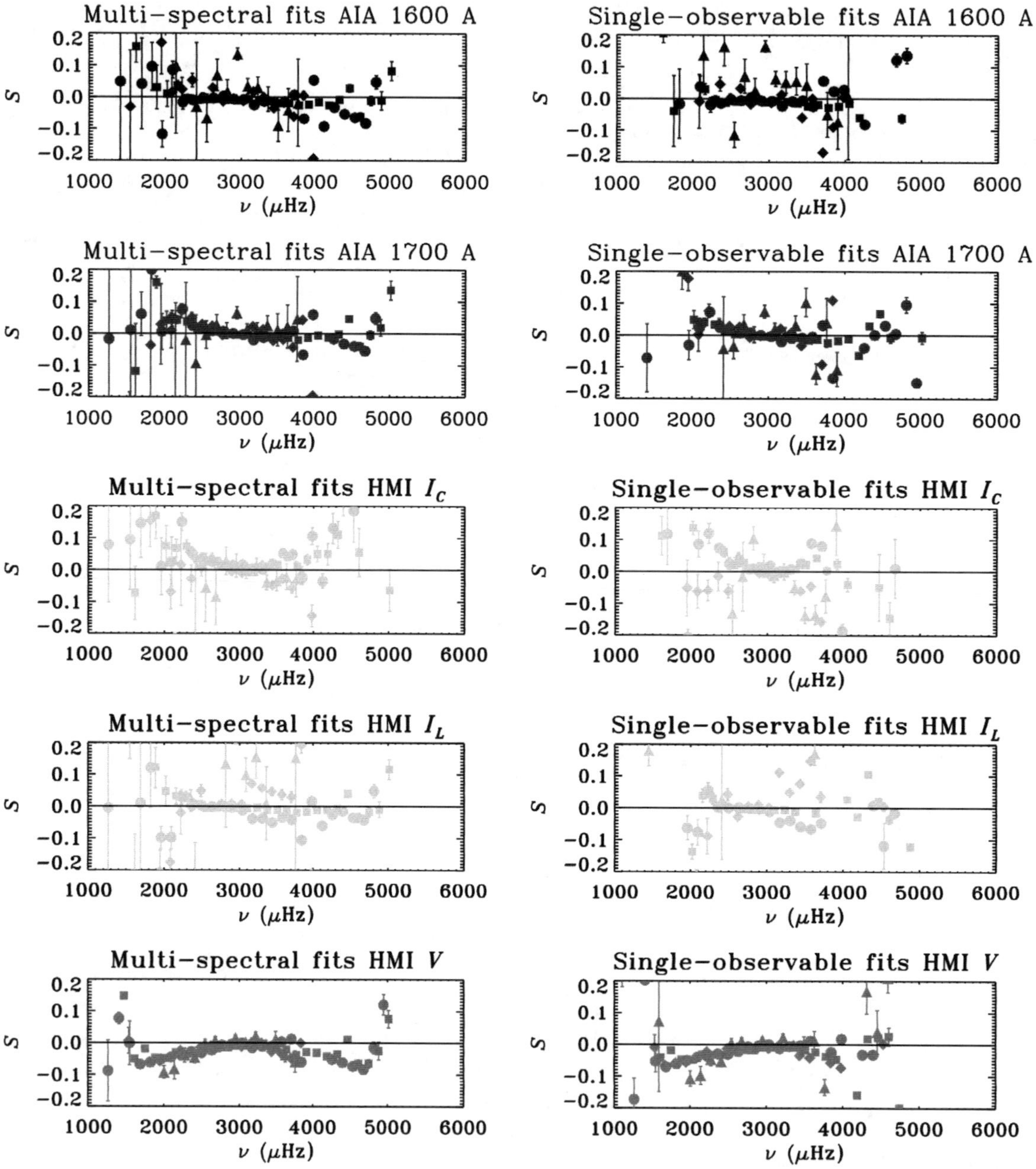

Figure 5. Comparison of fitting results from multi-spectral (left) and single (right) fits, showing asymmetry parameter for each observable. The different degrees are shown as circles ($\ell = 0$), squares ($\ell = 1$), diamonds ($\ell = 2$) and triangles ($\ell = 3$).

Journal of Physics: Conference Series **440** (2013) 012011 doi:10.1088/1742-6596/440/1/012011

modes in the intensity observables. This may be because the information from the HMI V helps to fix the frequency of these modes and ensure that the correct peak is being fitted even when it is buried in the noise. The asymmetry estimates are noisier, but the predominantly negative asymmetry for HMI V, and positive asymmetry for HMI Ic, are clearly seen, whereas the observables formed higher in the atmosphere – the AIA bands and HMI Lc – show a tendency towards positve asymmetry at low frequencies and negative at high frequencies. As the asymmetry depends on the interaction of the modes with the solar background noise, such as granulation, this may perhaps be understood as a lesser influence of granulation higher in the atmosphere. At the highest frequencies, close to the acoustic cut-off, where the modes are broad and not resolved, the fitting is less robust and the trend is less clear, but there are hints that the HMI Ic asymmetry decreases again, while those for the other observables may be increasing.

The results from this preliminary exercise are promising, but there is plenty of room to improve the method. A more sophisticated fitting algorithm might well give better results even for the single-spectrum case. To take full advantage of the multiple observables we need to use the phase and coherence information as well as the power spectra. We hope to do this in future work.

Acknowledgments

HMI data courtesy of SDO (NASA) and the HMI consortium. The National Solar Observatory is operated by the Association of Universities for Research in Astronomy (AURA), Inc. under a cooperative agreement with the National Science Foundation. This work was partly supported by NASA grant NNH12AT11I to NSO. RH acknowledges computing support from NSO.

References

[1] Fossum A and Carlsson M 2005 *Astrophys. J.* **625** 556–562

[2] Howe R, Hill F, Komm R, Broomhall A, Chaplin W J and Elsworth Y 2011 *Journal of Physics Conference Series* **271** 012058–+

[3] Norton A A, Graham J P, Ulrich R K, Schou J, Tomczyk S, Liu Y, Lites B W, López Ariste A, Bush R I, Socas-Navarro H and Scherrer P H 2006 *Solar Phys.* **239** 69–91 (*Preprint* arXiv:astro-ph/0608124)

[4] Fleck B, Couvidat S and Straus T 2011 *Solar Phys.* **271** 27–40 (*Preprint* 1104.5166)

[5] Nigam R and Kosovichev A G 1998 *Astrophys. J. Lett.* **505** L51–54

[6] Press W H, Flannery B P and Teukolsky S A 1989 *Numerical recipes. The art of scientific computing (FORTRAN Version)* (Cambridge University Press)

Journal of Physics: Conference Series **440** (2013) 012012 doi:10.1088/1742-6596/440/1/012012

Multi-spectral study of acoustic mode parameters and sub-surface flows

Kiran Jain[1], S Tripathy[1], S Basu[2], R Bogart[3], I González Hernández[1], F Hill[1] and R Howe[4]

[1]National Solar Observatory, Tucson, AZ 85719, USA
[2]Yale University, New Haven, CT 06520, USA
[3]Stanford University, Stanford, CA 94305, USA
[4]University of Birmingham, Birmingham, UK

E-mail: kjain@nso.edu

Abstract. Simultaneous measurements at different wavelengths offer the prospect of studying the sensitivity of helioseismic inferences to the choice of observing height both in quiet-Sun and magnetically active regions. In this context, we use observations from space-borne measurements from the Solar Dynamics Observatory and ground-based Global Oscillation Network Group to analyze high-degree acoustic mode parameters and sub-surface flows obtained with different observables. We also quantify differences and interpret results in the framework of the formation height in solar atmosphere.

1. Introduction

The helioseismic study at multi wavelength is important in order to obtain precise information about the excitation and damping of the oscillations. While this information can be used as a path to understand the transfer of energy through the solar interior and atmosphere, it is also useful as a tool to improve the fitting of the oscillation spectrum [1]. Here we study the sensitivity of helioseismic inferences to the choice of observing height and observables, and to clarify the effects of observing-line-related systematic errors. Since the physics of the solar oscillations cannot depend on the observed physical variable, here we test this assumption for high-degree modes obtained with measurements at different heights in the solar atmosphere. We also investigate whether there is any effect of observing height on the sub-surface flows.

2. Analysis and Results

We use multi-spectral data from two sources; the space-borne *Solar Dynamics Observatory (SDO)* and the ground-based Global Oscillation Network Group (GONG). The Helioseismic and Magnetic Imager (HMI) on board *SDO* provides simultaneous full-disk Doppler (HMI V) and continuum intensity (HMI Ic) observations with a resolution of 0.5 arcsec/pixel and a uniform cadence of 45 sec in spectral line Fe I 6173 Å. The Atmospheric Imaging Assembly (AIA) also onboard *SDO* provides continuum intensity measurements from 1700 Å and 1600 Å passbands with 0.6 arcsec/pixel resolution and the uniform cadence of 24 sec. The full-disk Doppler measurements from GONG (GONG V) have a resolution of 2.5 arcsec/pixel and a uniform cadence of 60 seconds in spectral line Ni I 6768 Å. These observables are formed at different

Journal of Physics: Conference Series **440** (2013) 012012 doi:10.1088/1742-6596/440/1/012012

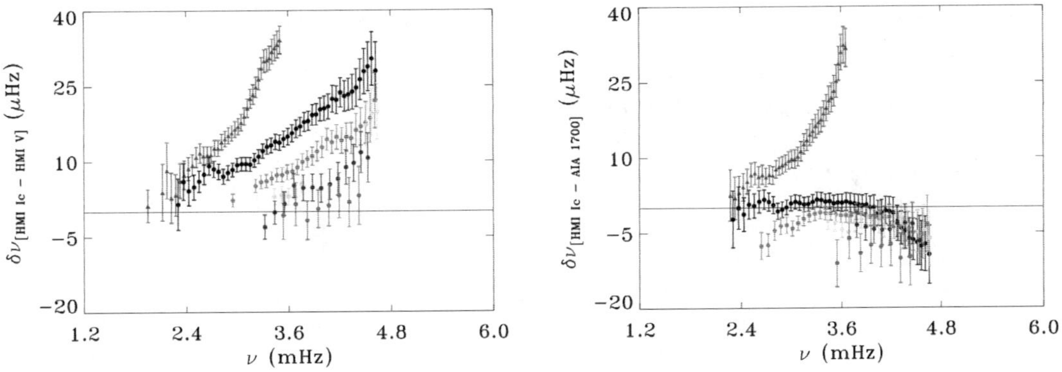

Figure 1. The frequency dependence of the difference between frequencies obtained using the symmetric profile model: (left) for HMI Ic and HMI V, and (right) HMI Ic and AIA 1700. Different colors are for different n ridges in the ℓ-ν diagram.

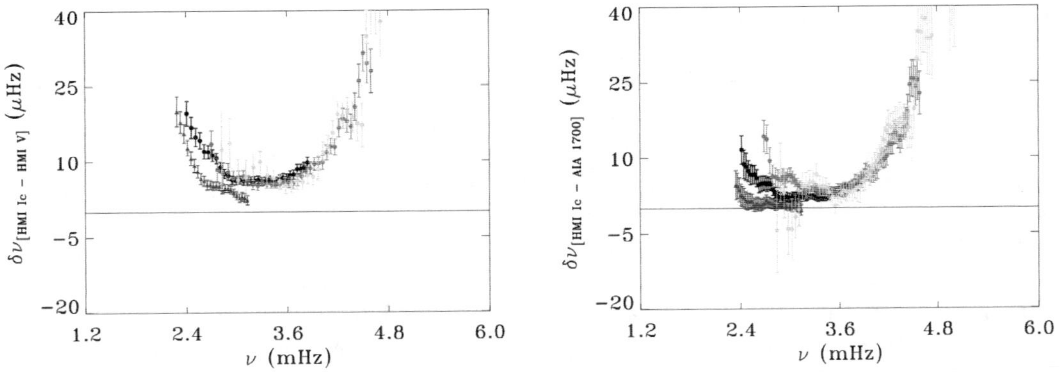

Figure 2. Similar to Figure 1 but for the asymmetric profile model.

heights in the solar atmosphere. While the HMI Ic, HMI V and GONG V measurements are taken at around 20 km, 100 km, 200 km respectively, the AIA 1700 and 1600 bands are centered at 360 km and 480 km respectively.

In order to assess the observing height related effects on high-degree mode parameters and sub-surface flows, we use simultaneous observations during a quiet period (June 24-25, 2010) and apply the technique of ring diagrams [2]. The regions of $15^\circ \times 15^\circ$ for each observable are analyzed where these are remapped and tracked for 1664 minutes. A 3-dimensional FFT is subsequently applied to obtain power spectra which are fitted to symmetric [3] and asymmetric profile models [4] to estimate mode parameters. Finally, the fitted velocities are inverted using RLS method to infer the depth dependence of horizontal velocity vectors.

In Figure 1, we show the frequency difference, $\delta\nu$, between modes obtained with the symmetric profile model. It is observed that frequencies from different observables do not agree with each other and also the difference increases with increasing frequency. This difference could be as large as 40 μHz or even more around 5 mHz. We notice the $\delta\nu$ between HMI V and HMI Ic is much more apparent as compared to two continuum measurements. We also find that the frequencies for two Doppler mesurements, e.g. HMI V and and GONG V, are comparable to each other. Frequency differences for modes fitted with the asymmetric profile model are plotted in Figure 2. For comparison, we have shown values for same type of observables as in Figure 1. In this case, $\delta\nu$ is significantly reduced, particularly for the HMI V and HMI Ic in the 3 mHz

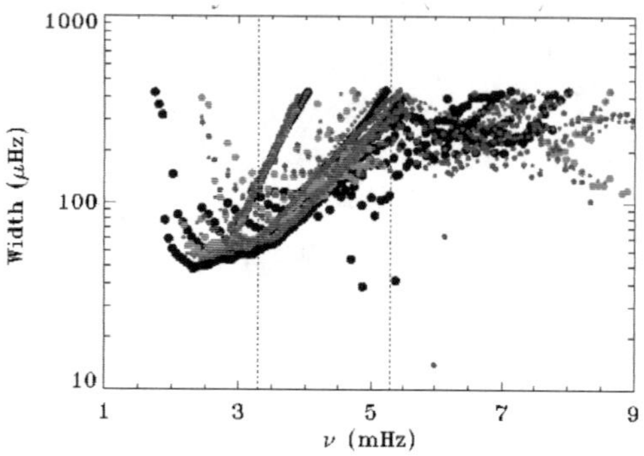

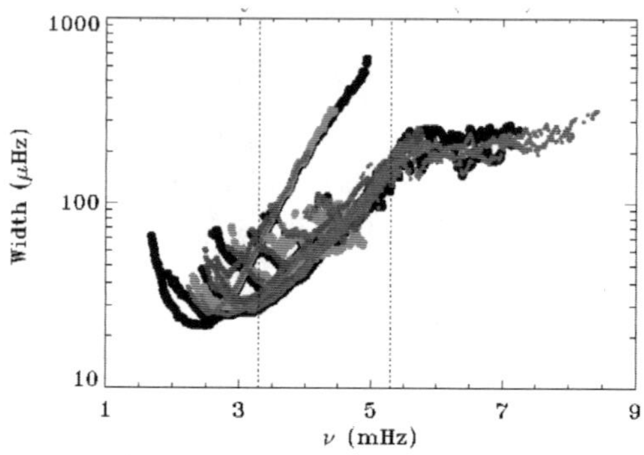

Figure 3. Half width of the fitted modes for SDO observables: HMI Ic (yellow), HMI V (blue), AIA 1700 (green) and AIA 1600 (red). The left panel shows values obtained with the symmetric profile model while the right panel is for the asymmetric profile model.

band. However, $\delta\nu$ tends to increase sharply on lower and higher ends of the spectra. It seems that the model used to define asymmetry in peaks is most suitable for the 3 mHz band and does not properly describe the asymmetry in a broader range of frequencies. We plan to revise the model to include asymmetry at frequencies below and above 3 mHz band.

In an earlier study, using data from GONG and Magneto-Optical filters at Two Heights (MOTH), the frequency differences were found relatively small as compared to the present study [5]. Note that the previous study considers only one type of observable, i.e. Doppler measurements, at different heights while present analysis is based on both Doppler and continuum measurements. In a different study, Tripathy et al. used Doppler and continuum measurements from the Michelson Doppler Imager (MDI) on board *Solar and Heliospheric Observatory (SOHO)* and found $\delta\nu$ for the asymmetric profile model are comparable to our analysis [6]. However, frequency differences in symmetric profile model are not as large as we obtain in the present study. We anticipate that different symmetric profile models used in both studies led to inconsistent results. A more detailed study is underway to understand the difference in frequencies in two symmetric profile models and will be published elsewhere.

In Figure 3, we further show the width of the fitted modes obtained using both fitting methods. It is noted that the width increases with frequency for all observables and appears to reach a constant value of about 200 μHz at high frequencies ($\nu > 6$ mHz). It is consistent with previous studies where similar results were found for a single observable [7,8]. We see a large scatter for modes fitted with the symmetric profile while values obtained with the asymmetric profile model for all observables are comparable. Since life time of the acoustic modes cannot depend on the type of the observable, our results clearly indicate that the symmetric profile does not adequately represent the true nature of peaks in power spectra, hence it is important to include proper asymmetry in mode fitting while dealing with different types of observables.

We also investigate the sensitivity of inferred subsurface flows to the choice of spectral line. Figure 4 exhibits the variation of zonal (V_x) and meridional (V_y) components of the horizontal flow with depth at the disk center for all three observables shown in Figures 1 and 2. While profiles for V_y in all cases show similar variations, there is a noticeable difference in profiles for V_x. Since different modes have different sensitivity to the averaging kernels and affect velocity profiles, we interpret this difference as resulted from the mode sets used in the inversion. In addition, the inaccurate dettermination of the asymmetry in different observables would also affect the inverted velocities. In order to check these assumptions, we analyze simultaneous Doppler measurements from GONG and HMI and the inversion is carried out only for those modes which are common in both data sets. Figures 5 and 6 show depth dependence of V_x and

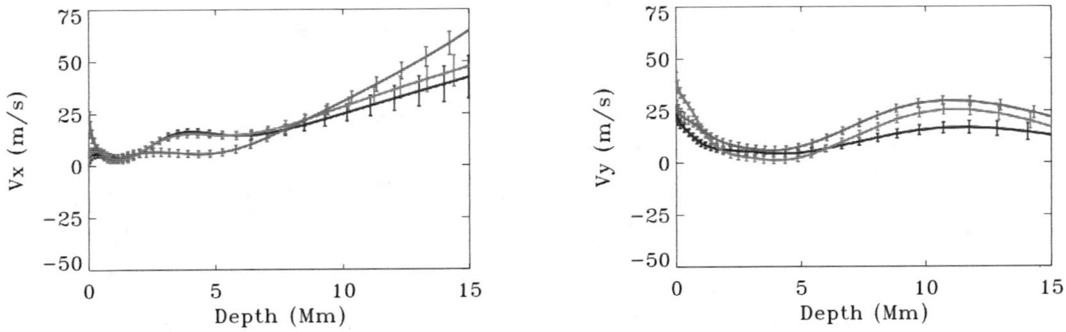

Figure 4. Variation of zonal (V_x) and meridional (V_y) components of the horizontal flow obtained for HMI V (red), HMI Ic (green) and AIA 1700 (blue) at the disk center.

V_y for several regions along the central meridian and the equator, respectively. It is evident that the inference of sub-surface flow is not much affected by the line formation height in the solar atmosphere when common modes are inverted for the same type of observable. We also notice that the agreement starts decreasing around 45° away from the disk center which could be attributed to the foreshortening as both data sets have different spatial resolution. This is consistent with earlier findings where similar flows were found for Doppler measurements in Ni and K spectral lines [9].

3. Summary

Our analysis shows that the estimated frequency for a particular mode is different for different observables. The use of symmetric profile model leads to a larger discrepancy while the asymmetric profile model reduces the difference significantly near 3 mHz. This implies that the model used to describe asymmetry in peaks is most suitable for the 3 mHz band, but needs to be revisited for a broader range of frequencies. It is clearly evident that the proper asymmetry should be included while fitting modes in different types of observables. We plan to perform a

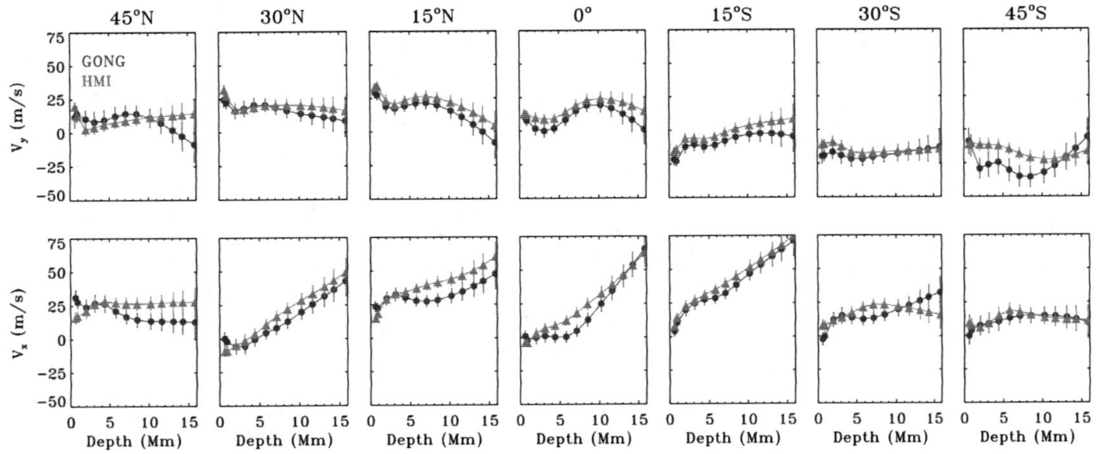

Figure 5. Variation of zonal (V_x) and meridional (V_y) components of the horizontal flow obtained using GONG (blue circles) and HMI (red triangles) Dopplergrams at different latitudes along the central meridian.

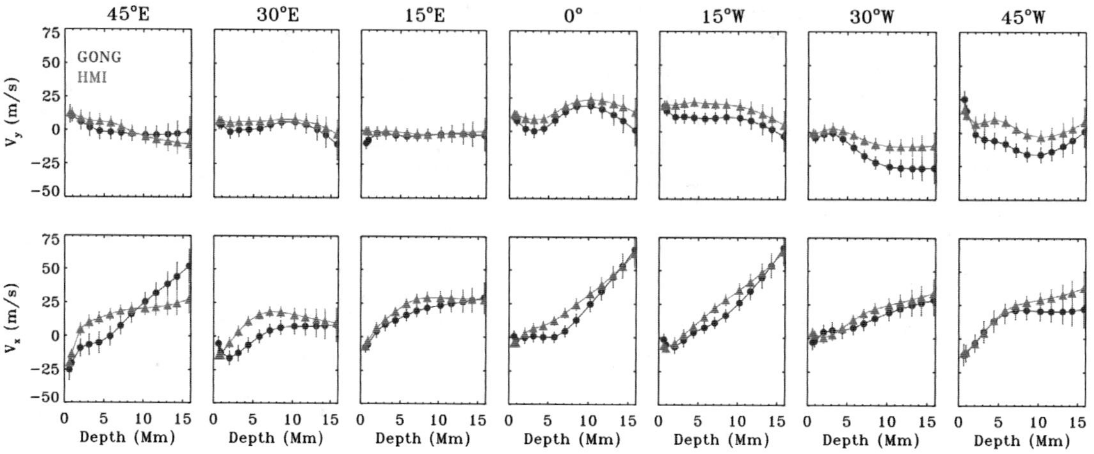

Figure 6. Same as Figure 5 but at different longitudes along the equator.

cross-spectral fitting to further improve the estimates of mode parameters. We also find that the simultaneous observations do not show any significant change in zonal and meridional flows with the choice of spectral line.

Acknowledgments

This work utilizes GONG data obtained by the NSO Integrated Synoptic Program (NISP), managed by the National Solar Observatory, which is operated by AURA, Inc. under a cooperative agreement with the National Science Foundation. The data were acquired by instruments operated by the Big Bear Solar Observatory, High Altitude Observatory, Learmonth Solar Observatory, Udaipur Solar Observatory, Instituto de Astrofisica de Canarias, and Cerro Tololo Interamerican Observatory. SDO data courtesy of SDO (NASA) and the HMI and AIA consortium. This work is partially supported by NASA grant NNH12AT11I.

References

[1] Braban C, Hill F and Kras S 2004 *Astrophys. J.* **602** 516-527
[2] Hill F 1988 *Astrophys. J.* **333** 996-1013
[3] Haber D, Hindman B W, Toomre J, Bogart R S, Larsen R M and Hill F 2002 *Astrophys. J.* **570** 855-864
[4] Basu S and Antia H M 1999 *Astrophys. J.* **525** 517-523
[5] Antia H M and Basu S 1999 *Astrophys. J.* **519** 400-406
[6] Duvall, T L Jr., Harvey, J W, Jefferies, S M and Pomerantz, M A 1991 *Astrophys. J.* **373** 308-316
[7] Jain K, Hill F, Tripathy S C, Antia H M, Armstrong J D, Jefferies S M, Rhodes Jr E J and Rose P J 2006 *Proc. Beyond the spherical Sun* vol 624 ed K Fletcher and M Thompson (Noordwijk: ESA) p 103
[8] Tripathy S C, Antia H M, Jain K and Hill F 2009 *Astrophys. J.* **691** 365-371
[9] Jain K, Hill F, González Hernández I, Toner C G, Tripathy S C, Armstrong J D and Jefferies S M 2006 *Proc. Beyond the spherical Sun* vol 624 ed K Fletcher and M Thompson (Noordwijk: ESA) p 127

Journal of Physics: Conference Series **440** (2013) 012013 doi:10.1088/1742-6596/440/1/012013

Observational consequences of the local re-acceleration thick-target model

M Varady[1], Z Moravec[1], M Karlický[2], J Kašparová[2]

[1]Faculty of Science, J.E. Purkyně University, Physics Department, České mládeže 8, 400 75 Ústí nad Labem, Czech Republic

[2]Astronomical Institute of the Academy of Sciences of the Czech Republic, v.v.i., 251 65 Ondřejov , Czech Republic

E-mail: mvarady@physics.ujep.cz

Abstract. In our contribution we compare the efficiency of the hard X-ray production and the vertical sizes and positions of the hard X-ray sources for the classical collisional thick-target model and for its recently proposed modification, the local re-acceleration thick-target model. The latter model has been proposed in order to ease some of the severe theoretical problems of the collisional thick-target model related to interpretation of the observational properties of the foot-point HXR sources in solar flares. The results are obtained using a relativistic test-particle approach for a fully ionised atmosphere with a converging magnetic field and a single (compact) flare loop.

1. Introduction

The Collisional Thick-Target Model (CTTM) of the impulsive phase of solar flares [1] presented for many years a quite successful tool to interpret the fundamental processes related to the energy deposition and hard X-ray (HXR) production at the foot-point regions of flare loops due to the interaction of chromospheric plasma with electron beams. On the other hand, in order to explain the energetics of flares and observed intensities of HXR, many serious theoretical problems emerged (see [2] and the Introduction in [3]), especially in combination with the idea that the beam acceleration site corresponds to the coronal reconnection site [4], [5], [6] and consequently with the inevitability to transport enormous fluxes of accelerated charged particles from the corona to the thick-target region. Moreover novel RHESSI measurements of HXR chromospheric albedo [7] or of the vertical sizes and positions of the HXR sources are inconsistent with the predictions given by the CTTM [8], [9].

To resolve the discrepancies a modification of the CTTM known as the Local Re-acceleration Thick-Target Model (LRTTM) has recently been proposed [3], [10], [11]. The model assumes an existence of a secondary re-acceleration region located within the thick-target region in the chromosphere. The non-thermal electrons initially accelerated at the primary coronal acceleration site experience, in addition to the energy losses and scattering due to the Coulomb collisions, also sequences of re-acceleration events at the secondary re-acceleration region due to the stochastic electric fields generated in the current sheet cascades [10], [11] excited by random photospheric motions.

In our contribution we concentrate on a comparison of outcomes of the chromospheric bombardment by electron beams with a power-law spectrum modelled in frame of the CTTM

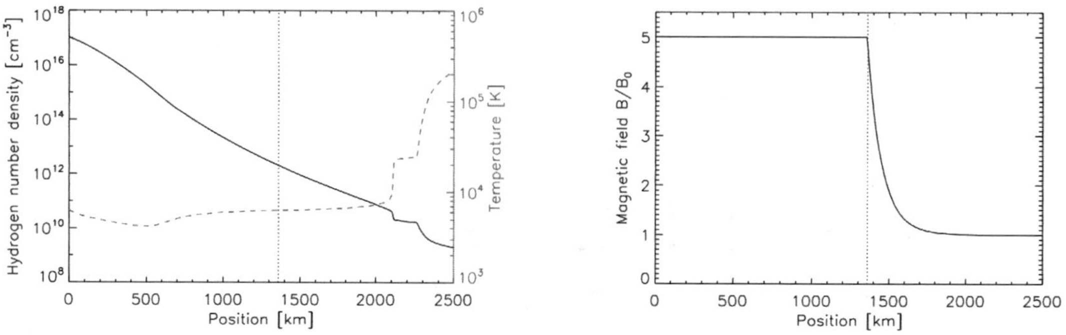

Figure 1. Left: Density (black solid line) and temperature (red dashed line) profiles of the VAL C atmosphere [12]. The dotted line indicates the bottom boundary of the magnetic mirror. Right: Relative magnetic field strength along the loop. Only the lower part of the atmosphere is plotted.

and LRTTM. We compare the non-thermal electron distribution functions in the vicinity of the thick-target region, corresponding chromospheric energy deposits, HXR production, spectra and vertical sizes and heights above the photosphere (corresponding to $\tau = 1$ at 5000 Å) of the HXR foot-point sources calculated for a fully ionised VAL C atmosphere [12] and a single (compact) loop with a converging magnetic field towards the photosphere. The results are obtained using a relativistic test-particle code [13], [14], [15].

2. Model description

The starting point of our simulations is a generation of an electron beam with a power-law spectrum and with parameters: power-law index $\delta = 3$, low-energy cutoff $E_0 = 10$ keV, high-energy cutoff $E_1 = 400$ keV, initial energy flux $\mathcal{F}_0 = 2.5 \times 10^9$ erg cm^{-2} s^{-1} (or 5×10^{10} erg cm^{-2} s^{-1} – the reference flux for the CTTM) and a normalised semi-uniform initial distribution of the pitch angle cosine $\mu_0 = \cos\vartheta \in (0.5, 1)$. In order to demonstrate the potential of secondary re-acceleration we selected a very small value of $\mathcal{F}_0$ in context of flare physics. As the target atmosphere in the presented approximation is static, by changing $\mathcal{F}_0$ the vertical structure of the distribution function remains the same and the energy deposits and HXR intensities scale linearly. In our model we assume that the beam is accelerated at the primary acceleration site localised in the corona. Using a relativistic test-particle code [13], [14], [15], we model the propagation of beam electrons through a fully ionised, static atmosphere with a converging magnetic field with the mirror ratio $R_m = 5$ (see right panel of Fig. 1). The temperature and density profiles of the atmosphere correspond to the VAL C model [12] (see left panel of Fig. 1) with an extension into the corona, the ionisation is artificially set to one in the whole atmosphere. The code calculates the corresponding Coulomb collisional energy losses and scattering and the effects of the magnetic mirroring in the converging field for the CTTM. For the LRTTM it also includes the effects of the secondary re-acceleration events due to the presence of the stochastic electric fields in the vicinity of the thick-target region. Only one-half of the loop is considered.

We assume normally distributed stochastic electric fields in the directions parallel and antiparallel relative to the loop axis, i.e. a simplified model of the electric field distribution due to the current sheet cascade in a randomly stressed magnetic loop in the chromosphere [3], [11]. We present results for zero mean value of the stochastic field and for its variances var$(E) = 0.5, 1$ and 5 V m^{-1}. The secondary re-acceleration region is placed into the chromosphere in range of

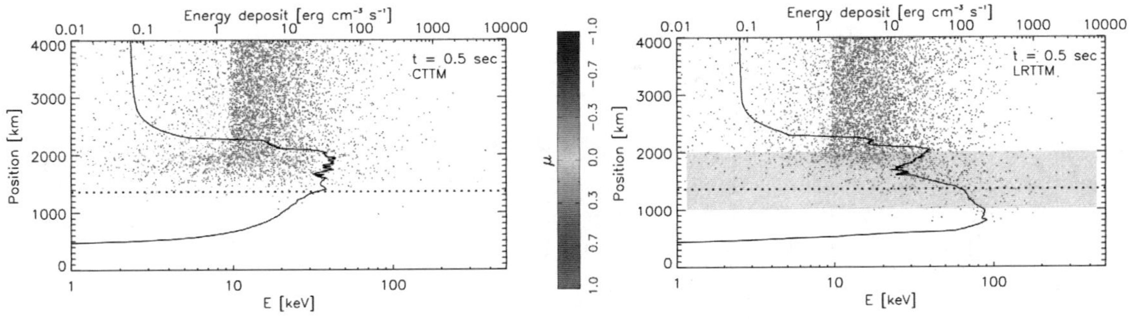

Figure 2. The distribution functions of non-thermal electron energies versus positions with a colour coded μ-distribution for the classical the CTTM (left panel) and LRTTM (right panel) for $\mathrm{var}(E) = 1$ V m^{-1} at time $t = 0.5$ s after the beam injection into the loop at its apex. The solid lines indicate the instantaneous energy deposit corresponding to $\mathcal{F}_0 = 2.5 \times 10^9$ erg cm^{-3} s^{-1} in the logarithmic scale. The dotted horizontal line indicates the bottom boundary of the magnetic mirror, the grey area corresponds the secondary re-acceleration region. Only the lower parts of the loop are captured.

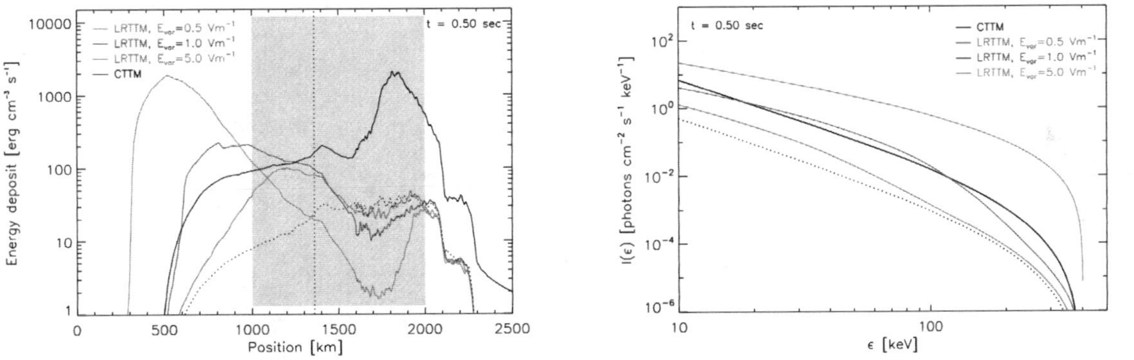

Figure 3. The energy deposits (left) and HXR spectra (right) at $t = 0.5$ s. The green, red and orange lines correspond to the LRTTM with the initial beam energy flux $\mathcal{F}_0 = 2.5 \times 10^9$ erg cm^{-2} s^{-1} and field variances $\mathrm{var}(E) = 0.5, 1$ and 5 V m^{-1}, respectively. The black lines indicate the energy deposits and HXR spectra for the CTTM, where the black solid and dotted lines correspond to the initial energy fluxes $\mathcal{F}_0 = 5 \times 10^{10}$ erg cm^{-2} s^{-1} and 2.5×10^9 erg cm^{-2} s^{-1}, respectively. The grey area indicates the secondary re-acceleration region, the vertical dotted line the bottom boundary of the magnetic mirror.

heights from 1000 km to 2000 km.

The HXR fluxes and spectra produced in the vicinity of loop foot-points are computed according to [1] using the semi-relativistic cross-section given by [16]. All the results are presented at time $t = 0.5$ s after the beam injection into the loop at its apex, when the model is close to a quasi-steady state.

3. Results

The distribution functions of non-thermal electrons for the CTTM and LRTTM are shown in left and right panel of Fig. 2, respectively. The electrons propagating from the loop apex towards the photosphere ($\mu \sim 1$) are coded with shades of red, the electrons with $\mu \sim 0$ are coded with shades of green and the electrons reflected by the magnetic mirror or backscattered ($\mu \sim -1$) are coded with shades of blue. For the CTTM case, it is evident that the majority of electrons arriving from the primary acceleration site (colour coded red to brown), into the higher density region between the upper chromosphere ($s \approx 2000$ km) and the bottom boundary of the magnetic mirror ($s = 1360$ km), quickly lose their energies and increase their pitch angles due to the Coulomb collisions and the effects of converging magnetic field. As a consequence a low energy tail is formed in the chromosphere above the bottom boundary of the magnetic mirror. The tail is dominated by particles of $\mu \sim 0$ and these particles either completely thermalise their kinetic energies or are reflected by the magnetic mirror back towards the second foot-point. Only few particles with higher energies penetrate behind the bottom boundary of the magnetic mirror.

The right panel of Fig. 2 shows the distribution function for the LRTTM. Two tails in the distribution function can be identified. A low energy tail corresponding to the tail identified in the distribution function for the CTTM, but less distinct, and a tail aiming towards high energies, spatially located within the re-acceleration region. This tail is formed by the particles re-accelerated by the stochastic electric fields. Another obvious effect of the stochastic fields is the enhancement of the number of backscattered electrons, particularly at higher energies. Comparing the corresponding energy deposits, it is apparent that the presence of the re-acceleration field substantially increases the energy deposit and shifts its maximum towards deeper layers.

The energy deposits for several variances of the stochastic field are shown in the left panel of Fig. 3. The model indicates that an increase of var(E) results in a dramatic increase of the energy deposit into the lower atmosphere and also the maxima of the energy deposits can be shifted to very deep layers (see the deposit for var(E) $= 5$ V m^{-1}). If the deposits are integrated over the chromosphere and photosphere we can conclude that the latter mentioned variance of stochastic field can increase the integrated deposit of almost one and half order in comparison with the deposit obtained using the CTTM and equal initial beam energy flux (dotted black line in the left panel of Fig. 3).

In flares the non-thermal electron distribution functions, the spatial position, and the extent of the energy deposits are influenced by the instantaneous state of the target flare atmosphere, in particular by the density profile of the chromosphere, which corresponds to the history of flare heating [14], [17], [18]. Due to the interaction of the low-energy electrons, which are predominant in the beam, with the evaporated flare plasma a shift of their penetration depth higher into the atmosphere can be expected. Similar effects would be also present using a static flare chromosphere (e.g. models F1 or F2 [19]).

The corresponding HXR spectra are shown in the right panel of Fig. 3. The HXR intensities and power-law indices are dependent on the values of var(E). With increasing variances $I_{10\,\mathrm{keV}}$ (the intensity at 10 keV) increases and the power-law indices harden. Comparing the intensity $I_{10\,\mathrm{keV}}$ for the LRTTM with var(E) $= 5$ V m^{-1} and for the CTTM with equal initial beam energy flux (dotted black line in the right panel of Fig. 3) we again obtain an increase in $I_{10\,\mathrm{keV}}$ of one and half order.

The HXR photon fluxes at various energies as a function of height above the photosphere for the CTTM and LRTTM are shown in Fig. 4. These profiles are used to calculate the HXR source heights and vertical sizes as the profile first and second moments, respectively. It is apparent that the placement and parameters of the magnetic mirror strongly influence the size and position of the CTTM HXR source. The results are shown in Fig. 5. The vertical sizes of

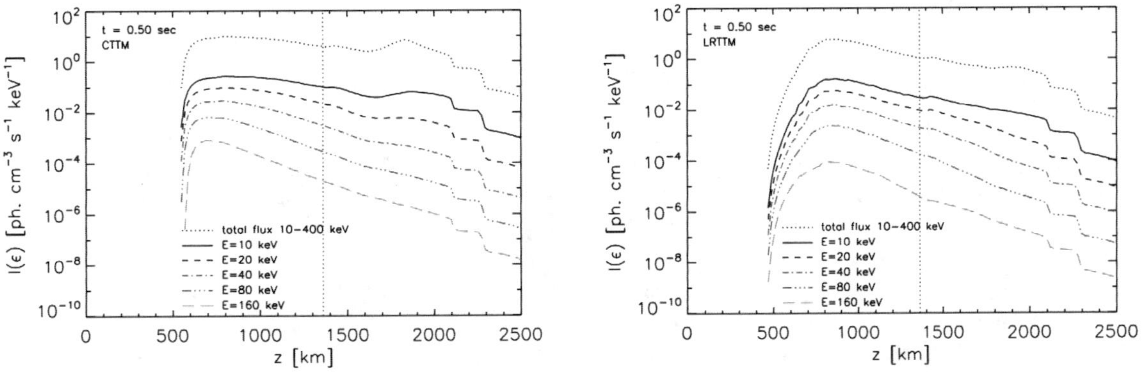

Figure 4. HXR photon flux emanating from the thick-target region as a function of height for selected photon energies (colour coded). Left for the CTTM, right for the LRTTM and $\text{var}(E) = 1\ \text{V}\,\text{m}^{-1}$. The dotted vertical line indicates the position of the bottom boundary of the magnetic mirror.

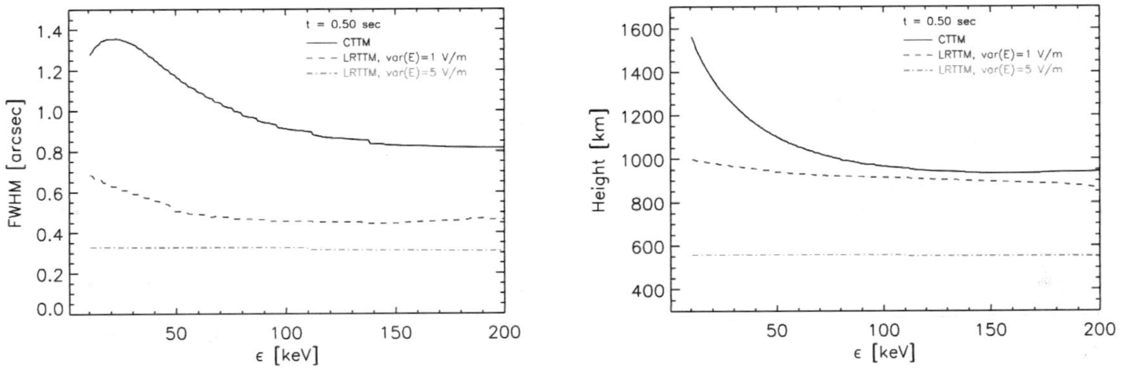

Figure 5. FWHM vertical sizes of the foot-point HXR sources (left panels) and the positions of the HXR emission maxima (right panels) versus photon energy for the CTTM (black solid lines) and the LRTTM (red dash-dotted lines) with $\text{var}(E) = 1\ \text{V}\,\text{m}^{-1}$ (upper panels) and $\text{var}(E) = 5\ \text{V}\,\text{m}^{-1}$ (lower panels).

HXR sources for the CTTM (left panel of Fig. 5 solid black lines) are about 1.2 arcsec for energies 10 – 30 keV. The source sizes decrease with increasing energy down to approximately 0.8 arcsec for energies above approximately 130 keV. For the LRTTM we found the HXR sources to be much more compact with increasing $\text{var}(E)$ on all energies (compare dashed blue and dashed-dotted red lines in the left panel of Fig. 5) with a typical vertical source size 0.5 – 0.7 arcsec for $\text{var}(E) = 1\ \text{V m}^{-1}$ and approximately 0.3 arcsec for $\text{var}(E) = 5\ \text{V m}^{-1}$ on all energies.

Also the heights of the HXR emission maxima for the LRTTM are sensitive to the variance of stochastic field. With increasing $\text{var}(E)$ the maxima tend to be placed deeper in the atmosphere (compare the blue dashed and red dash-dotted lines in the right panel of Fig. 5). The height of the HXR emission maximum for $\text{var}(E) = 1\ \text{V m}^{-1}$ is 1000 – 900 km for all energies. It corresponds to the height of the CTTM emission maximum for energies above 90 keV. The height for $\text{var}(E) = 5\ \text{V m}^{-1}$ is even only about 550 km for all energies and the source is thus much deeper than the CTTM source on all energies.

4. Conclusions

It has been shown that the presence of secondary re-acceleration in the region close to the flare loop foot-points reduces by one or one and half orders of magnitude (depending on the stochastic field parameters) the particle flux from the corona required to explain the observed HXR foot-point intensities. It also decreases the vertical size of the foot-point HXR sources. While the first findings agree with the results obtained by [3], [10], [11], the latter presents a problem.

Namely, the vertical sizes of HXR sources computed for the LRTTM are only 0.5 – 0.7 arcsec for var(E) = 1 V m^{-1}, and even only 0.3 arcsec for var(E) = 5 V m^{-1} on all energies. These values are much smaller than those obtained from the recent RHESSI observations (1.3 – 8 arcsec), see [9] and [8]. However, a comparison of the vertical sizes of X-ray sources for the LRTTM and CTTM models together with the multi-threaded chromosphere [20] offer a solution. Besides different densities in the multi-thread structure of the flaring chromosphere (which increases the vertical source size), in different threads a different distribution of the localised electric field can be expected; even with zero electric field (as in the case of the CTTM model). Therefore, integrating the vertical sizes of HXR sources for both considered models (LRTTM and CTTM) the observed vertical source sizes might be obtained. This suggestion will be tested in future models.

The heights of the HXR emission maxima for the LRTTM are in comparison with the CTTM substantially shifted towards the photosphere. The height above the photosphere, corresponding to height 0 km in the VAL C atmosphere model [12], decreases with the variance of the stochastic field. For var(E) = 5 V m^{-1} we obtained the source height only approximately 550 km, which is not far from the heights of the white-light sources (305 ± 170 km and 195 ± 70 km) found from observations of white-light flares [21].

5. Acknowledgments

This work was supported by grants P209/10/1680 and P209/12/0103 of the Grant Agency of the Czech Republic. The research at the Astronomical Institute, ASČR leading to these results has received funding from the European Commission's Seventh Framework Programme (FP7) under the grant agreement SWIFT (project number 263340).

References

[1] Brown J C 1971 *Sol. Phys.* **18** 489
[2] Brown J C, Karlicky M, MacKinnon A L and van den Oord G H J 1990 *ApJS* **73** 343–348
[3] Brown J C, Turkmani R, Kontar E P, MacKinnon A L and Vlahos L 2009 *A&A* **508** 993–1000
[4] Sturrock P A 1968 A Model of Solar Flares *Structure and Development of Solar Active Regions* (*IAU Symposium* vol 35) ed Kiepenheuer K O p 471
[5] Kopp R A and Pneuman G W 1976 *Sol. Phys.* **50** 85–98
[6] Shibata K 1996 *Advances in Space Research* **17** 9
[7] Kontar E P and Brown J C 2006 *ApJ* **653** L149–L152 (*Preprint* arXiv:astro-ph/0611170)
[8] Battaglia M, Kontar E P, Fletcher L and MacKinnon A L 2012 *ApJ* **752** 4
[9] Battaglia M and Kontar E P 2011 *ApJ* **735** 42 (*Preprint* 1104.2997)
[10] Turkmani R, Vlahos L, Galsgaard K, Cargill P J and Isliker H 2005 *ApJ* **620** L59
[11] Turkmani R, Cargill P J, Galsgaard K, Vlahos L and Isliker H 2006 *A&A* **449** 749–757
[12] Vernazza J E, Avrett E H and Loeser R 1981 *ApJS* **45** 635–725
[13] Varady M, Kašparová J, Moravec Z, Heinzel P and Karlický M 2010 *IEEE Transactions on Plasma Science* **38** 2249–2253
[14] Kašparová J, Varady M, Heinzel P, Karlický M and Moravec Z 2009 *A&A* **499** 923–934
[15] Varady M, Karlický M, Moravec Z and Kašparová J 2012 Influence of Static and Stochastic Electric Fields on Electron Beams Bombarding the Chromosphere *Fifth Hinode Science Meeting* (*Astronomical Society of the Pacific Conference Series* vol 456) ed Golub L, De Moortel I and Shimizu T p 203
[16] Haug E 1997 *A&A* **326** 417–418
[17] Abbett W P and Hawley S L 1999 *ApJ* **521** 906–919
[18] Allred J C, Hawley S L, Abbett W P and Carlsson M 2005 *ApJ* **630** 573–586

[19] Machado M E, Avrett E H, Vernazza J E and Noyes R W 1980 *ApJ* **242** 336–351
[20] Kontar E P, Hannah I G, Jeffrey N L S and Battaglia M 2010 *ApJ* **717** 250–256 (*Preprint* 1005.0599)
[21] Martínez Oliveros J C, Hudson H S, Hurford G J, Krucker S, Lin R P, Lindsey C, Couvidat S, Schou J and Thompson W T 2012 *ApJ* **753** L26 (*Preprint* 1206.0497)

Eclipse on the Coral Sea: Cycle 24 Ascending (GONG 2012, LWS/SDO-5, and SOHO 27) IOP Publishing
Journal of Physics: Conference Series **440** (2013) 012014 doi:10.1088/1742-6596/440/1/012014

Where is the solar dynamo ?

Paul Charbonneau

Département de Physique, Université de Montréal, C.P. 6128 Centre-Ville, Montréal, Qc H3C3J7, CANADA

E-mail: paulchar@astro.umontreal.ca

Abstract. In this paper I review results from recent global magnetohydrodynamical numerical simulations of solar convection, as a springboard to address the question "Where is the solar dynamo". I first describe and contrast similarities and differences in the large-scale flows and magnetic fields such simulations can produce, with emphasis on polarity reversals (or lack thereof) in the large-scale magnetic components they generate. On the basis of these simulation results, I argue that some of the significant differences in the spatiotemporal evolution of the large-scale magnetic field can be traced to the competing effects of turbulent electromotive forces and induction by large-scale flows, whose mutual near-cancellation in the nonlinearly saturated regime leads to a high sensitivity to the numerical/physical treatment of small scales. Some of these recent simulation results also reopen the possibility that dynamo action driving the solar activity cycle may reside entirely within the convection zone, with the tachocline perhaps playing a lesser role than has been assumed in the last two decades. On the other hand, other subsets of simulations suggest that magnetohydrodynamical processes taking place within the tachocline may have a significant impact on timescales comparable to or longer than the primary cycle.

1. Where the solar dynamo was

Helioseismology has changed forever the way we go about modelling the solar dynamo, as it provides us with vital information on the flows pervading the solar interior. These flows are the dominant energy source for the magnetohydrodynamical (MHD) dynamo process powering all of solar activity. Interestingly, helioseismology's initial impact on solar dynamo modelling was to stir trouble in what was then emerging as an almost consensual view. Instead of rotational isocontours constant on cylinders matching the observed surface latitudinal differential rotation, as expected on theoretical grounds, the first helioseismic inversions of the deep solar rotation (Brown & Morrow 1987; Brown et al. 1989; Dziembowski et al. 1989) revealed a shear that remains largely latitudinal from the subsurface layers down to the base of the solar convection zone ($0.71 \leq r/R_\odot \leq 1$), vanishing abruptly below its base across a thin shear layer now called the tachocline. At about the same time, serious doubts were beginning to be raised concerning the viability of the other cornerstone of the solar dynamo models developed in the 1970 and 1980, namely the turbulent electromotive force associated with cyclonic turbulence, as embodied in the so-called α-effect (see Cattaneo et al. 1996; Blackman & Brandenburg 2002; and references therein). Finally, the first pioneering global magnetohydrodynamical simulations of solar convection (Gilman 1983; Glatzmaier 1984, 1985) also challenged many prevailing theoretical tenets of the solar cycle as modeled through the dynamically- and geometrically-simplified dynamo models relying on mean-field electrodynamics.

This state of affair led to the development of dynamo models specifically designed to bypass some of these difficulties, e.g., the interface dynamo picture proposed by Parker (1993), as well as the revival of other classes of solar cycle models formerly eclipsed by the rise of mean-field electrodynamics. Most notable among these, perhaps, are the so-called Babcock-Leighton models (Wang & Sheeley 1991; Choudhuri et al. 1995; Durney 1995; Dikpati & Charbonneau 1999), in which the sun's large-scale poloidal field is regenerated through the decay of active regions. The study of the stability and rise of thin magnetic flux tubes, presumably producing sunspots upon emergence through the photosphere, also got underway and produced a number of key insights, most important perhaps the notion that such sunspot-producing magnetic flux ropes could only be stored in the weakly subadiabatic overshoot layer underlying the solar convection zone (Ferriz-Mas et al. 1994). The strong rotational shear therein was also seen as conducive to the production of intense toroidal magnetic fields, from which strongly magnetized flux ropes would presumably form. Various instabilities were later identified, which could drive the production of a poloidal magnetic component from a strong tachocline toroidal magnetic field, thus opening the possibility of closing the dynamo loop wholly within the tachocline. For more details on all these solar cycle models, see Charbonneau (2010).

This very brief historical survey, incomplete as it may be, nonetheless illustrates the main point I want to make here: as recently as a few years ago, if one were to be asked "where is the solar dynamo", the short answer would have likely been along the line "mostly in the tachocline". But remarkably, just as this almost consensual new picture was emerging, something else came around to stir trouble once again.

2. Magnetic cycles in MHD simulations of solar convection

It was clearly in the air; in the years 2009-2011, various groups working independently and using distinct numerical and modelling approaches managed to achieve turbulent MHD simulations of solar/stellar convection producing magnetic fields well-organized on large spatial scales, and undergoing polarity reversals which, in some cases, showed remarkable similarities with the solar cycle. In this section I review three such classes of numerical simulations, highlighting similarities and differences in the flows and magnetic fields they generate. In principle, all simulate the same thing: turbulent convection and MHD dynamo action in a thick, rotating, stratified, convectively unstable shell of electricaly conducting fluid subjected to thermal forcing sustaining convection.

The PENCIL code (`www.nordita.org/pencil-code`) is a public domain parallel high order finite-difference compressible MHD code, formulated in cartesian geometry. PENCIL uses constant values for dissipative coefficients such as viscosity, thermal and magnetic diffusivities, and so operates as a Direct Numerical Simulation (DNS). Pseudo-global simulations of magnetic cycles in a zonally periodic spherical wedge have been presented by Käpylä et al. (2010, 2012).

The ASH code (Anelastic Spherical Harmonics; see Clune et al. 1999) is a massively parallel spectral code in spherical geometry. Most published solar/stellar simulations (e.g., Brun et al. 2004; Browning et al. 2006; Brown et al. 2010, 2011, and references therein) use a Large-Eddy Simulation (LES) approach, where nonlinear stability is maintained through the use of enhanced values for dissipative coefficients such as viscosity and magnetic diffusivity. Some of the most recent ASH simulations (Nelson et al. 2011, 2013a, 2013b), including some discussed further below, make use of a dynamical subgrid model which yields significantly lower dissipation levels than prior ASH LES.

The EULAG code (EUlerian-LAGrangian; `www.mmm.ucar.edu/eulag`; see also Prusa et al. 2008) is a publicly available, robust hydrodynamical flow solver formulated in generalized curvilinear coordinates. Its MHD extension (not yet publicly available) is a relatively recent development and is documented in Smolarkiewicz & Charbonneau (2013). EULAG achieves strongly turbulent, low-dissipation flow states through the use of a non-oscillatory forward-in time advection algorithm, and thus belongs to the class of so-called Implicit Large-Eddy

Journal of Physics: Conference Series **440** (2013) 012014 doi:10.1088/1742-6596/440/1/012014

Simulations (ILES), where dissipation is delegated to the numerical scheme, rather than being formulated explicitly as in LES. Global simulations of solar-like cycles in the anelastic approximation have been reported in Ghizaru et al. (2010) and Racine et al. (2011).

Despite widely varying numerical approaches, presence or absence of an underlying stable fluid layer, treatment of small scales, resolution, and physical parameter regimes, the convective flows produced by these simulations, and the large-scale flows they power via inverse cascade, show remarkable similarities. Convective patterns take the form of broad upflows of hotter fluid, delineated by a network of narrow downflow lanes of colder fluid, typical of thermal convection in a gravitationally stratified background. This pattern is broken only in the equatorial regions, where larger, latitudinally-elongated convective cells are stacked longitudinally, with the flow showing little variations in the direction parallel to the rotation axis, as expected in rotationally-constrained thermal convection (cf. Fig. 2 in Käpylä et al. 2010, Fig. 1A,B,C in Nelson et al. 2013a; Fig. 1 in Ghizaru et al. 2010; §2.2.3 in Miesch & Toomre 2009). This rotational structuring of convection is also reflected in the large-scale zonal flow, in all cases showing equatorial acceleration but with a strong tendency for cylindrical isocontours throughout the convection zone (cf. Fig. 1A in Käpylä et al. 2010, Fig. 7A in Nelson et al. 2012, and Fig. 4 in Racine et al. 2011), unlike the primarily radial isocontours revealed by helioseismology (e.g., Howe 2009, Fig. 19). Moreover, in all three cases considered here the pole-to-equator angular velocity contrast turns out smaller than observed by a factor of about 3; since purely hydrodynamical but otherwise identical parent simulations tend to yield a contrast much closer to observed, this indicates that magnetic forces are playing an important role in zonal dynamics.

Few of these simulations yield a large-scale meridional flow profile (the axisymmetric flow component in $[r, \theta]$ planes) that resembles the smooth, steady single flow cell per quadrant commonly used in flux transport dynamos. Instead, the zonally-averaged meridional flow is typically very weak, highly time variable, and characterized by multiple cells in both radius and latitude (cf. Fig. 1C in Käpylä et al. 2012, Fig. 1D in Brown et al. 2010, and Fig. 4 in Racine et al. 2011).

Subtracting these axisymmetric large-scale flows extracted from the simulations yields the "small-scale" contribution $\mathbf{u}'$, from which the zonally-averaged kinetic helicity $h_u = \langle \mathbf{u}' \cdot \nabla \times \mathbf{u}' \rangle$, measuring the cyclonic character of the turbulence, is readily computed. Once again the kinetic helicity profiles turn out remarkably similar, with h_u negative (positive) in the Northern (Southern) hemisphere, peaking at high latitudes with a secondary peak in equatorial regions, often with a sign change near the base of the convection zone (cf. Fig. 1B in Käpylä et al. 2012, Fig. 8A in Brown et al. 2010, and Fig. 15C in Racine et al. 2011).

With such striking similarities in the flows powering dynamo action, one might rightfully expect that the large-scale magnetic fields produced by their associated dynamo action should show a comparable level of similarities. Interestingly, this is not the case, although some similarities do exist. Most prominent perhaps is the production of persistent and spatially well-organized toroidal magnetic flux structures in equatorial regions, undergoing polarity reversals on timescales of a few years, in a manner more or less regular and synchronous across hemispheres, depending on the simulation considered. Figure 1A shows an example taken from a EULAG simulation, in the form of a Mollweide projection of the toroidal magnetic field extracted at $r/R = 0.88$. The slightly tilted banded structures at low latitudes, maintaining their polarity over a significant longitudinal extent, are also seen in other MHD simulations (cf., e.g., Fig. 2 in Brown et al. 2010; Fig. 4 in Nelson et al 2013a).

Interestingly, zonal averaging of this non-axisymmetric magnetic field still yields a significant axisymmetric component, as shown on Figure 1B. This component, in turn, undergoes cyclic polarity reversals, here on a cadence of $\sim 2\,\mathrm{yr}$. Moreover, dynamo wave-like poleward propagation shows up upon computing time-latitude diagrams of the axisymmetric component, as shown on Fig. 2. A remarkably similar pattern is observed in ASH simulations rotating at 3

Journal of Physics: Conference Series **440** (2013) 012014 doi:10.1088/1742-6596/440/1/012014

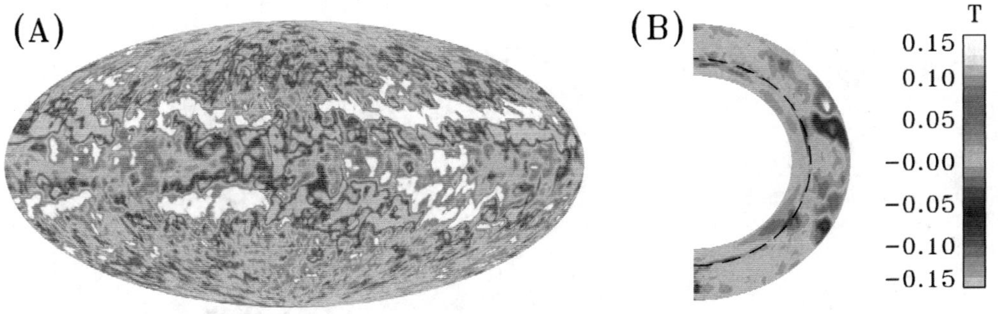

Figure 1. (A) Mollweide projection of the toroidal magnetic field at $r/R = 0.88$ in a EULAG simulation; (B) zonal average of the same, displayed in a meridional plane.

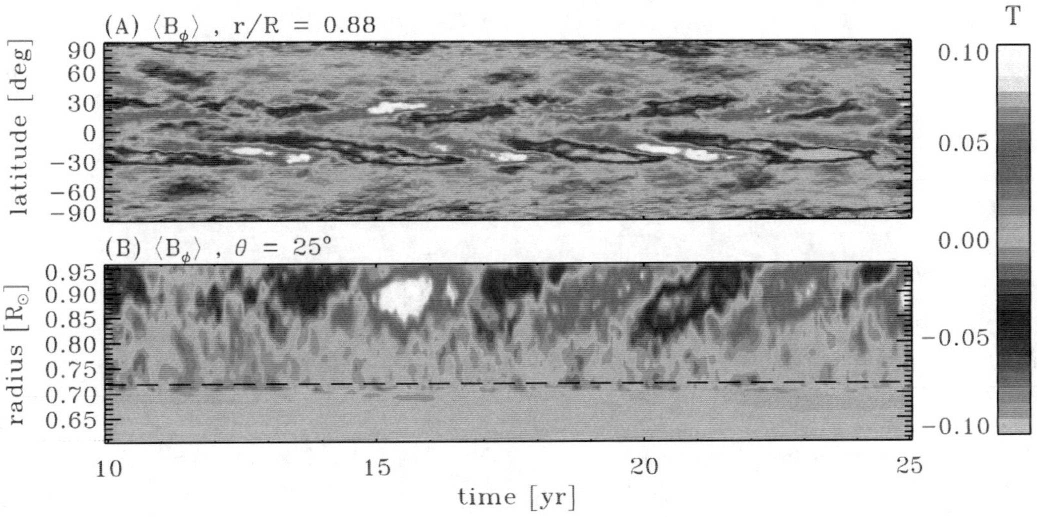

Figure 2. Time-latitude (top) and time-radius (bottom) cuts of the zonally-averaged toroidal magnetic component at $r/R = 0.88$ in the EULAG simulation of Fig. 1. The horizontal dashed line indicates the base of the convectively unstable fluid layer. This is a moderate resolution simulation ($N_\phi \times N_\theta \times N_r = 256 \times 128 \times 97$), including (relatively) small explicit magnetic dissipation ($\eta = 10^6\,\mathrm{m}^2\,\mathrm{s}^{-1}$). This simulation begins with an unmagnetized, convectively unstable static state, in which a very small flow and magnetic field perturbations are introduced to initiate convection and dynamo action.

times the solar rate (see Fig. 2 in Nelson et al. 2013a and Fig. 15 in Nelson et al 2013b). Now, with rotation increasing outwards on cylinders at low latitudes in these simulations, poleward propagation is what one would expect for an α-effect positive in the Northern hemisphere; which is what mean-field theory would predict for a kinetic helicity negative in the Northern hemisphere. Using the full α-tensor extracted by Racine et al. (2011) from a similar EULAG simulation, Simard et al. (2013) constructed kinematic $\alpha^2\Omega$ mean-field dynamo models that also exhibit an equatorially-concentrated, poleward-propagating dynamo mode, supporting — without strictly proving— a dynamo wave interpretation.

Journal of Physics: Conference Series **440** (2013) 012014 doi:10.1088/1742-6596/440/1/012014

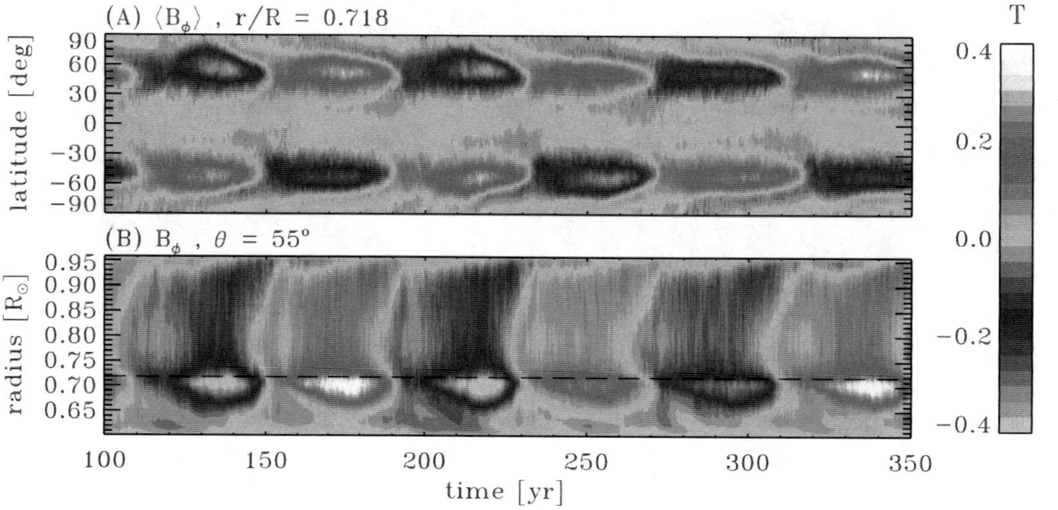

Figure 3. Time-latitude (top) and time-radius (bottom) cuts of the zonally-averaged toroidal magnetic component, extracted respectively at $r/R = 0.718$ (base of convection zone), and latitude 55°. The zonally-averaged toroidal component now peaks at around 0.5 Tesla. This is a lower resolution EULAG simulation ($N_\phi \times N_\theta \times N_r = 128 \times 64 \times 47$), covering a much longer time span than on Fig. 2.

Cycles on longer timescales also materialize in some of these MHD simulations, when a stably stratified fluid layer is included below the convecting fluid layers. Figure 3 shows an example, taken from a low resolution simulation segment spanning two and a half centuries of simulated time. Time-latitude and time-radius diagrams of the zonally-averaged toroidal magnetic component are shown, in same format as on Fig. 2. Here regular polarity reversals, essentially synchronous in both hemispheres, take place every $\sim 40\,$yr. Notice how the large-scale toroidal field now peaks at mid-latitudes and deep down, slightly below the base of the convecting layers.

Close examination of this simulation reveals a short period oscillation present at low latitudes within the convection zone (see Fig. 4A below). This appears to be the faster dynamo mode already encountered on Fig. 2, superimposed on the deep-seated, dominant mode and, as a consequence, enslaved to the latter's polarity reversals. Observational evidence for such "double-cycle" behavior actually abounds in indicators of solar activity (Benevolenskaya 1995), geomagnetic activity (Mursula et al. 2003), as well as in solar-cycle related shifts in p-mode frequency splittings (Fletcher et al. 2010; and references therein).

The ILES approach underlying these EULAG simulations makes it possible to reach strongly turbulent states on relatively coarse spatial grids[1], which in turns allows very long temporal

[1] A persistent and widely held belief is that low spatial resolution automatically implies high dissipation. It is certainly the case that *for a given numerical scheme and/or subgrid scale model*, dissipation scales with resolution; but it is also the case that different treatment of subgrid scales can yield different effective dissipation *at a given mesh size*. The set of ASH simulations recently published by Nelson et al. (2013a) offers a wonderful case in point. By switching from a static, eddy diffusivity-like LES scheme to a dynamical Smagorinsky model, these authors report a $\sim$ 50-fold decrease in dissipation on a fixed spatial mesh (cf. the quoted values for ν and η for simulations D3b and S3 in their Table 1). This is consistent with decades of community experience in computational fluid dynamics using ILES schemes; see the papers collected in Grinstein et al. (2007) for specific examples, including thermally-driven convection.

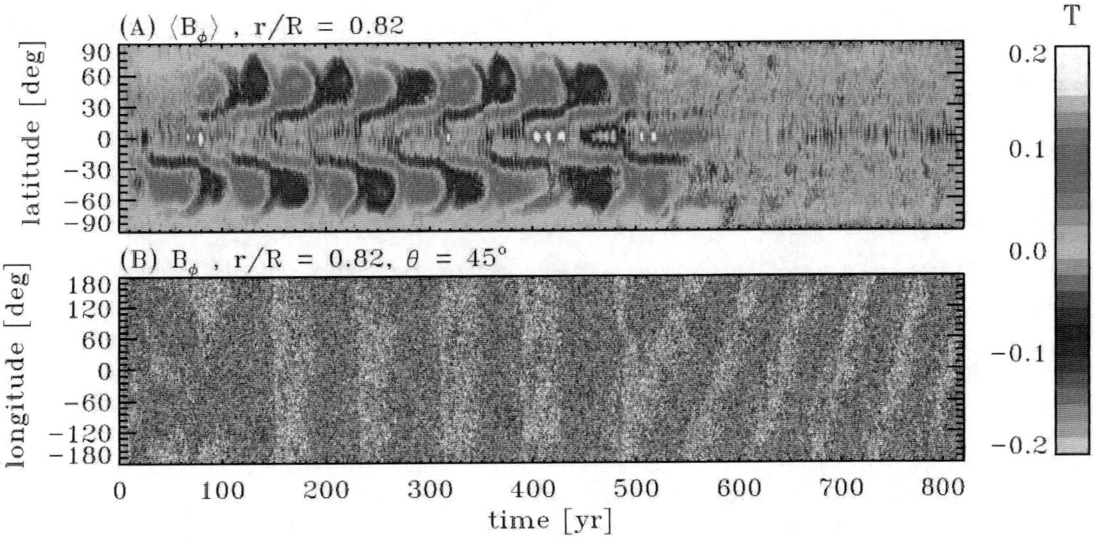

Figure 4. Extension of the EULAG simulation of Fig. 3 to 800yr. The top plot is the usual time-latitude diagram of the zonally-averaged toroidal magnetic component, now extracted at $r/R = 0.82$. The bottom plot is a longitude-time diagram of the non-averaged toroidal component, constructed at a location $(r/R, \theta) = (0.82, 45°)$ in the meridional plane.

integrations. Figure 4 shows the simulation of Fig. 3, now extended to 800 yr. The top plot is again a time-latitude diagram of the zonally-averaged toroidal field, now extracted from the inner third of the convecting fluid layer. The well-defined cycle settling in after initialisation, antisymmetric about the equatorial plane, undergoes a switch to a symmetric mode following the failure of the Southern hemisphere to inverse its polarity at around 390 yr. Some 150 years later the cycle seems to vanish, but this turns out to be an artefact of the zonal averaging used to produce this plot. The bottom plot is a time-longitude diagram of the toroidal field at $(r/R, \theta) = (0.82, 45°)$, and reveals that around 500 yr the dynamo switches now to a non-axisymmetric, tilted-dipole-like configuration, which continues to undergo polarity reversals, with a slight reduction in both period and amplitude. Ongoing analysis of this simulation suggests that the destablization towards the non-axisymmetric $m = 1$ mode is mediated by the runaway growth of a MHD instability triggered in the upper portion of the stable fluid layers underlying the convection zone (on this topic see, e.g., Gilman & Fox 1997; Dikpati et al. 2004; Strugarek et al. 2011; and references therein).

3. Old casks for new wines ? Or is it the other way around ?

It has been well understood for quite some time now that turbulent convective flows of electrically conducting fluid can be powerful dynamos (e.g., Cattaneo 1999). With regards to the solar cycle, the real challenge is to identify the mechanism(s) through which the resulting magnetic fields, generated primarily on temporal and spatial scales commensurate with those of the convective flow, can self-organize into temporally persistent and spatially coherent structures on the scale of the sun as a whole.

Traditionally, this self-organization has been viewed as a form of inverse cascade, similar to that powering large-scale flows via Reynolds stresses in purely hydrodynamical turbulence. This process can be formalized through mean-field electrodynamics in a manner such as to allow, under certain (physically restrictive) circumstances, the calculation of source terms for the large-scale magnetic field arising from the inductive action of small-scale flow and field. If a good separation of scales exists between the large-scale magnetic component and the turbulent flow, such source terms end up embodied in the so-called α-effect, refering to the first term in the tensorial development of the mean turbulent electromotive force $\boldsymbol{\xi}$ in terms of the mean magnetic field $\mathbf{B}$:

$$\boldsymbol{\xi} = \langle \mathbf{u}' \times \mathbf{B}' \rangle \approx \boldsymbol{\alpha}\mathbf{B} \ , \tag{1}$$

where $\boldsymbol{\alpha}$ is the α-tensor, whose calculation in terms of the statistical properties of the turbulent flow and field becomes the crux of the matter. Mathematically, representation of the turbulent electromotive force in terms of the large-scale magnetic field via eq. (1) is always possible, although it may not always be physically meaningful. Nonetheless, and with this caveat in mind, mean-field electrodynamics can also be turned around and used as an interpretive tool to analyze turbulent induction in MHD simulations. Accordingly, measurements of the turbulent emf and α-tensor components in turbulent MHD simulations have been carried out, and yielded results often in surprisingly good agreement with the predictions from mean-field theory, even though the simulations operate in physical regimes where many underlying assumptions under which the α-tensor can be computed analytically do not hold (see, e.g., Ossendrijver et al. 2001, 2002; Käpylä et al. 2006, 2009; Racine et al. 2011; but also Courvoisier et al. 2006). Chief among these restrictive assumptions is the requirement that the turbulence not be affected by the large-scale magnetic field, which is seldom the case in most simulations considered here (on this point see also Hubbard et al. 2009). In fact, in the context of large-scale dynamo action at high magnetic Reynolds number, the constraint of magnetic helicity conservation guarantees that the small and large scales will be coupled through a dual forward/inverse cascade of magnetic helicity of opposite sign (see Brandenburg & Subramanian 2005, and references therein).

A different angle on the inevitable coupling of small and large scales in the low dissipation limit can be directly obtained from the MHD induction equation governing the evolution of the large-scale magnetic field:

$$\frac{\partial \mathbf{B}}{\partial t} = \nabla \times (\mathbf{u} \times \mathbf{B} + \boldsymbol{\xi} - \eta \nabla \times \mathbf{B}) \ . \tag{2}$$

Assume now that because dissipation is very small, the last term on the RHS can be neglected. Then, for the mean field to remain quasi-stationary on the timecale of convection one must evidently have:

$$\mathbf{u} \times \mathbf{B} + \boldsymbol{\xi} \approx 0 \ , \tag{3}$$

implying the near-cancellation of the turbulent emf and induction by the large-scale flow. Remarkably, this is precisely what is found in the EULAG simulation analyzed by Racine et al. (2011; see their Fig. 14): the "long" cycle is driven by the small residual total emf resulting from this near-cancellation of the two inductive contributions. A similar situation was also uncovered in the ASH simulations presented in Brown et al. (2010), which produced steady large-scale magnetic fields within the convecting layers. From the modeling point of view, the implications of this state of affairs are worrisome. Minor changes in the treatment of small scales can alter slightly the turbulent emf (say), but these can translate into large changes in the residual total emf, and therefore large changes in global behavior.

Another exciting recent development has been the autonomous production of strongly magnetized (many times equipartition) flux rope-like structures in the ASH simulations of Nelson et al. (2011, 2013a, 2013b), a process by all appearance favored by the strong turbulent intermittency characterizing these simulations. Figure 5 shows an example, taken from Nelson et al. (2013a; Fig. 18). Some flux strands building up within toroidal wreaths develop enough of a density deficit in their cores for magnetic buoyancy to contribute to their rise through the convecting layers, after being pulled away from the wreaths by strong convective updrafts. Subsequent detailed analysis by Nelson et al. (2013b) has shown that most of these structures

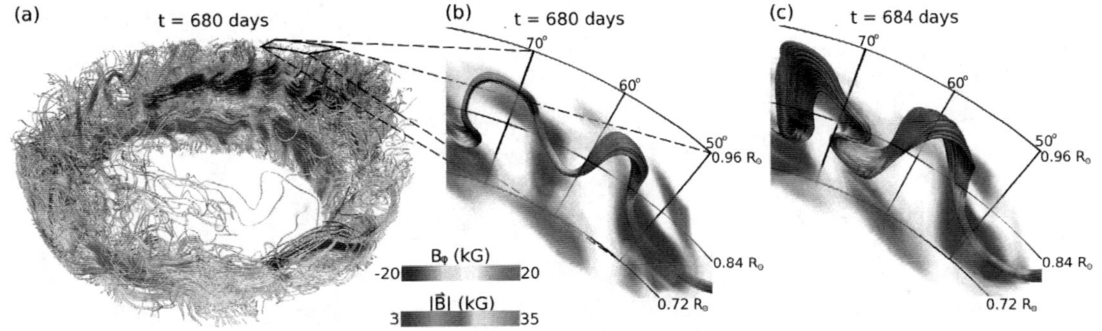

Figure 5. Formation and rise of rope-like magnetic flux systems in the toroidal wreaths produced by the ASH simulations of Nelson et al. (2013a). The superquipartition-strength magnetic flux strands form as substructures within the toroidal wreaths, and expand as they rise through the convecting layers.

maintain their approximately East-West orientation as they rise, and thus would comply with Hale's polarity rules upon emergence. This offers an intriguing alternative to the prevalent idea that sunspot-forming toroidal magnetic flux ropes must be formed, or at least stored, in the overshoot layer within the tachocline.

4. Coda: where is the solar dynamo?

Global three-dimensional MHD numerical simulations of solar-like magnetic cycles have landed. But what have we learned, and what can we still hope to learn, from such simulations? Perhaps the most significant lesson learned is that the buildup of a large-scale magnetic component, retaining spatial coherence over timescales much longer than the convective turnover time, is possible in simulations operating in strongly turbulent regimes. Now, it was not at all obvious *a priori* that this should be possible. In the turbulent regimes characterizing MHD simulations operating at low dissipation, enhancements of both turbulent induction (stretching and/or constructive folding of magnetic fieldlines) and turbulent dissipation (destructive folding of magnetic fieldlines) are expected. Until recently, whether the former could outpace the latter remained very much an open question. The fact that three distinct classes of MHD simulations, relying on different physical setups and numerical approaches, should all provide the same positive answer is noteworthy indeed.

The overall complexity of global dynamo action and magnetic self-organization in the strongly turbulent regime is also well established by these simulations; large-scale flows generated by convection itself are important players in the growth and evolution of the large-scale magnetic field. The turbulent emf is quite complex and attempts to relate it to the mean magnetic field yield an intricate picture where turbulent induction, turbulent pumping and perhaps even time lags all play a significant role. Further complications ensue in simulations including a stably stratified fluid layer underlying the convection zone. In the absence of magnetic fields, such simulations naturally develop a thin, tachocline-like shear layer immediately beneath this interface (e.g. Brun et al. 2011, Fig. 10; Racine et al. 2011, Fig. 3). This shear, although reduced in MHD simulations, can still play a significant role in global dynamo action, and in some cases lead to the buildup of distinct dynamo modes, peaking in different regions of the domain and interfering with one another.

What have these MHD simulations provided in terms of a (tentative) answer to the query raised in the title of this paper? The fact that spatially coherent, persistent magnetic flux systems can be sustained within the convection zone, and even produce buoyantly rising flux rope-like structures, certainly brings back to the fore the possibility that the dynamo driving the solar cycle may "live" entirely within the convection zone. Käpylä et al. (2012) even produced simulations which, in their nonlinearly saturated regime, exhibit a very solar-like equatorward propagation. Then again, the EULAG simulations producing deep cycles in the tachocline also show a number of solar-like secondary features, such as double cycles (Simard et al. 2013), magnetically-mediated modulation of convective energy transport (Cossette et al., preprint), of the meridional flow (Passos et al. 2012) and rotational torsional oscillation patterns that compare very favorably to observations (Beaudoin et al. 2013). Where is the solar dynamo? We still don't know, but there is a fair chance that the little light we are currently seeing, very far away at the end of the tunnel, for once may not be the proverbial incoming locomotive.

Acknowledgments

I wish to thank Nic Nelson and Mark Miesch for kindly providing Figure 5 herein, Patrice Beaudoin for producing all others "on order", and Piotr Smolarkiewicz and Sacha Brun for useful comments and suggestions on the original version of this paper. The development of EULAG-MHD is a joint venture with Piotr K. Smolarkiewicz (MMM/NCAR, Boulder, U.S.A.) and Mihai Ghizaru (UdeM), and has been supported by the Natural Science and Engineering Research Council, the Canada Research Chair program, and the Canadian Space Agency (grant 9SCIGRA-21).

References

Beaudoin, P, Charbonneau, P, Racine, É and Smolarkiewicz, P K 2013 *Solar Phys.* **282** 335–360
Benevolenskaya, E E 1995 *Solar Phys.* **161** 1–8
Blackman, E G, and Brandenburg, A 2002 *Astrophys. J.* **579** 359–373
Brandenburg, A, and Subramanian, K 2005 *Phys. Rep.* **417** 1–209
Browning, M K, Miesch, M S, Brun, A S, and Toomre, J 2006 *Astrophys. J.* **648**, L157–160
Brown, T M, and Morrow, C A 1987, *Astrophys. J. Lett.* **314** L21–26
Brown, T M, Christensen-Dalsgaard, J, Dziembowski, W A, Goode, P, Gough, D A, and Morrow, C A 1989 *Astrophys. J.* **343** 526–546
Brown, B P, Browning, M K, Brun, A S, Miesch, M S, and Toomre, J 2010 *Astrophys. J.* **711**, 424–438
Brown, B P, Miesch, M S, Browning, M K, Brun, A S, and Toomre, J 2011 *Astrophys. J.* **731** id69
Brun, A S, Miesch, M S, and Toomre, J 2004 *Astrophys. J.* **614** 1073–1098
Brun, A S, Miesch, M S, and Toomre, J 2011 *Astrophys. J.* **742** id79
Cattaneo, F, Hughes, D W, and Kim, R-J 1996 *Phys. Rev. Lett.* **76** 2057–2060
Cattaneo, F 1999 *Astrophys. J.* **515** L39–42
Charbonneau, P 2010 *Liv. Rev. Solar Phy.* **7** lrsp-2010-3
Choudhuri, A R, Schüssler, M, and Dikpati, M 2010 *Astron. Astrophys.* **303** L29–32
Clune, T L, Elliot, J R, Glatzmaier, G A, Miesch, M S and Toomre, J 1999, *Parallel Comput.* **25** 361
Courvoisier, A, Hughes, D W, and Tobias, S M 2006 *Phys. Rev. Lett.* **96** id034503
Dikpati, M, and Charbonneau, P 1999, *Astrophys. J.* **518** 508–520
Dikpati, M, Cally, P S, and Gilman, P A 2004, *Astrophys. J.* **610** 597–615
Durney, B R 1995, *Solar Phys.* **160** 213–235
Dziembowski, W A, Goode, P R, and Libbrecht, K G 1989 *Astrophys. J. Lett.* **337** L53–57
Ferriz-Mas, A, Schmitt, D, and Schüssler, M 1994 *Astron. Astrophys.* **289** 949
Fletcher, S T, Broomhall, A-M, Salabert, D, Basu, S, Chaplin, W J, Eslworth, Y, Garcia, R A, and New, R 2010 *Astrophys. J. Lett.* **718** L19–22
Ghizaru, M, Charbonneau, P, and Smolarkiewicz, P K 2010 *Astrophys. J. Lett.* **715** L133–137
Gilman, P A 1983, *Astrophys. J. Suppl.* **53**, 243–268
Gilman, P A, and Fox, P A 1997, *Astrophys. J.* **484**, 439–454
Glatzmaier, G A 1984 *J. Comput. Phys.* **55**, 461–484
Glatzmaier, G A 1985, *Astrophys. J.* **291**, 300–307
Howe, R 2009 *Liv. Rev. Solar Phy.* **6** lrsp-2009-1
Hubbard, A, DelSordo, F, Käpylä, P J, and Brandenburg, A 2009 *Month. Not. Roy. Astr. Soc.* **398** 1981–1899
Käpylä, P J, Korpi, M J, Ossendrijver, M, and Stix, M 2006 *Astron. Astrophys.* **455** 401–412

Käpylä, P J, Korpi, M J and Brandenburg, A 2009, *Astron. Astrophys.* **500** 633–646
Käpylä, P J, Korpi, M J, Brandenburg, A, Mitra, D, and Tavakol, R 2010 *Astr. Nach.* **331** 73
Käpylä, P J, Mantere, M J, and Brandenburg, A 2012 *Astrophys. J. Lett.* **755** idL22
Grinstein, F F , Margolin, L G, and Rider, W J (eds.) 2007 *Implicit Large Eddy Simulation: Computing Turbulent Fluid Dynamics*, Cambridge: Cambridge University Press
Miesch, M S, and Toomre, J 2009 *Ann. Rev. Fluid Mech.* **41** 317–340
Mursula, K, Ziegler, B, Vilppola, J H 2003 *Solar Phys.* **212** 201–207
Nelson, N J, Brown, B P, Brun, A S, Miesch, M S, and Toomre, J 2011 *Astrophys. J. Lett.* **739** idL38
Nelson, N J, Brown, B P, Brun, A S, Miesch, M S, and Toomre, J 2013a *Astrophys. J.* **762** id73
Nelson, N J, Brown, B P, Brun, A S, Miesch, M S, and Toomre, J 2013b *Solar Phys.* preprint
Ossendrijver, M A J H, Stix, M, and Brandenburg, A 2001 *Astronom. Astrophys.* **376** 713–726
Ossendrijver, M A J H, Stix, M, Brandenburg, A, and Rüdiger, G 2002 *Astron. Astrophys.* **394**, 735–745
Parker, E N 1993, *Astrophys. J.* **408** 707–719
Passos, D, Charbonneau, P and Beaudoin, P 2012, *Solar Phys.* **279** 1–22
Prusa, J M, Smolarkiewicz, P K, and Wyszogrodzki, A A 2008 *Comp. Fluids* **37**, 1193–1207
Racine, É, Charbonneau, P, Ghizaru, M, Bouchat, A and Smolarkiewicz, P 2011 *Astrophys. J.* **735** id46
Simard, C, Charbonneau, P and Bouchat, A 2013 *Astrophys. J.* submitted
Smolarkiewicz, P K and Charbonneau, P 2013 *J. Comput. Phys.* **236**, 608–623
Strugarek, A, Brun A S, and Zahn, J-P 2011 *Astronom. Astrophys.* **532** A34
Wang, Y-M, and Sheeley, N R Jr 1991 *Astrophys. J.* **375** 761–770

Eclipse on the Coral Sea: Cycle 24 Ascending (GONG 2012, LWS/SDO-5, and SOHO 27) IOP Publishing
Journal of Physics: Conference Series **440** (2013) 012015 doi:10.1088/1742-6596/440/1/012015

Mode Frequencies from 17, 15 and 2 Years of GONG, MDI, and HMI Data

S G Korzennik[1], **A Eff-Darwich**[2,3]

[1] Harvard-Smithsonian Center for Astrophysics, Cambridge, MA, USA
[2] Dept. Edafologia y Geologia, Univ. La Laguna, 38206, Tenerife, Spain
[3] Instituto de Astrofisica de Canarias, 38205, Tenerife, Spain

E-mail: `skorzennik@cfa.harvard.edu, adarwich@ull.es`

Abstract. We present results from fitting all the available data to date from the three major helioseismic instruments: MDI, GONG and HMI. These data were fitted using an innovative and independent methodology devised a few years ago. The mode fitting was carried out on time series of varying lengths (1×, 2×, 4×, 8×, 16×, 32×, 64 × 72 day-long), using co-eval epochs for all three data sets. By fitting time series of varying lengths, one trades off some temporal resolution for a better precision. We present a comparison of these results, and discuss the potential sources of the residual small discrepancies. We also present inferences from these results on the determination of the solar internal rotation and changes with epoch and thus activity levels.

1. Introduction

We present results from fitting all the available data sets acquired to date by the GONG, MDI and HMI instruments, namely: 171 × 36 days of GONG, 76 × 72 days of MDI and 10 × 72 days of HMI data[1]. The time series of spherical harmonic coefficients for all three instruments were fitted using the same procedure, described in Korzennik [3, 4]. This method fits all the individual singlets (n, ℓ, m), for a given multiplet (n, ℓ). It uses an optimal sine-multitaper spectral estimator, the complete leakage matrix, and fits an asymmetric mode profile. The spherical harmonic coefficients for MDI we used are the ones computed using an *improved* spatial decomposition, *i.e.*, the one that includes our best knowledge of both components of the image distortion. The spherical harmonic coefficients for HMI we used were still *preliminary* values.

As a trade-off between precision and temporal resolution, we fitted time series of different lengths instead of the canonical 72 or 108 days used by either the MDI or GONG projects, respectively. Instead, we fitted 1×, 2×, 4×, 8×, 16×, 32×, and 64 × 72 day long time series. For all but the 1 × 72 day-long ones, the epochs we fitted overlap by half their length. Figure 1 illustrates the temporal coverage of the available data, the length of the fitted segments and the corresponding mean solar activity level.

To investigate the influence of the leakage matrix on the results, we also fitted some time series using different leakage matrices: one computed by Schou for $B_o = 0$, one computed by Korzennik also for $B_o = 0$, and one computed by Korzennik for $B_o = \bar{B}_o$ (*i.e.*, the mean heliographic latitude at disk center, B_o, over the time series of each epoch).

[1] Two 72 day-long MDI epochs are missing due to the loss of contact with the SOHO spacecraft.

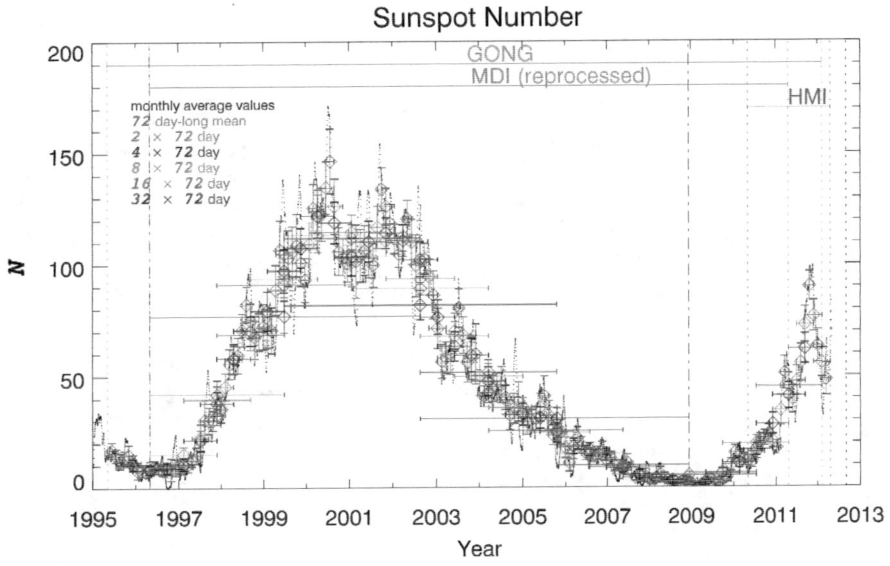

Figure 1. Temporal extent of the available data (top bars), length of the fitted segments (horizontal error bars) and the corresponding solar activity level, in terms of mean sunspot number.

This paper presents comparisons of the resulting fitted frequencies and the solar internal rotation rate inferred from the frequency splittings. These are comparisons of results obtained with the same fitting method and for identical time intervals, by contrast to comparing table of frequencies produced by the respective projects where neither the fitting method nor the time interval are similar.

2. Comparisons

Figure 2 compares GONG results to MDI's, complemented by HMI's, in terms of scaled changes of the multiplet frequencies, (*i.e.*, $\delta\nu/\sigma_\nu$), as a function of time and for epochs of different lengths. The frequency changes with activity, as measured by both sets of observations, are quite similar, but not identical.

Figure 3 compares scaled multiplet differences between GONG and MDI results, as a function of epoch and for epochs of different lengths. The upgrade of the GONG instrument, during the summer of 2001, from a 256×242 to a 1024×1024 pixels CCD camera, is visible in the frequency differences. Frequency comparison, also in terms of scaled multiplet differences, for the 360 days of available overlap between MDI and HMI data is shown in Fig. 4. This comparison shows no systematics, but differences at the 1 to 2 σ level.

The effects of the leakage matrix are illustrated in Figs. 5 and 6. the scaled differences for the singlets are plotted, as well as the corresponding distribution. The comparison between two independent leakage matrix calculations shows almost no bias, but systematic differences for some ridges, as large as 2σ. The inclusion of the mean heliographic latitude at disk center is barely affecting the low and intermediate degrees results (*i.e.*, remaining mostly at or below the 0.1σ level).

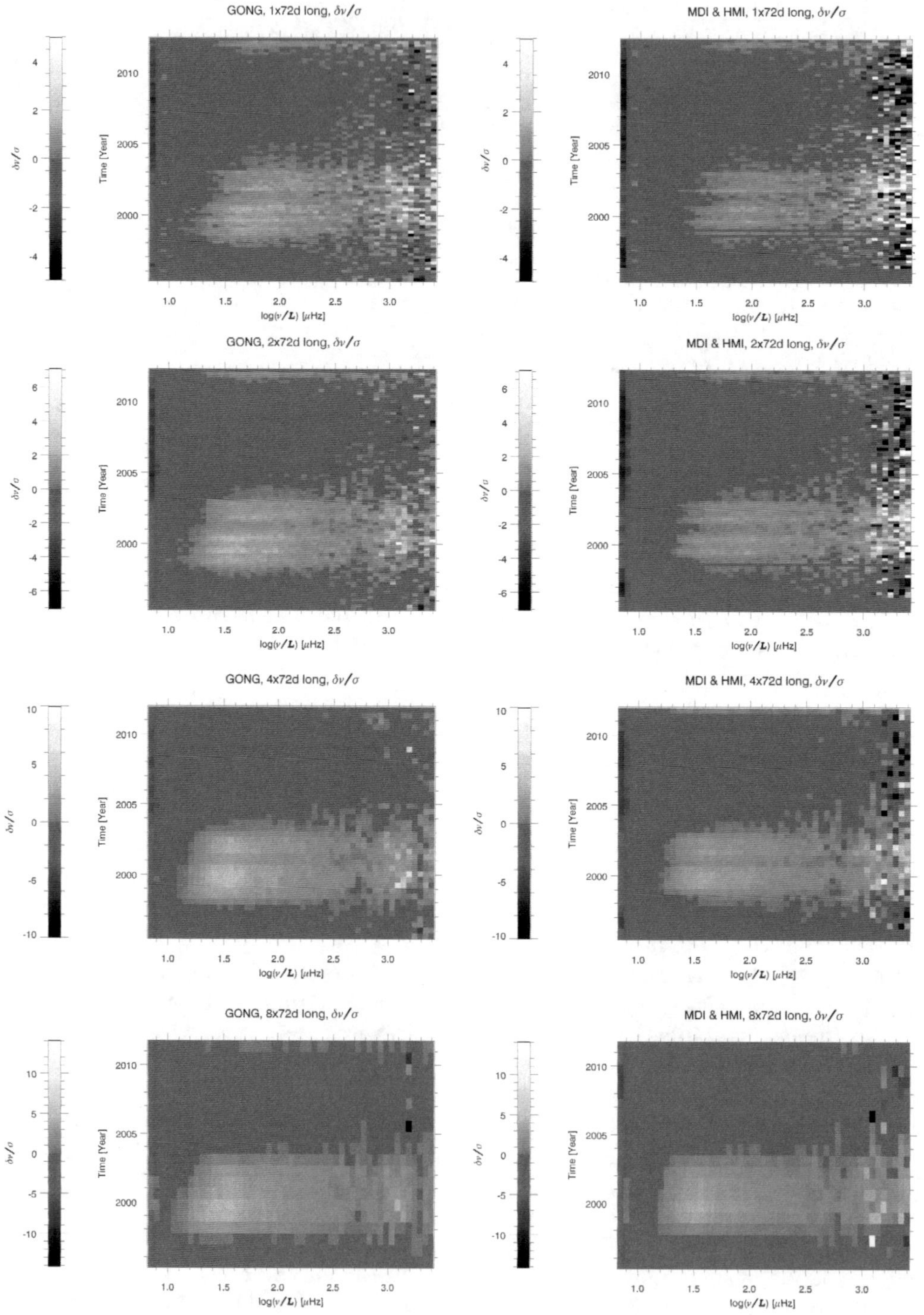

Figure 2. Scaled changes of multiplet frequencies, (*i.e.*, $\delta\nu/\sigma_\nu$), as a function of time, and for fitting epochs of different lengths (1, 2, 4, and 8 $\times$ 72 day-long, top to bottom), between fitting GONG data (left) and MDI augmented by HMI data (right). The scaled differences are computed with respect to the first fitted epoch, and averaged over bins equispaced in $\log(\nu/L)$.

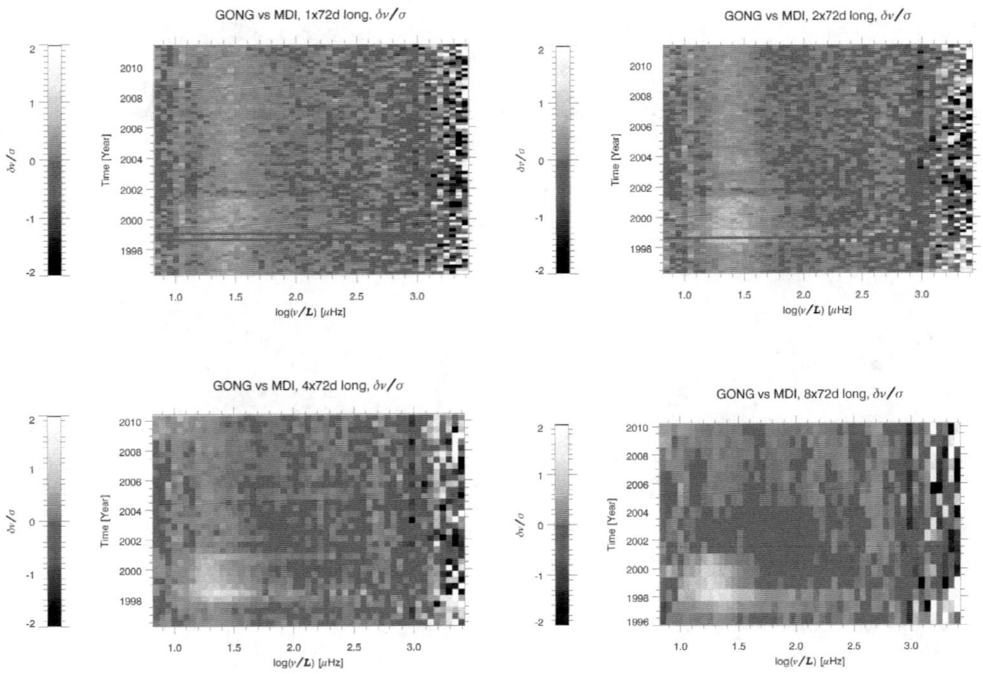

Figure 3. Scaled differences of multiplet frequencies, ($i.e.$, $\delta\nu/\sigma_\nu$), as a function of time, and for fitting epochs of different lengths (1, 2, 4, 8×72 day-long), using GONG data or MDI data. The horizontal dash line corresponds to the summer of 2001, when the GONG instrument was upgraded from a 256×242 to a 1024×1024 pixels CCD camera. As for Fig. 2, The scaled differences are averaged over bins equispaced in $\log(\nu/L)$.

3. Variation of the Solar Rotation

We inverted the resulting rotational frequency splittings for all the data and all the fitted epochs, using the inversion methodology described in [1, 2]. Note that this method uses the actual frequency splittings, not the coefficients of a polynomial expansion in m of the splittings. This method optimizes the model grid used for the inversion based on the precision of the inverted data set.

Figure 7 shows the resulting variation of the solar rotation, at four depths spanning most of the convection as a function of time and latitude, derived from fitting epochs of different lengths and either GONG data or MDI augmented by HMI data. The figure also shows the difference of the inferred rotation between the two data sets. Animations of the resulting rotation changes as a function of depth or time, and for other lengths of fitted time series, are available in the supplementary material.

4. Conclusions

We have derived sets of frequencies[2], using a single fitting methodology and identical epochs, that are consistent across the three instruments (GONG, MDI and HMI). Detailed comparisons show: (a) a systematic offset at the 1σ level (relatively high) between GONG and MDI that corresponds to the upgrade of the GONG instrument; (b) no systematic differences when comparing MDI

[2] Tables of frequencies are available at `www.cfa.harvard.edu/~sylvain/research/tables/MediumL/`.

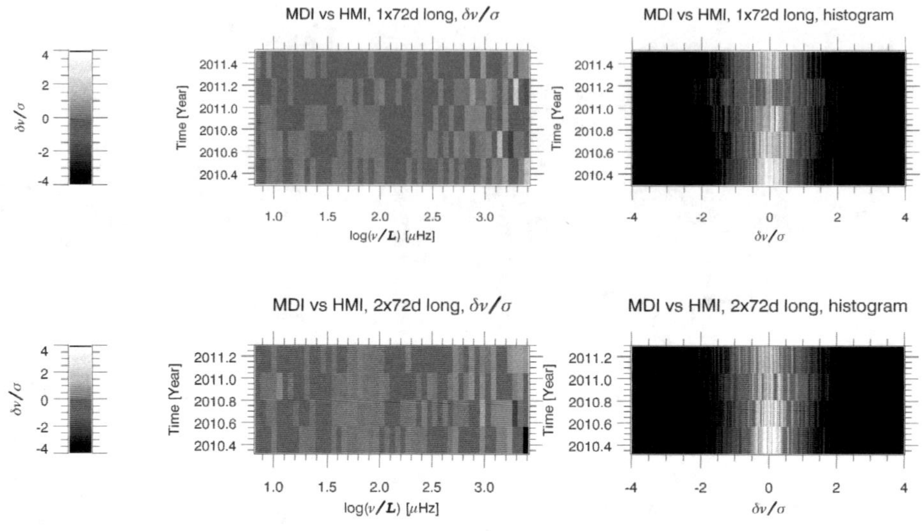

Figure 4. Scaled differences of multiplet frequencies, (*i.e.*, $\delta\nu/\sigma_\nu$), between MDI and HMI data, as a function of time, and for 1 & 2 × 72 day-long fitting epochs. The left panels show averaged scaled differences, averaged over bins equispaced in $\log(\nu/L)$, as in Figs. 2 and 3. The right panels show the distributions of the differences.

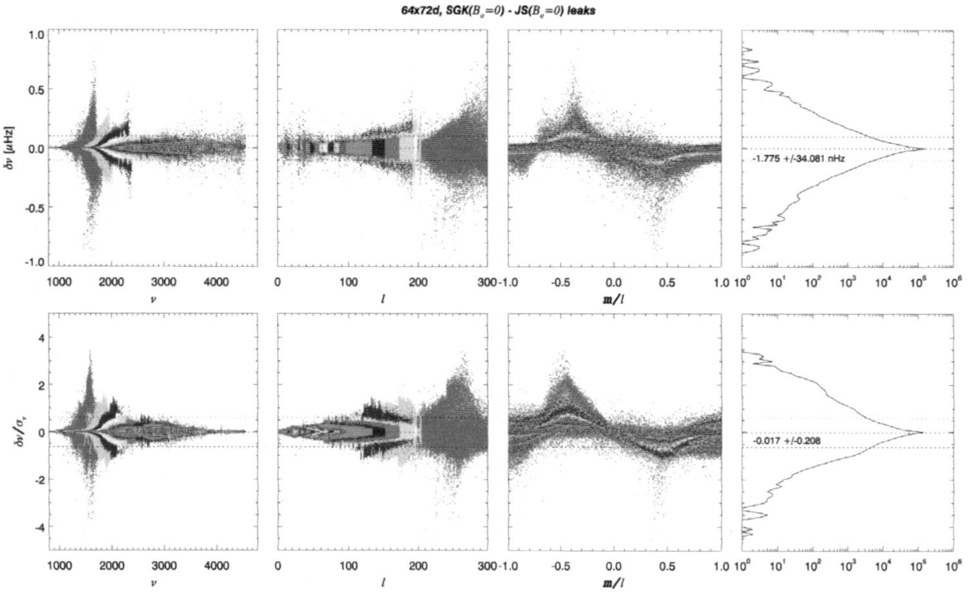

Figure 5. Scaled differences of singlet frequencies, (*i.e.*, $\delta\nu/\sigma_\nu$), when fitting a 64×72 day-long time series but using two independent leakage matrix calculations. The color coding corresponds to the mode's order, n. This comparison shows almost no bias, but systematic differences for some ridges, as large as 2σ.

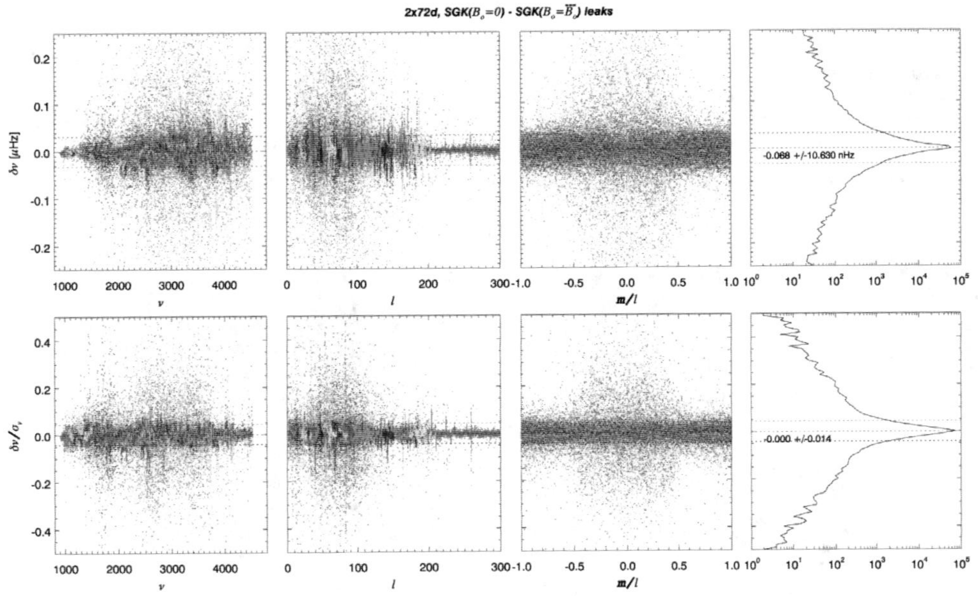

Figure 6. Scaled differences of singlet frequencies, (*i.e.*, $\delta\nu/\sigma_\nu$), when fitting a 2×72 day-long time series but using leakage matrix calculations that account for the mean heliographic latitude at disk center.

and HMI results, except at the 2σ level (again surprisingly high); and (c) that the effect of using different leaks results in small effects, at the 0.1σ to 2σ level, when comparing singlets, with almost no bias.

Our rotation inversions show how the high latitude branch becomes more visible and significant when using longer time series, while differences of the inferred rotation using different instruments illustrate the difficulty of deriving significant information on potential changes at the deeper layers.

References

[1] Eff-Darwich, A., Korzennik, S. G., & Jiménez-Reyes, S. J. 2002, ApJ, 573, 857
[2] Eff-Darwich, A., Korzennik, S. G., Jiménez-Reyes, S. J., & García, R. A. 2008, Ap.J., 679, 1636
[3] Korzennik, S. G. 2005, Ap.J., 626, 585
[4] Korzennik, S. G. 2008, Journal of Physics Conference Series, 118, 012082

Acknowledgments

The Solar Oscillations Investigation - Michelson Doppler Imager project on SOHO is supported by NASA grant NAG5–8878 and NAG5–10483 at Stanford University. SOHO is a project of international cooperation between ESA and NASA. This work utilizes data obtained by the Global Oscillation Network Group (GONG) program, managed by the National Solar Observatory, which is operated by AURA, Inc. under a cooperative agreement with the National Science Foundation. The data were acquired by instruments operated by the Big Bear Solar Observatory, High Altitude Observatory, Learmonth Solar Observatory, Udaipur Solar Observatory, Instituto de Astrofísica de Canarias, and Cerro Tololo Interamerican Observatory. SGK was supported by NASA grants NAG5–9819 and NNG05GD58G.

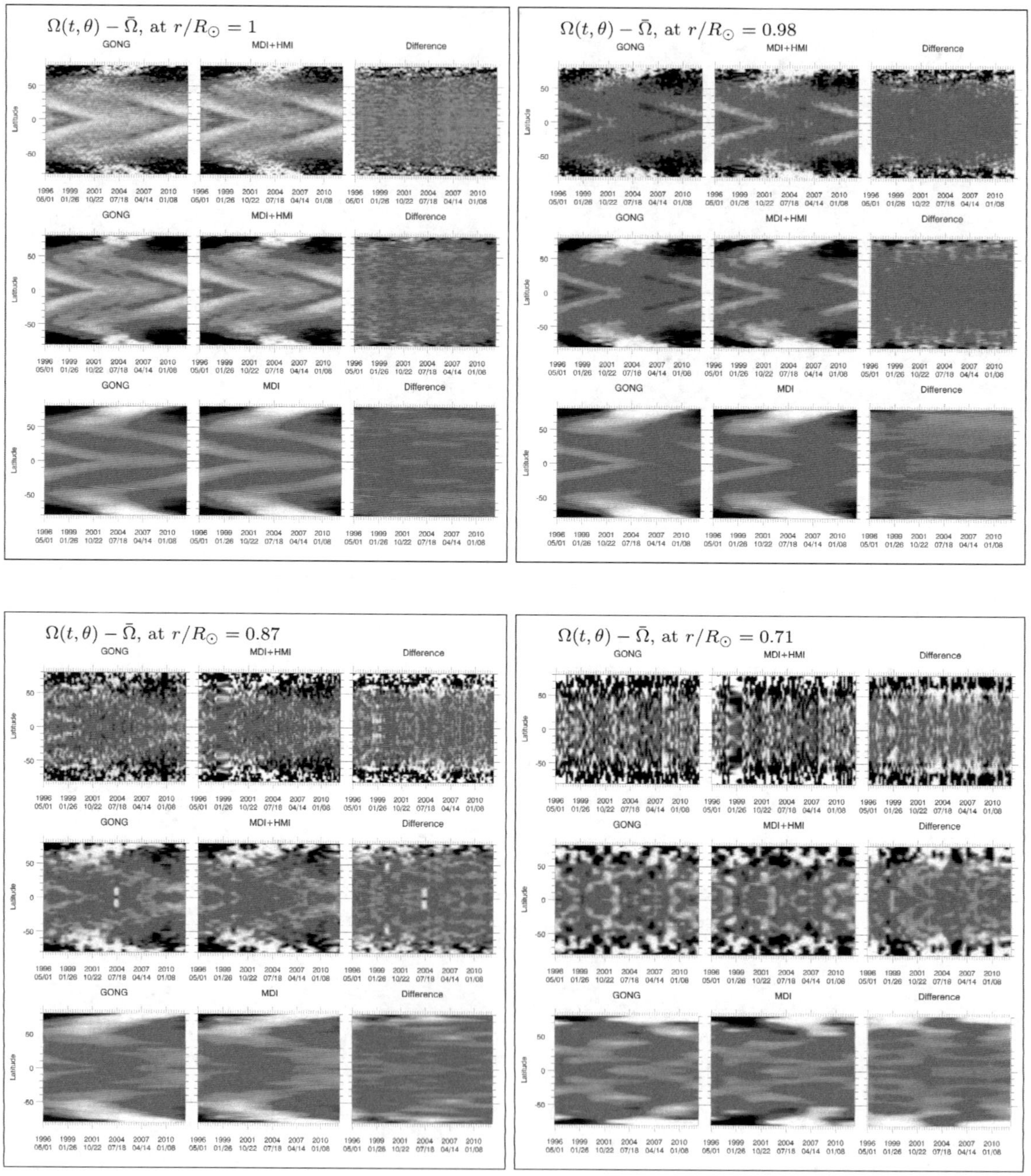

Figure 7. Variation of the solar rotation as a function of time and latitude, inferred from fitting epochs of different lengths (rows in each box: $1\times$, $4\times$ & 16×72 day-long) at four depths (each box: $r/R_\odot = 1, 0.98, 0.87, 0.71$). The columns in each box show results from GONG, MDI augmented by HMI data, and their difference (left to right). The residual variations of the differences are an indication of the remaining uncertainties of these inferences, especially for the deeper layers.

Eclipse on the Coral Sea: Cycle 24 Ascending (GONG 2012, LWS/SDO-5, and SOHO 27) IOP Publishing
Journal of Physics: Conference Series **440** (2013) 012016 doi:10.1088/1742-6596/440/1/012016

Accurate characterization of high-degree modes using MDI data

S G Korzennik[1], M C Rabello-Soares[2,3], J Schou[2] and T Larson[2]

[1] Harvard-Smithsonian Center for Astrophysics, Cambridge, MA, USA
[2] W.W. Hansen Experimental Physics Laboratory, Stanford University, CA, USA

E-mail: skorzennik@cfa.harvard.edu

Abstract. We present the first accurate characterization of high-degree modes (i.e., ℓ up to 1000), using the best MDI full-disk full-resolution data set available (90-day long time series, acquired in 2001). The Dopplergrams were spatially decomposed using our best estimate of the image scale and the known components of MDI's image distortion. Multi-tapered power spectra were fitted for all degrees and all azimuthal orders, between $\ell = 100$ and $\ell = 1000$, and for all orders with substantial amplitude. Fitting at high degrees generates ridge characteristics, characteristics that do not correspond to the underlying mode characteristics. We used a sophisticated forward modeling to recover the best possible estimate of the underlying mode characteristics. We have derived a final set of corrected mode characteristics (frequencies, line widths, asymmetries and amplitudes) and their uncertainties.

1. Introduction

It has been shown that the inclusion of high degree modes ($\ell \leq 1000$) has great potential to further constrain the outer 4% of the solar interior (see [4] for the stratification and [1] for the internal rotation). Unfortunately, modes merge into ridges at high degrees (*i.e.*, $\ell > 300$ for f-modes, $\ell > 200$ for low order p-modes) and while the ridge can be fitted, the resulting ridge characteristics ($\tilde{\nu}$, $\tilde{\Gamma}$, $\tilde{A}$, $\tilde{\alpha}$) are not equal to the underlying mode characteristics. It is thus not enough to fit ridges, but one needs to correct the ridge characteristics using a precise and sophisticated ridge modeling procedure to derive accurate mode characteristics.

2. Data Analysis and Ridge Fitting

We have analyzed MDI's *Dynamics 2001*: 90 days of full disk images, with a 97% fill factor, after gap filling (95.8% prior). The full-disk two arc-second per pixel resolution Dopplergrams were spatially decomposed using our *best* spatial decomposition, including both components of the image distortion [2]. We used a 61-term sine multi-taper spectral estimator, since the ridges are wide (no need for high spectral resolution) and the resulting widening causes modes to blend into ridges at lower degrees (hence some overlap between mode fitting and ridge fitting). The resulting power spectra, for the zonal modes, is shown in Fig.1. All the power spectra, between $\ell = 100$ and $\ell = 1000$ and for all the azimuthal orders, were fitted with a sum of modified (asymmetric) Lorentzians, using a multi-step least-squares minimization. The number of orders included in the fitting was contrast limited (1:500 power ratio followed by rejecting the highest

[3] Present address: Physics Department, Universidade Federal de Minas Gerais, Minas Gerais, Brazil.

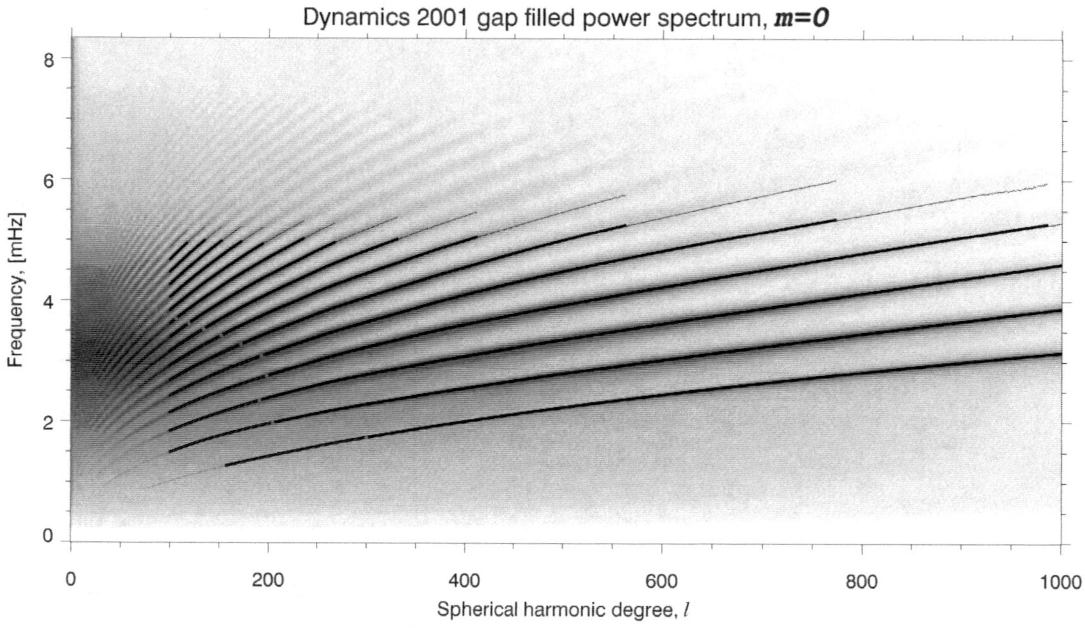

Figure 1. Gap-filled *Dynamics* 2001 power spectrum, displayed on a logarithmic scale, for the zonal modes ($m = 0$). The dots represent the fitted ridges (black) or modes (green).

n) to limit the dynamical range of the fitting. This produced some 5.2×10^6 singlets (n, ℓ, m), or 5,780 multiplets (n, ℓ)[4].

3. Ridge Modeling

Over the years, we have developed a precise and sophisticated model of the ridge [2, 5, 3]. This model produces a simulated ridge, based on a mode characteristics input set. The simulated ridge is fitted to derive the resulting ridge characteristics, and thus a correspondence between the mode and ridge characteristics. The input model is refined by adjusting iteratively most of the parameters to produce a model that matches the observations. The input set was also perturbed to derive the sensitivity and precision of the correction scheme (see [3] for additional details.)

4. Results

Mode frequencies were estimated from ridge frequencies as follows:

$$\nu_{n,\ell,m} = \tilde{\nu}_{n,\ell,m} - \Delta^{\nu}_{n,\ell,m} \tag{1}$$

where $\tilde{\nu}$ represents the ridge frequency and Δ^{ν} the ridge-to-mode offset:

$$\Delta^{\nu}_{n,\ell,m} = \tilde{\nu}^{M}_{n,\ell,m} - \nu^{M}_{n,\ell,m} \tag{2}$$

where ν^M is the input mode frequency and $\tilde{\nu}^M$ the resulting ridge frequency of our model.

[4] The tables are available at `https://www.cfa.harvard.edu/~sylvain/research/tables/HiL/`.

Journal of Physics: Conference Series **440** (2013) 012016 doi:10.1088/1742-6596/440/1/012016

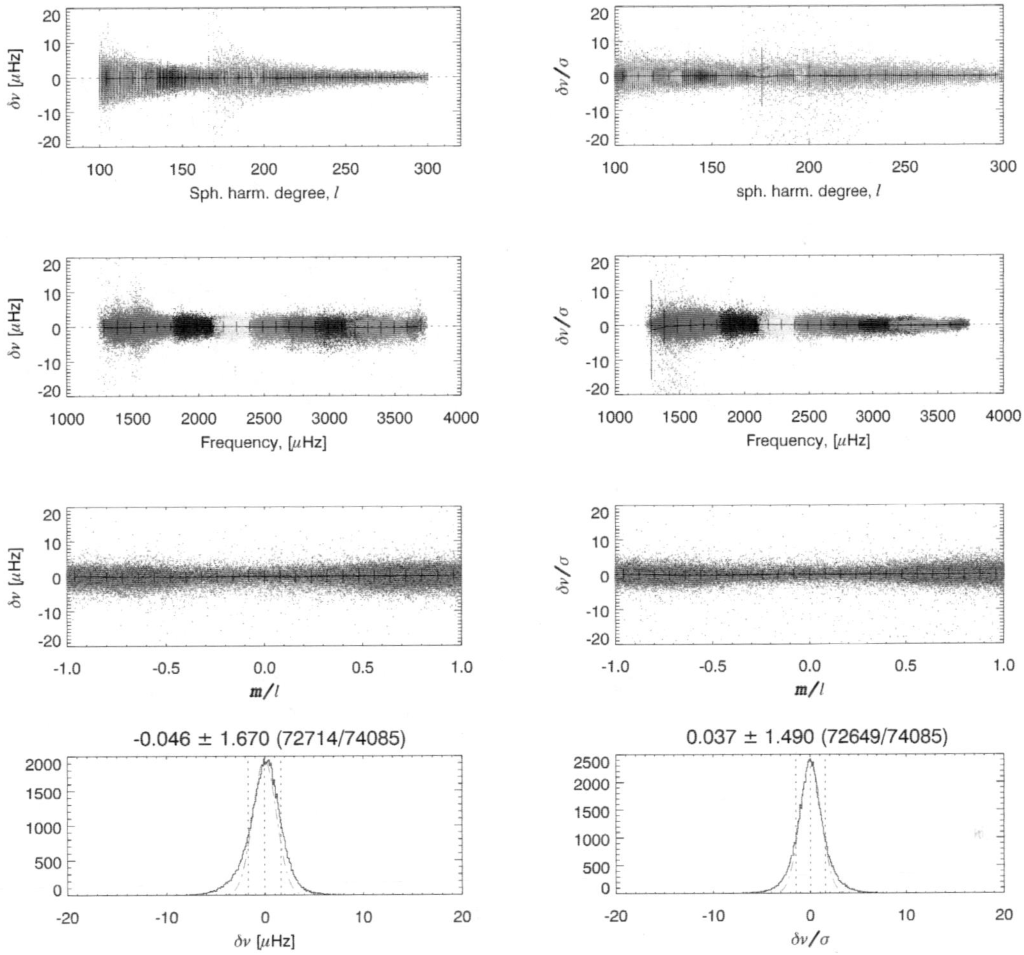

Figure 2. Comparison of estimate of mode singlet frequencies derived from ridge fitting to actual mode frequency computed by fitting resolved modes at intermediate degrees. The frequency differences (left panels), and the scaled differences (differences divided by their respective uncertainties. σ, right panels) are plotted as a function of degree, frequency, and m/ℓ, and as a histogram (top to bottom). The color coding in the top three rows corresponds to, n. The red curves in the histograms are the Gaussians corresponding to the mean and standard deviation of the distributions (indicated by the vertical dotted lines), drawn to show the excess and skewness in the wings of the actual distribution.

The overlap between mode and ridge fitting, *i.e.*, $100 \leq \ell \leq 300$ for f-modes and $100 \leq \ell \leq 200$ for p-modes, provides validation of the methodology. Figure 2 compares mode frequencies measured at low and intermediate degrees[5] to corrected mode frequencies derived from ridge fitting, for all the overlapping singlets. This figure demonstrates clearly that there is almost no bias in the corrected mode frequencies derived from ridge fitting, except for some excess and skewness in the wings of the distribution of the differences (bottom panels of Fig. 2.)

[5] Where the modes are resolved.

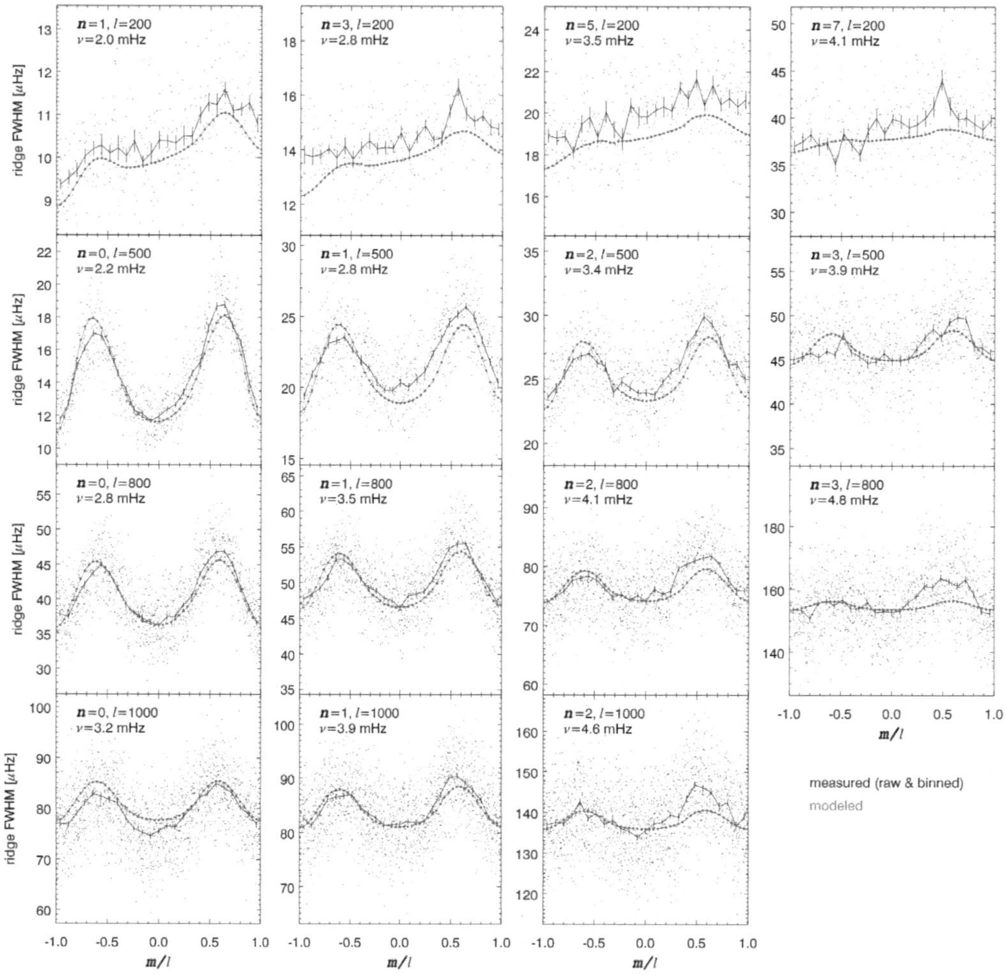

Figure 3. Comparison of ridge FWHM for a selection of multiplets, as a function of the ratio m/ℓ. The dots correspond to individual ridge width measurements, and the black curves the corresponding binned values. The red curves correspond to our best model of the ridge width, as evaluated at 51 equispaced m values. In most cases, our model reproduces rather well the variation of ridge width with the azimuthal order, m, although the input mode width of the model is constant with m.

Figure 3 shows that the ridge width is a function of m, although the input mode width is assumed constant. That variation is well modeled and is thus taken out to estimate a ridge width for each multiplet, that is then corrected using the ridge modeling to estimate a mode width for that multiplet.

The correction scheme we developed to estimate mode widths uses a Gaussian widening model:

$$\tilde{\Gamma}^2_{n,\ell} = (\Gamma^E_{n,\ell})^2 + \mathcal{W}^2_{n,\ell} \quad (3)$$

where $(\Gamma^E_{n,\ell})^2 = \Gamma^2_{n,\ell} + W^2_{N,T}$ and $W_{N,T}$ is the spectral resolution of the Nth order sine multi-taper ($N = 61$).

Figure 4 illustrates how the ridge amplitude is also a function of m. It shows also that the model ridge amplitude does not match the observations very closely. We suspect that this mismatch results from our poor knowledge of the point-spread function (PSF) of the MDI instrument.

5. Conclusions

We have derived the most accurate high-ℓ mode characteristics to date, using *Dynamics 2002* MDI observations. The limiting factor remains our poor knowledge of MDI's PSF. Since we have co-eval HMI & MDI data, we may be able to better characterize MDI's PSF. We anticipate that the method will be extended to more MDI *Dynamics* segments and ported to HMI and GONG observations.

References

[1] Korzennik, S. G., & Eff-Darwich, A. 1999, "Solar Rotation Near the Surface: The Outer 1%", SOHO-9 Workshop on Helioseismic Diagnostics of Solar Convection and Activity, 9,

[2] Korzennik, S. G., Rabello-Soares, M. C., & Schou, J. 2004, "On the Determination of Michelson Doppler Imager High-Degree Mode Frequencies", Ap.J., 602, 481

[3] Korzennik, S. G., Rabello-Soares, M. C., Schou, J., & Larson, T.P. 20128, " Accurate Characterization of High-Degree Modes using MDI data", *submitted* to Ap.J., arXiv:1209.6414

[4] Rabello-Soares, M. C., Basu, S., Christensen-Dalsgaard, J., & Di Mauro, M. P. 2000, Sol.Phys., 193, 345

[5] Rabello-Soares, M. C., Korzennik, S. G., & Schou, J. 2008, "Analysis of MDI High-Degree Mode Frequencies and their Rotational Splittings", Sol.Phys., 251, 197

Acknowledgments

The Solar Oscillations Investigation - Michelson Doppler Imager project on SOHO is supported by NASA grant NAG5–8878 and NAG5–10483 at Stanford University. SOHO is a project of international cooperation between ESA and NASA. SGK was supported by Stanford contract PR–6333 and NASA grants NAG5–9819 and NNG05GD58G.

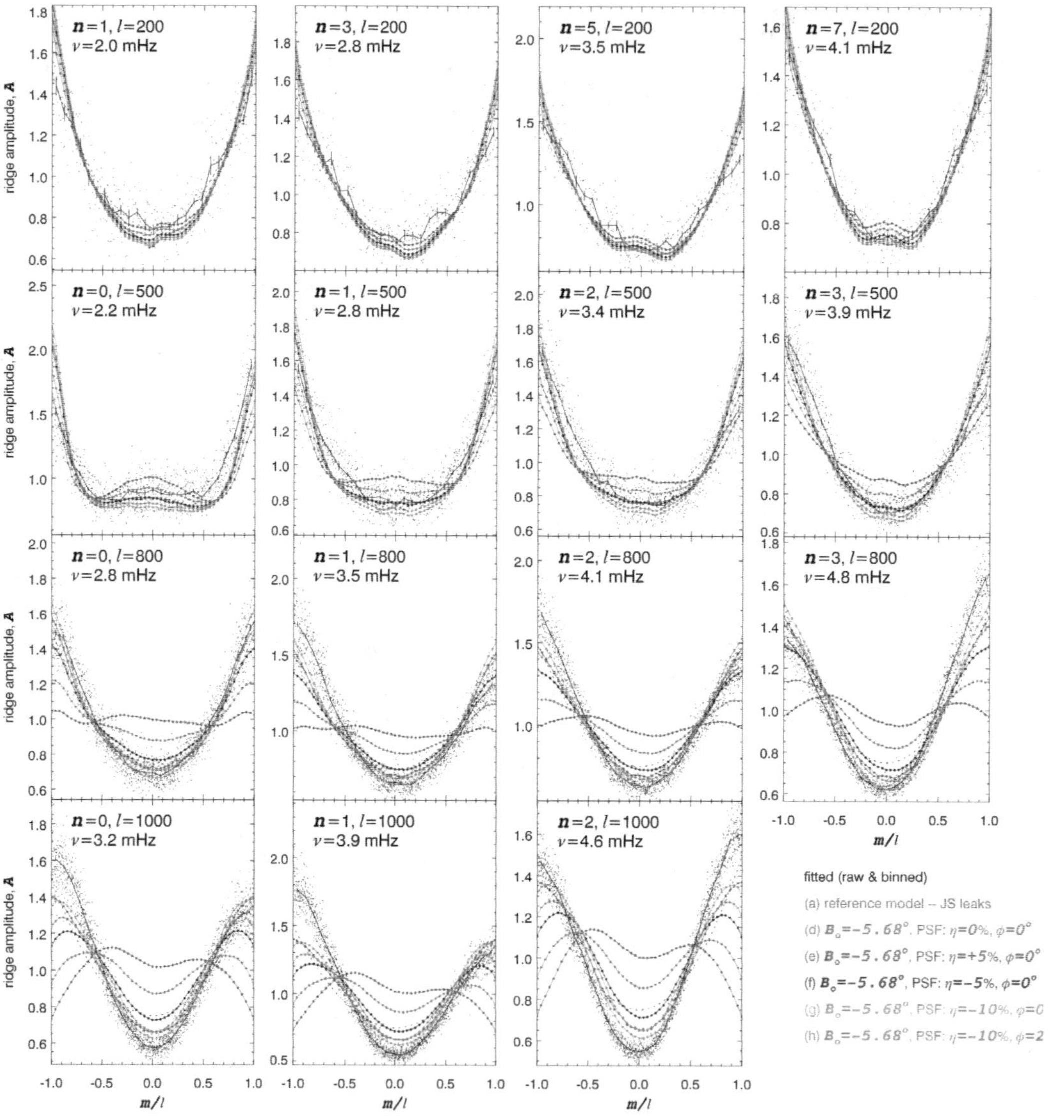

Figure 4. Comparison of ridge amplitude, as a function of the ratio m/ℓ. The dots correspond to individual ridge power amplitude measurements, and the black curves the corresponding binned values. The colored curves correspond to different models of the ridge amplitude, resulting from different models computed for different leakage matrices.

Eclipse on the Coral Sea: Cycle 24 Ascending (GONG 2012, LWS/SDO-5, and SOHO 27) IOP Publishing
Journal of Physics: Conference Series **440** (2013) 012017 doi:10.1088/1742-6596/440/1/012017

Revisiting the Issue of Solar Abundances

Sarbani Basu[1] **and H. M. Antia**[2]

[1] Department of Astronomy, Yale University, P. O. Box 20801, New Haven CT 06520-8101, U. S. A.

[2] Tata Institute of Fundamental Research, Homi Bhabha Road, Mumbai 400005, India

E-mail: sarbani.basu@yale.edu, antia@tifr.res.in

Abstract. We revisit the issue of solar abundances and examine whether the updated abundances result in solar models that have structures that agree with the structure of the Sun as determined by helioseismology. We quantify the changes in opacity required to bring the models constructed with the newer solar abundances in agreement with the Sun.

1. Introduction

The solar heavy element abundance, both total and relative, is still an unsettled quantity. For a considerable period of time, solar heavy element abundances as determined by Grevesse & Sauval (1998; henceforth GS98) were used. Solar models constructed with these abundances, that yield $Z/X = 0.023$, matched the helioseismically determined structure of the Sun quite well (see e.g., Basu & Antia 2008 and references therein).

Following updated analysis techniques that included the use of 3D model atmosphere, the solar metallicity was revised downwards by Asplund et al. (2005; henceforth AGS05) to $Z/X = 0.0165$. These changes resulted in solar models that did not match the structure of the Sun (Basu & Antia 2008 and references therein). With further improvement in analysis, abundances were updated by Asplund et al. (2009; henceforth AGSS09) to $Z/X = 0.018$. These abundances were quickly adopted by some members of the stellar astrophysics community since the 3D model atmospheres were considered to be better than the 1D atmospheres used earlier. Additionally, the oxygen and carbon abundances derived from different lines were consistent with each other, and brought the solar abundance in line with some of the young stars in the solar neighbourhood. However, others in the stellar astrophysics community did not adopt these abundances because they could not be used to model the Sun properly; additionally they argued that the difference between solar abundances and those in the young nearby stars could be a result of the fact that the Sun was not born in what is the current solar neighbourhood and hence a difference in abundances is not surprising.

An independent calculation using a different 3D model atmosphere found $Z/X = 0.0209$ (Caffau et al. 2010, 2011; henceforth CAF10). Another compilation by Lodders (2010; henceforth LOD10) resulted in $Z/X = 0.019$

We have constructed models with the different heavy element abundances and compared the results with the helioseismically determined structure of the Sun. We have relied on a mix of full solar models obtained by evolving the Sun from the zero-age main sequence, as well as envelope models of the Sun. Envelope models of the Sun have the advantage that they can be constructed with a specified position of the convection-zone (CZ) base and a specified helium abundance.

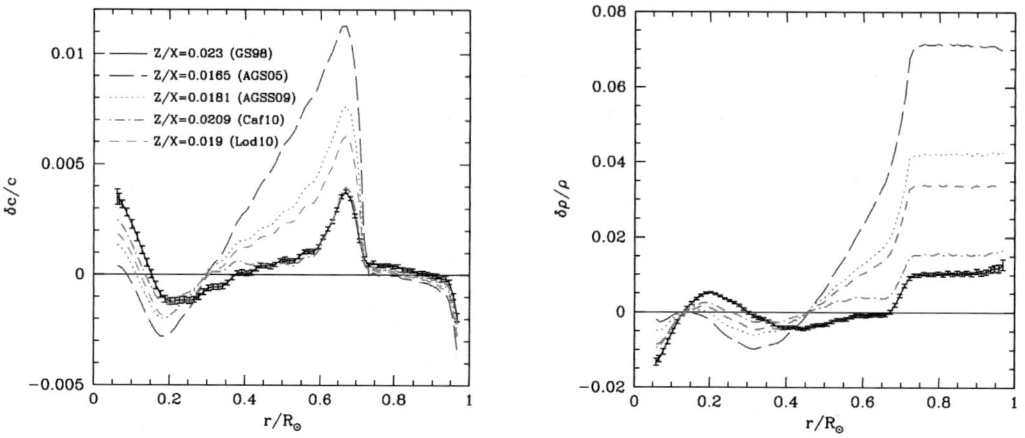

Figure 1. The sound-speed and density differences between the Sun and standard solar models constructed with different heavy-element abundances. The differences are shown by the lines and the ticks show 1σ errorbars. For the sake of clarity, errorbars are shown only for one result.

These envelope models were used to estimate the extent by which current opacities need to be changed to make models constructed with different metallicities yield the correct convection zone depth.

2. Results

In Fig. 1 we show the sound-speed and density differences between the Sun and solar models constructed with the different heavy element abundances mentioned above. These models were constructed with the Yale Stellar Evolution Code, YREC (Demarque et al. 2008) with OPAL equation of state (Rogers & Nayfonov 2002) and OPAL opacities (Iglesias & Rogers 1996) that were supplemented by low-temperature opacities of Ferguson et al. (2005). Nuclear reaction rates from Adelberger (1998) were used, except for the $^{14}\mathrm{N}(p,\gamma)^{15}\mathrm{O}$ reaction for which the rates of Formicola et al. (2004) were used. Note from the figure that the lower-metallicity models do not match the Sun very well. The GS98 and CAF10 models show a similar degree of agreement with the Sun despite the fact that the CAF10 abundance is lower, this is because the heavy element mixture of CAF10 is such that the opacity of the CAF10 model is similar to that of the GS98 model. Table 1 lists main properties of models. Note that the lower-abundance models have shallower convection zones compared with the Sun. They also have lower helium abundances.

We have also constructed a model with the CAF10 abundances, but with the abundance of Ne increased by a factor of 1.4. An increased Ne abundance has been suggested as a means to get low-Z models in concordance with the Sun (Antia & Basu 2005; Bahcall et al. 2005). In Fig. 2 we compare the sound-speed and density differences between the Sun and the CAF10 model with the Ne abundance enhanced by a factor of 1.4. Note that increasing Ne by this factor improves the density profile of the model, but over-corrects the sound-speed profile. A smaller Ne enhancement, say by a factor of 1.2 will be better as far as the sound-speed profile is concerned.

In Fig. 3 we have compared the dimensionless sound-speed gradient of the models with that of the Sun. This quantity, W(r) defined as

$$W(r) = \frac{r^2}{Gm}\frac{dc^2}{dr}, \tag{1}$$

Table 1. Properties of models with different heavy-element mixtures

Mixture	Z/X	$R_{\rm CZ}$	$Y_{\rm CZ}$	Y_0
Helioseismic	-	0.713 ± 0.001^a	0.2485 ± 0.0034^b	0.273 ± 0.006^c
GS98	0.023	0.7139	0.2456	0.2755
AGS05	0.0165	0.7259	0.2286	0.2586
AGSS09	0.018	0.7205	0.2352	0.2650
CAF10	0.0209	0.7150	0.2415	0.2711
LOD10	0.019	0.7136	0.2412	0.2665

[a] Basu & Antia (1997); [b] Basu & Antia (2004); [c] Serenelli & Basu (2010)

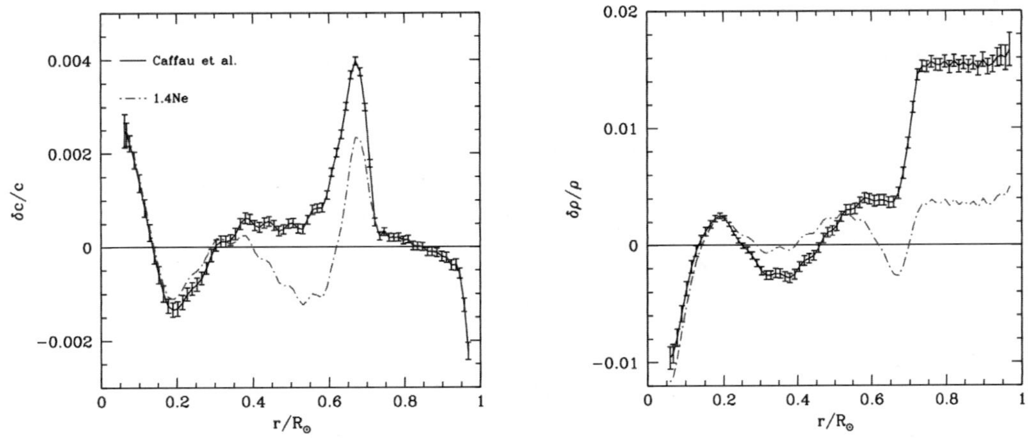

Figure 2. The sound-speed and density differences between the Sun and standard solar models. One model was constructed with the CAF10 heavy element abundance and the other with the CAF10 et al. abundances, but with the Neon abundance increased by a factor of 1.4. Again the difefrences are shown by the lines and the ticks show 1σ errorbars.

is sensitive to the changes in the adiabatic index in ionization zones of various elements. Here we use solar envelope models that were constructed to have the observed solar CZ base and the observed helium abundance. The envelope models were used to ensure that differences in convection zone depth does not contribute to the difference in $W(r)$. The solar value was obtained by inverting solar oscillation frequencies obtained by the GONG project (Hill et al. 1996) to determine the solar sound-speed and density profiles. As can be seen in the figure, the GS98 model shows the best agreement. The CAF10 model does not fare badly either, though enhancing Ne does not help as far as $W(r)$ is concerned. The other, lower-Z, models are not in agreement with the Sun.

The envelope models are constructed with the observed CZ depth and helium abundance and hence the sound speed profile in these models agrees well with that in the Sun, but the density profile in general doesn't match (see Fig. 4). By modifying the opacity near the base of the convection zone, it is possible to make the density profiles of the models also agree with that of the Sun (Basu & Antia 2004). We used the models shown in Fig. 4 to determine the factor by which opacity must change for any given composition so that the density profiles match that of the Sun. We modified opacity by a constant factor till the density of the model matched the inverted density profile in the lower part of the CZ. There is an error of about 2% in estimating

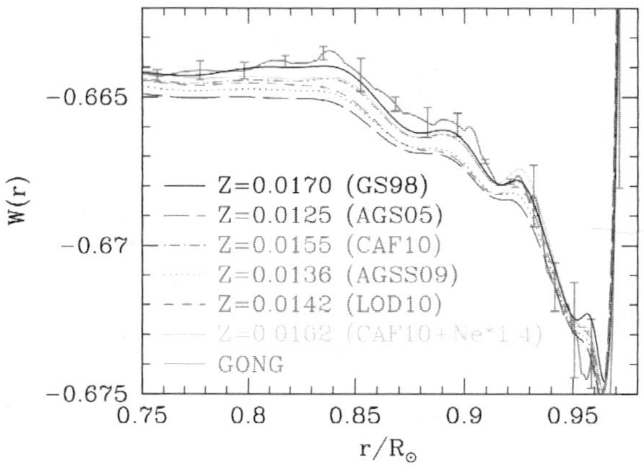

Figure 3. The dimensionless sound-speed gradient of models constructed with different compositions compared with the solar value as determined using GONG data.

this factor. The factors are listed in Table 2. We find that GS98 and CAF10 models can be made to agree with the Sun with very little change in opacities. The AGS05 models need a large change (26.5%) while even the updated AGSS09 models need a fairly large change (16.5%).

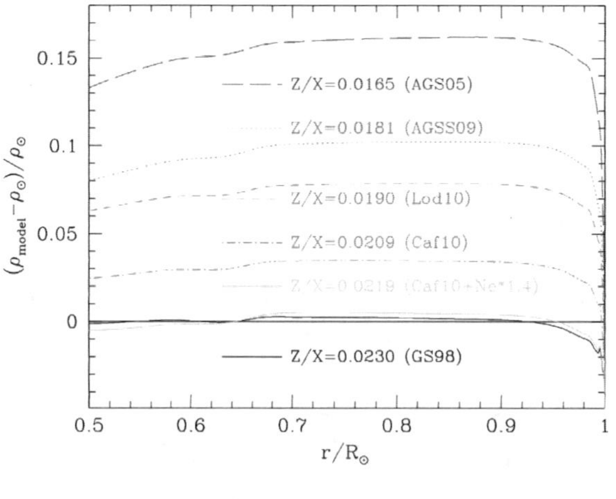

Figure 4. The density difference between solar envelope models and the Sun. All models have the correct CZ base and helium abundance and hence the sound-speed profiles match that of the Sun by construction. The density profiles can be made to match the Sun's by modifying the opacity.

3. Summary

We find that solar models constructed with the GS98 abundances continue to have the best match with helioseismically determined solar properties. Of the more recent solar abundance determinations, the CAF10 abundances result in models that have reasonable properties in terms of the position of the convection-zone base, the convection-zone helium abundances and initial helium abundances. The sound-speed and density profiles are comparable to those of the GS98 models. We also find that enhancing the Neon abundance by a factor of 1.4 in the CAF10

Table 2. Factor by which OPAL opacities have to be multiplied to bring density profiles of models shown in Fig. 4 to match that of the Sun

Model	Opacity factor
GS98	1.005
AGS05	1.265
AGSS09	1.165
CAF10	1.060
LOD10	1.125
CAF10+1.4Ne	1.010

models improves the density profile and other properties, however, the sound-speed profile in the radiative zone is over-corrected. A smaller enhancement, by a factor of about 1.2 will probably work better as far as the sound-speed profile is concerned. GS98 and CAF10 models can be made to match solar properties better with very little change in opacities. The other abundances require a larger change. Increasing the amount of Neon in the CAF10 mixture decreases the need to modify opacities.

Acknowledgements
This work utilizes data obtained by the Global Oscillation Network Group (GONG) project managed by the National Solar Observatory, which is operated by AURA, Inc. under a cooperative agreement with the National Science Foundation. The data were acquired by instruments operated by the Big Bear Solar Observatory, High Altitude Observatory, Learmonth Solar Observatory, Udaipur Solar Observatory, Instituto de Astrofisico de Canarias, and Cerro Tololo Inter-American Observatory. This work was supported by NSF grant AST 1105930 to SB.

References

[1] Adelberger, E. G., Austin, S. M., Bahcall, J. N. et al. 1998 *Rev. Mod. Phys.* **70**, 1265
[2] Antia H. M. & Basu S. 2005 *ApJ* **620**, L129
[3] Asplund M., Grevesse N. & Sauval A. J. 2005 *ASPCS* **336**, 25 (astro-ph/0410214)
[4] Asplund M., Grevesse N., Sauval A. J. & Scott, P. 2009 *ARAA* **47**, 481
[5] Bahcall J. N., Basu S. & Serenelli A.M. 2005 *ApJ* **631**, 1281
[6] Basu S. & Antia, H. M. 1997 *MNRAS* **287**, 189
[7] Basu, S. & Antia, H. M. 2004 *ApJ* **606**, L85
[8] Basu, S. & Antia, H. M. 2008 *Phys. Rep.* **457**, 217
[9] Caffau E. et al. 2010 *A&A* **514**, 92
[10] Caffau E., Ludwig, H.-G. Steffen, M. Freytag, B. & Bonifacio, P. 2011 *Solar Phys.* **268**, 255
[11] Demarque P., Guenther D. B., Li L. H., Mazumdar A., & Straka C. W. 2008, *ApSS* **316**, 31
[12] Ferguson J. W. et al. 2005 *ApJ* **623**, 585
[13] Formicola A. et al. 2004 *Phys. Let. B* **591**, 61
[14] Grevesse N., & Sauval A. J. 1998 *Space Sci. Rev* **85**, 161
[15] Hill F. et al. 1996 *Science* **272**, 1292
[16] Iglesias C. A., & Rogers F. J. 1996 *ApJ* **464**, 943
[17] Lodders K. 2010, in Principles and Perspectives in Cosmochemistry, Astrophysics and Space Science Proceedings., p 379
[18] Rogers F. J. & Nayfonov A. 2002 *ApJ* **576**, 1064
[19] Serenelli A.M. & Basu S. 2010 *ApJ* **719**, 865

Journal of Physics: Conference Series **440** (2013) 012018 doi:10.1088/1742-6596/440/1/012018

Contrasting the solar rotation rate of cycles 23 and 24

H. M. Antia[1] and Sarbani Basu[2]

[1] Tata Institute of Fundamental Research, Homi Bhabha Road, Mumbai 400005, India
[2] Department of Astronomy, Yale University, P. O. Box 20801, New Haven CT 06520-8101, U. S. A.

E-mail: antia@tifr.res.in, sarbani.basu@yale.edu

Abstract. The minimum between solar cycles 23 and 24 was quite unusual compared with other minima for which detailed data are available and this pointed to the possibility that cycle 24 will be unusual. Cycle 24 is almost at its maximum now and we take this opportunity to compare and contrast the solar rotation rate and zonal flows between the two cycles. We find that the rotation rate during cycle 24 is slightly lower than that during cycle 23. Additionally we find that the poleward branch of the zonal flow that is believed to be the harbinger of the next solar cycle is very week in cycle 24.

1. Introduction

The period of solar activity minimum before the onset of cycle 24 was unusual in its duration and depth. The cycle 24 minimum was longer that of cycle 23; it was quieter as measured by the 10.7 cm flux; the polar fields were weaker (Lo et al. 2010) and the solar winds were slower (Manoharan 2012). The differences were present inside the Sun too, particularly with regards to solar zonal flows (Howe et al 2009; Antia & Basu 2010) as well as meridional flows (Basu & Antia 2010).

The Global Oscillation Network Group (GONG) and the MDI instrument on board SOHO have been collecting helioseismic data since the minimum preceding solar cycle 23. The data sets span cycle 23 in its entirety as well has the ascending part of cycle 24. These data sets provide the means of determining the rotation rate in the solar interior, and we use these data to compare and contrast the solar rotation rate during cycles 23 and the ascending part of cycle 24. We use odd order splitting coefficients from GONG data (Hill et al. 1996) and MDI data (Schou 1999) for inferring the rotation rate in the solar interior.

In this work we use 170 data sets obtained by GONG. Each set covers a period of 108 days with a spacing of 36 days between consecutive data sets; the sets thus overlap in time. The data cover a period from May 1995 to April 2012. Additionally we use 73 sets from MDI. Each MDI set covers a period of 72 days and the sets do not overlap. The MDI data sets cover a period from May 1996 to April 2011, with a gap around 1998 when contact with the SOHO satellite was lost. We use the 2D Regularized Least Squares (2D RLS) technique as described by Antia et al. (1998) for determining the rotation rate.

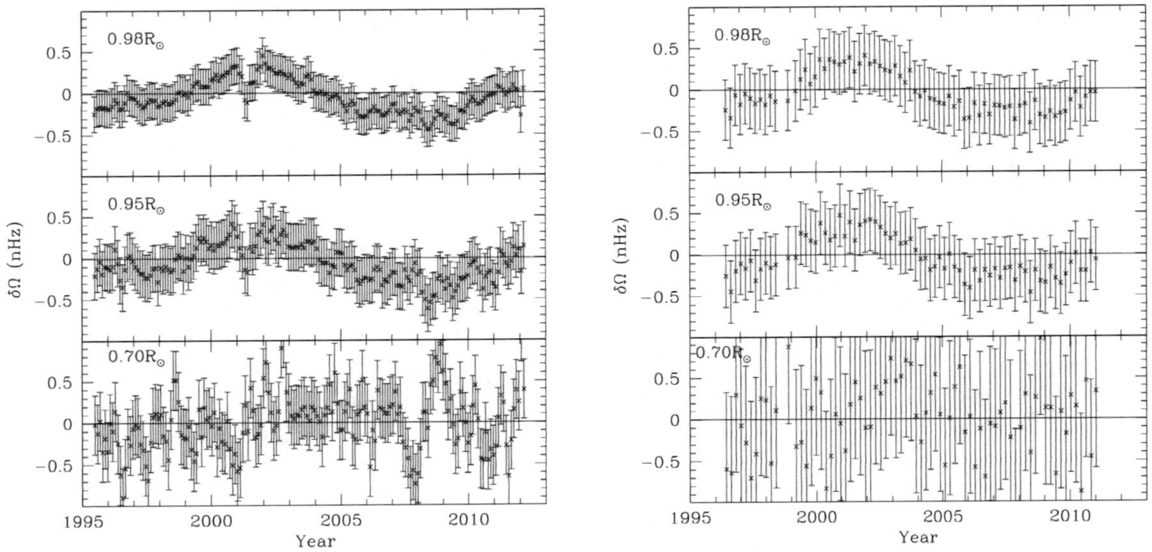

Figure 1. The change in the latitudinally independent component of the solar rotation rate plotted as a function of time obtained with GONG (left) and MDI (right) data.

2. Results

In Fig. 1 we show the change in the latitude-independent part of the solar rotation rate at three different depths. The latitude-independent part of the rotation rate was obtained by inverting the splitting coefficient a_1. The changes were obtained by subtracting the average rotation rate from the rotation rate at each epoch. Note that there appears to be a solar-cycle related change, with the fastest rotation at the cycle 23 solar maximum and the slowest during the two minima. Also note that the rotation rate is slower during cycle 24 than it was in cycle 23, in particular the solar rotation rate as shown by the latest data point which is close to the cycle 24 maximum is lower than the rate at cycle 23 maximum, at least in the outer layers in the Sun. The data are too noisy in deeper layers to determine whether or not the rotation rate in the deeper layers changed too. The results with GONG data show some large variations at the cycle 23 maximum which the MDI data do not, that could be due to some systematic differences introduced during the upgrade of GONG instruments. Although, $\delta\Omega$ is comparable to errorbars at individual points, if we take average over neighbouring points the value will be significant. Further, the changes in the latitude-independent part of the solar rotation rate correlate well with the 10.7 cm flux, which is a proxy for solar activity. This can be seen in Fig. 2.

The pattern of the solar zonal flows is also different between the two cycles. We define the zonal flow rate as

$$\delta\Omega(r,\theta,t) = \Omega(r,\theta,t) - \langle\Omega(r,\theta,t)\rangle, \tag{1}$$

where Ω is the rotation rate, r, the radius, θ the latitude and t time. $\langle\Omega(r,\theta,t)\rangle$ is the time average of the rotation rate. From the zonal flow rate we can define the zonal flow velocity as

$$\delta v_\phi = \delta\Omega r \cos\theta. \tag{2}$$

Thus the zonal flow is simply the residual of the rotation when the average rotation is removed.

In Fig. 3 we plot the solar zonal flows at $0.98R_\odot$ as a function of time. We show two versions of the flow velocity. The results in the top panel were obtained using Eqs. 1 and 2 with $\langle\Omega(r,\theta,t)\rangle$ defined as the average over cycle 23, the lower panel shows results with $\langle\Omega(r,\theta,t)\rangle$ defined as the average over cycle 24. The higher rotation rate of cycle 23 washes out some of the weaker

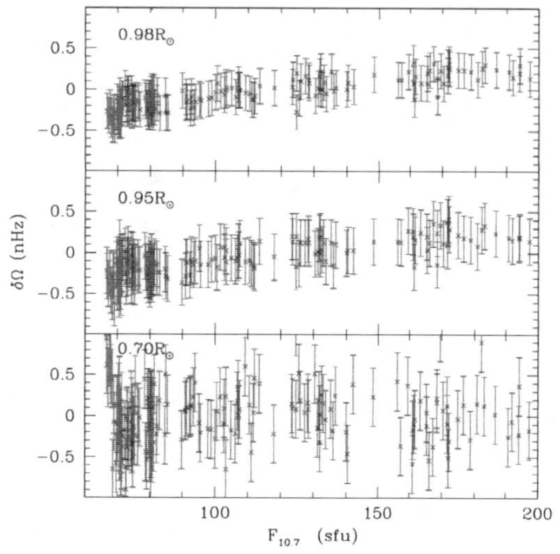

Figure 2. The change in the latitude-independent part of the solar rotation rate plotted as a function of the 10.7 cm flux. The blue points are results for cycle 23 and the red points for cycle 24. Only GONG results are shown.

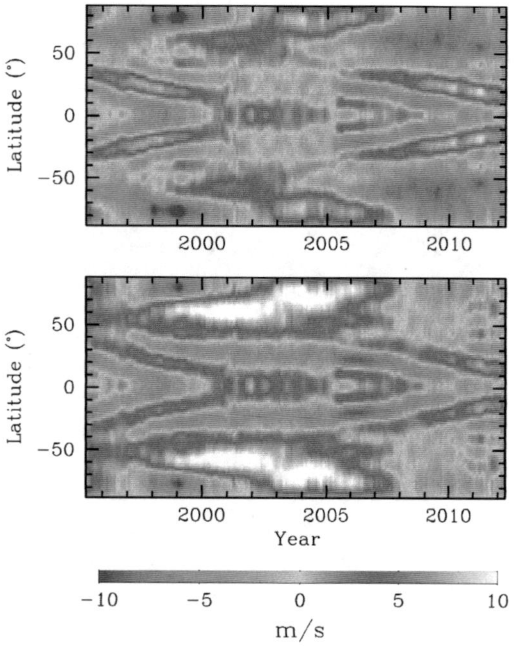

Figure 3. The zonal flow velocity at $0.98R_\odot$ as determined using GONG data plotted as a function of latitude and time. MDI results are similar. The top panel was obtained by subtracting the average of the cycle 23 rotation velocities, while the lower panel is that obtained by subtracting the cycle 24 average rotation.

features of the zonal flow during cycle 24, in particular we cannot see the high-latitude poleward branch of the zonal flow during cycle 24, this had earlier led to speculation that this might mean that cycle 25 may be delayed (Hill et al. 2011). The weak polar branch is seen more clearly in the lower panel of Fig. 3. As is clear, the flow is much weaker than the corresponding flow during cycle 23. It should be noted that we have data for only a part of cycle 24 and hence the average rotation rate during this period is heavily biased by the values during the deep and extended minimum. This will also affect the pattern in lower panel of Fig. 3.

Plots of the zonal flow seen at different latitudes plotted as a function of radius and time show other differences between the flows during cycles 23 and 24, as can be seen in Fig. 4. Basu & Antia (2003) found that during cycle 23, the low latitude flows appears to rise from the deeper layers to the surface at a speed of about $0.05R_\odot$ per year. The signature of this rise is different in cycle 24, in particular, at 10° latitude, the flow stalled between 0.8 and $0.9R_\odot$ before beginning

Journal of Physics: Conference Series **440** (2013) 012018 doi:10.1088/1742-6596/440/1/012018

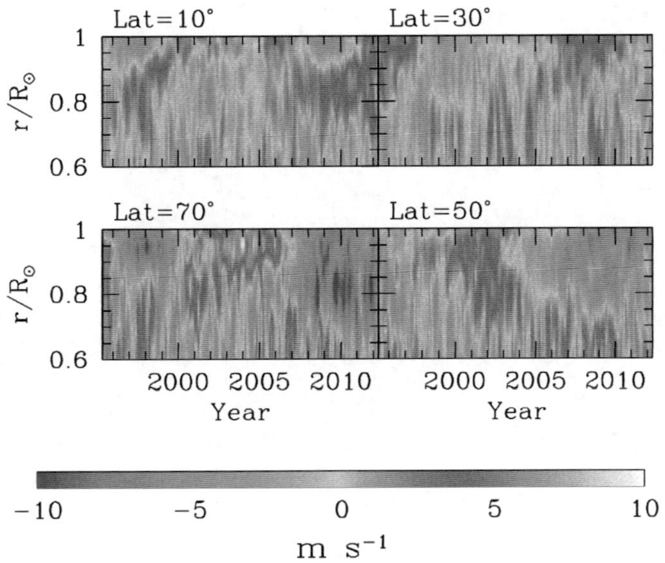

Figure 4. The zonal flow velocity plotted as a function of radius and time for a few latitudes. Only GONG results are shown.

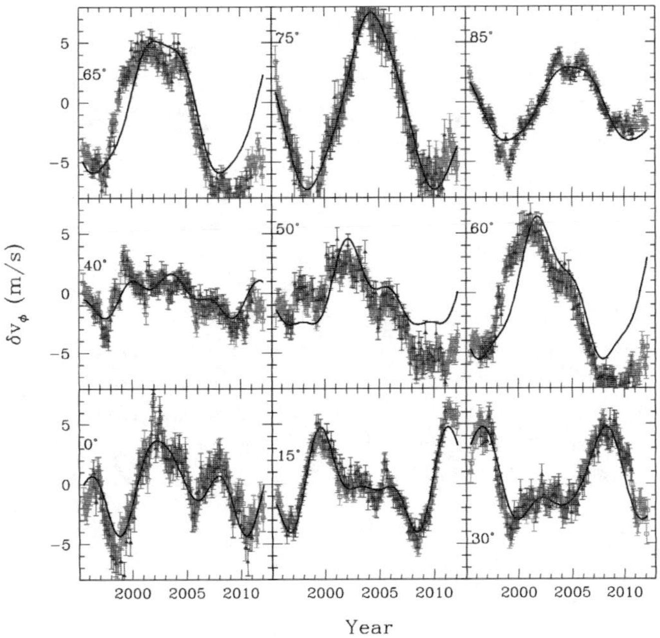

Figure 5. The zonal flow velocity at $0.98R_\odot$ plotted as a function of time for different latitudes. The blue points are MDI results, the red are GONG results and the black line is the fit to the GONG data using the function shown in Eq. 3. Note that while the low and high latitude flows fit well to the function over both cycle 23 and 24, at mid latitudes, we cannot fit cycles 23 and 24 simultaneously.

its rise.

We can examine differences in the zonal flows between the two cycles in more detail by fitting the zonal flow (i.e. rotation rate residuals) with a periodic function and examining if the same form fits both cycles. Antia & Basu (2010) had found that the cycle 23 zonal flows could be fitted to a form given by

$$\delta v(r, \theta, t) = \sum_{k=1}^{k=3} A_k \sin(k\omega_0 t + \phi_k), \qquad (3)$$

with $\omega_0 = 11.7$ years. In the above expression A_k is the amplitude of the kth component and ϕ_k a phase. We show the same fit to the zonal flows in Fig. 5. We can see that while the same value of ω_0 fits the low and high latitude flows in both cycles 23 and 24, at intermediate latitudes there are deviations from this form during cycle 24 and the two cycles cannot be fitted simultaneously. Further, in the intermediate latitude range (50°–65°) the rotation rate during cycle 24 is lower than that during the same phase of cycle 23.

3. Summary

We have used available rotational splitting data from the GONG and MDI projects to compare the solar rotation rate and zonal flows during solar cycle 24 with that of solar cycle 23. We find that latitude-independent part of the solar rotation rate of cycle 24 is lower than that of cycle 23, at least in the outer layers of the Sun. This component of the rotation rate shows a correlation with the 10.7 cm flux, which is a solar activity indicator. While examining the zonal flows we find that the poleward, prograde flow that is the precursor of cycle 25 is very weak compared with the corresponding flow that was the precursor of the current cycle. While the zonal flows at all latitudes in cycle 23 can be fitted with three harmonics of an 11.7 year period, we cannot simultaneously fit the mid-latitudes of cycle 24. However, the low and high latitudes of cycle 23 and 24 can be fitted simultaneously

Acknowledgements

This work utilises data obtained by the Global Oscillation Network Group (GONG) project, managed by the National Solar Observatory, which is operated by AURA, Inc. under a cooperative agreement with the National Science Foundation. The data were acquired by instruments operated by the Big Bear Solar Observatory, High Altitude Observatory, Learmonth Solar Observatory, Udaipur Solar Observatory, Instituto de Astrofisico de Canarias, and Cerro Tololo Inter-American Observatory. This work also utilises data from the Solar Oscillations Investigation/ Michelson Doppler Imager (SOI/MDI) on the Solar and Heliospheric Observatory (SOHO). SOHO is a project of international cooperation between ESA and NASA. SB acknowledges support from NASA grant NXX10AE60G.

References

[1] Antia H M and Basu S 2010 *ApJ* **720**, 494
[2] Antia H. M., Basu S. and Chitre S. M. 1998 *MNRAS* **298**, 543
[3] Basu S and Antia H M 2003 *ApJ* **585**, 553
[4] Basu S and Antia H M 2010 *ApJ* **717**, 488
[5] Hill, F. et al. 1996 *Science* **272**, 1292
[6] Hill, F. et al. 2011 *BAAS* **43**, 16.10
[7] Howe R et al. 2009 *ApJ*, **701**, L87
[8] Lo L, Hoeksema J T and Scherrer P 2010 *ASPCS* **428**, 109
[9] Manoharan P K 2012 *ApJ* **751**, 128
[10] Schou, J. 1999 *ApJ* **523**, L181

Journal of Physics: Conference Series **440** (2013) 012019 doi:10.1088/1742-6596/440/1/012019

Can we detect local helioseismic parameter shifts in coronal holes?

R Howe[1], D A Haber[2], R S Bogart[3], S Zharkov[4,5], D Baker[4], L Harra[4] and L van Driel-Gesztelyi[4]

[1]School of Physics and Astronomy, University of Birmingham, UK
[2]University of Colorado, Boulder, CO, USA
[3]Stanford University, USA
[4]UCL/MSSL, United Kingdom
[5]Department of Physics and Mathematics, University of Hull, UK

E-mail: `rhowe@nso.edu`

Abstract. Changes in helioseismic mode parameters in active regions and across the solar disk are well documented, but local magnetic activity and geometric effects may not account for all of the scatter seen in the results. We use results from the *Helioseismic and Magnetic Imager* ring-diagram pipeline for Carrington rotation 2113 to look for differences in mode amplitude and frequency between coronal holes and other quiet-Sun regions. While we do not find a systematic difference, the results do suggest that the correlation between magnetic activity index and mode parameters shows less scatter in coronal hole regions than in general quiet Sun.

1. Introduction

Helioseismic mode parameters in the five-minute frequency range have long been known to correlate (frequency, linewidth) and anticorrelate (amplitude) with surface magnetic activity indices at a wide range of scales. However, the indices do not account for all of the variation and scatter. Here we investigate the possible contribution from coronal holes effects to the local mode properties.

2. Data

2.1. Helioseismic Data

We use ring-diagram parameters (specifically mode frequency and the logarithmic mode amplitude) from the *Helioseismic and Magnetic Imager* (HMI) ring diagram pipeline [1] for 5-degree tiles. These tiles, each tracked at the Carrington rate for an 8-hour period, are processed for the full solar disk using both the "rdfitf" [2] (symmetric) and "rdfitc" [3] (asymmetric) peak-fitting algorithms. Each patch has a magnetic activity index [MAI] calculated from the total unsigned magnetic flux in the patch as measured by HMI, averaged across the observing interval. The available data cover Carrington rotations 2096 to 2126, but for this report we concentrate on the rdfitf fits for C.R. 2113, which shows a good mixture of coronal holes and active regions. For the purposes of this analysis, we first correct the mode parameters for disk-position effects by subtracting a disk-position average (excluding patches with magnetic index greater than 10 G) over the whole Carrington rotation from the parameters at each position; we then form a

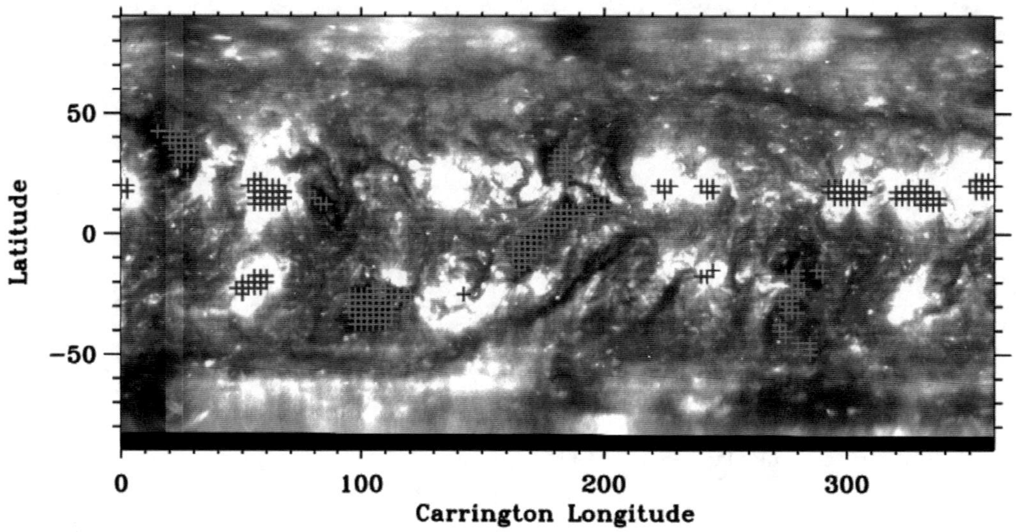

Figure 1. The "coronal hole" areas in red and the areas with MAI $\geq$ 30 G in blue, superimposed on the EIT 195 Å intensity, for Carrington rotation 2113.

synoptic chart by averaging the parameters for each location in heliographic latitude/longitude across the whole disk passage of the location.

2.2. Coronal Hole Identification

The identification of coronal holes is a research topic in itself and beyond the scope of this work. We use the NSO GONG[1] and SOLIS[2] preliminary synoptic maps of open-field regions based on PFSS modeling of synoptic magnetograms, together with EIT synoptic charts[3] of EUV 195 Å intensity, to locate regions that can be considered as coronal holes. We designate a ring-diagram patch as being in a coronal hole if it has an MAI less than 30 G, has non-zero values in either of the two PFSS maps, and has an EIT 195 Å intensity less than the median value for the rotation. Because of projection difficulties with the high-latitude measurements, we concentrate on features within ±50 deg of the Equator. Figure 1 shows the active (MAI $\geq$ 30 G) and coronal-hole regions superimposed on the EIT 195 Å intensity map for C.R. 2113.

3. Results

For illustrative purposes, we select two peaks in the ring-diagram spectrum; $l = 660, n = 1$, which lies close to the center of the five-minute p-mode spectrum, and $l = 840, n = 3$, which has a frequency close to the acoustic cutoff. Figures 2 and 3 show synoptic charts of the peak frequency and amplitude for each case over C.R. 2113, and Figure 4 shows scatter plots of the same data as a function of MAI in the quiet-Sun range, with results from the coronal-hole patches indicated by color.

As seen in earlier work [4–7], and in many global-mode studies, for example [8–10], in the five-minute band the tendency is for frequencies to be positively, and amplitudes negatively,

[1] http://gong.nso.edu/data/magmap/
[2] ftp://solis.nso.edu/synoptic/
[3] http://lasco-www.nrl.navy.mil/carr_maps/eit/_synomap/

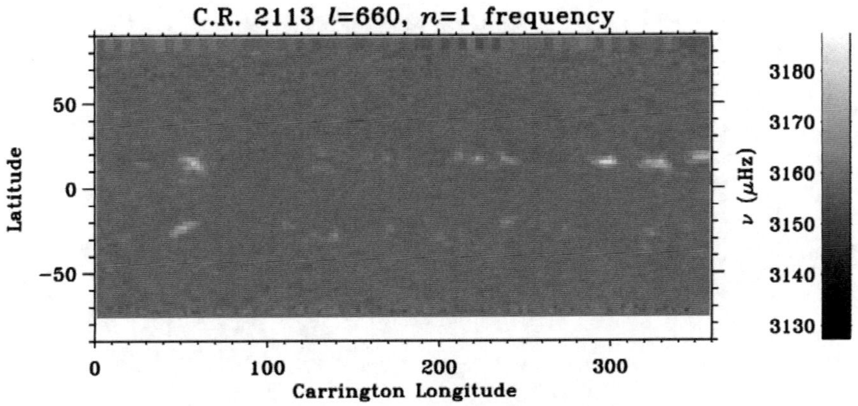

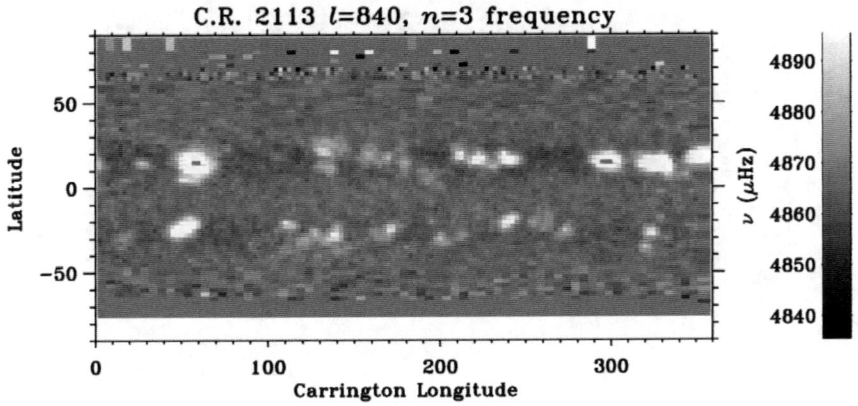

Figure 2. Maps of frequency shifts for rdfitf fits to two sample peaks – one (top) in the middle of the 5-min band and one (bottom) close to the acoustic cutoff.

correlated with the magnetic activity index. In the case of mode amplitude, this trend also applies to power maps at the pixel-by-pixel level, *e.g.* [11; 12]. Closer to the acoustic cutoff, while the frequency continues to increase in the range covered by the fits to the 5-degree tiles, the trend in amplitude reverses; detailed examination of power maps reveals a "halo" of enhanced power surrounding an active region. These haloes may arise from the interaction of waves in the solar atmosphere with inclined magnetic fields around active regions. This halo effect is not easily seen in ring-diagram measurements with the conventional 15-degree tiling, but the five-degree tiles give a fine enough spatial resolution to reveal the halo. For the higher-frequency case, therefore, the direct anti-correlation between activity level and amplitude breaks down to some extent and the plots show more scatter. In each case, the coronal-hole data points lie within the "cloud" of general quiet-Sun measurements; however, there does seem to be a tendency for the coronal-hole measurements to be somewhat less scattered.

4. Discussion

The five-degree ring-diagram tiles are useful for seeing the fine details of changes to local parameters around active regions, such as the high-frequency haloes. The improved quality of HMI magnetograms compared to those from MDI allows us to look at the effects of relatively weak fields. So far we have found no obvious signature of coronal holes in the frequency and

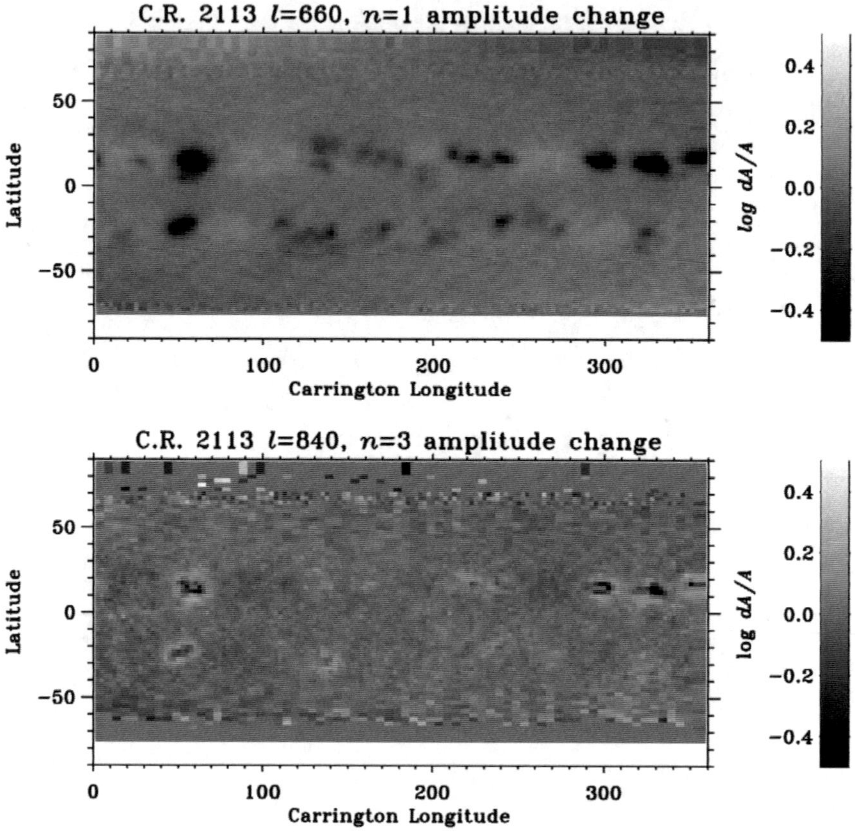

Figure 3. Maps of amplitude shifts for rdfitf fits to two sample peaks – one (top) in the middle of the 5-min band and one (bottom) close to the acoustic cutoff.

mode-amplitude parameters. However, the scatter in the coronal hole areas does seem to be marginally smaller than for general quiet Sun. This would make sense in terms of simpler field geometry and less emerging flux in coronal hole regions. In future work we hope to examine this question in more detail using data from many solar rotations.

Acknowledgments

RH acknowledges computing support from NSO. We acknowledge the Leverhulme Trust for funding the "Probing the Sun: inside and out" project upon which this research is based. HMI data courtesy of SDO (NASA) and the HMI consortium. This work utilizes GONG and SOLIS data obtained by the NSO Integrated Synoptic Program (NISP), managed by the National Solar Observatory, which is operated by the Association of Universities for Research in Astronomy (AURA), Inc. under a cooperative agreement with the National Science Foundation.

References

[1] Bogart R S, Baldner C, Basu S, Haber D A and Rabello-Soares M C 2011 *Journal of Physics Conference Series* **271** 012008

[2] Haber D A, Hindman B W, Toomre J, Bogart R S, Thompson M J and Hill F 2000 *Solar Phys.* **192** 335–350

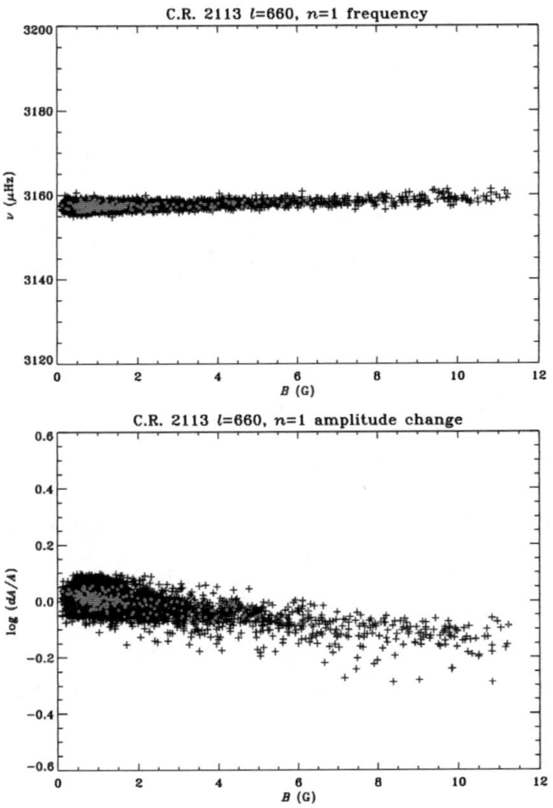

Figure 4. Scatter plots of frequency (top) and amplitude (bottom) change against MAI for low-activity regions in C.R. 2113. The red symbols represent patches in the coronal hole areas and the black symbols other patches in the same MAI range, for sample peaks in the 5-min band (left) and at the acoustic cutoff (right). Both sets of data are restricted to latitudes below 50 degrees.

[3] Basu S and Antia H M 1999 *Astrophys. J.* **525** 517–523 (*Preprint* arXiv:astro-ph/9906252)

[4] Hindman B, Haber D, Toomre J and Bogart R 2000 *Solar Phys.* **192** 363–372

[5] Rajaguru S P, Basu S and Antia H M 2001 *Astrophys. J.* **563** 410–418 (*Preprint* astro-ph/0108227)

[6] Howe R, Komm R W, Hill F, Haber D A and Hindman B W 2004 *Astrophys. J.* **608** 562–579

[7] Howe R, Komm R W and Hill F 2002 *Astrophys. J.* **580** 1172–1187

[8] Libbrecht K G and Woodard M F 1990 *Nat.* **345** 779–782

[9] Elsworth Y, Howe R, Isaak G R, McLeod C P, Miller B A, New R, Speake C C and Wheeler S J 1994 *Astrophys. J.* **434** 801–806

[10] Komm R, Howe R and Hill F 2002 *Astrophys. J.* **572** 663–673

[11] Jain R and Haber D 2002 *Astron. Astrophys.* **387** 1092–1099

[12] Howe R, Jain K, Bogart R S, Haber D A and Baldner C S 2012 *Solar Phys.* **281** 533–549 (*Preprint* 1208.1644)

Journal of Physics: Conference Series **440** (2013) 012020 doi:10.1088/1742-6596/440/1/012020

Towards solar activity maximum 24 as seen by GOLF and VIRGO/SPM instruments

R. A. García[1], D. Salabert[2], S. Mathur[3,4], C. Régulo[5,6], J. Ballot[7,8], G.R. Davies[1,9], A. Jiménez[5,6], R. Simoniello[1]

[1] Laboratoire AIM, CEA/DSM-CNRS-Université Paris Diderot; CEA, IRFU, SAp, F-91191, Gif-sur-Yvette, France

[2] Laboratoire Lagrange, UMR7293, Université de Nice Sophia-Antipolis, CNRS, Observatoire de la Côte d'Azur, Bd. de l'Observatoire, 06304 Nice, France

[3] High Altitude Observatory, 3080 Center Green Drive, Boulder, CO, 80302 USA

[4] Space Science Institute, 4750 Walnut Street, Suite 205, Boulder, Colorado 80301 USA

[5] Instituto de Astrofísica de Canarias, E-38200 La Laguna, Tenerife, Spain

[6] Dept. de Astrofísica, Universidad de La Laguna, E-38206 La Laguna, Tenerife, Spain

[7] CNRS, Institut de Recherche en Astrophysique et Planétologie, 14 avenue Edouard Belin, 31400 Toulouse, France

[8] Université de Toulouse, UPS-OMP, IRAP, 31400 Toulouse, France

[9] School of Physics and Astronomy, University of Birmingham, Edgbaston, Birmingham, B15 2TT, UK

E-mail: rgarcia@cea.fr, salabert@oca.eu, smathur@SpaceScience.org, crr@iac.es, Jerome.ballot@irap.omp.eu, davies@bison.ph.bham.ac.uk, ajm@iac.es, Rosaria.Simoniello@cea.fr

Abstract. All p-mode parameters vary with time as a response to the changes induced by the cyclic behavior of solar magnetic activity. After the unusual long solar-activity minimum between cycles 23 and 24 –where the p-mode parameters have shown a different behavior than the surface magnetic proxies– we analyze the temporal variation of low-degree p-mode parameters measured by GOLF (in velocity) and VIRGO (in intensity) Sun-as-a-star instruments on board SoHO. We compare our results with other activity proxies.

1. Introduction

GOLF and VIRGO/SPM instruments have been observering since 1996 and are well suited to the study of low-degree, solar-acoustic mode characteristics [1]. Thanks to their high-quality observations, we are able to measure all the p-mode properties with very high precision not previously attainable, including mode asymmetries [2]. Moreover, the temporal variations of the p-mode parameters during cycle 23 can be studied [3, 4, 5]. The unexpected long activity minimum between cycles 23 and 24 [6] demonstrates that the physical processes governing the magnetic activity in the Sun are not yet well understood.

2. Observations and Data Analysis

We analyzed observations collected by the space-based GOLF [7] and VIRGO [8] instruments onboard SoHO. A total of 6000 days were analyzed covering nearly 16.5 years between 1996 and 2012. These datasets were split into contiguous 365-day subseries, with a one-fourth overlap. The power spectrum of each subseries was fitted to extract the mode parameters using a standard likelihood maximization function (power spectrum with a χ^2 with 2 d.o.f. statistics). Each mode component was parameterized using an asymmetric Lorentzian profile [9]. The temporal variations of the frequency shifts were defined as the difference between reference values (taken as the average over 1996-1997) and the parameters of the corresponding modes observed at different dates. Subseries with duty cycles less than 90% (around the SoHO vacation) were not taken into account for this analysis. The weighted averages over the central part of the 5-min oscillation power –from 2200 to 3400 μHz– of the temporal variations of the mode parameters were then calculated. Mean values of daily measurements of the 10.7-cm radio flux were used as a proxy of the solar surface activity. Linear regressions were performed between the temporal variations of the mode parameters and the radio flux using independent points only.

3. Frequency Variations of Individual Low-Degree p Modes

In Fig. 1 we show the average temporal variations of the l=0,1, and 2 p-mode frequencies observed by GOLF and VIRGO following [10]. The 11-year solar cycle is clearly visible with a quasi-biennial superimposed, originally described by [11], and fully discussed in [12, 13]. Mode frequency shifts are anticorrelated with their amplitude and no significant temporal variations have been found in the rotational splittings, in agreement with [14, 15].

The Pearson linear correlation coefficients and the probabilities of having no correlation between the temporal variations of the p-mode parameters measured in the photometric VIRGO data and the 10.7-cm radio flux are given in Table 1. These variations are linked to the cyclic changes of Sun's magnetic activity.

Table 1. Pearson correlation coefficients (r) and probabilities of having no correlation ($P_{r=0}$) between the temporal variations of the mode frequencies ($\langle\Delta\nu\rangle$), heights ($\langle\Delta h\rangle$), linewidths ($\langle\Delta\gamma\rangle$), and asymmetries ($\langle\Delta b\rangle$) measured from the blue channel VIRGO data and the corresponding $F_{10.7}$ radio flux used as a magnetic activity proxy.

$\langle\Delta\nu\rangle$		$\langle\Delta h\rangle$		$\langle\Delta\gamma\rangle$		$\langle\Delta b\rangle$	
r	$P_{r=0}$	r	$P_{r=0}$	r	$P_{r=0}$	r	$P_{r=0}$
0.97	4.76×10^{-9}	0.94	3.56×10^{-7}	0.93	1.10×10^{-6}	0.69	1.26×10^{-3}

4. Mode Excitation and Damping with Solar Activity

The temporal variations of the mode amplitudes, $<\Delta h>$, and linewidths, $<\Delta\gamma>$, are shown in Fig. 2 computed using the three VIRGO/SPM channels. The Pearson linear correlation coefficients with the 10.7-cm radio flux are given in Table 1. Note that due to absolute calibration problems and the changes of the observing wings [16], the GOLF amplitudes and linewidths are not exploitable in this analysis. A proper calibration is currently underway.

In Fig. 3, we show the variations of A_{max} [17] as it is usually done in asteroseismology to track down activity cycles [18]. The maximum power at ν_{max} was computed by fitting a Gaussian on the p-mode envelope, as explained in [19], and converted to bolometric amplitude following [20]. The results are comparable to the average of the individual mode amplitudes as shown in Fig.2a.

Journal of Physics: Conference Series **440** (2013) 012020 doi:10.1088/1742-6596/440/1/012020

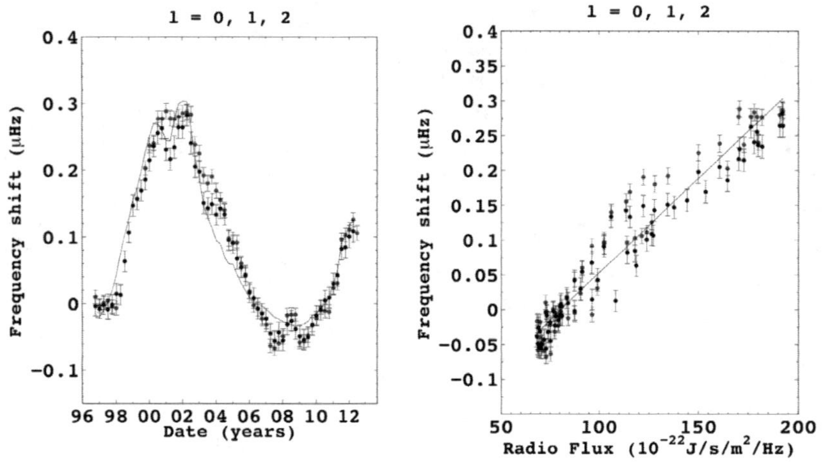

Figure 1. Left panel: Temporal variations of low-degree modes as described in the text; GOLF(black) and VIRGO (blue). The solid lines correspond to the scaled radio flux. Right panel: Frequency shifts as a function of the radio flux at 10.7 cm.

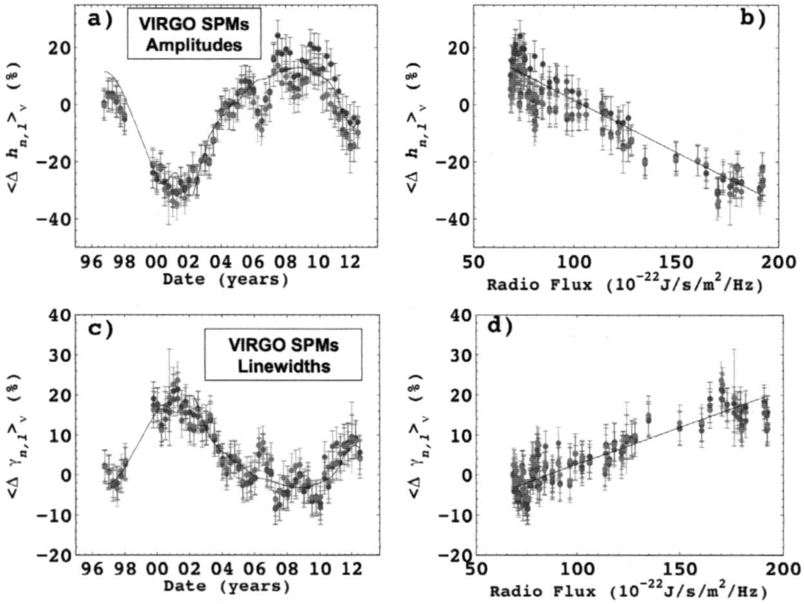

Figure 2. Average of the amplitudes (a,b) and linewidths (c,d) of the modes l = 0, 1 and 2 obtained using the three independent VIRGO channels: blue, green, and red as a function of time (left panels) and the radio flux (right panels). The solid lines correspond to the scaled radio flux.

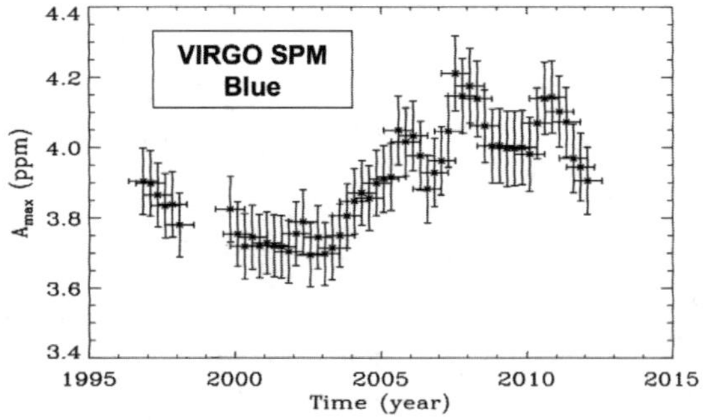

Figure 3. Temporal variation of A_{max} using the averaged three VIRGO/SPM channels as explained in the text.

5. Variation of Asymmetry

In Fig. 4, the temporal variations of the peak asymmetry $<\Delta b>$ of the modes observed by GOLF and VIRGO/SPM are shown. Due to the change in the GOLF observing configuration between the blue and the red wing [16], the variations (gradient) with the cycle are different (see Fig.4b). Pearson correlation coefficients of the VIRGO/SPM asymmetries are included in Table 1.

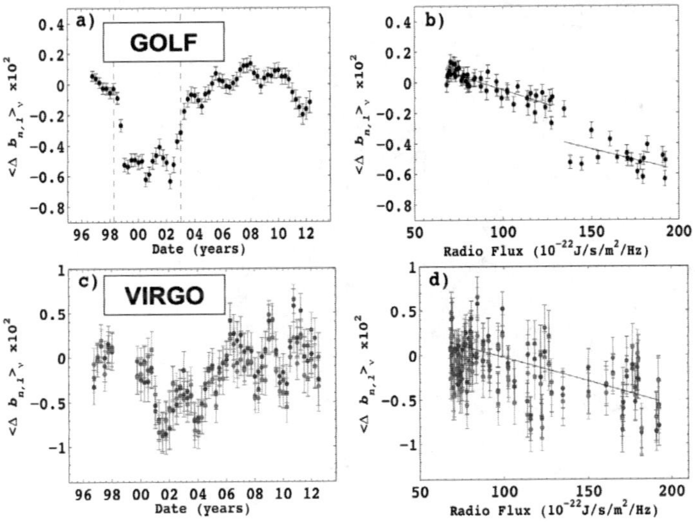

Figure 4. Average of the mode asymmetries computed using GOLF (a,b) and VIRGO/SPM (c,d) data as a function of time (left panels) and radio flux (right panels). The solid lines correspond to the scaled radio flux.

6. Solar Activity Proxy with GOLF and VIRGO/SPM

By correcting the raw VIRGO/SPM averaged data with the algorithms developed to process *Kepler* light curves [21], we are able to measure the temporal evolution of the rotation signature produced by the sunspots crossing the visible solar disk. Indeed standard VIRGO/SPM time series are filtered from periods longer than around 3 days. With this new procedure, we can keep longer periods up to 140 days. Therefore, we are not sensitive to changes in the faculae during the cycle. This effect produces a smooth increase of the solar brightness correlated with the cycle that cannot be disentangled from the ageing of the phototubes that we have filtered out.

We performed a time-frequency analysis of the time series using a Morlet wavelet [22]. It consists of calculating the correlation between a sliding mother wavelet and the time series where we change the period of the wavelet. This produces the wavelet power spectrum (Fig. 5, top panel).

The projection of the wavelet power spectrum in the range 6 to 60 days on the time axis (Fig. 5, bottom panel), provides information of the temporal evolution of the magnetic structures crossing the visible solar disk at the frequencies of the rotation and its first harmonics. We use here a conservative approach by selecting a wide range between 6 and 60 days. This is equivalent to computing a moving variance of the time series. However, our methodology has the advantage of being only sensitive to structures that have a periodicity around the rotation period. Therefore, we obtain a magnetic activity proxy linked to magnetic spots. This methodology could be used to track magnetic activity cycles in other stars.

Acknowledgments

SoHO is a space mission of international cooperation between ESA and NASA. R.A.G and D.S. thank the support from CNES. This work was partially supported by the NASA grant NNX09AE59G, by the White Dwarf Research Corporation through the Pale Blue Dot project, and by the grant AYA2010-17803 from the Spanish National Research Plan. NCAR is supported by the National Science Foundation. RAG acknowledges the support of the European

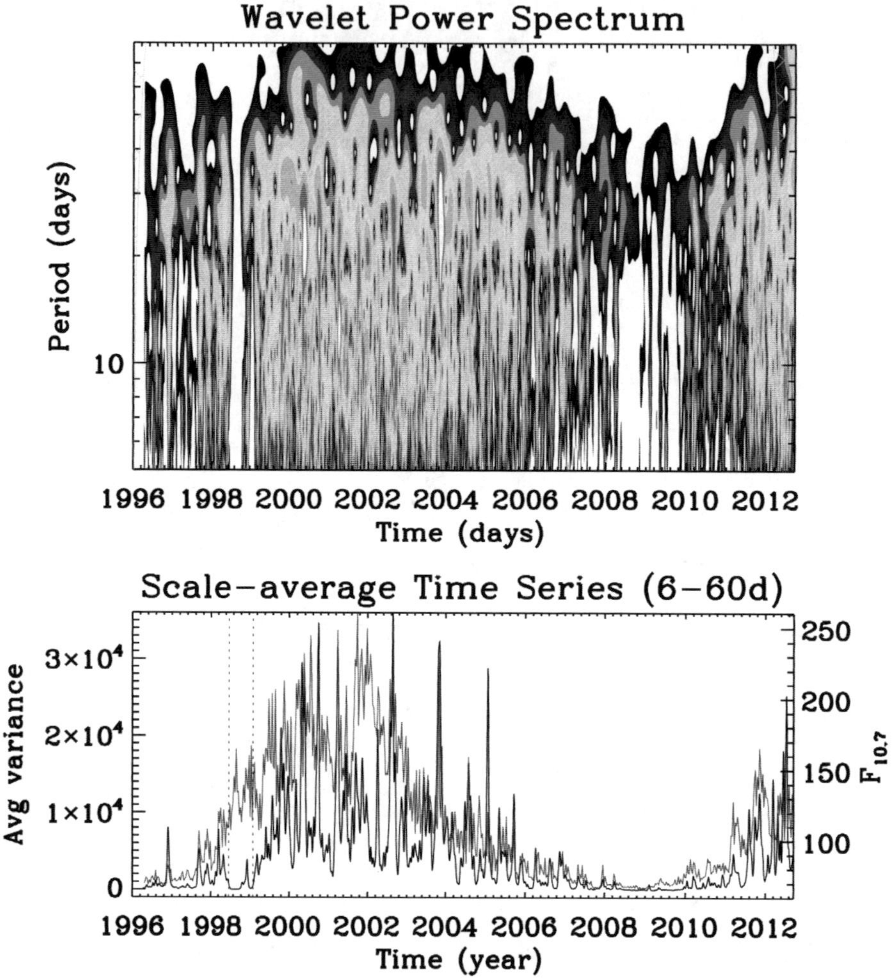

Figure 5. Top: Wavelet power spectrum of VIRGO/SPM computed using [22, 19]. Bottom: Projection onto the time axis as explained in the text. The dotted vertical lines correspond to the starting of the SoHO summer vacation period and the end of the winter vacation period. The blue line corresponds to the radio flux used as a proxy of the magnetic activity cycle averaged over 10 days.

Community's Seventh Framework Program (FP7/2007-2013) under grant agreement no. 269194 (IRSES/ASK).

References

[1] Toutain T, Appourchaux T, Baudin F et al. 1997 *Solar Phys.* **175** 311
[2] Thiery S, Boumier P, Gabriel A H et al. 2000 *Astron. & Astrophys.* **355** 743
[3] Gelly B, Lazrek M, Grec G, et al. 2002 *Astron. & Astrophys.* **394** 285
[4] Jiménez-Reyes S J, García R A, Jiménez A, Chaplin W J 2003 *Astrophys. J.* **595** 446
[5] Jiménez-Reyes S J, Chaplin W J, Elsworth Y et al. 2007 *Astrophys. J.* **654** 1135
[6] Salabert D, García R A, Pallé P L, Jiménez-Reyes S J 2009, *Astron. & Astroph.* **504** L1
[7] Gabriel A H, Grec G, Charra J et al. 1995 *Solar Phys.* **162** 61
[8] Frohlich C, Romero J, Roth H et al. 1995 *Solar Phys.* **162** 101

[9] Nigam R & Kosovichev A G 1998 *Astrophys. J.* **505** L51
[10] Salabert D, Chaplin W J, Elsworth Y, New R, Verner G A 2008, *Astron. & Astrophys.* **463** 1181
[11] Broomhall A M, Chaplin W J, Elsworth Y, Fletcher S T, New R 2009 *Astrophys. J.* **700** 162
[12] Fletcher S T, Broomhall A M, Salabert D et al. 2010 *Astrophys. J.* **718** 19
[13] Simoniello R, Finsterle W, Salabert D et al. 2012 *Astron. & Astroph.* **539** 135
[14] García R A, Mathur S, Ballot J, et al. 2008 *Solar Phys.* **251** 119
[15] Broomhall A M, Salabert D, Chaplin W J et al. 2012 *MNRAS* **422** 3564
[16] García R A, Turck-Chièze S, Boumier P et al. 2005 *Astron. & Astroph.* **442** 385
[17] Kjeldsen H, Bedding T R, Arentoft T et al. 2008 *Astrophys. J.* **682** 1370
[18] García R A, Mathur S, Salabert D et al. 2010 *Science* **329** 1032
[19] Mathur S, García R A, Régulo C et al. 2010 *Astron. & Astroph.* **511** 46
[20] Michel E, Baglin A, Auvergne M et al. 2008 *Science* **322** 558
[21] García R A, Hekker S, Stello D, et al. *MNRAS* **414** L6
[22] Torrence C & Compo G P 1998 *BAMS* **79** 61

Journal of Physics: Conference Series **440** (2013) 012021 doi:10.1088/1742-6596/440/1/012021

The Butterfly Diagram Internal Structure

Maurizio Ternullo

INAF – Osservatorio Astrofisico di Catania, Via S. Sofia 78, I-95123 Catania, Italy

E-mail: `maurizio.ternullo@oact.inaf.it`

Abstract. A time-latitude diagram, where the spotgroup area is taken into account, is presented for cycles 12 through 23. The results show that the spotted area is concentrated in few, small portions (*"knots"*) of the Butterfly Diagram (BD). *The BD may be described as a cluster of knots.* Knots are distributed in the butterfly wings in a seemingly randomly way. A knot may appear at either lower or higher latitudes than previous ones, in spite of the prevalent tendency to appear at lower and lower latitudes. Accordingly, the spotted area centroid, far from continuously drifting equatorward, drifts poleward or remains stationary in any hemisphere for significant fractions ($\approx 1/3$) of the cycle total duration. In a relevant number of semicycles, knots seem to form two roughly parallel, oblique "chains", separated by an underspotted band. This picture suggests that two (or more) "activity streams" approach the equator at a rate higher than the spot zone as a whole.

1. Introduction

It is widely accepted that the Butterfly Diagram (BD) represents the scattering of spots around a mean latitude *steadily* drifting equatorward. Nevertheless, this description of the spot zone structure is hardly reconciliable with others, given by many authors. Bell [3] described the spots as appearing in distinct *non shifting* belts; Antalova and Gnevyshev [1] stated that *"there is no steady equatorward progression of a single maximum of activity"*, Gnevyshev [9; 10] recognized that the cycles from 12 to 19 show two activity peaks, a couple of years distant from each other. This important feature has been confirmed by further works [2; 14; 16] involving various indices of the solar activity. Ternullo [18] described the activation of new long-living activity centers, on the poleward side of the pre-existing spot zone, mimicing the inversion of the familiar equatorward drift of the spot zone centroid [19; 20]; Norton and Gilman [15] observed that *"the sunspot belt migration toward the equator does not proceed as smoothly as is commonly perceived"*. Similar conclusions have been reached by Cho and Chang [7]. The examination of the cycles 20 to 23 [21] showed that, in any hemisphere, the trace of the spot zone centroid results from the quasi-biennnal alternation of high-speed prograde phases with stationary or even retrograde phases; the average duration of the latter phases amounts to $\approx 35\%$ of the cycle total duration. A detailed picture of the BD internal structure has shown [24; 25] the striking lack of homogeneity of the spotted area distribution inside the butterfly wings: indeed, most of the spotted area is concentrated in some small portions (*"knots"*) of the BD, so as to give it a *"leopard skin"* aspect. The present work extends the study of the BD stucture and evolution to the whole set of cycles 12 to 23. The Royal Greenwich Observatory data – integrated with data from the US Air Force and the US National Oceanic and Atmospheric Administration from 1976 onward – have been used. These data are available at the site `http://solarscience.msfc.nasa.gov/greenwch.shtml`.

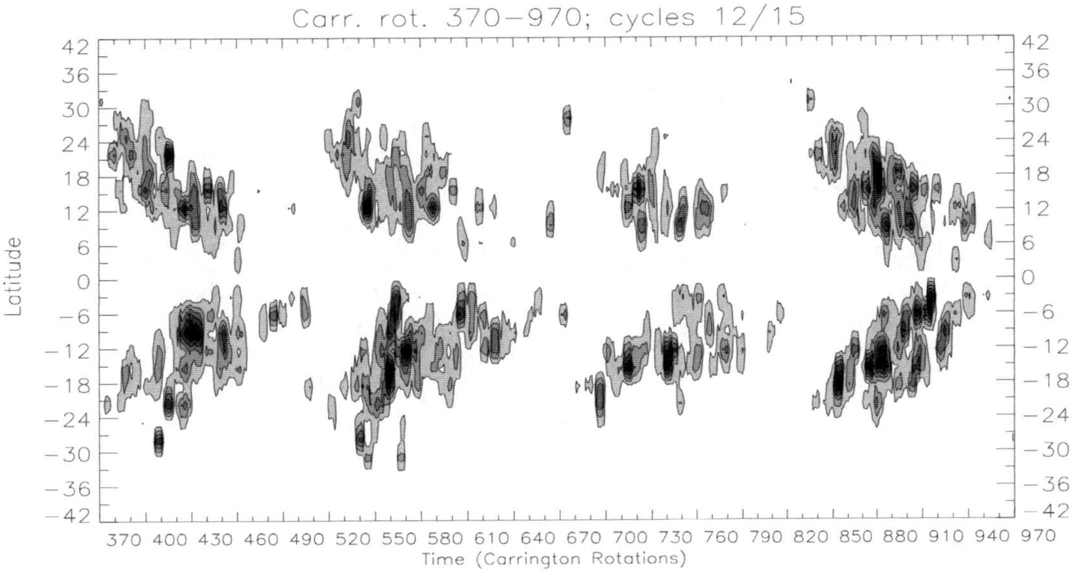

Figure 1. Butterfly Diagram for Carrington rotations 370-970 (years 1881–1926)

2. Results

For each of the 1741 Carrington rotations (CR's) from the 370th (June 1881) to the 2110th (May 2011) and for each of the 84 1°-wide latitude strips in the interval −42, +42°, the average value of daily spotted area has been determined. The resulting figures are the elements of an 84 × 1741 array, which is the quantitative counterpart of the diagram drawn by Maunder. This array has been smoothed by a running window covering 5 CR's, to suppress short period events and give visibility to features lasting longer than 5 CR's, as *complexes of activity* [4; 8] or *sunspot nests* [5]. The resulting smoothed array is visualized in Figures 1 to 3 by means of level curves, for which the term *"isospotted"* is suggested. For any semicycle (that is, for any "wing" of the BD), the first (the most external) isospotted excludes all the array elements spotted at a level lower than 10% of the spotted area maximal value registered in that wing. Inner isospotted are plotted for levels 20, 30, ... 90% of the maximal value. The area of the BD portions lying between the two most external lines typically amounts to ≈ 70% of the wing area, but host only ≈ 20% of the total spotted area. On the other hand, the 3rd level lines circumscribe less than ≈ 15% of the wing area but contain ≈ 50% of the total spotted area. These data support the description of the BD as a cluster of small, highly concentrated spot aggregations (*knots*).

A first-approach description of the spot zone evolution may be given by involving the most external level curves only. Their examination reveals that the first spots of any cycle appear at latitudes usually not higher than 25 ≈ 30°. Afterward, the spot zone rapidly expands both equatorward and poleward. It should be stressed that the maximal latitude is reached a couple of years after the cycle commencement. After reaching the minimal latitude, the spot zone equatorward boundary frequently drifts poleward (*e.g.*, see the cycles 13, northen and southern hemispheres (n.h. and s.h.), 14 (n.h. and s.h.), 15 (n.h. and s.h.), 16 (s.h.)); this phenomenon is visible with increasing sharpness in the second- and higher-level isospotted (*e.g.*, see the cycles

18 (n.h. and s.h.), 19 (n.h. and s.h.), 20 (n.h.), 21 (n.h.), 22 (n.h. and s.h.) and 23 (n.h.)). It should be emphasized that – in further phases of the same cycle – the spot zone again gains the low latitudes from which it retired. On the other hand, after that the spot zone poleward boundary has reached its maximal latitude and begun to retire, spots may reappear at high latitudes, even after a couple of years of quiescence (the semicycles 14 s.h., 14 n.h., 15 s.h., 16 s.h. and 17 s.h. represent cogent exempla).

We may gain a deeper insight on this phenomenology by taking the innermost isospotted, namely the ones delimiting *knots*, into consideration. Figures 1 to 3 show that, in spite of the knots tendency to appear at lower and lower latitudes as the cycle goes on, a knot may actually appear at latitudes either higher or lower than that of previous ones, in a seemingly random way. In some semicycles (for example, cycle 15 s.h., 16 s.h., 17 s.h., 20 n.h., 22 n.h. and 23 s.h.) knots seem arranged into two oblique, roughly parallel *chains*, or *streams*, leaving an underspotted band between them; in these cases, the cycle begins with the activation of a knot at latitude usually not higher than $25 \approx 30°$; this is the first element of a chain of knots which will appear at lower and lower latitudes, forming the butterfly wing equatorward boundary. The last of them appears roughly $50 \approx 60$ CR's after the first one, close to the equator. Accordingly, the slope of the butterfly wing equatorward boundary may be assumed as the rate of the equatorward progression of these knots; its order of magnitude is $\approx 0.5°\mathrm{CR}^{-1} \approx 6°\mathrm{year}^{-1}$. The remnant part of the familiar butterfly wing will be formed by knots belonging to another stream (or to other streams).

3. Discussion and Conclusions

This paper confirms the results described in previous papers by the present author.

The first, relevant result is the discovery of the butterfly wings fragmentation into knots

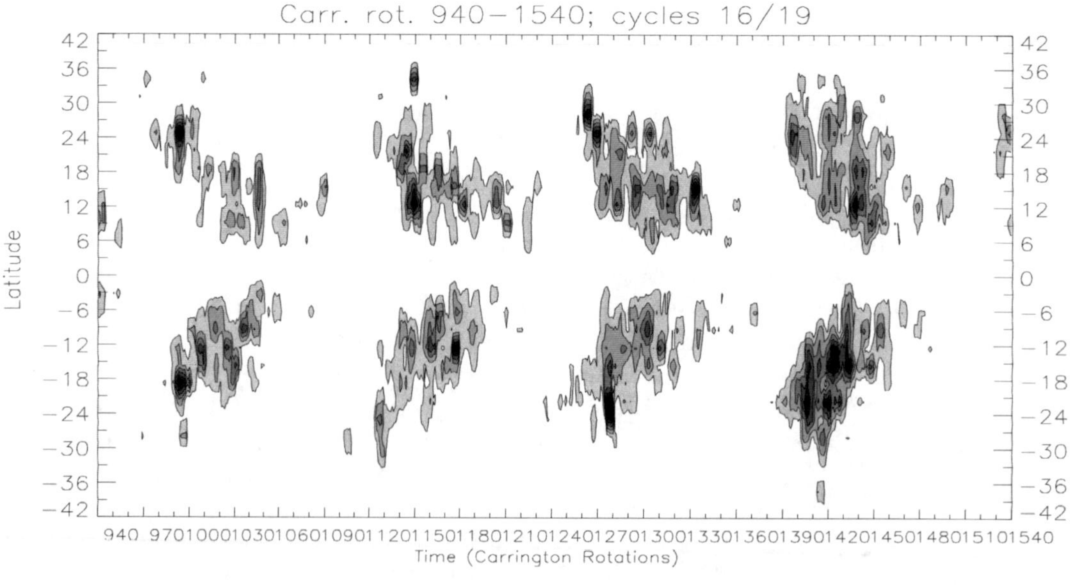

Figure 2. Butterfly Diagram for Carrington rotations 940-1540 (years 1923–1968)

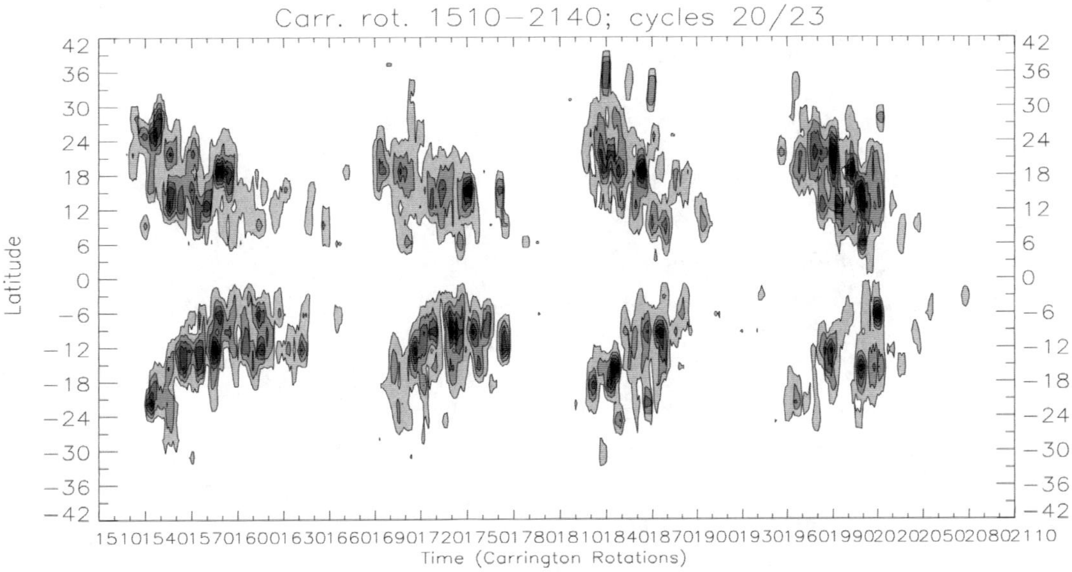

Figure 3. Butterfly Diagram for Carrington rotations 1510-2110 (years 1967-2011)

[24; 25]. This fragmentation is fully compatible with the spotgroups well-known tendency to appear at preferred locations, where they form *complexes of activity* [4; 8] or *sunspot nests* [5]. As a consequence of this new picture of the BD, no continuous, steadily drifing equatorward line can be drawn to represent the "mean latitude of spots": a line of this kind, indeed, rather than denoting the photospheric zone hosting the spot largest concentration, around which spots are taken to scatter, should be considered but an arithmetic artefact, to which it would be hard to attach any physical significance. If these conclusion will be confirmed, much theoretical work, aimed at predicting the location and evolution of such a mean latitude, should be reconsidered.

The second relevant result is that we begin to understand that the knot distribution in the BD wings obeys some regularities, in spite of the complexity of the processes stochastically acting on the solar dynamo. Describing the spot cycle as a two-wave process throws light on previous results concerning cycles 21 to 23 [21–23]; these works show that the latitudinal drift of the spot zone centroid may be decomposed into genuine equatorward segments and stationary or even poleward ones. The alternance of these prograde/retrograde segments is compatible with the sequence of activations and extinctions of knots belonging to either stream. Indeed, the activation of the second wave, occurring a couple of years after the first one, mimics the first poleward drift of the spot zone centroid; afterwards, other retrograde drifts of the spot zone centroid are the effect of the extinction of a low latitude knot, followed by the activation of a high latitude one; on the other hand, whenever a new, lower-latitude knot becomes active, a high speed prograde phase results. Finally, the extinction of the first wave mimics the last poleward drift of the spot zone centroid.

The pulse-like character of the magnetic flux emergence [8; 17; 23] is probably related to the tachocline oscillating rotation rate [12; 26]. On the basis of the reasonable assumption that, the faster the tachocline rotation, the more efficient the magnetic flux production, one could conjecture that any sudden augmentation of the photospheric flux is the effect of the tachocline increased rotation rate.

Acknowledgments

Helpful discussions with Paolo Romano are gratefully acknowledged. This work has been supported by the Inaf (Prin Inaf 2010).

References

[1] Antalová A. & Gnevyshev M. 1983 *Contr. Astron. Obs. Skalnaté Pleso* **11** 63
[2] Bazilevskaya M. et al. 2003 *Adv. Space Res.* **31** 895
[3] Bell B. 1960 *Smithsonian Contrib. Astrophys.* 5, no 3
[4] Bumba V. & Howard R. 1965, *Astrophys. J.* **141** 1502
[5] Castenmiller M. J.M. et al. 1986 *Sol. Phys.* **105** 237
[6] Charbonneau P. 2005, *Living Rev. Solar Phys.* 2, 2. [Online Article]: http://www.livingreviews.org/lrsp-2005-2
[7] Cho Il-Hyun & Chang Heon-Young 2011, *J. Astron. Space Sci.* 28(1), 1-7 (2011) DOI: 10.5140/JASS.2011.28.1.001
[8] Gaizauskas V. et al. 1983 *Astrophys. J.* **265** 1056
[9] Gnevyshev M. N. 1967 *Sol. Phys.* **1** 107
[10] Gnevyshev M. N. 1977 *Sol. Phys.* **51** 175
[11] Harvey K. L. & Zwann C. 1993 *Solar Phys.* **148** 85
[12] Howe R. "Solar Interior Rotation and its Variation", *Living Rev. Solar Phys.* 6, (2009), 1. URL (cited on September 2009): http://www.livingreviews.org/lrsp-2009-1
[13] Jiang J. & Wang J. X. 2007 *Mon. Not. R. Astron. Soc.* **377** 711
[14] Kopecký M. & Kuklin G. V. 1969 *BAICz* **20** 22
[15] Norton A. A. & Gilman P. A. 2004 *Astrophys. J.* **603** 348
[16] Norton A. A. & Gallagher J. C. 2010 *Sol. Phys.* **261** 193
[17] Rabin D. et al. 1991 in *Solar Interior and Atmosphere*, edited by Cox et al., The University of Arizona Press, 781
[18] Ternullo M. 1990 *Solar Phys.* **127** 29
[19] Ternullo M. 1997 *Solar Phys.* **172** 37
[20] Ternullo M. 2001 *Mem. S.A.It.* **72** 565
[21] Ternullo M. 2007a *Solar Phys.* **240** 153
[22] Ternullo M. 2007b *Mem. S.A.It.* **78** 596
[23] Ternullo M. 2007c *Astron. Nachr.* **328** 1023
[24] Ternullo M. 2008 in *Electronic Proceedings - 12th European Solar Physics Meeting, 8-12 September 2008 - Freiburg, Germany* http://espm.kis.uni-freiburg.de
[25] Ternullo M. 2010 *Astrophys. Sp. Science* **328** Issue 1-2, 301
[26] Toomre J. et al. 2003 in *Proceedings of SOHO 12/GONG+2002 on Local and Global Helioseismology: The Present and Future* Sawaya-Lacoste (ed.), ESA SP-517, ESA publication division, Noordwijk, Netherlands
[27] Van Driel-Gesztelyi L., van der Zalm E.B.J. & Zwann C. 1992 in *The solar cycle*, Asp Conference Series Vol. 27. Harvey. K.L. ed.

Journal of Physics: Conference Series **440** (2013) 012022 doi:10.1088/1742-6596/440/1/012022

Are subsurface flows and coronal holes related?

R. Komm[1], R. Howe[2], I. González Hernández[1], L. Harra[3], D. Baker[3], L. van Driel-Gesztelyi[3]

[1] National Solar Observatory, Tucson, AZ 85719, USA

[2] School of Physics and Astronomy, University of Birmingham, Edgbaston, Birmingham B15 2TT, UK

[3] Mullard Space Science Laboratory, University College London, Dorking RH5 6NT, UK

E-mail: `rkomm@nso.edu`

Abstract. We study synoptic maps of solar subsurface flows covering six Carrington rotations (2050 to 2055). The subsurface flows are determined with a ring-diagram analysis of GONG high-resolution Doppler data. We identify the locations of coronal holes in synoptic maps of EUV images at 195Å from the EIT instrument and determine the characteristics of associated subsurface flows. We study two long-lived coronal holes that are present during this epoch. We find that large-scale patterns are present in the subsurface flows but appear to be unrelated to these coronal holes. The horizontal subsurface flows associated with the two long-lived coronal holes are weakly divergent (upflows) with small cyclonic vorticity. These flows are thus similar to subsurface flows of quiet regions with regard to the vertical flows and similar to flows of active regions with regard to vorticity.

1. Introduction

We study solar subsurface flows measured with a ring-diagram analysis of high-resolution Doppler data obtained by *Global Oscillation Network Group* (GONG) instruments. In previous studies, we have focused on the relationship between active regions and subsurface flows associated with them [1, 2]. Here, we explore whether there is a relationship between subsurface flows and coronal features. Synoptic subsurface flow maps show large-scale patterns that are not obviously associated with active regions. It is unknown whether these flow patterns correlate with any large-scale magnetic features. We also want to determine whether subsurface flows associated with coronal holes are different from subsurface flows of active or quiet regions.

We analyze synoptic maps of subsurface flows covering six Carrington rotations during the years 2006 and 2007. Long-lived coronal holes are present during this epoch at latitudes that are accessible by ring-diagram analysis of GONG data. To compare the subsurface flows with coronal images, we use synoptic observations in the EUV from the *EUV Imaging Telescope* (EIT) onboard the *Solar and Heliospheric Observatory* (SOHO) spacecraft.

2. Data and analysis

We analyze observations obtained during Carrington Rotations 2050 – 2055 (14 November 2006 – 27 April 2007). To derive solar subsurface flows, we use high-resolution full-disk Doppler data from the GONG network (`http://gong.nso.edu/data`). We determine the horizontal components of subsurface flows with a ring-diagram analysis using the dense-pack technique

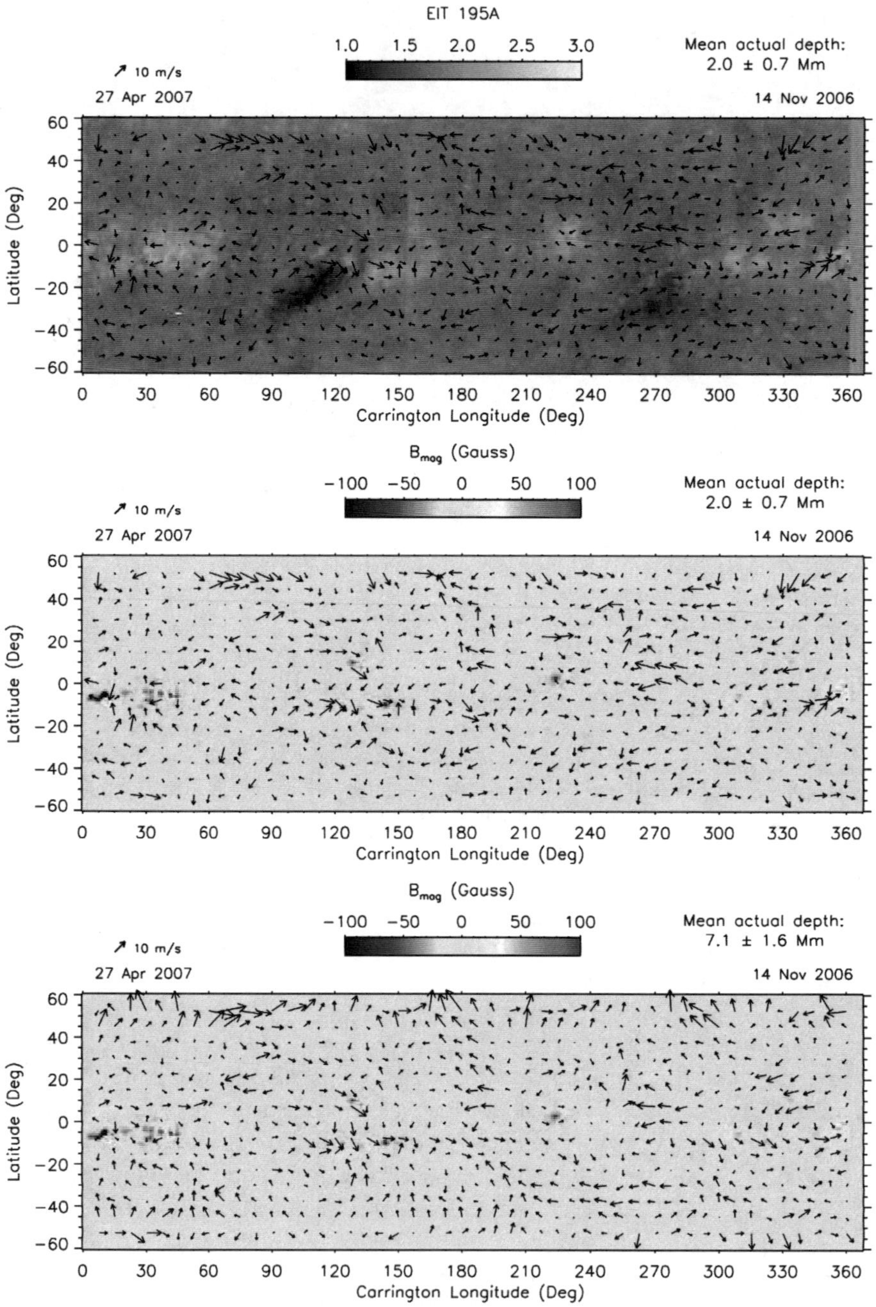

Figure 1. Top: The subsurface flows at a depth of 2 Mm (arrows) averaged over Carrington rotation 2050 – 2055 are superposed on the corresponding average EIT map (log-scale). Middle: The flows at 2 Mm and the corresponding photospheric magnetogram. Bottom: The subsurface flows at 7 Mm are superposed on the magnetogram.

[3] adapted to GONG data [4]. The full-disk Doppler images are divided into 189 overlapping regions with centers spaced by 7.5° ranging over ±52.5° in latitude and central meridian distance.

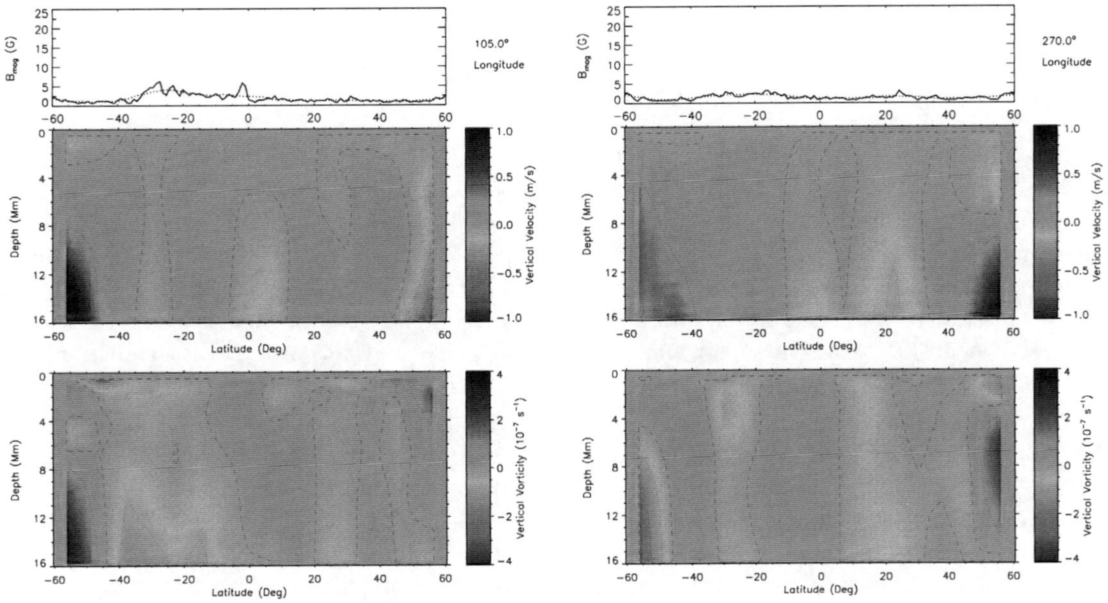

Figure 2. The unsigned magnetic activity (top), the vertical velocity (middle), and the vertical vorticity (bottom) averaged over Carrington rotation 2050 – 2055 as a function of latitude and depth at 105° (left) and 270° longitude (right). Positive vertical-velocity values imply upflows, negative values imply downflows. The coronal holes are centered at −22.5° (left) and −30° latitude (right), as seen in Figure 1.

Each region is tracked for 1664 minutes and a three-dimensional power spectrum is calculated for each tracked region. Horizontal flows advect the acoustic modes and shift their temporal frequency, which appears as a shift of the rings in spatial frequency. The shift of these rings in spatial frequency is a measure of the subsurface flows. We derive daily flow maps of horizontal velocities from the 189 dense-pack patches and combine them to form synoptic flow maps at 16 depths from 0.6 to 16 Mm. To remove the large-scale trends in latitude of the differential rotation and the meridional flow, we subtract a low-order polynomial fit in latitude of the longitudinal average of the flows for each Carrington rotation and study the resulting residual synoptic flow maps [1, 2]. We estimate the vertical-velocity component from the divergence of horizontal flows using mass conservation [1, 5] and also calculate maps of the vertical component of the vorticity vector (curl of the horizontal velocities).

To compare synoptic subsurface flow maps with observations of the solar corona, we use synoptic observations in the EUV at 195Å from the EIT instrument onboard the SOHO spacecraft (`http://sun.stanford.edu/synop/EIT/`). The synoptic maps of EUV images represent the values of the line intensity centered on the central meridian and they have a resolution of 1° in both longitude and latitude [6]. As a measure of photospheric magnetic activity, we use NSO SOLIS magnetograms (`http://solis.nso.edu/solis_data.html`).

3. Results

We average the synoptic flow maps of the six Carrington rotations to improve the signal-to-noise ratio and to study long-lived structures (Figure 1). Two long-lived coronal holes are present in the southern hemisphere near 105° and 270° longitude. The subsurface residual flows are large and complicated near active regions but small and simple near coronal holes. Small subsurface residual flows is a characteristic that coronal holes have in common with quiet regions. Large-scale flow patterns are present near 180° to 210°, for example. These large-scale flows are present to varying degree in the six synoptic flow maps and do not seem to be associated with coronal holes or active regions.

We also derive the vertical velocity and vorticity averaged over the six rotations (Figure 2). The large values at high latitudes are unreliable [1, 5]. At the locations of the coronal holes near $-22.5°$ latitude and 105° longitude (left) and $-30°$ latitude and 270° longitude (right), the average vertical velocity shows weak upflows and the vertical vorticity is small and mainly negative (cyclonic or clockwise in the southern hemisphere). The signal-to-error ratio is greater than three at shallow depths less than about 6 Mm and smaller at greater depths. The long-lived active region near 30° shows downflows and mainly negative vorticity values (not shown).

4. Summary and conclusions

Large-scale flow patterns are present in the average synoptic flow maps. They are unrelated to coronal holes or active regions; they seem to almost avoid or separate regions of magnetic activity. The horizontal subsurface flows associated with the two long-lived coronal holes are weakly divergent (upflows) with small cyclonic vorticity. Quiet regions are characterized by weakly divergent subsurface flows with small anticyclonic vorticity, while active regions are characterized by convergent subsurface flows with cyclonic vorticity [7]. The sign of subsurface-flow vorticity thus distinguishes these coronal holes from quiet regions, while the sign of the vertical-flow component distinguishes them from active regions.

These are encouraging results. However, we have studied only two long-lived coronal holes so far. We clearly need to analyze a larger data set. For this purpose, we plan to analyze synoptic maps of GONG and EIT data covering Carrington rotation 1979 to 2118 and characterize each dense-pack patch by their dominant magnetic feature.

Acknowledgments

This work utilizes GONG data obtained by the NSO Integrated Synoptic Program (NISP), managed by the National Solar Observatory, which is operated by the Association of Universities for Research in Astronomy (AURA), Inc. under a cooperative agreement with the National Science Foundation. GONG data were acquired by instruments operated by the Big Bear Solar Observatory, High Altitude Observatory, Learmonth Solar Observatory, Udaipur Solar Observatory, Instituto de Astrofísica de Canarias, and Cerro Tololo Interamerican Observatory. SOHO is a project of international cooperation between ESA and NASA. This work was supported by NSF/SHINE Award No. 1062054 and by NASA grant NNX11AQ57G to the National Solar Observatory. RH acknowledges computing support from NSO. We acknowledge the Leverhulme Trust for funding the Probing the Sun: inside and out project upon which this research is based.

References

[1] Komm R, Corbard T, Durney B R, González Hernández I, Hill F, Howe R and Toner C 2004 *Astrophys. J.* **605** 554
[2] Komm R, Howe R, Hill F, González Hernández I, Toner C and Corbard T 2005 *Astrophys. J.* **631** 636
[3] Haber D A, Hindman B W, Toomre J, Bogart R S, Larsen R M and Hill F 2002 *Astrophys. J.* **570** 855
[4] Corbard T, Toner C, Hill F, Hanna K D, Haber D A, Hindman B W and Bogart R S 2003 *Local and Global Helioseismology: The Present and Future* ed H Sawaya-Lacoste (Noordwijk: ESA) **SP-517** 255
[5] Komm R 2007 *Astron. Nachr.* **328** 269
[6] Benevolenskaya E E, Kosovichev A G and Scherrer P H 2001 *Astrophys. J.* **554** L107
[7] Komm R, Howe R, Hill F, Miesch M, Haber D A and Hindman B 2007 *Astrophys. J.* **667** 571

Journal of Physics: Conference Series **440** (2013) 012023 doi:10.1088/1742-6596/440/1/012023

Solar oscillations in cycle 24 ascending

Kiran Jain[1], Sushanta Tripathy[1], Frank Hill[1] and Timothy Larson[2]

[1]National Solar Observatory, Tucson, AZ 85719, USA
[2]Stanford University, Stanford, CA 94305, USA

E-mail: kjain@nso.edu

Abstract. Solar oscillation frequencies are known to follow the trend of solar cycle and show a strong correlation with various activity indices. However, the extended minimum between cycles 23 and 24 has raised several questions on the correlation between frequencies and solar activity where frequencies with different mode sets sensed different minima. In this paper, we analyze intermediate-degree mode frequencies as the Sun emerges from the unusually long period of minimal magnetic activity to study their behaviour with activity indices and compare results with the corresponding phase of cycle 23. We show that a model based on the rising phase of cycle 23 is a good predictor for behaviour in the rising phase of cycle 24.

1. Introduction

A precise knowledge of the solar interior and its variation with time is important to understand the origin of solar activity. The most simple and reliable way to infer variability of the interior is by examining its resonant modes. It is now well established that mode frequencies closely follow the trend of solar cycle and exhibit strong correlation with various proxies of solar activity [e.g. 1,2]. However, using uniform data for cycle 23, it has been shown that the degree of correlation depends on phase of the solar cycle [3]. While ascending and descending phases are tightly correlated with the surface activity, the maximum and minimum phases have relatively poor correlation. In addition, the analysis of frequencies during the recent extended minimum has demonstrated that different mode sets may exhibit different scenario; the most interesting finding is the disagreement in minima as sensed by low- and intermediate-degree modes [4,5]. In this paper, we analyze intermediate-degree mode frequencies for the ascending phase of cycle 24 to study how well these frequencies follow the solar activity after a long spell of extremely low activity.

2. Data

The p-mode frequencies used here are taken from three sources: the Global Oscillation Network Group (GONG) covering a period of 17 years from 7 May 1995 to 24 May 2012; the Michelson Doppler Imager (MDI) onboard *Solar Heliospheric Observatory (SoHO)* covering 15 years of data from 1 May 1996 to 24 Apr 2011; and the Helioseismic Magnetic Imager (HMI) onboard *Solar Dynamics Observatory (SDO)* covering a period of more than 2 years from 30 Apr 2010 to 9 Sep 2012. We have used MDI frequencies processed through the HMI pipeline [6]. While the GONG frequencies are calculated using overlapping 108-day timeseries spaced by 36 days, the MDI and HMI frequencies are for non-overlapping 72-day timeseries. In total, we use 171, 74 and 12 data sets from GONG, MDI and HMI, respectively with simultaneous observations

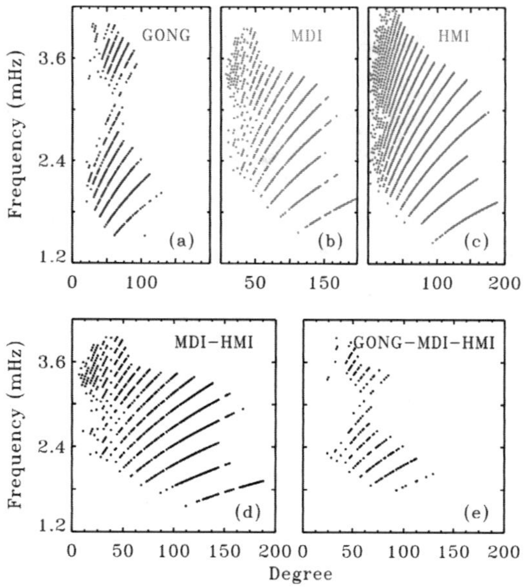

Figure 1. Comparison of common p modes observed in (a) GONG, (b) MDI and (c) HMI datasets. The bottom row shows p modes common between (d) MDI and HMI datasets, and (e) all three datasets. The number of modes fitted in all GONG data are significantly smaller than those in the MDI and HMI data.

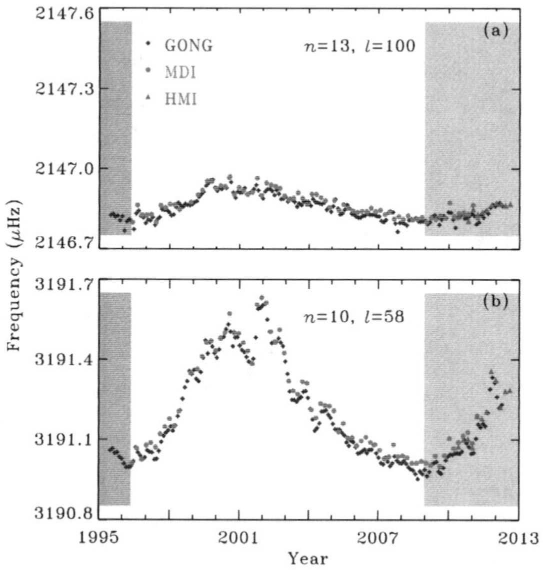

Figure 2. Variation of oscillation frequencies for two modes observed in all three data sets at (a) low frequency and (b) high frequency. The uncertainties in frequencies are smaller than the size of symbols. Shaded areas on left and right sides in these plots represent solar cycles 22 and 24, respectively.

from all three instruments for a year from 30 April 2010 to 25 April 2011. In order to study variation of frequencies with time, we analyze modes that are common in all data sets and are shown in Figure 1. It is seen that there is a significant difference in fitted modes using data from different instruments. We interpret the different mode sets to arise due to the difference in lengths of time series and data reduction techniques.

3. Results

In Figure 2 we show two examples representing the low- and intermediate-frequency modes as a function of time for all three data sets. The GONG frequencies are found to be lower than the MDI and HMI frequencies. However, as MDI and HMI frequencies are calculated using identical fitting method and 72-day timeseries, the frequencies of these two data sets are in better agreement.

To study temporal variation in oscillation frequencies, we calculate average frequency shift, $\delta\nu$, from the relation

$$\delta\nu(t) = \sum_{n,\ell} \frac{Q_{n,\ell}}{\sigma^2_{n,\ell}} \delta\nu_{n,\ell}(t) / \sum_{n,\ell} \frac{Q_{n,\ell}}{\sigma^2_{n,\ell}}, \tag{1}$$

where $\delta\nu_{n,\ell}(t)$ is the change in the measured frequency with respect to a reference frequency, $\sigma_{n,\ell}$ is the uncertainty in the frequency measurement for a given n, ℓ multiplet and $Q_{n,\ell}$ is the inertia ratio. The reference frequency is calculated using a temporal mean over all the available data sets. In this analysis, we have used two sets of reference frequencies; the first set is calculated using modes shown in Figure 1(d) and second set is for Figure 1(e). The resultant frequency

Journal of Physics: Conference Series **440** (2013) 012023 doi:10.1088/1742-6596/440/1/012023

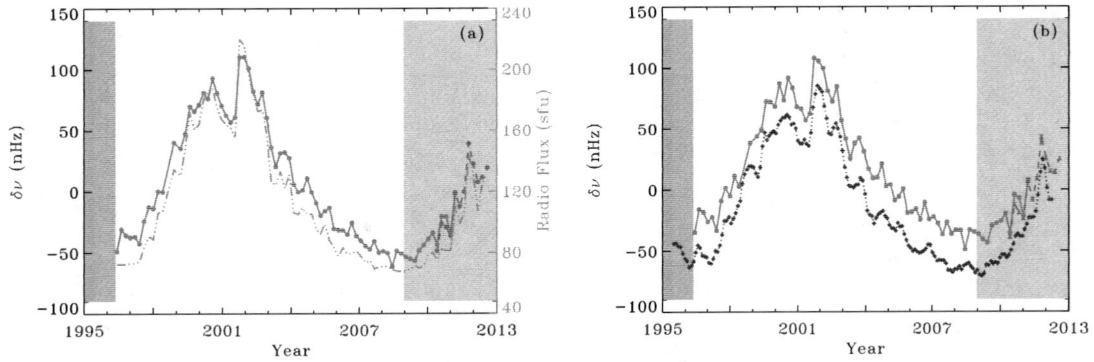

Figure 3. Temporal variation of frequency shifts calculated using (a) common modes in all MDI (green solid) and HMI (red dashed) datasets, and (b) common modes in all MDI (green solid), HMI (red dashed) and GONG (blue dotted) datasets. The yellow dashed-dot-dot-dot line in figure (a) represents the variation of the 10.7 cm radio flux. Shaded areas have the same meaning as in Figure 2.

shifts in the frequency range of 1.5 mHz $\leq \nu \leq$ 4.0 mHz for p modes are plotted in Figure 3. It is clearly seen that the frequencies vary in phase with the solar activity as discussed in earlier studies, however there is a significant variation in its magnitude depending on the modes used in the analysis. It is evident from Figure 3 that the MDI and HMI frequency shifts are consistently higher than the GONG but the overall shifts from low- to high-activity periods in all cases are approximately similar. In addition, the MDI and HMI shifts are in close agreement in the period of simultaneous observations.

To compare relative variation in frequencies with the change in solar activity δi during ascending phases of cycles 23 and 24, we assume a linear relationship of the form $\delta\nu$ = a δi + b, where coefficients a and b are obtained by performing a linear least square. Examples of the linear regression of GONG frequencies in the ascending phase of solar cycle 23 with 10.7 cm radio flux and sunspot numbers are shown in Figure 4(a-b). The solid line is the best regression

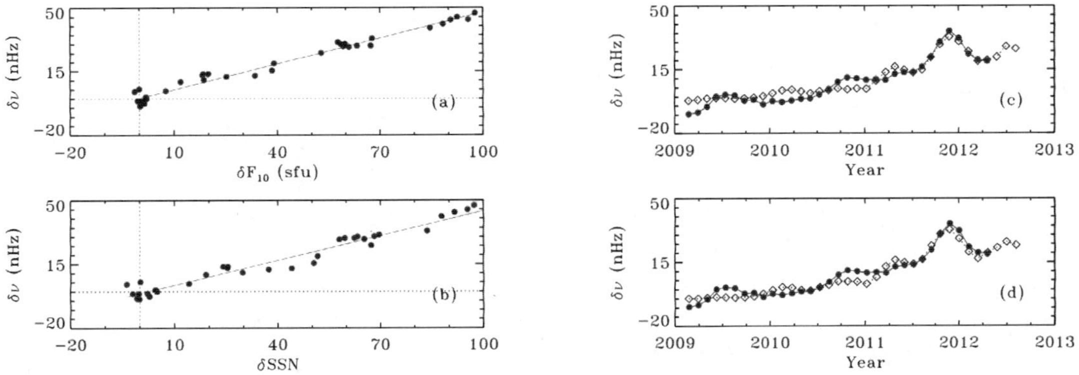

Figure 4. (Left) Linear regression of the GONG frequency shifts and the change in activity proxies (a) 10.7 cm radio flux (δF_{10}) and (b) the international sunspot number (δSSN) for the ascending phase of solar cycle 23. The solid line represents the least-square fit. Models for cycle 23 are used to predict frequency variation in cycle 24. (Right) Temporal variation of modeled (black diamond) and observed frequency shifts in cycle 24 (blue circle) using (c) 10.7 cm radio flux and (d) international sunspot numbers.

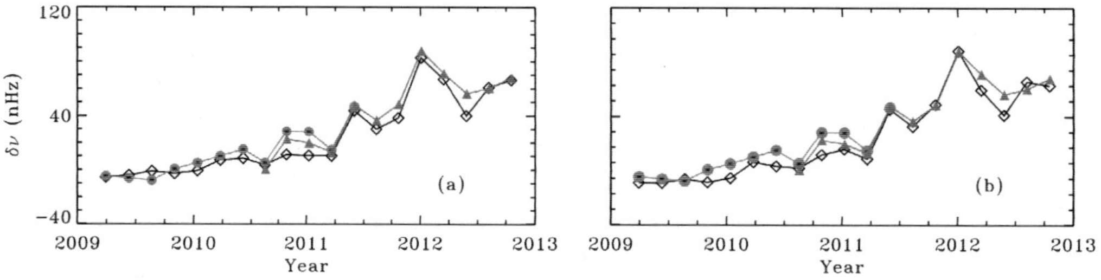

Figure 5. Temporal variation of modeled (black diamond) and observed frequency shifts in cycle 24 for MDI (green circle) and HMI data (red triangle); The models are developed using MDI frequencies for the ascending phase of cycle 23 with (a) 10.7 cm radio flux and (b) international sunspot numbers.

fit and confirms that the data sets are consistent with the assumption of a linear relationship. This assumption is further tested by calculating Pearson's correlation coefficients which are found to be > 0.98 in both cases. These regression coefficients are further used to model mean shifts in the ascending phase of cycle 24.

Figures 4(c-d) and 5(a-b) compare observed and modeled variation of frequency shifts for ascending phase of cycle 24. We find a good correlation, i.e. > 0.96 between the observed and modeled shifts, thus the change in frequency per unit solar activity obtained for the ascending phase of cycle 23 is a good estimate for frequency shifts in the same phase of cycle 24. In addition, we find a relatively higher correlation between observed and modeled frequencies when we use F_{10} as an input to the regression model. This is in agreement with earlier results where frequency shifts are found to be better correlated with F_{10} than SSN [3]. These findings indicate that the sensitivity of oscillation frequencies to the change of magnetic activity does not vary significantly for the same phase of solar cycles.

In summary, the continuous and improved measurements of oscillation frequencies for more than a solar cycle provide a deeper insight on the variability of solar interior. The obtained high linear correlation between frequency shifts and solar activity in the rising phase of cycle 24 is consistent with the previous findings. However, longer data sets covering the maximum phase are needed to provide better understanding on the role of magnetic field in the variation of oscillation frequencies.

Acknowledgments

This work utilizes GONG data obtained by the NSO Integrated Synoptic Program (NISP), managed by the National Solar Observatory, which is operated by AURA, Inc. under a cooperative agreement with the National Science Foundation. The data were acquired by instruments operated by the Big Bear Solar Observatory, High Altitude Observatory, Learmonth Solar Observatory, Udaipur Solar Observatory, Instituto de Astrofisica de Canarias, and Cerro Tololo Interamerican Observatory. SOHO is a mission of international cooperation between ESA and NASA. SDO data courtesy of SDO (NASA) and the HMI and AIA consortium. This work is partially supported by NASA grant NNH12AT11I .

References

[1] Tripathy S C, Hill F, Jain K and Leibacher J W 2007 *Solar Phys.* **243** 105-120
[2] Chaplin W J, Elsworth Y, Isaak G R, Miller B A and New R 2004 *MNRAS* **352** 1102-1108
[3] Jain K, Tripathy S C and Hill F 2009 *Astrophys. J.* **695** 1567-1676
[4] Jain K, Tripathy S C and Hill F 2011 *Astrophys. J.* **739** 6
[5] Salabert D, García R A, Pallé P L, Jiménez-Reyes S J 2009 *A&A* **504** L1-L4
[6] Larson T and Schou J 2011 *J. Phys. Conf. Ser.* **271** 012062

Time–distance inversions for horizontal and vertical flows on supergranular scales applied to MDI and HMI data

M Švanda[1,2], H Schunker[3] and R Burston[3]

[1] Astronomical Institute, Academy of Sciences of the Czech Republic (v. v. i.), Fričova 298, CZ-25165 Ondřejov, Czech Republic
[2] Charles University in Prague, Faculty of Mathematics and Physics, Astronomical Institute, V Holešovičkách 2, CZ-18000 Prague 8, Czech Republic
[3] Max-Planck-Institut für Sonnensystemforschung, Max-Planck-Strasse 2, D-37191 Katlenburg-Lindau, Germany

E-mail: michal@astronomie.cz

Abstract. We study the possibility of consistent extension of MDI full-disc helioseismic campaigns with the growing data set of HMI observations. To do so, we down-sample and filter the HMI Dopplegrams so that the resulting spatial power spectrum is similar to the spatial power spectrum of MDI full-disc Dopplergrams. The set of co-spatial and co-temporal datacube pairs from both instruments containing no missing and no bad frames were processed using the same codes and inverted independently for all three components of the plasma flow in the near surface layers. The results from the two instruments are highly correlated, however systematically larger (by $\sim 20\%$) flow magnitudes are derived from HMI. We comment that this may be an effect of the different formation depth of the Doppler signal from the two instruments.

1. Motivations

Understanding the details of the solar dynamo is inevitably linked to measurements of plasma flows inside the Sun. Methods of helioseismology [1, 2] can measure such flows, one of them being time–distance local helioseismology [3]. It is based on the measurements of travel times of seismic waves, which propagate through the solar interior and carry information about the inhomogeneities (flows in particular) affecting their travel time. By solving the inverse problem one can learn about these inhomogeneities.

A proper assessment of the character and structure of solar flows requires a long term study. Recently, the Michelson Doppler Imager (MDI [4], on board of SOHO spacecraft), which covered 15 years of observations, was replaced by its successor Helioseismic and Magnetic Imager (HMI [5], on board of SDO spacecraft). Our study is thus driven by the need to fully exploit *all* of the data. Here, we specifically ask the question: *Can MDI and HMI data be possibly utilised together for inverse modelling for time–distance local helioseismology?*

In local helioseismology, two methods are usually used to solve the inverse problem. The regularised least squares (RLS; in time–distance helioseismology used for the first time by [6]) minimisation seeks to find models of solar interior that minimise the difference between the predicted and measured travel times. The inverse problem must be solved separately for each

Journal of Physics: Conference Series **440** (2013) 012024 doi:10.1088/1742-6596/440/1/012024

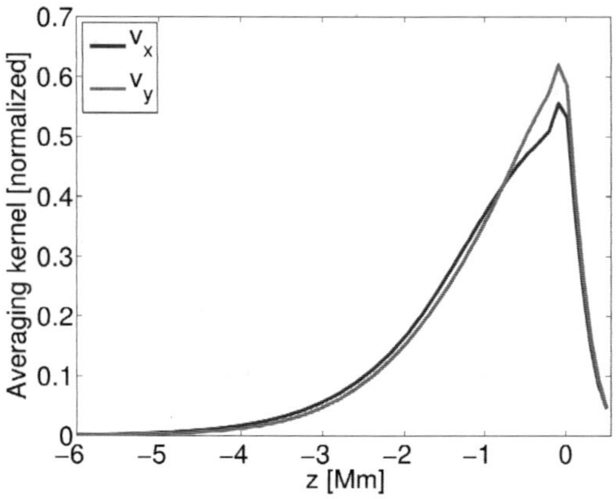

Figure 1. Cuts through averaging kernels of inversion for horizontal v_x and vertical v_z flow in the middle point as a function of depth. Note that $z = 0$ at an optical depth of $\tau_{500} = 1$, negative z values refer to solar interior and positive ones refer to atmosphere.

set of travel times. On the contrary, optimally localised averaging (OLA; [7, 8, 9]) method is based on explicitly constructed spatially confined averaging kernels, while bounding the noise propagation at the same time. The inverse problem in the OLA approach must be solved only once and does not directly involve the measurements, except for the knowledge of statistical properties of the travel-time noise. Estimates for inverted quantities are then obtained by taking linear combination of travel times utilising a set of weights resulting from the inversion. When the measurements from various instruments are combined to form a homogeneous data set, the OLA is a natural approach to be used for inverse modelling from such data set.

To test inverse modelling, we constructed a well-understood inversion for all three flow components. The inversion follows the Fourier-space based solution [10] using the validated inversion code [11] implemented within the German Science Center for SDO at Max-Planck-Institut für Sonnensystemforschung in Katlenburg-Lindau, Germany. This code introduces additional terms in the cost function of the inverse problem, which allows the minimisation of the contributions from cross-talks. This property is crucial in order to be able independently invert also for a weak vertical flow. The inversion utilised a combination of wave sensitivity kernels computed in the Born approximation [12] for the surface gravity wave (f) mode, assuming the application to the MDI full-disc Dopplergrams. The travel-time noise covariance matrix used in the inverse problem was measured from the data using $1/T$ fitting approach [13]. The inversion combined sensitivity kernels for travel times for three point-to-annulus and point-to-quadrant geometries (making the measurements sensitive to divergence of the flow and vertical velocity, and to waves travelling in west–east and north–south directions) with radii of the annuli from 7.3 to 29.2 Mm. The inversion targets only near sub-surface layers (see the averaging kernels in Fig. 1).

2. Data

To do a proper comparison between the two instruments, we required the consistent analysis of the data, which are both co-spatial and co-temporal. MDI delivered data suitable for local helioseismology only two months a year (during Dynamics Campaigns), with the last one (June and July 2011) overlapping with routine operations of HMI. In this two-month period we prepared 64 consecutive pairs (one for MDI, one for HMI) of 12-hour long Dopplergram datacubes capturing the centre of the solar disc using the mapping and tracking pipeline at the GDC-SDO. Additionally, we did not want to have results affected by missing or bad frames in datacubes, thus we required a 100% duty cycle. A down-selection was done on a frame-by-frame basis manually. Only 4 (!) out of 64 co-spatial and co-temporal datacube pairs fulfilled our requirements.

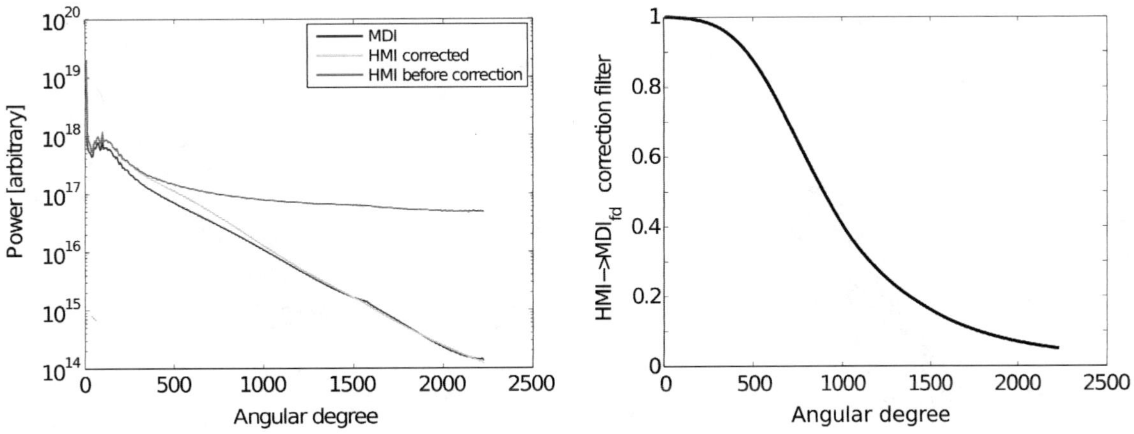

Figure 2. Left: Spatial power spectra of MDI_{fd} Dopplergrams compared to HMI Dopplergrams down-sampled to MDI_{fd} resolution before and after correction. Right: The wave-number filter designed for the correction.

The sensitivity kernels used in the inverse problem described above were designed to be consistent with MDI full-disc (MDI_{fd} hereafter) Dopplergrams. HMI is a higher-resolution instrument, thus we needed to make sure that HMI data looked like MDI_{fd} ones. This was achieved by a Postel-projection of both MDI and HMI Dopplergrams with the same map scale (corresponding to MDI_{fd} pixel size) using bicubic interpolation. HMI Dopplergrams were thus down-sampled to MDI_{fd} resolution. Even in such data the contribution from large wave-number was much higher than in the case of true MDI_{fd} data (see Fig. 2 left) and we wanted to correct for it. This may be done by an application of a filter in the wave-number space (Fig. 2 right), designed so that the spatial power spectra of both instruments decay similarly in the large wave-number region. This filter approximates the correction to the different modulation transfer functions (MTF) of the two instruments. Note that despite the filtering, spatial power spectrum of corrected HMI Dopplergrams still have an excess in power at angular degrees l of 50 to 300. We will return to this issue in the next section.

Apart from the power-spectrum correction, the rest of the analysis (filtering, travel-time measurements, application of inversion weights) was identical for both MDI and degraded HMI datacubes.

3. Comparison of helioseismic observables

Consistently with the setup of the inversion, we measured wave travel times from f-mode-pass filtered Dopplergram datacubes with point-to-annulus and point-to-quadrants geometries with distances 7.3 to 29.2 Mm. Travel-time maps measured from corresponding MDI and HMI datacubes are highly correlated (correlation coefficients of 0.8 or higher), but the travel times measured from HMI datacubes are systematically larger by 20% (e.g., Fig. 3 left). This statement holds for all pairs of travel-time maps. The excess is mostly prominent on supergranular and similar scales (angular degree between 50 and 300) and it seems to be a consequence of the excess power discussed above (see Fig. 2 left).

Since the estimate of the inverted flows is just a linear combination of travel times and the inversion weights are the same for both MDI and HMI measurements, the magnitudes of flow estimates are also systematically larger, when the flow is inverted from the HMI datacubes than from the MDI datacubes (see example in Fig. 3 right). Apart from this systematic effect (there is an excess of 20% for horizontal components and around 30% for vertical component) the flows

Journal of Physics: Conference Series **440** (2013) 012024 doi:10.1088/1742-6596/440/1/012024

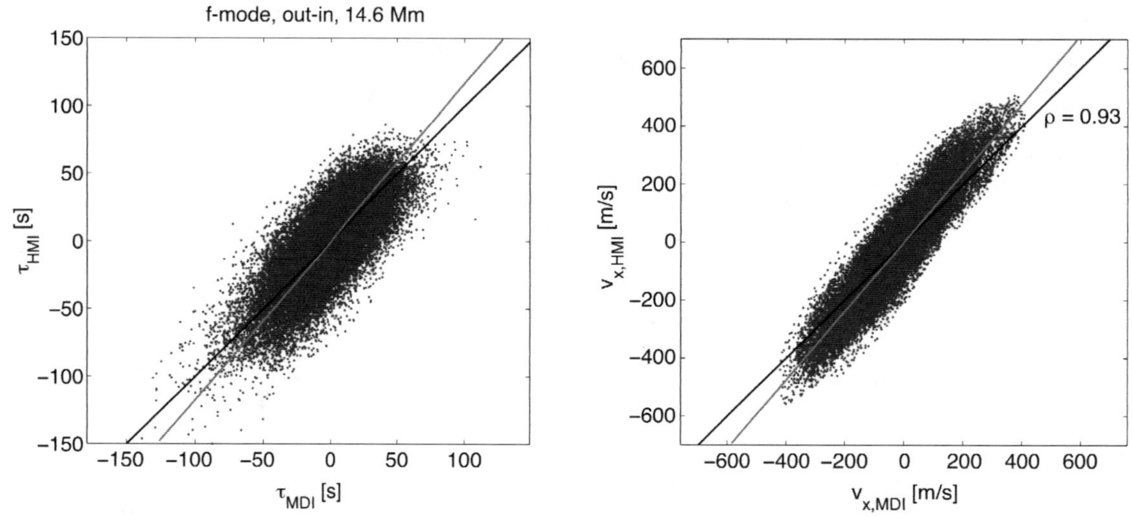

Figure 3. Left: Scatter plot of the travel times measured for the f-mode with point-to-annulus geometry with distance 10 px from $\mathrm{MDI_{fd}}$ (horizontal axis) and degraded HMI (vertical axis). Right: Equivalent scatter plot to compare the inversion estimate of v_x with the correlation coefficient indicated. In both cases, the red solid line represents the best least-squares fit of the points and the black solid line has a unity slope to increase the visibility of the systematically larger values derived from HMI Dopplergrams.

derived from MDI and HMI measurements are almost identical (correlation coefficient larger than 0.9 for horizontal components and around 0.8 for the vertical component). The lower correlation in the case of vertical component is probably caused by lower signal-to-noise ratio of the vertical flow estimate.

An example of full-vector flow maps for HMI and MDI is displayed in Fig. 4. One can see that except for the systematically larger magnitudes in the HMI flow map, both maps are practically identical. In these maps, supergranular cells (rosette-like features in the horizontal components) are dominant. Also note that the upflows (red) in the vertical component in most cases correspond to outflows in the horizontal component. The correlation between the vertical component and the horizontal divergence of the horizontal flow is higher than 0.8. This is one indication that the inversion for the vertical flow is reasonable.

4. Final remarks

We showed that with some drawbacks, HMI and MDI helioseismic datasets may be analysed together. Long-term studies of flows in the solar convection zone employing combined datasets are therefore possible. However, the noticed bias must be understood and corrected for.

It is likely that the different flow magnitudes are an effect of the different spectral line used in the two instruments. The formation depth of HMI Doppler signal measured in Fe I (617.3 nm) line is some 25 km deeper in the photosphere than the formation depth of the MDI Doppler signal measured in Ni I (676.8 mm) line [14]. Thus the convection (especially supergranules) is better pronounced when observed in the Fe I line, hence the power spectrum is expected to have more power on scales where convection is dominant, including the supergranular scales around angular degree of 120. That is the case and it can be clearly seen in Fig. 2. Increased sensitivity to the convection affects the sensitivity kernels, which slightly differ for the two spectral lines (however, we did not compute the proper sensitivity kernels for Fe I line, so we did not model

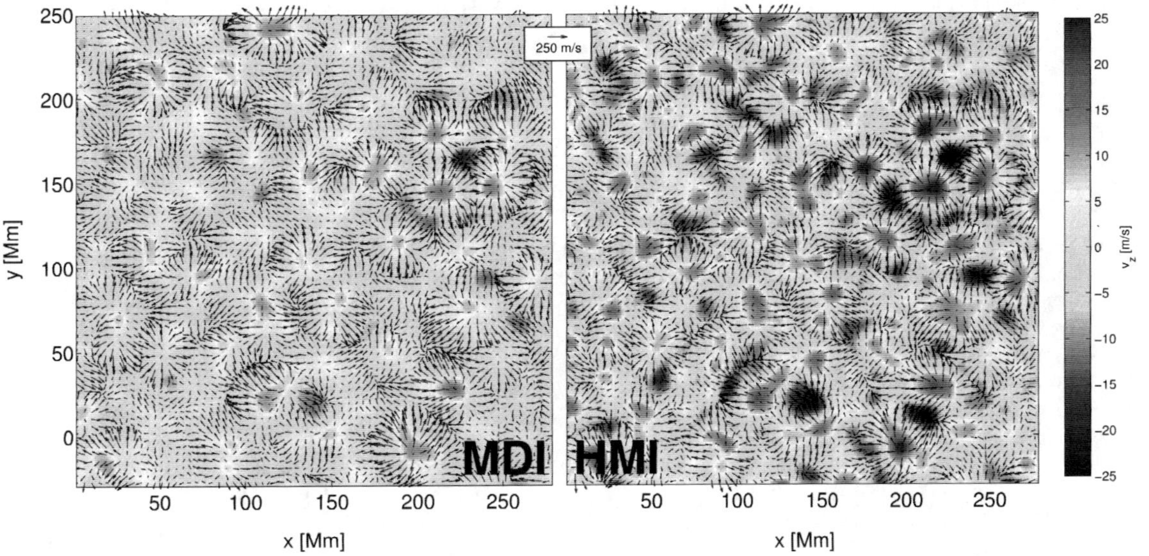

Figure 4. Comparison of full-vector flow maps in the same region on the Sun, inverted from MDI (left) and HMI (right) Dopplergrams. Horizontal components in the plane of solar surface are indicated by arrows, the vertical component by colours. The level of random error in the flow estimates is 49 m/s for horizontal flow components and 4 m/s for the vertical one.

the effect properly). As a consequence, we expect that when the weights from the inversion performed with Ni I-line sensitivity kernels are applied to travel-time maps measured from Fe I line, the true inversion averaging kernel is slightly different, with the peak sensitivity located probably a little deeper (the averaging kernels for the Ni I-line inversion are plotted in Fig. 1). If we assume that the magnitude of the plasma flow increases with height in the last 1 Mm of the convection zone and then drops gradually around $\tau_{500} = 1$ level, the averaging kernel with a deeper peak averages the flow with slightly higher amplitudes and therefore the resulting flow estimate will also have larger magnitude. That is observed as a bias when comparing the flow estimates from the two instruments. The higher bias in the vertical flow ($\sim 30\%$) than in the horizontal components ($\sim 20\%$) might indicate a steeper gradient of the vertical flow below the surface.

Other effects may also be responsible for the excess power in the spectrum, such as a more complicated instrumental MTF shape which was not corrected for. After the bias is understood, it would be possible to correct for it, for example by application of an additional non-monotonic wave-number filter.

This study is the first to compare inverse modelling using the OLA technique for MDI and HMI data. Similar direct comparisons of flow and sound speed inversions was performed on early HMI data (J Zhao, private communication) using the RLS technique, but it showed larger discrepancies not only in magnitude, but also in structure of the flows than presented here. This higlights the necessity to first test and understand the systematics due to both the instrument and the analysis technique.

Acknowledgments

MŠ acknowledges the support of the Czech Science Foundation (grant P209/12/P568). This work utilised the resources and helioseismic products dispatched within the German Science Center at MPS in Katlenburg-Lindau, Germany, which is supported by German Aerospace

Center (DLR). The data were kindly provided by the HMI consortium. The HMI project is supported by NASA contract NAS5-02139. Tato práce vznikla s podporou na dlouhodobý koncepční rozvoj výzkumné organizace (RVO:67985815).

[1] Gizon L, Birch A C and Spruit H C 2010 *ARA&A* **48** 289–338 (*Preprint* arXiv:1001.0930)
[2] Gizon L and Birch A C 2005 *Living Reviews in Solar Physics* **2** 6
[3] Duvall Jr T L, Jefferies S M, Harvey J W and Pomerantz M A 1993 *Nature* **362** 430–432
[4] Scherrer P H, Bogart R S, Bush R I, Hoeksema J T, Kosovichev A G, Schou J, Rosenberg W, Springer L, Tarbell T D, Title A, Wolfson C J, Zayer I and MDI Engineering Team 1995 *Sol. Phys.* **162** 129–188
[5] Schou J, Scherrer P H, Bush R I, Wachter R, Couvidat S, Rabello-Soares M C, Bogart R S, Hoeksema J T, Liu Y, Duvall T L, Akin D J, Allard B A, Miles J W, Rairden R, Shine R A, Tarbell T D, Title A M, Wolfson C J, Elmore D F, Norton A A and Tomczyk S 2012 *Sol. Phys.* **275** 229–259
[6] Kosovichev A G 1996 *ApJ* **461** L55–L57
[7] Backus G E and Gilbert J F 1968 *Geophysical Journal* **16** 169–205
[8] Backus G E and Gilbert J F 1970 *Royal Society of London Philosophical Transactions Series A* **266** 123–192
[9] Pijpers F P and Thompson M J 1992 *A&A* **262** L33–L36
[10] Jackiewicz J, Birch A C, Gizon L, Hanasoge S M, Hohage T, Ruffio J B and Švanda M 2012 *Sol. Phys.* **276** 19–33 (*Preprint* arXiv:1109.2712)
[11] Švanda M, Gizon L, Hanasoge S M and Ustyugov S D 2011 *A&A* **530** A148 (*Preprint* arXiv:1104.4083)
[12] Birch A C and Gizon L 2007 *Astron. Nachr.* **328** 228
[13] Gizon L and Birch A C 2004 *ApJ* **614** 472–489
[14] Fleck B, Couvidat S and Straus T 2011 *Sol. Phys.* **271** 27–40 (*Preprint* arXiv:1104.5166)

Journal of Physics: Conference Series **440** (2013) 012025 doi:10.1088/1742-6596/440/1/012025

Helioseismology with PICARD

T Corbard[1], D Salabert[1], P Boumier[2], T Appourchaux[2], A Hauchecorne[3], P Journoud[2], A Nunge[2], B Gelly[4], J F Hochedez[3], A Irbah[3], M Meftah[3], C Renaud[1], S Turck-Chièze[5]

[1] Laboratoire Lagrange, UMR7293, Université de Nice Sophia-Antipolis, CNRS, Observatoire de la Côte d'Azur, Bd. de l'Observatoire, 06304 Nice, France

[2] Institut d'Astrophysique Spatiale, CNRS-Université Paris XI, UMR 8617, 91405 Orsay Cedex, France

[3] Laboratoire Atmosphères, Milieux, Observations Spatiales,CNRS, Université Paris VI & Université de Versailles Saint-Quentin-en-Yvelines, IPSL, F-78280 Guyancourt, France

[4] Themis, UPS 853 du CNRS, c/o IAC, Via Lactae s/n, 38200 La Laguna, Tenerife, Spain

[5] CEA/IRFU/Service d'Astrophysique, AIM, CE Saclay, 91191 Gif sur Yvette, France

E-mail: `Thierry.Corbard@oca.eu`

Abstract. PICARD is a CNES micro-satellite launched in June 2010 [17]. Its main goal is to measure the solar shape, total and spectral irradiance during the ascending phase of the activity cycle. The SODISM telescope onboard PICARD also allows us to conduct a program for helioseismology in intensity at 535.7 nm [5]. One-minute cadence low-resolution full images are available for a so-called medium-l program, and high-resolution images of the limb recorded every 2 minutes are used to study mode amplification near the limb in the perspective of g-mode search. First analyses and results from these two programs are presented here.

1. Medium-l program in intensity

Full continuum images at 535.7 nm (bandwidth 0.5 nm) are recorded by SODISM [8] every minute with a spatial resolution of about 1 square arcsecond on a 2048^2 CCD. For telemetry reasons, these images are downgraded to 256^2 pixels (about 8x8 arcsecond) by a simple onboard binning before being transmitted to ground. We have analyzed a dataset covering 209 consecutive days (2011/04/16-2011/11/10) with a duty cycle of 74.4%. The gaps are mostly due to interruptions for the other PICARD program (astrometric measurements at different wavelengths) and to the crossing of the South Atlantic Anomaly, which is the location of transient signals generated by trapped protons exposing low-altitude, Sun-synchronous orbiting satellites, like PICARD, to strong radiations.

Figure 1 (top) shows the l-ν diagram obtained from this 209-day dataset for angular degree up to l=300. As expected from the instrumental resolution, a spatial aliasing for spherical harmonics above l=200 is present. We can also see a temporal aliasing for frequencies above 4.5 mHz. This is the signature of a spurious cycle in the 2-min sampling slots corresponding to the astrometric program sequences. Figure 1 (bottom) illustrates a m-ν diagram for l=90 clearly showing the S-shape signature of the differential rotation. Figure 2 shows the mode parameters modelled using the asymmetrical line profile formula of Nigam and Kosovichev [10] and fitted using a Maximum Likelihood Estimator minimization [1, 7]. For a given mode (n,l), the $2l+1$ m-components of the multiplets are fitted simultaneously using an expansion on orthogonal

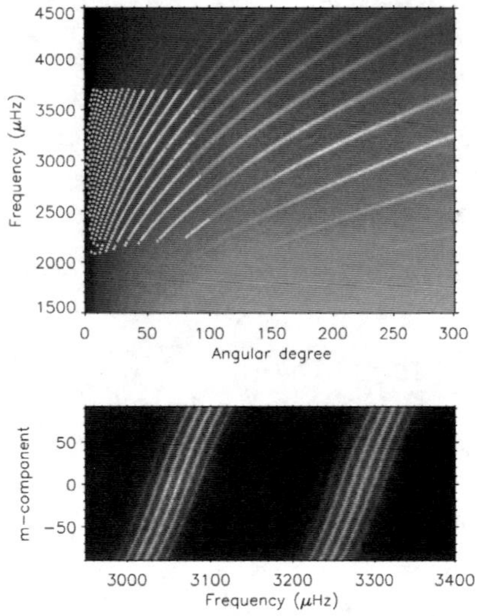

Figure 1. Top: l-ν diagram from 209-day PICARD observations. The fitted modes are shown by the green dots. Bottom: m-ν diagram for l=90 and radial orders n=7 (left) and n=8 (right). The l-leaks from l=90±3 are clearly seen.

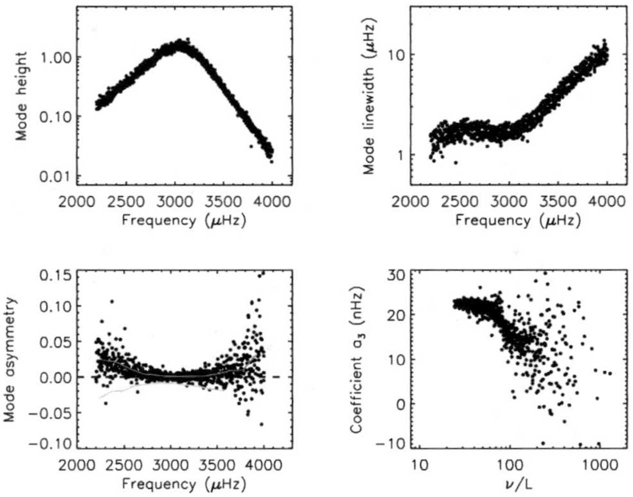

Figure 2. Fitted mode parameters: amplitude (upper left), linewidth (upper right), peak asymmetry (lower left), and a_3 splitting coefficients (lower right). The red line is a fit of SODISM positive asymmetries in intensity, while the green line shows the negative asymmetries obtained from the SOHO/GOLF velocity signal.

polynomials [12] with up to 9 so-called a-coefficients. The l and m mode leakages are included in the fitting process. Positive peak asymmetries are obtained as expected from previous analyses of intensity signal. The a_3 splitting coefficients plotted as a function of the mode turning point proxy ν/L, where $L = \sqrt{l(l+1)}$, exhibit a gradient which is a typical signature of the solar tachocline. Improvements of the peak-fitting process are currently underway. We need for instance to better model all the components of the leakage matrix in order to fit higher angular degrees.

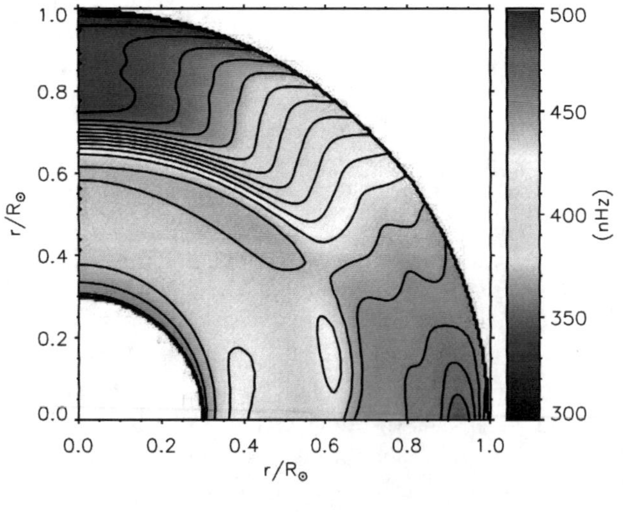

Figure 3. Symmetric part of the internal solar rotation rate with respect to the equator. Isocontours are shown every 10 nHz from 300 nHz up to 480 nHz. This was obtained by inverting the frequency splittings estimated from PICARD/SODISM intensity images over the period from April 16 to November 10 2011. The solution below $0.3R_\odot$ is poorly constrained by the data and is not shown.

Figure 3 shows a Regularized Least-Square (RLS) inversion [6] of the measured rotational splittings of 769 modes from l=1 to l=99. The steep gradient of the tachocline at the base of the convection zone around 0.7 $R_\odot$ is clearly seen. The inferred rotation rate is similar to the one previously obtained from velocity data [6, 11]: a radial gradient is obtained close to the surface followed by a latitudinal differential rotation persisting through the whole convection zone. The radiative interior is roughly rigid at a rate of 435 ± 5 nHz which corresponds to the surface rate for latitudes between 30 and 40 degrees. Finally, we note that a local maxima is found close to the equator near $0.95R_\odot$. These preliminary results are very encouraging. We are currently working in order to extend the peak-fitting analysis towards higher l and to increase the number of fitted a-coefficients thus improving the resolution of our inversion. The reversal of peak asymmetry observed between velocity and intensity power spectra is linked to the effect of a background component correlated to the modes [9]. The complete two years of PICARD observations available to date should then be used to confront the details of internal rotation studies obtained using simultaneous velocity and intensity data from SDO/HMI or GONG. By fitting intensity, velocity, their phase difference and coherence signals simultaneously, we hope to nail down systematics in either observables and better understand the correlated background noise [14, 4]. This will help to better interpret the fitted frequencies and frequency splittings in terms of internal structure and dynamics.

2. Limb seismology

Every 2 minutes, the full resolution of PICARD/SODISM images are kept in an annulus of 22 pixels around the solar limb (increased to 31 pixels since April 2012). In April 2011 and April 2012, 3 consecutive days were dedicated to helioseismic measurements at 535.7 nm without interruption for the other PICARD objectives. These datasets with a duty cycle above 95% have been used in order to study the expected mode amplification at the extreme limb [2, 16, 15]. Intensity data from SOHO/MDI and SDO/HMI were also used for comparison.

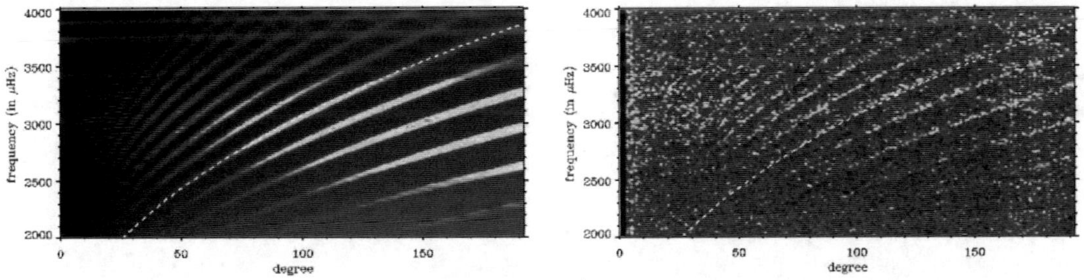

Figure 4. Classical spherical harmonic transform of full images (l-ν diagram) (left) and limb transform of limb pixels (l'-ν diagram) (right). The theoretical n=6 ridge is indicated by a dashed line for both cases clearly showing a frequency shift between the two diagrams.

In order to analyse the limb signal, we calculated the Fourier transform $e^{il'\rho}$ along the limb using the polar coordinates, where ρ=1 at the disk center and ρ=0 at the limb. Figure 4 shows the comparison between the l-ν diagram obtained from a classical spherical harmonic transform of full images (left) and the l'-ν diagram (right) resulting from the limb transform as described above. A frequency shift between ridges of the same radial order is clearly seen. We carried out simulations in order to understand this shift and the link between the two diagrams [3]. For a given l', it can be shown that the limb response are non zero for value of l higher than l'. Since the mode frequencies increase with the degree, the leaks will appear at higher frequencies than the original anticipated frequency. The leaks will then appear as an asymmetry at higher frequency with a decreasing amplitude when l increases. This is indeed what is shown in Figure 5, where

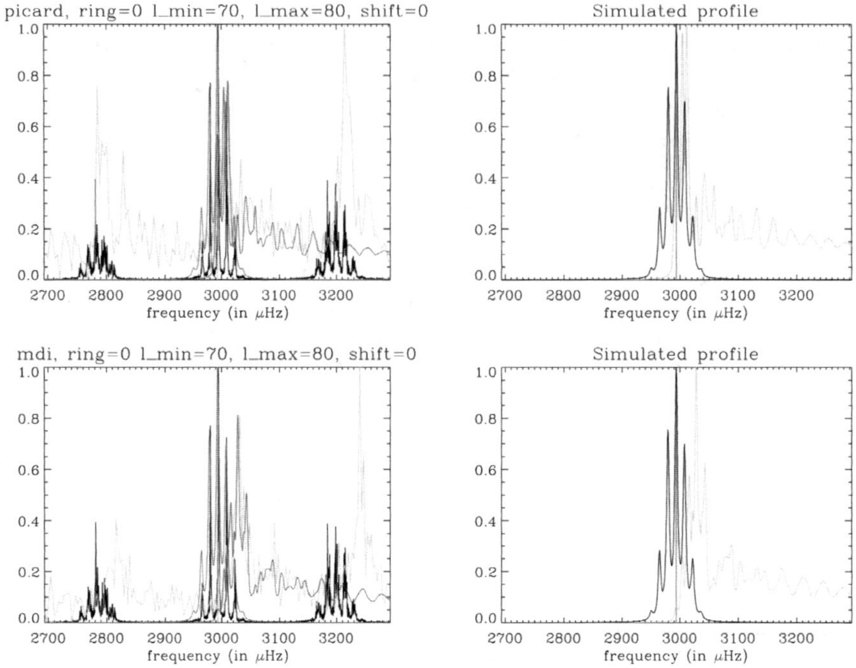

Figure 5. Comparison between observed and simulated profiles. On the left panels, the black lines are ridges in SOHO/MDI l-ν diagrams while the green lines are ridges in the PICARD/SODISM (top) and SOHO/MDI (bottom) l'-ν diagrams. The two panels on the right show the simulated profiles for both the l-ν (black) and l'-ν (green) diagrams which are also shown with an appropriate scalling on the left panels as the red and blue lines respectively.

the observed ridge profiles obtained from PICARD/SODISM and SOHO/MDI intensity data are compared to the simulated profiles. From these simulations, we are able to understand:

- why the peak in the l'-ν diagram is shifted with l (the shift with the original l-ν ridge increases as l increases);
- why the peak gets closer to the original ridge as we get closer to the limb;
- why the l'-ν ridge leaks towards higher frequencies.

Figure 6 (left) shows that the energy in the p-mode range peaks at the extreme limb, as was already seen using MDI intensity images [15] or the guiding pixels of SOHO/LOI [2]. If we normalize this by a proxy of the noise level obtained by integrating the energy in the low-frequency band (right panel), we can see that we effectively have an amplification of the signal-to-noise ratio at the extreme limb for the three instruments. This peak is however slightly shifted inward for SODISM. This may be an instrumental effect but this needs further investigations.

3. Conclusions

The first helioseismic analyses of SODISM full-continuum images have been carried out for both the medium-l and limb programs. We fitted mode parameters for a wide range of eigenmodes. The fitted peak asymmetries and frequency splittings are consistent with what is known from previous studies. We performed a first inversion of the rotational splittings and infered the internal rotation rate down to the radiative zone. The steep gradient of the tachocline is clearly seen at the base of the convection zone. The signature of p-modes can also be observed in limb

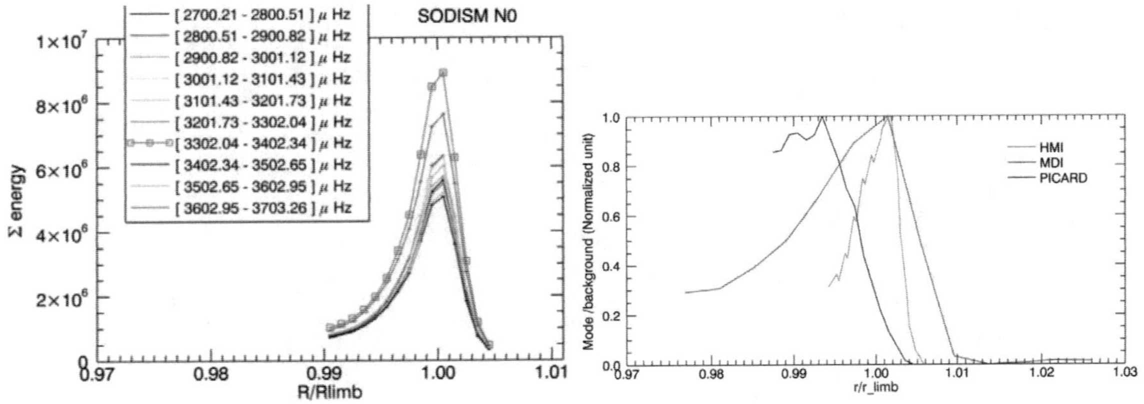

Figure 6. Left: spectral energy level integrated in several frequency ranges of the 5-min oscillations envelope. Right: energy of the 5-min envelope normalized by the low frequency energy, for SODISM, SOHO/MDI and SDO/HMI. As far as SODISM is concerned, the amplification curve is displaced toward the inner part of the limb. This could be due to an instrumental effect not well taken into account in the data reduction pipeline. Note the narrower profile in the case of HMI, consequence of a better PSF.

data and a signal amplification was detected at the extreme limb. By computing the response of any mode to the limb transform, we explain most of the features of the power spectra observed at the limb when compared to spectra built using full images. However, we have to face strong instrumental and orbital effects that affect both the photometric signal and the geometry of the solar images. We know from the past that our capacity in assessing the reliability of the details of many features analyzed using helioseismology (tachocline shape, polar jet, meridional flows, surface effects for structure inversions, etc. see e.g. [13]) greatly benefit from the cross-analysis made using different instruments (e.g. GONG and SOHO/MDI). We are therefore still working on improving our calibration procedure and we hope that PICARD will be extended at least one year in order to gather more intensity data that can be compared and cross-calibrated with SDO/HMI intensity and velocity images.

Acknowledgments

The data were provided by the PICARD/SODISM, SOHO/MDI, and SDO/HMI instrument teams. PICARD is a mission supported by the Centre National d'Etudes Spatiales (CNES), the CNRS/INSU, the Belgian Space Policy (BELSPO), the Swiss Space Office (SSO) and the European Space Agency (ESA). SOHO is a project of international cooperation between ESA and NASA. D. Salabert and P. Journoud acknowledge financial support from CNES. The authors wish to thank S. J. Jiménez-Reyes for useful discussions and for providing his codes at the early stage of the project.

References

[1] Appourchaux A, Gizon L and Rabello-Soares C 1998a *A&AS* **132** 107
[2] Appourchaux A and Toutain T 1998b *IAU Symposium 181* (Poster volume) Ed J Provost and F X Schmider (Nice: Nice University) p 5
[3] Appourchaux A and Nunge A 2012 *Internal IAS report*
[4] Barban C, Hill F, Kras S 2004 *ApJ* **602** 516
[5] Corbard T, Boumier P, Appourchaux T, Jiménez-Reyes S J, Gelly B and the PICARD Team 2008 *AN* **329** 508
[6] Corbard T, Berthomieu G, Morel P, Provost J, Schou J, Tomczyk S 1997 *A&A* **324** 298

[7] Jiménez-Reyes S J 2001 *PhD Thesis* (La Laguna University, Spain)
[8] Meftah M et al. 2013 *Sol. Phys.* submitted
[9] Nigam R, Kosovichev A G , Scherrer P H and Schou J 1998 *495* L115
[10] Nigam R and Kosovichev A G 1998 *ApJ* **505** L51
[11] Howe R 2009 *Living Rev. Solar Phys.* **6** 1. URL: http://www.livingreviews.org/lrsp-2009-1
[12] Schou J *PhD Thesis* 1992 (Stanford University , USA)
[13] Schou et al. 2002 *ApJ* **567** 1234
[14] Severino G, Magrí M, Oliviero M, Strauss T and Jefferies S M 2001 *ApJ* **561** 444
[15] Toner C, Jefferies S and Toutain T 1999 *ApJ* **518** L127
[16] Toutain T, Berthomieu G and Provost P 1999 *A&A* **344** 188
[17] Thuillier G, Dewitte S, Schmutz W and The PICARD Team 2006 *Adv. Space Res.* **38** 1792

Multi-wavelength analysis of active regions using ring-diagram technique

S C Tripathy[1], K Jain[1], R Howe[2], R Bogart[3], S Basu[4], M C Rabello-Soares[5], F Hill[1]

[1] National Solar Observatory, Tucson, AZ, 85719 USA
[2] School of Physics and Astronomy, Edgbaston, Birmingham B15 2TT, UK
[3] HEPL, Stanford University, Stanford, CA 94305, USA
[4] Yale University, New Haven, CT 06520, USA
[5] Departamento de Física, Universidad Federal de Minas Gerais, Belo Horizonte, Brazil

E-mail: `stripathy@nso.edu`

Abstract. With the availability of high-cadence and high-resolution Doppler and intensity images from the *Solar Dynamics Observatory's* Helioseismic Magnetic Imager (HMI) and Atmospheric Imager Assembly (AIA), we analyze the characteristics of high-degree solar acoustic modes in active regions. We apply the ring-diagram technique to Doppler, continuum intensity measurements from HMI, and intensity images from AIA 1600 Å and 1700 Å passband to examine the high-degree mode parameters and sub-surface horizontal flows.

1. Introduction

Helioseismic studies using many different observables, which provides information at different heights in solar atmosphere, are important to determine the mode parameters precisely [1] as well as to characterize the systematic errors. With this goal in mind, we have previously analyzed [2] a few active regions using high-resolution observations of the Sun from Helioseismic Magnetic Imager (HMI). Here, we extend these analysis to include the intensity observations from Atmospheric Imager Assembly (AIA) on board *Solar Dynamics Observatory* (SDO) as well as the Doppler observations from Global Oscillation Network Group (GONG).

2. Data Analysis

The data used in this work consist of Doppler velocity (V) and continuum intensity (C) from HMI and 1600 Å and 1700 Å passband images from AIA with a cadence of 45 seconds and 48 seconds, respectively. For consistency checks, we also use GONG's Doppler (V_g) measurements where velocity observations are taken at a 1 minute cadence. It may be noted that the C, V, and V_g observables are formed around 20 km, 100 km, and 200 km while the AIA 1700 Å and 1600 Å bands are centered at 360 km and 480 km, respectively above the bottom of the photosphere [3].

The active regions that we select for this study consist of NOAA active regions (AR) 11092, 11093 and 11330 out of which the first two AR's consist of isolated and simple axisymmetric sunspots while the last AR is more complex consisting of many sunspots. The full disk images corresponding to these ARs are processed through the ring-diagram pipeline [4] where we select

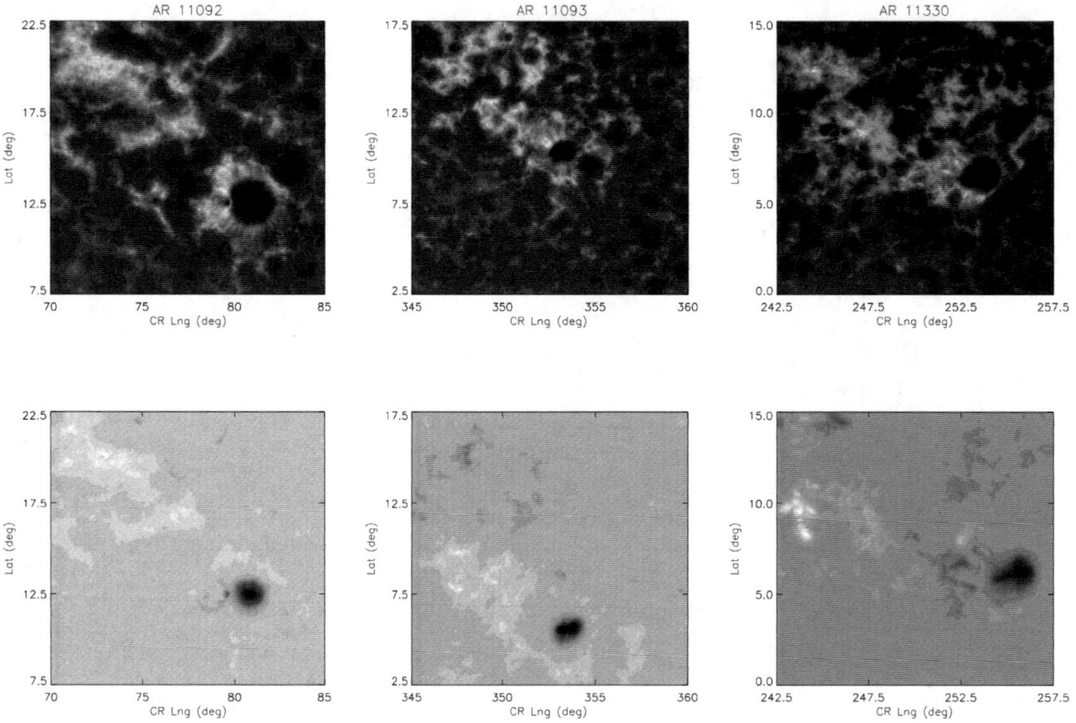

Figure 1. Mean values of active region data cubes as seen in AIA 1600 Å band (top panels) and HMI magnetograms (bottom panels).

Table 1. Details of the active regions used in this work. CR refers to the Carrington rotation. The values of the magnetic activity index (MAI) is provided by the HMI ring-diagram pipeline.

NOAA No.	CR No.	Latitude	Carrington Longitude	MAI in Gauss	Start Time in UT
11092	2099	15.0	077.5	40.78	2010.08.03_03:40
11093	2100	10.0	352.5	28.68	2010.10.08_13:57
11330	2116	07.5	250.0	91.00	2011.10.27_17:32

an area of 384 × 384 pixels around the active region and track it for 1728 minutes at the Carrington rate. Figure 1 shows the mean data cube of ARs as seen in the AIA 1600 Å band and the corresponding magnetic field data taken from HMI. More information on the data is summarized in Table 1.

The tracked cubes are then Fourier transformed in space and time to obtain the power spectra which are fitted to asymmetric profile model [5] to estimate the mode parameters. The fitted velocities are finally inverted to infer the depth dependence of horizontal flows [6].

3. Results

The frequency differences between modes obtained from different spectra are shown in Figure 2. We find that the differences are minimum between GONG and HMI Doppler observations while those between continuum and other observables are significantly large specially in the frequency

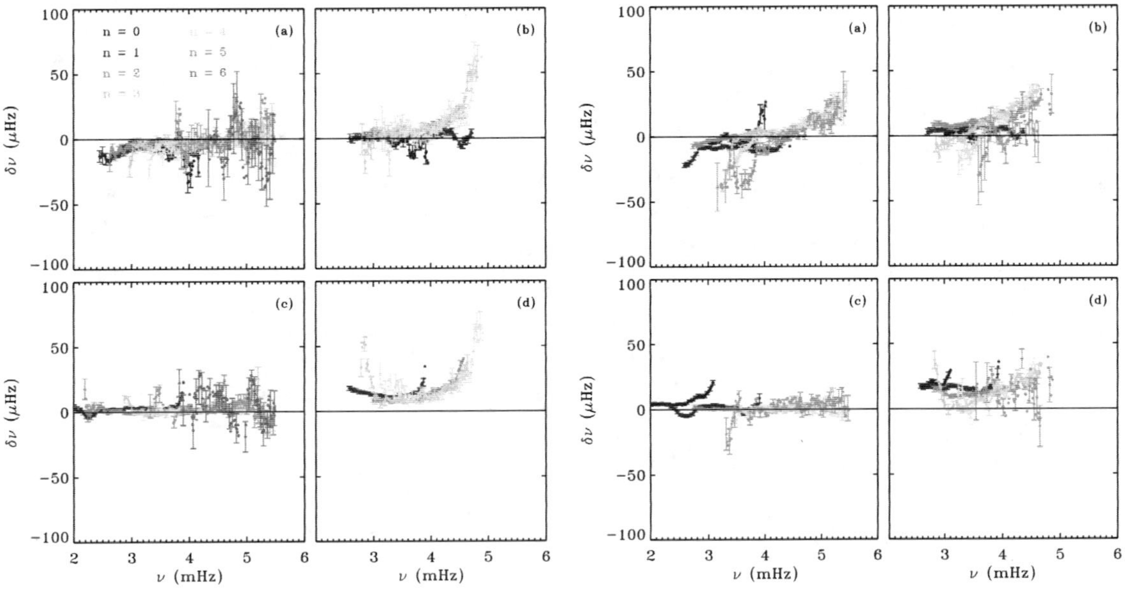

Figure 2. Frequency differences between different observables (a) HMI V - AIA 1600 (b) HMI C- AIA 1600 (c) HMI V - GONG V (d) HMI C - GONG V for AR 11092 (left) and AR 11330 (right)

range of 4.5−5 mHz. From the study of quiet regions it is known that the frequency differences are significantly reduced when an asymmetric profile is used for fitting the spectra from different observables as compared to the use of a symmetric model profile [7]. However, a recent study using HMI and AIA data has shown that the frequencies measured from different observables in quiet regions do not agree with each other [8]. Thus there appears to be disagreement between results reported between [7] and [8] which used data from different instruments. Thus we postulate that the simplified asymmetry profiles used in ring-diagram studies do not adequately describe the correlation between the background power and peaks in active regions.

The horizontal and vertical components of the sub-surface flows for different ARs are shown in Figure 3. Each plot has 5 curves representing the flows from different observables. In general, the flow shows the two-layer structure increasing inward with depth. The flows near the surface are negative implying a retrograde motion while deeper down these are positive indicating prograde motion. These flows are consistent with measurements of flows in active regions made previously using Michelson Doppler Imager data [6] and our analysis using HMI data [2]. We also find that the zonal flows measured from different observables agree reasonably except for the one measured from AIA 1600 Å band. On the other hand, the meridional flows show larger variations between different observables and similar to the zonal flows, the discrepancy is largest for AIA 1600 Å band. We also find a large discrepancy for AIA 1700 Å band at greater depths for AR 11092.

4. Summary

The ring-diagram analysis of active regions using data from AIA and HMI observables show that the magnitude of the differences between mode frequencies computed from different spectra are significantly large at the higher end of the frequency range while the differences between the Dopplergrams from two different instruments are minimized. The sub-surface zonal flows, in

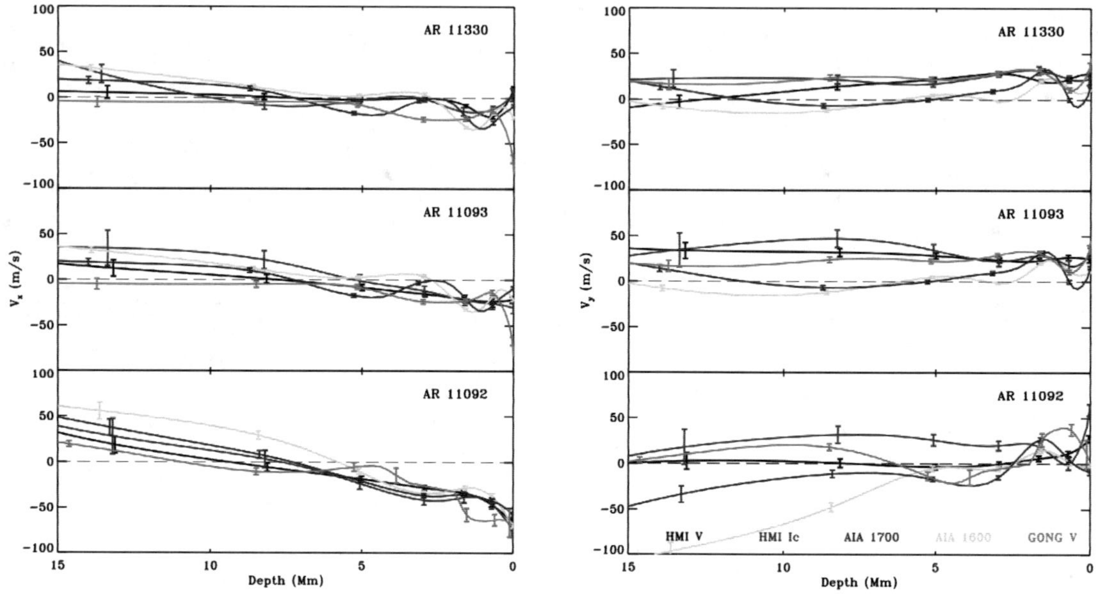

Figure 3. The horizontal velocity V_x (zonal) and V_y (meridional) for different observables and active regions as marked in the Figure.

general, are found to be consistent between different observables although the discrepancies are marginally larger for the meridional flows as compared to the zonal flows. Thus, it appears that the inference of sub-surface flows do not have a significant dependence on the choice of observables. A statistical study of many active regions is required to confirm these findings.

Acknowledgments

This work utilizes GONG data obtained by the NSO Integrated Synoptic Program (NISP), managed by the National Solar Observatory, which is operated by AURA, Inc. under cooperative agreement with the National Science Foundation. SDO data courtesy of SDO (NASA) and the HMI and AIA consortium. This work is partially supported by NASA grant NNH12AT11I.

References

[1] Harvey J, Hill F, Komm R, Leibacher J, Pohl B and the GONG 1998 *Proc. IAU Sump. 185: New Eyes to See Inside the Sun and Stars* eds. F-L Deubner, J Christensen-Dalsgaard and D Kurtz (Kluwer: Dordecht) p49-50
[2] Tripathy S C, Jain K, Howe R, Bogart R S and Hill F 2013 *Astron. Nachr.* **333** 1013-1017
[3] Fleck B, Couvidat S and Straus T 2011, *Sol. Phys.* **271** 27
[4] Hill F 1988 *Astrophys. J.* **333** 996-1013
[5] Basu S and Antia H M 1999, *Astrophys. J.* **525** 517-523
[6] Haber D A, Hindman B W, Toomre J and Thompson M J 2004, *Sol. Phys.* **220** 371-380
[7] Basu S, Antia H M and Bogart R S 2001 *Proc. Helio- and Asteroseismology at the Dawn of the Millennium* ed. A Wilson (Noordwijk:ESA) **464** p 183-186
[8] Jain K, Tripathy S C, Basu S, Bogart R S, González Hernández I, Hill F and Howe, R 2013 *J. Phys.: Conf. Series*, these proceedings

Journal of Physics: Conference Series **440** (2013) 012027 doi:10.1088/1742-6596/440/1/012027

Temporal Variations of High-Degree Solar p-Modes using Ring-Diagram Analysis

Olga Burtseva[1], Sushant Tripathy[1], Richard Bogart[2], Kiran Jain[1], Rachel Howe[3], Frank Hill[1], and Maria Cristina Rabello-Soares[4]

[1] National Solar Observatory, Tucson, AZ, USA
[2] HEPL, Stanford University, CA, USA
[3] School of Physics and Astronomy, Edgbaston, Birmingham, UK
[4] Departamento de Física, Universidade Federal de Minas Gerais, Belo Horizonte, Brazil

E-mail: burtseva@noao.edu

Abstract. We study temporal variations in the amplitudes and widths of high-degree acoustic modes by applying the ring-diagram technique to the GONG+, MDI and HMI Dopplergrams during the declining phase of cycle 23 and rising phase of cycle 24. The mode parameters from all three instruments respond similarly to the varying magnetic activity. The mode amplitudes and widths show consistently lower variation due to smaller magnetic activity in cycle 24 as compared to the previous solar cycle.

1. Introduction

The structural and dynamical properties of the Sun, changing with the solar cycle, are reflected in variations of solar oscillation parameters. In this context, we study temporal variations of the amplitudes and widths of high-degree solar acoustic modes. Temporal variations in these parameters can give us insight into the processes of excitation and damping of the solar acoustic oscillations, and mechanisms regulating the solar magnetic cycle. The global and local analyses of solar acoustic modes have shown that the mode amplitudes and lifetimes are anti-correlated with the solar activity level, and strongly depend on the local magnetic flux [1, 2, 3, 4]. In this work, we derive the amplitudes and widths of high-degree acoustic modes by applying the ring-diagram technique [5] to the GONG (Global Oscillation Network Group), MDI (Michelson Doppler Imager) and HMI (Helioseismic and Magnetic Imager) Dopplergrams during the declining phase of cycle 23 and rising phase of cycle 24. We compare the results with level of magnetic activity in the analyzed regions.

2. Data analysis

The mode parameters of the solar acoustic oscillations in this work are obtained from GONG (for the period from 2001 to 2012), MDI (for the period from 2001 to 2010) and HMI (for the period from 2010 till 2013) Dopplergrams using the standard ring-diagram technique of the GONG and HMI pipelines. The size of a standard patch in the ring analysis is $15^\circ \times 15^\circ$. The GONG pipeline routine uses a symmetric Lorentzian profile [6] to fit the power spectrum of the solar oscillations. The HMI pipeline [7] applies both symmetric and asymmetric [8] profiles to

Journal of Physics: Conference Series **440** (2013) 012027 doi:10.1088/1742-6596/440/1/012027

derive the mode parameters. The results from both procedures are generally consistent for the parameters of interest here; thus, we show results from the symmetric profile fitting only.

The Magnetic Activity Index (MAI) values are computed from MDI 96-minute [9] and HMI 45-second magnetograms for the same location and times as the Doppler data.

The amplitudes and widths fitted from the ring analysis need to be corrected for center-to-limb and duty-cycle dependencies [4]. In this work, we discuss results from disk center patches where the foreshortening effect is minimal. GONG data were restricted to time sequences with a duty cycle of 70% or higher and corrected for the duty-cycle dependence using a linear regression technique [2]. MDI data during Dynamics Runs usually have a high duty cycle. However the Dynamics Runs are restricted to about two months of observations per year. In this work, we also use MDI observations between the Dynamics Campaigns when available with a sufficiently high duty cycle in order to get a better temporal MDI coverage of the mode parameters. Most of these data have a lower duty cycle, so we also restrict the MDI data to times with a duty cycle of 70% or higher, and correct the mode parameters for the duty-cycle dependence. We analyzed 453 days from MDI, improving MDI data coverage in comparison with previous work [10]. HMI amplitudes and widths were corrected for the low duty-cycle related offsets during the HMI semi-annual daily eclipse periods.

3. Results and discussion

The results for amplitudes and widths shown here are for a multiplet $\ell = 440$, $n = 2$ ($\nu = 3.2$ mHz). The correlation of its variations over a solar activity cycle with other multiplets in the 2.5−3.5 mHz frequency range is above 70%.

The variations of amplitude and width as a function of time and MAI, computed from GONG, MDI and HMI data, are shown in Figure 1. The quietest period during the last solar minimum, according to the Mt.Wilson sunspot index data (`http://obs.astro.ucla.edu/150_data.html`) and MAI from MDI, was observed during most of the year 2008 and the beginning of 2009. However, HMI observations started in April of 2010; thus for consistency we plot the mode parameters from all three instruments relative to the mean value for the year 2010.

The amplitudes from GONG increase by $\sim 4\%$ from the high solar activity period in 2001 till 2004. From 2004 through 2008, we find a long-term variation of a non-magnetic nature (see, [10]). A co-temporal systematical problem was found in GONG velocity measurements, especially those from the Learmonth station. We believe this caused the long-term variations seen in the mode amplitudes. The mode widths are affected to a much smaller degree. They increase by an amount comparable with the error of the width measurement and are thus not obvious in the plot. We plan to explore the problem and correct it in a subsequent analysis. Ignoring these long-term variations, we conclude that the mode amplitudes obtained from GONG increased from 2001 to 2008-2009 by $\sim 18\%$, and then decreased from 2009 to 2012 by $\sim 10\%$. The widths decreased from 2001 to 2008 by $\sim 7\%$. A rising trend in the widths also starts from $\sim$ 2009 and reaches $\sim 2\%$ by mid-2012. After correction for duty cycle, the amplitude of the solar cycle variations for cycle 23 is somewhat lower than that reported in [10]. The overestimation of solar cycle variations was due to a lower temporal coverage of GONG high-resolution observations (and consequently lower amplitudes and higher widths) in the early years and then a gradual increase.

The variations in amplitude and width from MDI data show a solar cycle trend similar to that from GONG data. The mode amplitudes obtained from MDI from 2001 to 2008 increased by $\sim 15\%$ and the mode widths decreased by $\sim 4\%$, both 3% less than the same quantities from GONG. The amplitudes from MDI then decreased from 2008 to 2010 by $\sim 8\%$, while the mode widths for the same period of time do not show a clear trend.

The mode amplitudes and widths from HMI nicely merge and continue the solar cycle trend shown by MDI, and are similar to those from GONG. The mode amplitudes obtained from HMI

Journal of Physics: Conference Series **440** (2013) 012027 doi:10.1088/1742-6596/440/1/012027

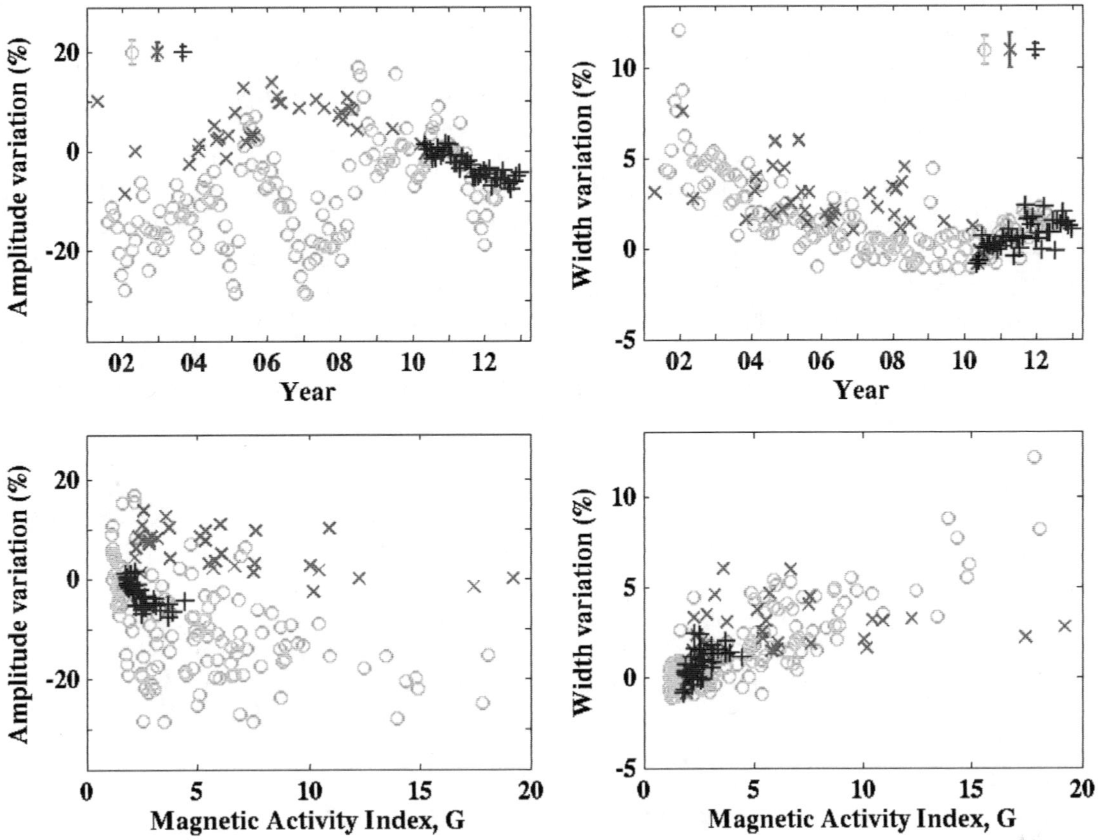

Figure 1. Variation of amplitude and width (relative to the mean value of 2010) as a function of time and MAI, computed from GONG (open green cicrcles), MDI (red crosses) and HMI (blue pluses). The mode parameters are derived from disk center patches. Each point is an average over one Carrington rotation. The typical error bars for the measurements are shown in the top panels.

decreased from 2010 to 2013 by $\sim 7\%$ while the mode widths increased by $\sim 1.5\%$.

In comparison with studies using global mode analysis (*e.g.*, [11, 12]), the mode parameters from local analysis in this work show a smaller variation over the solar cycle. This is mostly due to our analysis being limited to the central area of the Sun, which is usually outside the activity belt. For a complete picture of solar-cycle variations and a better comparison, we plan to analyze the temporal variations of the mode amplitudes and widths at different latitudes.

The decrease in amplitudes and increase in widths with increase in MAI (bottom panels of Figure 1) are consistent with previous results from global and local analysis. The long-term non-magnetic variations in GONG amplitudes from around mid-2004 to mid-2008, mentioned earlier, introduce large scatter in the values of the mode amplitudes in the range MAI < 8 G. After excluding the data points from GONG for this period, we find the linear correlation coefficient between amplitude and MAI is -0.80, and between width and MAI is 0.86. The correlation coefficients for MDI amplitudes and widths are -0.58 and 0.49, respectively. This is consistent with lower values of temporal variations with solar activity in the amplitudes and widths from MDI in comparison with those from GONG, discussed above. The correlation coefficients for HMI amplitudes and widths are −0.70 and 0.51, respectively. Note however that the HMI results so far only sample a small range in MAI.

According to some predictions and solar activity proxies (*e.g.*, [13], `http://solarscience.msfc.nasa.gov/predict.shtml`, `http://obs.astro.ucla.edu/150_data.html`), we are almost at the maximum of cycle 24, or may even have reached it. If the predictions are right, than the current solar activity maximum is certainly much lower than that of cycle 23, and the mode parameters clearly reflect this. Due to the lower magnetic activity, the mode amplitudes and widths in this study show consistently smaller variation, about two and three times, respectively, in the rising phase of cycle 24 as compared with the declining phase of cycle 23. It must be noted, however, that solar activity in the rising phase starts at higher latitudes and thus the equatorial part of the Sun, where these comparisons were made, exhibits less magnetic activity during the rising phase than the declining phase of the cycle. The actual differences are likely to be smaller than reported, if we compare the mode parameters near the equator at the same phase of the two cycles.

Acknowledgments
This work utilizes GONG data obtained by the NSO Integrated Synoptic Program (NISP), managed by the National Solar Observatory, which is operated by AURA, Inc. under a cooperative agreement with the National Science Foundation. The data were acquired by instruments operated by the Big Bear Solar Observatory, High Altitude Observatory, Learmonth Solar Observatory, Udaipur Solar Observatory, Instituto de Astrofísica de Canarias, and Cerro Tololo Interamerican Observatory. SOHO is a mission of international cooperation between ESA and NASA. The HMI data are provided by NASA/SDO and the HMI Science Team.

References
[1] Chaplin W J, Elsworth Y, Isaak G R, Miller B A and New R 2000 *Mon. Not. Royal Astron. Soc.* **313** 32
[2] Komm R W, Howe R, Hill F 2000 *Astrophys. J.* **531** 1094
[3] Rajaguru S P, Basu S and Antia H M 2001 *Astrophys. J.* **563** 410
[4] Howe R, Komm R W, Hill F, Haber D A and Hindman B W 2004 *Astrophys. J.* **608** 562
[5] Hill F 1988 *Astrophys. J.* **333** 996
[6] Haber D A, Hindman B W, Toomre J, Bogart R S, Thompson M J and Hill F 2000 *Solar Phys.* **192** 335
[7] Bogart R S, Baldner C, Basu S, Haber D A and Rabello-Soares M C 2011 *Journal of Physics: Conference Series* **271** 012008
[8] Basu S and Antia H M 1999 *Astrophys. J.* **525** 517
[9] Basu S, Antia H M and Bogart R S 2004 *Astrophys. J.* **610** 1157
[10] Burtseva O, Tripathy S, Howe R, Jain K, Hill F, Bogart R and Rabello-Soares M C 2011 *Journal of Physics: Conference Series* **271** 012012
[11] Salabert D and Jimenez-Reyes S J 2006 *Astrophys. J.* **650** 451
[12] Garcia R A, Salabert D, Mathur S, Regulo C, Ballot J, Davies G R, Jimenez A and Simoniello R 2013 *J. Phys. Conf. Proc.* this issue
[13] Lockwood M, Owens M, Barnard L, Davis C and Thomas S 2012 *Astron. Geophys.* **53** 3.09

Journal of Physics: Conference Series **440** (2013) 012028 doi:10.1088/1742-6596/440/1/012028

Enhanced sources of acoustic power surrounding AR 11429

Alina Donea, Christopher Hanson

Monash Centre for Astrophysics, School of Mathematical Sciences, Monash University, Victoria 3800, Australia

E-mail: alina.donea@monash.edu

Abstract. Multi-frequency power maps of the local acoustic oscillations show acoustic enhancements ("acoustic-power halos") at high frequencies surrounding large active region. Computational seismic holography reveals a high-frequency "acoustic-emission halo", or "seismic glory" surrounding large active regions. In this study, we have applied computational seismic holography to map the seismic seismic source density surrounding AR 11429. Studies of HMI/SDO Doppler data, shows that the "acoustic halos" and the "seismic glories" are prominent at high frequencies 5–8 mHz. We investigate morphological properties of acoustic-power and acoustic emission halos around an active region to see if they are spatially correlated. Details about the local magnetic field from vectormagnetograms of AR 11429 are included.

We identify a $15''$ region of seismic deficit power (dark moat) shielding the white-light boundary of the active region. The size of the seismic moat is related to region of intermediate magnetic field strength. The acoustic moat is circled by the halo of enhanced seismic amplitude as well as enhanced seismic emission. Overall, the results suggest that features are related. However, if we narrow the frequency band to 5.5 – 6.5 mHz, we find that the seismic source density dominates over the local acoustic power, suggesting the existence of sources that emit more energy downward into the solar interior than upward toward the solar surface.

1. Introduction

Regions of enhanced high-frequency oscillation power are usually observed at the boundaries of active regions and are referred to as "acoustic halos". They show a local enhancement of oscillatory amplitude at the Sun's surface that could be due to enhanced local generation of acoustic noise or to a greater local sensitivity of the surface to acoustic noise arriving from distant sources. Properties of halos were studied by Braun et al. [1], Brown et al. [2], Hindman and Brown [3], Jain and Haber [4], Donea at al. [5, 6], more recently by Schunker and Braun [7] and Rajaguru et al. [8]. Khomenko and Collados [9] summarized the halos' properties including a vast list of references related to this topic. Schunker and Braun [7] have examined the power distribution for velocity observations from MDI and found that the excess high-frequency power corresponds to regions with the magnetic-field inclination (as deduced from potential-field source-surface extrapolation) in the 40–60 degree range. Jain and Haber [4] showed that halos seen in Dopplergrams from the MDI/SOHO satellite tend to be prominent in intermediate magnetic-field strengths of 50–250 G. The origin of these high-frequency halos remains unclear.

Jacoutot et al. [10] simulated the enhanced acoustic emission in active regions by investigating how magnetic field affected convective motion and excitation of solar oscillations. They concluded that localized enhancements seen in acoustic halos are caused by high-frequency turbulent convective motions in the presence of moderate magnetic field. Hanasoge [11] argued that the enhanced acoustic powers were a result of scattering by magnetic regions' low-l to high-l modes. It has been suggested [2] that acoustic emission from the solar granulation should be relatively localized and episodic, emanating largely as relatively discrete wavepackets emitted from convective plumes falling into the solar interior from near-surface layers at which granular convection takes place (see also [12]).

A computational seismic holography technique has been also used to search for additional emission of acoustic waves relative to the quiet Sun. The seismic source power computations consist in regressing the acoustic field a single skip, from the observed surface disturbance back to the surface point at which it is supposed to have originated. The resulting 6 mHz emission power of outward propagating waves reveal a seismic source excess surrounding the active region called the "seismic glory" ([6, 13]). The seismic glories contain some of the most intense seismic emitters, emitting as much as 150–250% more acoustic power than the quiet Sun.

Brown et al. [2] suggested that seismic emission originates from enhanced episodic source activity, as a characteristic of wave generation by turbulence. Donea et al. [6] found no evidence of this in temporal and spatial distributions of seismic-source episodes, either in acoustic-power halos or acoustic-emission halos. Recently Lindsey and Donea (2013, this proceedings) have applied seismic holography to high-frequency p-modes in the quiet Sun to study the statistical relationship between seismic sources and the structure of the solar granulation. For details on the general technique of seismic holography, we refer the reader to [13].

Khomenko and Collados [9] proposed that seismic halos can be caused by additional energy injected by high-frequency fast mode waves refracted in the higher atmosphere due to the rapid increase of the Alfvén speed. Acoustic waves trapped under magnetic canopy have been also suggested as a source of seismic power [14].

In this paper, we explore spatial properties of acoustic halos and seismic glories, as a function of wave frequency, inclination and strength of magnetic field, in order to understand the origin of seismic excess emission surrounding large active regions.

We use 24-hours of Doppler velocity and vector magnetogram observations from the Helioseismic and Magnetic Imager (HMI) on the Solar Dynamics Observatory (SDO) of the active region AR 11429 which generated a X5.4 solar flare on March 7, 2012.

2. Comparison of helioseismic observables

Tracked Doppler data cubes from HMI were Fourier transformed in time and acoustic power maps calculated in the frequency ranges 2.5–4.5, 5.5–6.5, 6.5–7.5, and 7.5–8.5 mHz. Seismic holography is also applied to the Fourier transforms to derive source power maps (left panels in Figure 1). For details on the general technique of seismic holography, we refer to [13]. For details on the application of the holography technique to episodic phenomena, we refer to [6]. The panels of Figure 1 show seismic source power maps (left) and acoustic power maps (right) in the frequency bands centered at 9 (top row), 8, 6, and 3.5 mHz (bottom row) for March 7, 2012. Both seismic-source power (left column) and local seismic power (right column) are seen to be suppressed in the active region at all frequencies; this is coincident with the location of moderate-to-strong magnetic fields, and includes plages. A high-frequency enhancement of power in both local seismic amplitude and local seismic emission surrounds the active region. The regions of enhanced local acoustic amplitude have been called "acoustic power halos". The regions of enhanced seismic emission have been called "acoustic glories". The seismic emission maps have a spatial resolution of 2.5 Mm, a diffraction limit attained by applying the diagnostic with an annular pupil (see Fig 4, p. 269 of [13]) whose radial range was 7–28 Mm. We found that the 6 mHz acoustic halo and the 6 mHz seismic-source glory streched over a region approximately 80–100 Mm thick.

The seismic source density is $|H_+|^2$. The mean of this for the quiet Sun is signified by $|H_{+,0}|^2$. Figure 2 shows the l-o-s magnetic field map with yellow contours indicating where normalized 6 mHz seismic-source power reverts to its quiet-Sun value, $|H_+|^2/|H_{+,0}|^2 = 1$. Green contours indicate the most seismic areas of the glories (exceeding 150% of $H_{+,0}$). Similarly, these zones also show the greatest enhancement in local acoustic power (50% above the quiet-Sun power). Zone 2 was found to be the most acoustically active at high frequencies. Figure 3 shows a slice of the acoustic emission maps centred on zone 2. The multifrequency acoustic and seismic emissions from zones 1 and 3 are shown in Figure 4. Table 1 shows a summary of the total acoustic power averaged over the defined areas, along with estimates of the mean magnetic field strength in the areas, and mean inclination angle of **B**.

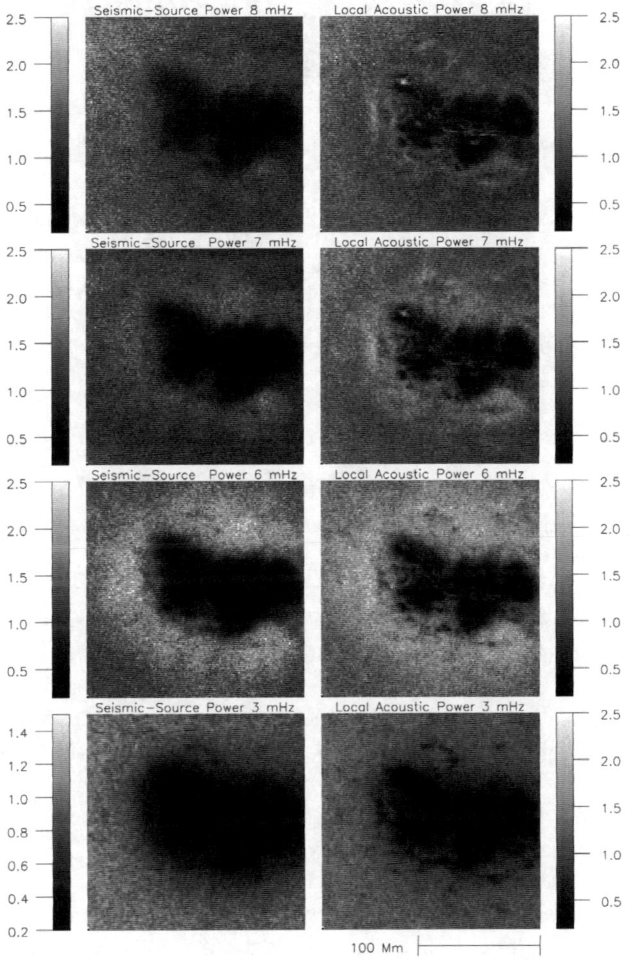

Figure 1. Multi-frequency oscillation power maps (as labelled) from seismic source emission (left column) and local Doppler velocity measurements (acoustic-power maps; right column). Maps are normalized to one in the quiet Sun.

We have identified a region of 6 mHz seismic emission deficit power ("quiet seismic moat") just beyond the white-light boundary of the active region (dashed yellow contours in Figure 2), surrounding even the most ragged areas of its boundary. It has a width of $\approx$ 20–30 Mm, and a

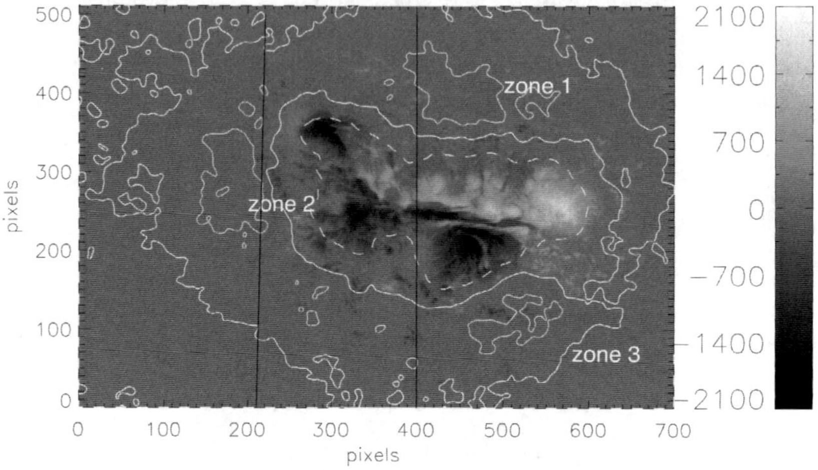

Figure 2. Contours of the 6 mHz seismic emission overplotted on the HMI l-o-s magnetic field strength map. Yellow contours show the seismic glory boundary, at the values of the quiet sun $|H_+|^2/|H_{+,0}|^2 = 1$. Dashed yellow curves show the white-light boundary of the active region. Green contours ($1.5\times|H_{+,0}|^2$) show the most intense 6 mHz seismic areas, labelled as zone 1,2 and 3. Vertical lines define the submaps of the next figure.

highly inclined magnetic field (averaged magnetic field strength $B = 230$ G). The local seismic power in the moat ranges from 0.5–1 (in normalized units). The moat contains most of the plages surrounding the active region, clearly showing patchy features in acoustic/seismic emission due to the magnetic field acoustic attenuation. The magnetic canopy of the active region [16] may also have a role in changing the properties of solar convection in the moat, reducing the seismic emission. The seismic moat is surrounded by the seismic glory seen in all high frequency seismic power maps. We also found that the deficit in the seismic emission at the white-light boundary of the active region stays at a constant value of 0.5 (dashed yellow contours) of the quiet-Sun power everywhere around the active region.

Cospatial maps of 6 mHz local acoustic power, seismic source emission, continuum intensity, magnetic field strength from a vector magnetogram and field inclination angles γ are shown in Figure 5. These images are centred upon zone 2 in Figure 2, the region of most intense seismic activity. We noticed an increased acoustic and seismic power in regions of approximately horizontal magnetic field whose strengths are between 50–200 G.

We are particularly interested in spatial correlations between the powerful discrete seismic emitters seen in glories and in the halos. Seismic holography is specifically designed for optimum local discrimination of acoustic sources, both temporally and spatially. Contours in Figure 5 show kernels in the 6 mHz local acoustic power (white) and 6 mHz seismic emission (black), within which the respective field value exceeds 180% of the quiet-Sun mean. Although in some areas we have identified a very good spatial correlation between the excess-local-seismic-power kernels and the excess-seismic-emission kernels, more than 60% of the respective kernels do not overlap. The highly emissive seismic sources have an average size of 2.0 Mm×3.5 Mm. To study the episodic nature of seismic emission in the glories we are preparing another paper, Hanson and Donea (2013).

We noticed a preference for the 7, 8 and 9 mHz enhanced acoustic power to shrink to a much

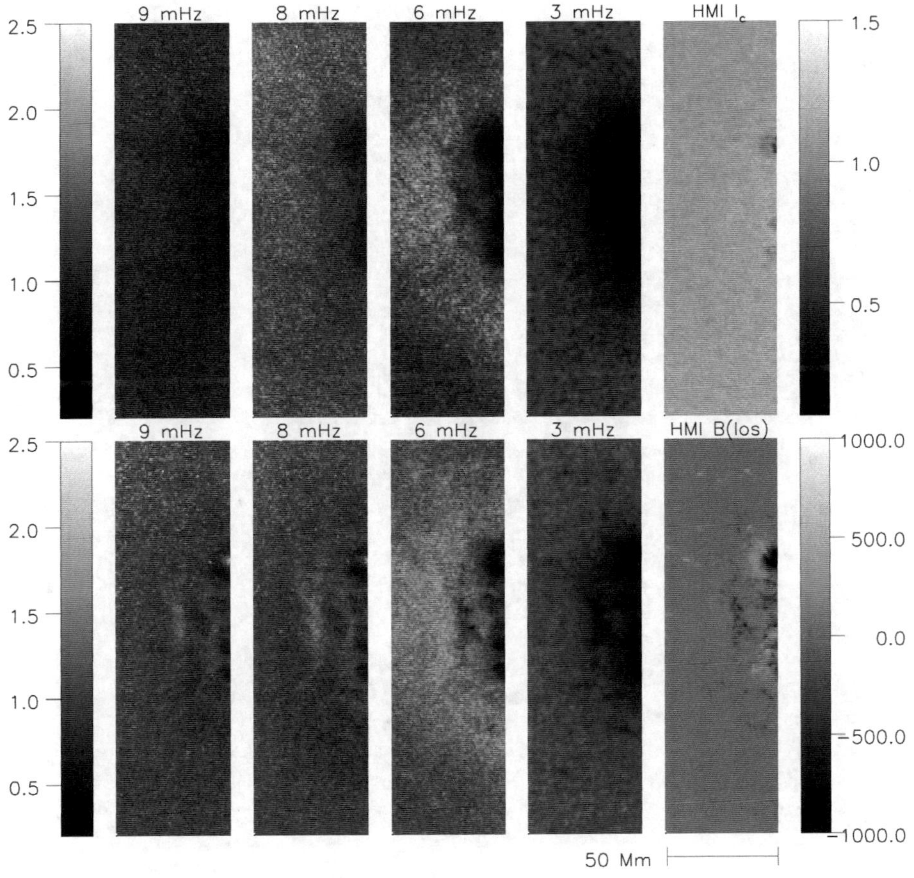

Figure 3. Seismic emission and acoustic local power maps at various frequencies in the most acoustically active part of the active region (zone 2). For reference, a normalized-to-the-quiet-Sun map of the intensity continuum and of the l-o-s magnetic field are included too.

narrower area outlining the active region and magnetic-field concentrations such as plages. Howe et al. [15] found similar behaviour for a different active region. We find that the corresponding regions of enhanced seismic emission at these frequencies become rather broad and diffuse by comparison (Figure 3), fading beyond 9 mHz. This finding is reflected also in Figure 6, where histogram plots emphasizing the difference in density distribution of the acoustic power halo and the seismic power glory at various frequencies are shown. The seismic emissivity dominates over the local acoustic oscillations at 6 mHz, reaching power values of 2.4×quiet-Sun.

The correlation of the 6-mHz local seismic power and seismic-source power with the strength and inclination of the magnetic field is expressed in zones 1, 2, and 3 by scatter plots in Figure 7. The 6 mHz seismic source power increases up to 240% of the quiet-Sun power at 6 mHz and is strongest at horizontal intermediate magnetic-field strength (120G).

We also computed the linear Pearson correlation coefficient between the 6 mHz acoustic power and 6 mHz seismic power in various frequency bands, for different zones of the halo. We find that there is a very good $R_{6\mathrm{mHz}} = 0.979$ correlation between the 6 mHz acoustic power and seismic emission for the whole seismic halo.

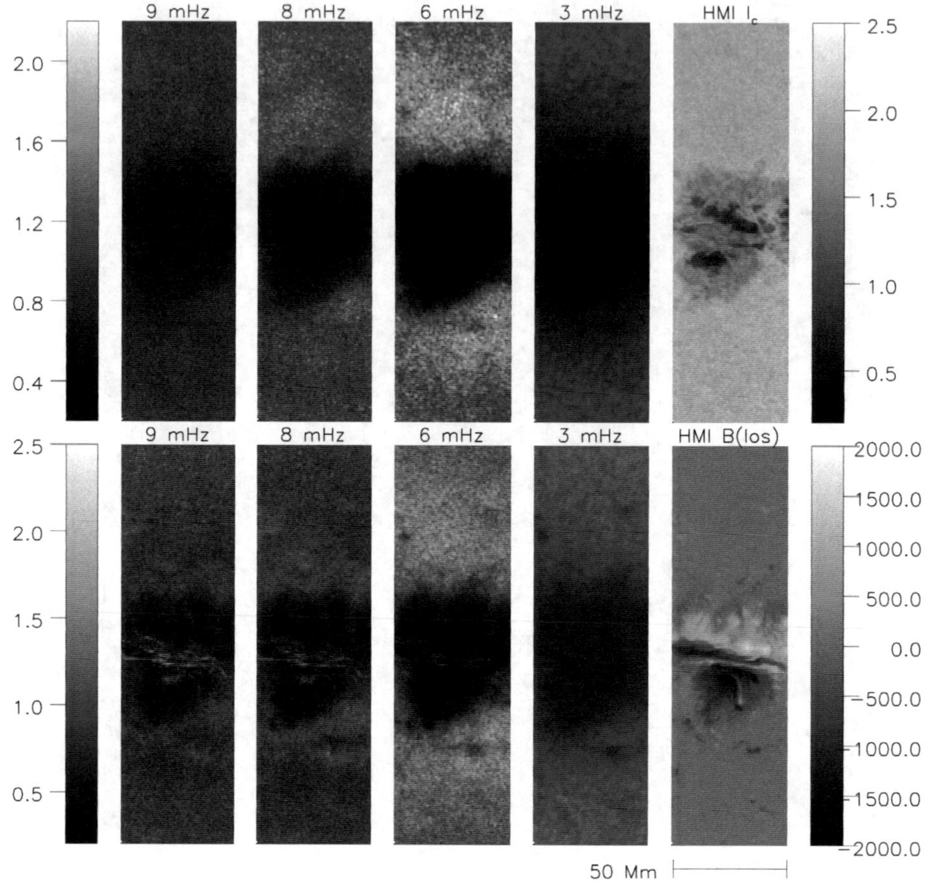

Figure 4. Same as the Figure 3, but for zone 1 (top) and zone 3 (below) the active region).

Table 1. Mean values of the Normalized Acoustic and Seismic Power of the most important areas of in the seismic halo. An average value for the total magnetic field strength and the inclination angle is also estimated.

Region	*Zone* 1	*Zone* 2	*Zone* 3	*Halo*	*QuietSun*	*Moat*
Egression Power 3mHz	0.72	0.96	0.80	0.88	1	0.44
Acoustic Power 3mHz	0.79	1.00	0.79	0.88	1	0.55
Egression Power 6mHz	1.91	9.91	1.73	1.55	1	1.27
Acoustic Power 6mHz	1.73	1.80	1.53	1.53	1	1.33
Egression Power 8mHz	1.14	1.34	1.00	1.07	1	0.83
Acoustic Power 8mHz	1.10	1.28	0.97	1.10	1	0.90
HMI B_{total} (G)	148	139	141	147	116	230
γ (°)	92	85	98	90	89	80

3. Final remarks

The purpose of this study is to examine the morphology of the enhanced seismic activity at high frequencies surrounding the region AR11429. We have found a very good spatial correlation

Journal of Physics: Conference Series **440** (2013) 012028 doi:10.1088/1742-6596/440/1/012028

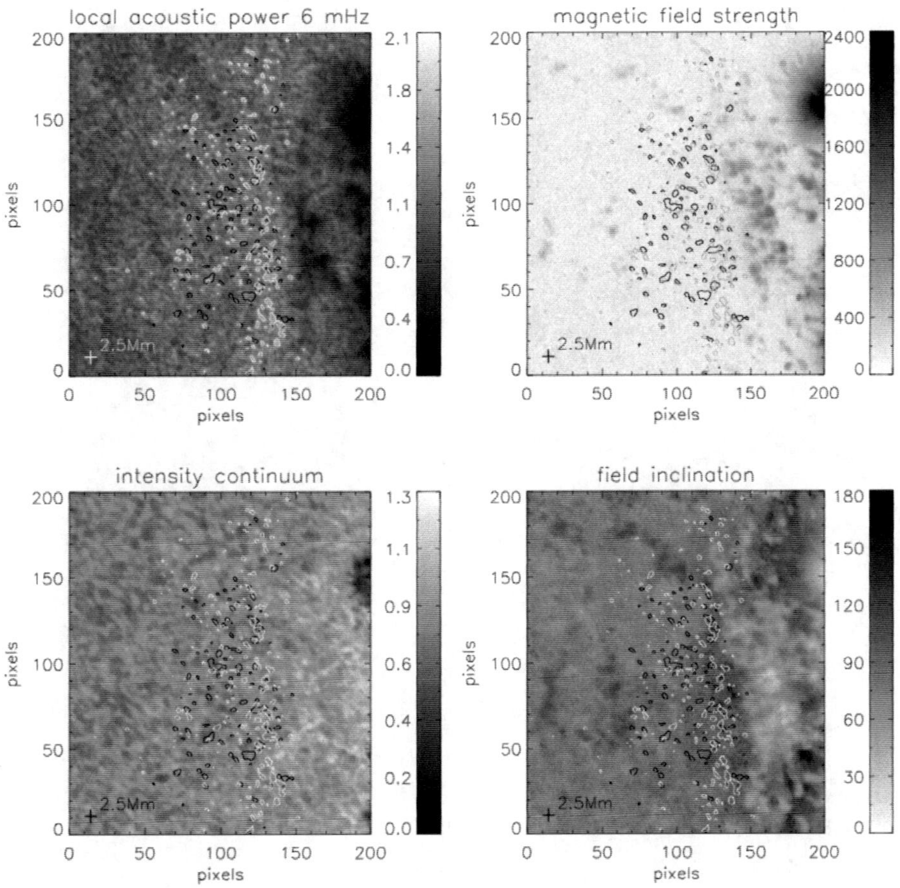

Figure 5. Co-spatial maps of zone 2 showing: a normalized 6 mHz local acoustic power, HMI/SDO vector magnetic field strength, normalized intensity continuum and field inclination γ. A weak spatial correlation exists between the acoustic power averaging roughly 180% in excess of the mean quiet Sun local acoustic power (white contours) and the seismic power at 180% in excess of the quiet Sun seismic power (black contours). Most of high power seismic emitters thrive in the areas free of plages, surrounding the active region. The length of each side of the maps is $\approx$ 70 Mm (1 pixel = 0.3475 Mm). The cross shows the spatial resolution of 2.5 Mm of the 6mHz seismic source power maps.

between the 6 mHz glories and the 6 mHz power in the local oscillations of the surface. However, the spatial correlation weakens when one compares the compact kernels of the most intense local oscillation and seismic emission (180% of the respective quiet Sun values). The latter are dependent of the diffraction resolution of the seismic-emission maps, as rendered by helioseismic holography. The 7–28 Mm pupil collects Doppler signal from magnetic areas of the active region, where the acoustic oscillations are attenuated. This "showerglass effect" [16, 17] will have an impact on the assessment of acoustic emissivity and its relationship to local acoustic amplitude. Efforts to correct the "shower glass" effect could answer some of these issues. Our conclusions are:

- As expected, the local oscillation power and any seismic source activity are suppressed in the active region at all frequencies.

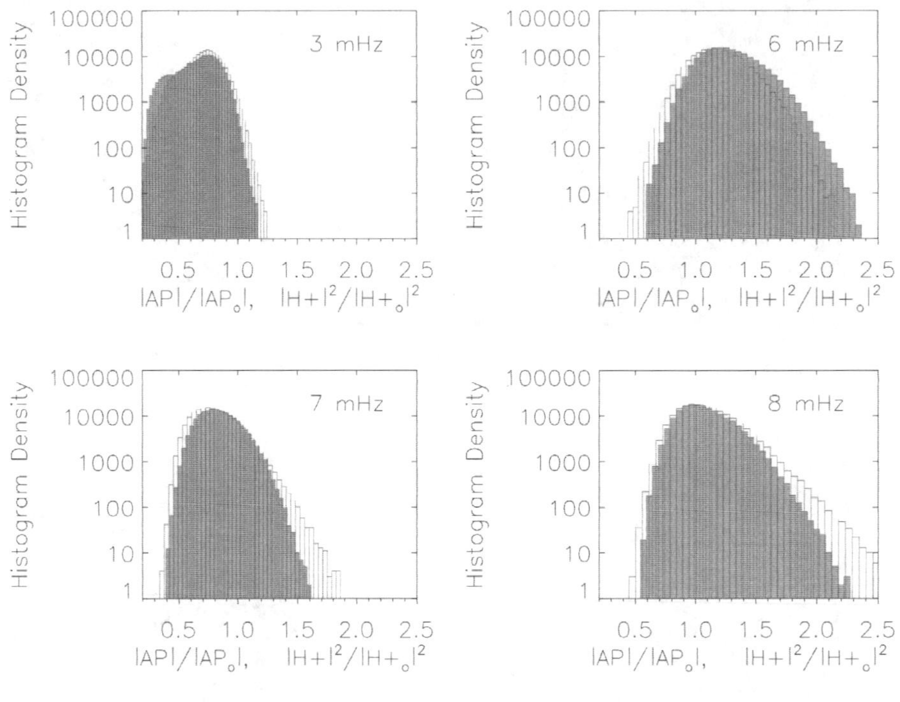

Figure 6. Histogram plots of the normalized acoustic power $|AP|/AP_o|$ (red) and seismic power $|H_+|^2/|H_{+,0}|^2$ (blue). At 6 mHz the seismic source emission dominates over the local acoustic oscillations.

- These regions of suppression are surrounded by halos of enhanced local-seismic-amplitude/power as well as enhanced seismic emission.
- The seismic-power halo contracts and weakens with increasing frequency, becoming indistinct by about 10 mHz. The seismic-emission halo (i.e., the "glory"), likewise weakens with increasing frequency, but, rather than shrinking, becomes more diffuse at higher frequencies.
- Seismic emission from areas surrounding the active region reaches up to 240% of the quiet Sun power in compact kernels.
- The patchiness of the seismic source density distribution in acoustic glories (and halos) appears to be related to plages.
- The seismic source power in the glory dominates over the local surface oscillations at frequencies of 6 mHz by a factor of 25%. This dominance appears to change with frequency.
- The white-light boundary of the active region (sometimes where the penumbra of large sunspots reaches into the quiet Sun) sustains a local seismic power and seismic emissivity of 0.5 of the respective mean-quiet-Sun power and emissivity.
- We recognise a "moat" of 6 mHz seismic emission deficit extending 20–30 Mm wide outside of the active region boundary.

There appears to be a strong relationship between the enhanced acoustic power seen at high frequencies and the enhanced seismic emissivity. However, differences between the distributions of the two phenomena suggest that this relationship may be somewhat oblique. If the sensitivity

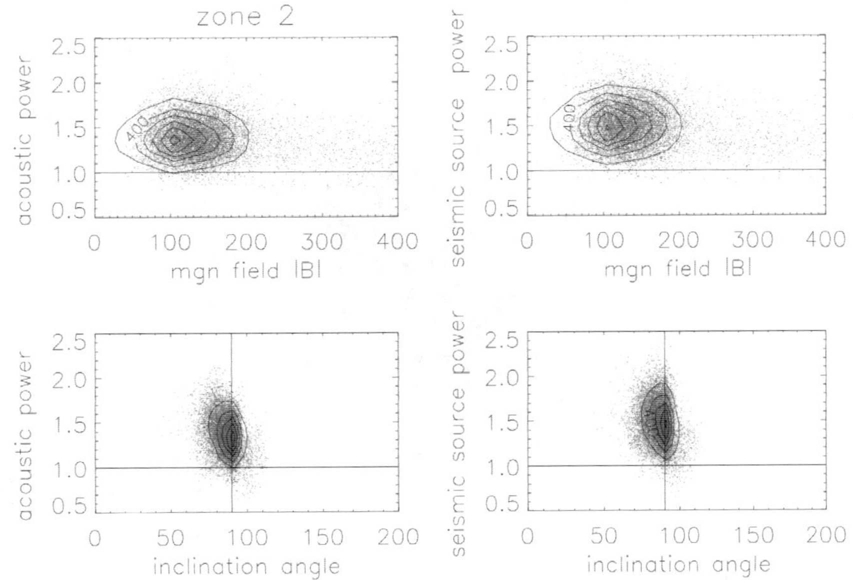

Figure 7. Scatter plot of normalized 6 mHz acoustic and seismic source power in zone 2, versus the magnetic field strength and magnetic field γ inclination angles. Vertical lines indicate $\gamma = 90°$. Power maps are normalized to one in the quiet-Sun.

of the photosphere to upcoming seismic radiation were the same in seismic halos as in the quiet Sun, we might expect a closer relationship, especially at high frequencies. If, for example, local emitters submerged just beneath the photosphere emitted the same power upward, into the local overlying photosphere, as downward (to arrive into the distant pupil of a holographic diagnostic), then we could expect regions of enhanced seismic power (directly above acoustic emitters) to coincide closely with regions of enhanced acoustic emission (into distant pupils). We understand that this would be the case for simple dipole or quadrupole emitters. However, it is possible to contrive coherently related combinations of dipole and quadrupole emitters that violate this relationship.

To complicate the issue, it is also possible that regions of quasi-horizontal magnetic fields with strength 50–200 Gauss have a different sensitivity to seismic noise arriving in their environments. Simulations of the effects of inclined magnetic field on helioseismic signatures could shed some light on this question.

We thank the referee for very useful comments and suggestions.

References

[1] Braun D C, Lindsey C, Fan Y and Jefferies S M 1992 *Ap.J.* **392** 739–745
[2] Brown T M, Bogdan T J, Lites B W and Thomas J H 1992 *Ap.J.* **394** L65–L68
[3] Hindman B W and Brown T M 1998 *Ap.J.* **504** 1029
[4] Jain R and Haber D 2002 *Astron. and Astrop.* **387** 1092–1099
[5] Donea A and Newington M 2011 *Journal of Physics Conference Series* **271** 012004
[6] Donea A C, Lindsey C and Braun D C 2000 *Sol.Phys.* **192** 321–333
[7] Schunker H and Braun D C 2011 *Sol.Phys.* **268** 349–362 (*Preprint* 0911.3042)
[8] Rajaguru S P, Couvidat S, Sun X, Hayashi K and Schunker H 2012 *Sol.Phys.* (*Preprint* 1206.5874)
[9] Khomenko E and Collados M 2009 *Astron. and Astroph.* **506** L5–L8 (*Preprint* 0905.3060)
[10] Jacoutot L, Kosovichev A G, Wray A A and Mansour N N 2008 *Ap.J.* **682** 1386–1391

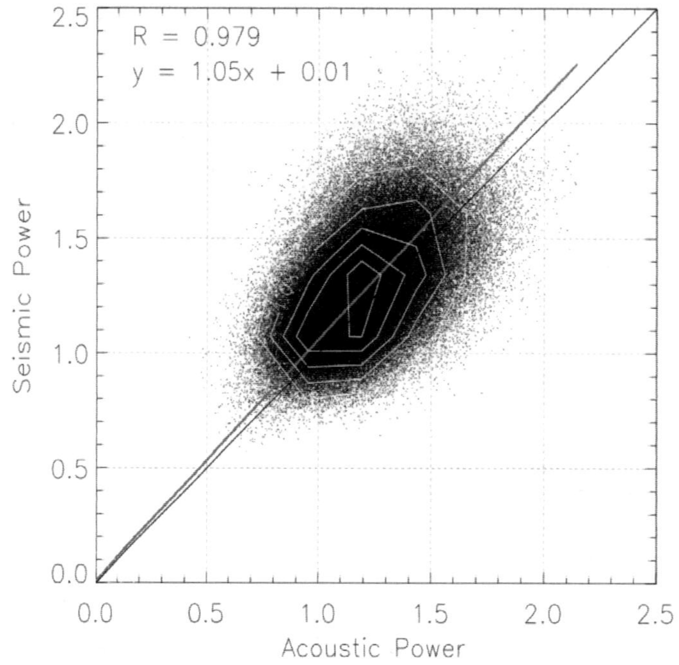

Figure 8. Scatter plot of normalized 6 mHz acoustic and seismic source power in the entire seismic glory. Contours show the density distribution of points, with a higher density at position (1.3,1.3). The Pearson correlation coefficient and the equation of the fitted line (thin red line) is displayed on the scatter plot. The black diagonally line is drawn for the purpose of seeing the slight deviation of the scatter plot from a perfect linear correlation.

[11] Hanasoge S M 2009 *Astron. and Astrop.* **503** 595–599 (*Preprint* 0906.4671)
[12] Rast M P 1999 *Ap.J.* **524** 462–468
[13] Lindsey C and Braun D C 2000 *Sol.Phys.* **192** 261–284
[14] Kuridze D, Zaqarashvili T V, Shergelashvili B M and Poedts S 2008 *Ann. Geophys.* **26** 2983–2989
[15] Howe R, Jain K, Bogart R S, Haber D A and Baldner C S 2012 *Sol.Phys.* **281** 533–549
[16] Lindsey C and Braun D C 2005 *Ap.J.* **620** 1107–1117
[17] Lindsey C and Braun D C 2005 *Ap.J.* **620** 1118–1131

Eclipse on the Coral Sea: Cycle 24 Ascending (GONG 2012, LWS/SDO-5, and SOHO 27) IOP Publishing
Journal of Physics: Conference Series **440** (2013) 012029 doi:10.1088/1742-6596/440/1/012029

Far-side helioseismic maps: the next generation

Irene González Hernández[1], Charles Lindsey[2], Douglas C. Braun[2], Richard S. Bogart[3], Philip H. Scherrer[3], and Frank Hill[1]

[1]National Solar Observatory, Tucson, Arizona, USA
[2]NorthWest Research Associates, Boulder, Colorado, USA
[3]Stanford University, Stanford, California, USA

E-mail: `irenegh@nso.edu`

Abstract.
For more than a decade, far-side seismic maps of medium-to-large active regions have proven their capability as a space weather forecasting tool. In the last few years, these maps have started to serve another purpose: complementing the front side observations that are input to different solar models. Photospheric flux transport as well as solar spectral irradiance models have been shown to produce improved results when incorporating the far-side seismic maps as well as providing better forecasting. The challenge for the future is twofold: Far-side seismic monitoring needs to be more sensitive, and it needs to offer more information. We present here initial steps towards fulfilling these goals using higher resolution input images, adding extra *skips* to the analysis and changing the presentation of the maps.

1. Introduction

Since 2000, when Lindsey and Braun [1] applied seismic holography to reconstruct large active regions at the non-visible hemisphere of the Sun, maps of the far hemisphere have been obtained from helioseismic analysis of Earth side observations of the solar surface velocity field. Currently far-side maps using data from both the Global Oscillation Network Group (GONG) and the Helioseismic and Magnetic Imager (HMI) are calculated daily and presented on-line for space weather forecasters (*farside.nso.edu* and *sdo.gsfc.nasa.gov/data/*). The location of the active regions is highlighted and the seismic signature is calibrated in terms of the magnetic field strength [2]. Figure 1 presents an example of a calibrated helioseismic far-side map for October 7, 2012 showing three active regions. The on-line maps include an associated file detailing the position, maximum seismic signature (phase shift), probability of eventual appearance on the front side [3], date of arrival at the East limb and the NOAA number if the active region was previously on the front side.

Fontenla et al. [4], in a case study, explored the possibility of using a combination of far-side maps from backscattered Ly-α observed by the SWAN instrument on board of SOHO and helioseismic far-side maps to complement the front-side observations used to model and forecast the solar spectral irradiance. The forecast of Ly-α was significantly improved by adding the far-side information. Very recently, Arge et al [5] used far-side helioseismic maps as input to the Air Force Data Assimilative Photospheric Flux Transport (ADAPT) model, finding that both the solar wind forecast and the F10.7 forecast from ADAPT maps improve with far-side information. This paper describes initial steps we are taking towards improving the look of the maps to make them more usable for space weather forecasters as well as to provide more accurate values for the maps to be used as input to solar models.

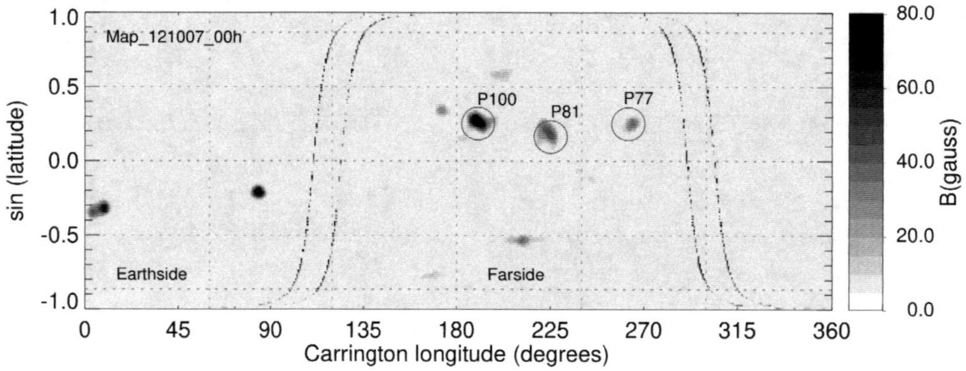

Figure 1. Calibrated GONG seismic far-side map for October 7, 2012 showing three active regions. Signatures with more than 70% probability of appearance upon arrival to the near hemisphere are highlighted in red.

2. Enhancing the seismic signature of active regions

Once the far-side seismic maps have been calibrated into magnetic field strength [2], they can complement front-side observations of the photospheric magnetic field for a full view of the Sun [5]. The information from the far side is particularly useful in case of emergence of new active regions or re-growth of decaying flux. Several lines of research are open to compute the maps more accurately. We have had successful results with two of them: the addition of extra *skips* to the computations and a higher spatial resolution in the input data.

2.1. Adding more skips to the calculations

The current far-side maps correlate helioseismic signatures from waves that skip twice beneath the solar surface (2×2-skip holography) to cover the central far side of the Sun and signatures from waves that skip once with waves that skip three times (1×3-skip holography) to cover the outer far side and limb. J. Zhao [6] developed 2×3-skip time-distance correlations for far-side monitoring. This has optimum sensitivity in the intermediate region where the 2×2-skip and the 1×3-skip correlations are sub-optimum. Figure 2 shows a comparison between calibrated far-side maps using the standard 2×2-skip and 1×3-skip combination (left) and the new 2×3-skip maps (right). Because the calibration is based on a relationship between the phase shift and the magnetic field found by González Hernández et al. [2] for the traditional maps, and the 2×3-skip maps present a different phase shift distribution, the new maps were first divided by a factor of 2.5 before applying the calibration. These maps shows a clear enhancement of the seismic signal of active region NOAA 11476 three days before it appears on the front side of the Sun. A preliminary analysis of three similar cases show improvements in two of them and no improvement in the other one. A statistical analysis of 4 years of 2×3-skip maps to quantify the improvement and to calibrate the phase shift in terms of magnetic field strength is underway.

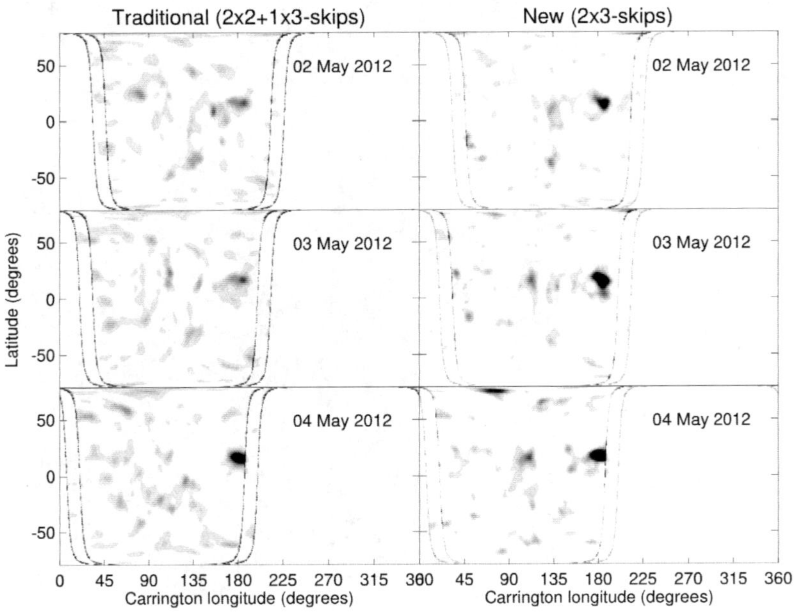

Figure 2. Far-side maps calculated with the traditional 2×2 and 1×3-skip schema (left) compared to the simultaneous maps calculated using 2×3-skip combinations. This series shows a particular example of far-side maps with an "ad-hoc" calibration where the 2×3-skip combination clearly enhances the signature of active region NOAA 11476 three days before it appears on the front side of the Sun. These maps were calculated from GONG Doppler observations.

2.2. Higher resolution input data

Due to technical limitations of several types, the current far-side maps are calculated with binned-down 200x200pixel^2 Dopplergrams. The near-real time availability of high resolution Dopplergrams enhances the quality of the far-side maps. Application of a Postel projection to the high-resolution image before it is binned down circumvents seismic signal lost approaching the limb due to foreshortening. Due to the heavy load of the computations, a faster algorithm for parallel processing of the data has been developed and tested, and is nearing implementation. Preliminary comparisons between far-side maps obtained from the low resolution (binned down) versus the high resolution HMI Dopplergrams show an increased signal-to-noise ratio in the far-side maps. A full statistical study is needed to properly quantify the improvement.

3. A new look for space weather forecasters

The higher resolution and inclusion of waves that have taken more skips offer the prospect of more sensitive maps with finer spatial resolution. Far-side seismic maps are a direct input for space weather forecasters, because they show the location and strength of active regions that will rotate to face the Earth. Currently, the synoptic maps present the far-side helioseismic signature along with the front side magnetograms. The front side represents the absolute value of the magnetic field smoothed to helioseismic resolution (see Figure 1 or Figure 3 left hand panels) to give a sense of continuity as the active region crosses from far side to near side. This has proven confusing for some users. The new look will present the front-side observations with a higher resolution and will preserve polarity, while the far-side calculations will use a color palette that will more intuitively represent helioseismic signatures, which cannot discriminate polarity. Preliminary new-look maps are shown in Figure 3. The same sequence of three consecutive days (May 22-24 2012) including active region NOAA 11476 on its second full passage across the farside is shown with GONG data and using the traditional wave combination and color palette (left) and calculated with HMI data, including the new 2×3-skip signature with the new palette (right). The inclusion of the new 2×3-skip signature seems to localize the active region better. However a full statistical evaluation of this is needed.

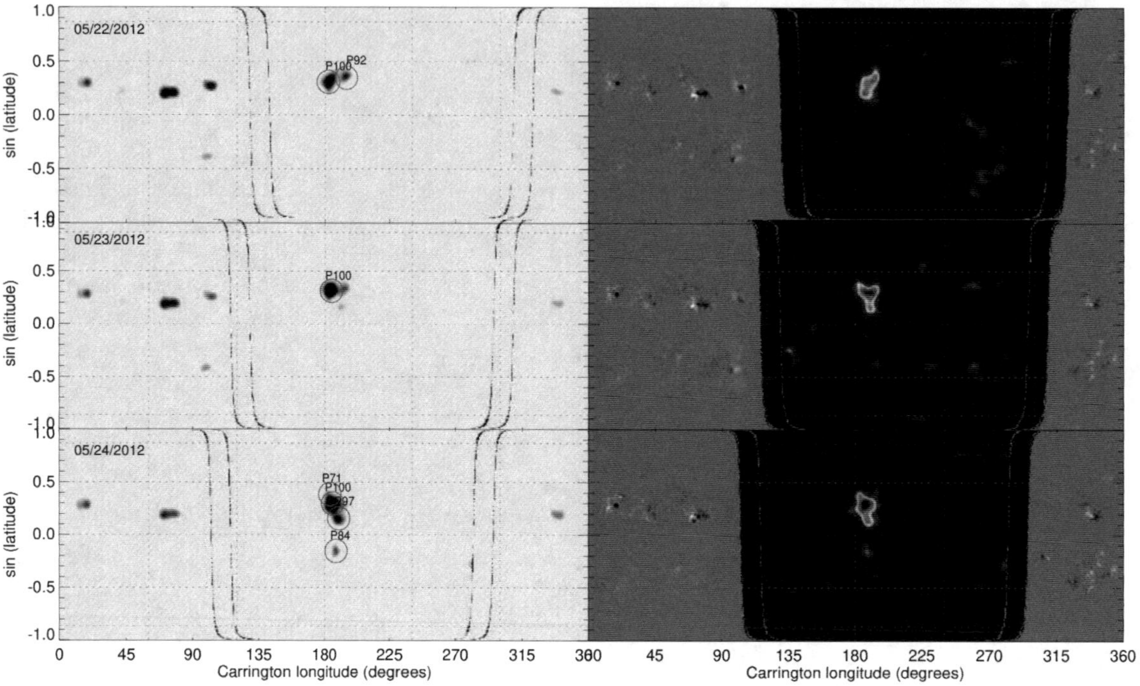

Figure 3. Comparison of far-side seismic maps for May 22-24 2012 calculated with GONG data and the traditional approach (left) and the new higher-resolution, more skips and new color palette calculated with HMI data (right).

4. Discussion

Far side helioseismic maps allow us to locate and track active regions over the whole solar surface. Preliminary case analysis has extended their usage for solar forecasting models. This paper presents initial steps towards improving the maps by including 2×3-skip correlations and applying Postel projections to a high-resolution version of the helioseismic data before binning it down to a more compact datacube. A full statistical analysis of the 2×3-skip maps is needed to quantify the improvement provided by incorporating this extra information, as well as an update in the calibration in terms of magnetic field strength. Future work includes a proper computation of intrinsic errors as well as an assessment of the effective area of the far-side active regions. The intended new look will make the maps more user-friendly for direct use by space weather forecasters.

Acknowledgments

This work utilizes data obtained by the GONG instrument of the NISP Program. NISP is managed by the National Solar Observatory, which is operated by AURA, Inc. under a cooperative agreement with the National Science Foundation. The data were acquired by instruments operated by the Big Bear Solar Observatory, High Altitude Observatory, Learmonth Solar Observatory, Udaipur Solar Observatory, Instituto de Astrofísica de Canarias, and Cerro Tololo Interamerican Observatory. This work has been supported by the NASA Living with a Star - Targeted Research and Technology program under grant NNH11AQ24I. Usage of the SDO/HMI data courtesy of NASA/SDO and the HMI science team.

Journal of Physics: Conference Series **440** (2013) 012029 doi:10.1088/1742-6596/440/1/012029

References

[1] Lindsey C and Braun D C 2000a *Science* **287** 1799
[2] González Hernández I, Hill F and Lindsey C 2007 *Astrophys. J.* **669** 1382
[3] González Hernández I, Hill F, Scherrer P, Lindsey C and Braun D C 2010 *Space Weather* **669** 806002G
[4] Fontenla J M, Quémerais E, González Hernández I, Lindsey C and Haberreiter M 2009 *AdSpR* **44** 457
[5] Arge C N, Henney C J, González Hernández I, Toussaint W A, Koller J and Godinez H C 2012 *Proceedings of the Thirteen International Solar Wind Conference, AIP, submitted*
[6] Zhao J 2007 *Astrophys. J. Lett.* **664** 113

Eclipse on the Coral Sea: Cycle 24 Ascending (GONG 2012, LWS/SDO-5, and SOHO 27) IOP Publishing
Journal of Physics: Conference Series **440** (2013) 012030 doi:10.1088/1742-6596/440/1/012030

Solar-like oscillations in distant stars as seen by CoRoT : the special case of HD 42618, a solar sister

C. Barban[1], S. Deheuvels[2,3], M.J. Goupil[1], Y. Lebreton[4,5], S. Mathur[6,7], E. Michel[1], Th. Morel[8], J. Ballot[2,3], F. Baudin[9], K. Belkacem[1], O. Benomar[10], P. Boumier[9], G.R. Davies[7,11], R.A. García[7], M.P. Hall[12], B. Mosser[1], E. Poretti[13], C. Régulo[14,15], I. Roxburgh[16], R. Samadi[1], G. Verner[16] and the CoRoT team.

[1]LESIA, CNRS UMR 8109, Observatoire de Paris, Université Pierre et Marie Curie, Université Denis Diderot, France
[2] CNRS, Institut de Recherche en Astrophysique et Planétologie, Toulouse, France
[3] Université de Toulouse, UPS-OMP, IRAP, France
[4] Observatoire de Paris, GEPI, CNRS UMR 8111, France
[5] Institut de Physique de Rennes, Université de Rennes 1, CNRS UMR 6251, France
[6] High Altitude Observatory, NCAR, P.O. Box 3000, Boulder, CO 80307, USA
[7] Laboratoire AIM, CEA/DSM-CNRS-Université Paris Diderot; IRFU/SAp, Centre de Saclay, France
[8] Institut dAstrophysique et de Géophysique, Université de Liège, Belgium
[9] Institut dAstrophysique Spatiale, UMR8617, CNRS, Université Paris XI, France
[10] Sydney Institute for Astronomy (SIfA), School of Physics, University of Sydney, Australia
[11] School of Physics and Astronomy, University of Birmingham, UK
[12] Jeremiah Horrocks Institute of Astrophysics, University of Central Lancashire, Preston PR1 2HE
[13] INAF Osservatorio Astronomico di Brera, Merate, Italy
[14] Instituto de Astrofísica de Canarias, Tenerife, Spain
[15] Departamento de Astrofísica, Universidad de La Laguna, Tenerife, Spain
[16] Queen Mary University of London, Astronomy Unit, UK

E-mail: `Caroline.Barban@obspm.fr`

Abstract. We report the observations of a main-sequence star, HD 42618 ($T_{\rm eff}$ = 5765 K, G3V) by the space telescope CoRoT. This is the closest star to the Sun ever observed by CoRoT in term of its fundamental parameters. Using a preliminary version of CoRoT light curves of HD 42618, p modes are detected around 3.2 mHz associated to $\ell = 0$, 1 and 2 modes with a large spacing of 142 μHz. Various methods are then used to derive the mass and radius of this star (scaling relations from solar values as well as comparison between theoretical and observationnal frequencies) giving values in the range of $(0.80 - 1.02)M_\odot$ and $(0.91 - 1.01)R_\odot$. A preliminary analysis of $\ell = 0$ and 1 modes allows us also to study the amount of penetrative convection at the base of the convective envelope.

1. Introduction

Solar-like oscillations are stochastically excited by convection in the outer layers of the star. They are produced in all stars with an external convective envelope. In the case of main-sequence stars, mode periods are typically a few minutes with amplitudes of the order of cm/s

Journal of Physics: Conference Series **440** (2013) 012030 doi:10.1088/1742-6596/440/1/012030

in spectroscopic data or of ppm in photometric data. Space photometry proved to be an efficient technique for measuring and characterizing these oscillations.

CoRoT (COnvection ROtation and planetary Transits) (Baglin et al. 2006) is a space mission built and operated by the French space agency (CNES) with the participation of ESA, Austria, Belgium, Brazil, Germany and Spain. It was designed to do ultra high-precision relative photometry. It has two scientific objectives: to detect stellar oscillations to probe the interior of stars and to detect small planets around distant stars through their transit. CoRoT was launched on Dec. 27, 2006 and since then it has observed about 150 bright stars and about 150 000 faint ones (11<V<16).

One of the scientific objectives of CoRoT is to detect and interpret solar-like oscillations in main-sequence stars (see e.g. Baglin et al. 2012 and Michel et al. 2012 for recent reviews of CoRoT results). We will focus here on HD 42618 a star observed by CoRoT with fundamental parameters very close to the solar ones.

2. Spectroscopic parameters of HD 42618

HD 42618 is one of the main CoRoT targets for the search for solar-like oscillations. It is a main-sequence star (G3V) whose properties are very similar to the Sun except for its metallicity (Morel et al. 2013), see table 1.

Table 1. Some fundamental parameters of HD 42618 obtained by Morel et al. (2013).

$T_{\rm eff}$	5765 ± 17 K	logg	4.48 ± 0.04
[Fe/H]	-0.10 ± 0.02	vsini	< 3 km/s

3. Analysis of CoRoT seismic data of HD 42618

HD 42618 has been observed by CoRoT during two long observing runs spanning 79 and 94 days, respectively. Data analysis of a preliminary light-curve version is presented here.

A Fourier analysis of the light curves shows an excess power around $\nu_{\rm max}$=(3.2 ± 0.1) mHz as well as a clear series of peaks in the power spectrum, very similar to the solar spectrum (figure 1). By cutting the power spectrum in several pieces and stacking them in order to build an échelle diagram, $\ell = 0$, 1 and 2 modes are detected (figure 2) showing a large spacing of $\Delta\nu$=(142 ± 1) μHz.

4. First interpretations of seismic data of HD 42618

Seismic data can be used to derive some global properties of the star such as its mass and radius. These two parameters are derived from the large spacing ($\Delta\nu$) and the frequency at maximum oscillation power ($\nu_{\rm max}$) using scaling relations (e.g. Kjeldsen & Bedding 1995, Kallinger et al. 2010, Chaplin et al. 2011, Belkacem et al. 2011). Using $T_{\rm eff,\odot} = 5777$ K, $\nu_{\rm max,\odot}$=(3.09 ± 0.03) mHz and $\Delta\nu_\odot$=(135.1 ± 0.1) μHz as in Huber et al. (2011), we found $M = (0.9 \pm 0.1)\ M_\odot$ and $R = (0.94 \pm 0.03)\ R_\odot$.

Another way to derive global properties of a star is by comparing theoretical oscillation mode frequencies to the observed ones. By using the Asteroseismic Modeling Portal (AMP, Metcalfe et al. 2009, Mathur et al. 2012), a mass of $M = (0.99 \pm 0.01)\ M_\odot$ and a radius of $R = (0.96 \pm 0.01)\ R_\odot$ with an age of (3.84 ± 0.12) Gyr have been found for HD 42618. Using Cesam2k (Morel et Lebreton 20008), we found $M = (1.00 \pm 0.02)\ M_\odot$ and $R = (0.99 \pm 0.02)\ R_\odot$. These values are in agreement with the ones found from scaling relations. We stress that these

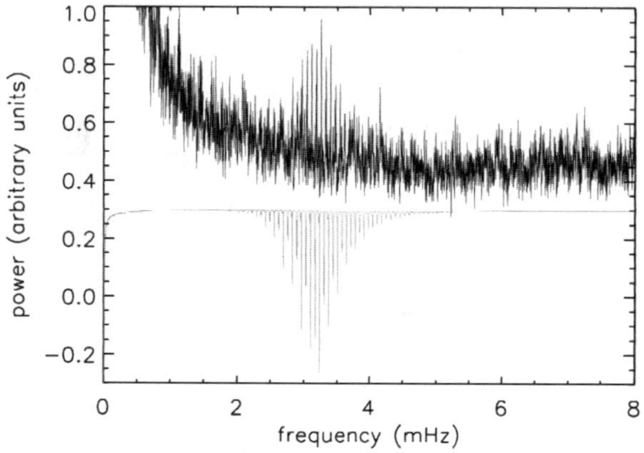

Figure 1. Top: Power spectrum of 175 days of COROT data obtained on HD 42618. Bottom: Power spectrum (upside down) of 175 days of solar data obtained by SOHO/GOLF. The two power spectra have been smoothed over 100 bins ($\sim 7\mu$Hz).

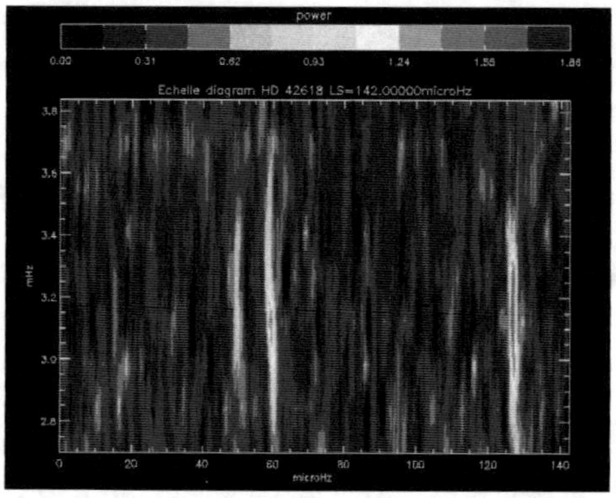

Figure 2. Echelle diagram computed from a smoothed version of the power spectrum of COROT data obtained on HD 42618 assuming a large spacing of 142 μHz.

results have been obtained for a given set of input physics; more detailed modelling studies are in progress.

Other preliminary results have been obtained regarding the penetrative convection at the base of the convective envelope. Using Cesam2k evolutionary code (ESTA version, Morel et Lebreton 2008) and the ADIPLS pulsation code (Christensen-Dalsgaard 2008), theoretical frequencies have been computed for HD 42618.

Figure 3 shows the seismic indices $rr01$ and $rr10$ frequency combination based on $\ell = 0$ and $\ell = 1$ modes as defined in Roxburgh (2009). The observations reveals a clear oscillation of these quantities with the frequency. The same seismic indices for two stellar models are also plotted for comparison. One model includes convective penetration below the convective envelope computed following Zahn (1991)'s prescription. In that approach, the amount (i.e. radial extent) of convective penetration is represented by an adimensionned free parameter ξ. When no penetrative convection is included ($\xi = 0$), the corresponding model in Fig.3 does not show any oscillation unlike what is observed for HD42618. On the other hand, when enough penetration is included ($\xi = 1.5$), the model does show an oscillation similar to that of HD42618. No effort has been made here to fit the data. Rather we illustrate the fact that the presence of some connvective penetration is able to reproduce the $rr01, rr10$ oscillation for HD52618. Neither the phase nor the amplitude and the period are really reproduce by the $\xi = 1.5$ model. A minimisation must be performed to optimise the modelling and fit the data (work in progress). These results nevertheless tend to indicate that a larger amount of convective penetration is necessary, larger than for the Sun and another star HD52265 (Lebreton, Goupil, 2012).

5. Conclusions and perspectives

Oscillations, like the famous 5 min oscillations of the Sun, have been observed in many stars thanks to the space missions COROT (CNES) and *Kepler* (NASA, Borucki et al. 2010). Among them is HD42618 ($T_{\rm eff} = 5765$ K, G3V), a solar sister observed by CoRoT. P modes are detected with a frequency at maximum power of 3.2 mHz and a large spacing of 142 μHz. These data obtained on HD 42618 are already very promising to test stellar models.

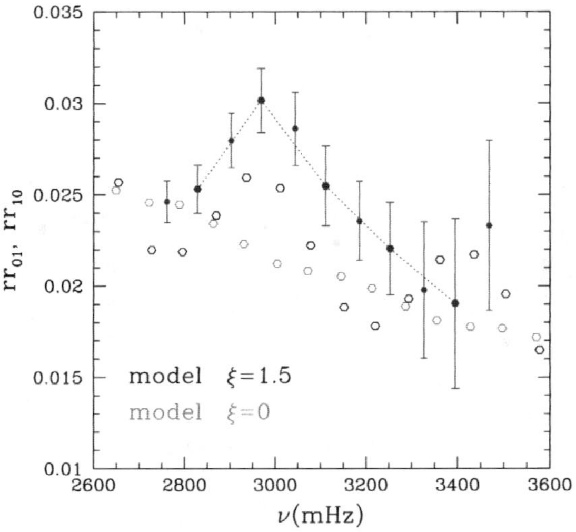

Figure 3. In blue, observed ratios $rr_{01/10}$, a seismic indicator based on $\ell = 0$ and $\ell = 1$ modes (Roxburgh 2009), as a function of mode frequency for the CoRoT data obtained on HD 42618. In black and magenta, modelled ratios $rr_{01/10}$ without ($\xi = 0$) and with penetrative convection ($\xi = 1.5$). See Lebreton & Goupil (2012) for more details.

Futur space missions are foreseen to study solar-like oscillations in main sequence stars. Among them, PLATO (PLAnetary Transits and Oscillations of stars) is a a medium class (M class) mission studied in the framework of the ESA Cosmic Vision 2015-2025 program (e.g. Rauer & Catala 2012). The objectives of this mission are to detect Earth Analogue systems around bright stars, and to reveal the interior structure of planets and their host stars.

5.1. Acknowledgments
The CoRoT space mission, launched on December 27, 2006, has been developed and is operated by CNES, with the contribution of Austria, Belgium, Brasil, ESA, Germany and Spain.

References
[1] Baglin A., Michel E., and the COROT team, 2012, *ASPC* **462** 492
[2] Baglin A., Auvergne M., Boisnard L., et al. 2006, in COSPAR, Plenary Meeting, **Vol. 36**, 36th COSPAR Scientific Assembly, 3749
[3] Belkacem K. Goupil M.J., Dupret M.A., 2011 *A&A* **530**, 142
[4] Borucki W.J. et al., 2010, *Science* **327**, 977
[5] Christensen-Dalsgaard, 2008 *Ap&SS* **316**, 113
[6] Huber D., Bedding T. R., Stello, D. et al. 2011 *ApJ* **743**, 143)
[7] Kallinger Th., Weiss W., Barban C., et al., 2010 *A&A* **509**, 77
[8] Kjeldsen H. and Bedding T., 1995 *A&A* **293**, 87
[9] Lebreton Y. and Goupil M.J., 2012 *A&A* **549**, L13
[10] Mathur S., Metcalfe T. S., Woitaszek M. et al., 2012 *ApJ* **749**, 152
[11] Metcalfe T.S., Creevey O.L., Christensen-Dalsgaard J., 2009 *ApJ* **699**, 373
[12] Michel E., Baglin A., and the COROT team, 2012, *arXIv* **1202** 1422
[13] Morel Th. et al. 2013 *A&A* submitted
[14] Morel P. & Lebreton Y., 2008 *ApSS* **316**, 61
[15] Rauer U. & Catala C., 2012, *EGUGA*, 14.7033
[16] Roxburgh I.W., 2009, *A&A*, **493**, 185-191
[17] Zahn J.P., 1991 *A&A* **252**, 179

Journal of Physics: Conference Series **440** (2013) 012031 doi:10.1088/1742-6596/440/1/012031

Solar-stellar connection: the frequency of maximum oscillation power from solar data

C. Barban[1], M. Beuret[1], F. Baudin[2], K. Belkacem[1], M.J. Goupil[1], R. Samadi[1]

[1]LESIA, CNRS UMR 8109, Observatoire de Paris, Université Pierre et Marie Curie, Université Denis Diderot, France

[2] Institut dAstrophysique Spatiale, UMR8617, CNRS, Université Paris XI, France

E-mail: Caroline.Barban@obspm.fr

Abstract. Stellar oscillations provide powerful tools to derive stellar fundamental parameters such as the mass and radius. These global quantities are derived from scaling relations linking seismic quantities (ν_{max} and $\Delta\nu$) to global stellar parameters. These relations use the Sun as a reference. In this work, we used VIRGO and GOLF data to study how the solar frequency at the maximum oscillation power (ν_{max}) varies with time along the solar cycle. We show that these variations imply differences of about 4% in radius and 12% in mass. We showed also that the observational method based on intensity or velocity data has also an impact, implying differences in mass of about 22% and 7% in radius.

1. Introduction

Stellar oscillations provide powerful tools to derive stellar model independant estimates of the mass and the radius of a pulsating star. These global quantities are derived from scaling relations linking seismic quantities to global stellar parameters (e.g. Kallinger et al. 2010 ; Belkacem et al. 2011, 2012). Two properties of the oscillation spectrum are used : the large separation, $\Delta\nu$, the frequency separation between two consecutive radial orders and the frequency at the maximum power in the oscillation spectrum, ν_{max}.

The scaling relations are normalised to the solar values and therefore depend on some seismic parameters of the Sun, among them the frequency at maximum power, $\nu_{\mathrm{max},\odot}$. The level of precision of these scaling relations is such that the accuracy on the solar reference maximum frequency becomes non negligible.

In this work, we investigate the possible biases and the precision of the determination of the solar frequency of the maximum oscillation power. For this purpose, we used data from different instruments: VIRGO and GOLF onboard SOHO.

2. Data and method

We used GOLF and VIRGO data from 04/11/1996 to 09/15/2004 cut into 6 months long time series. ν_{max} is derived by fitting the smoothed power spectrum by a gaussian-like model (figures 1 and 2). We can see clearly in VIRGO data the granulation background at low frequencies while it is less prevalent in GOLF data.

Journal of Physics: Conference Series **440** (2013) 012031 doi:10.1088/1742-6596/440/1/012031

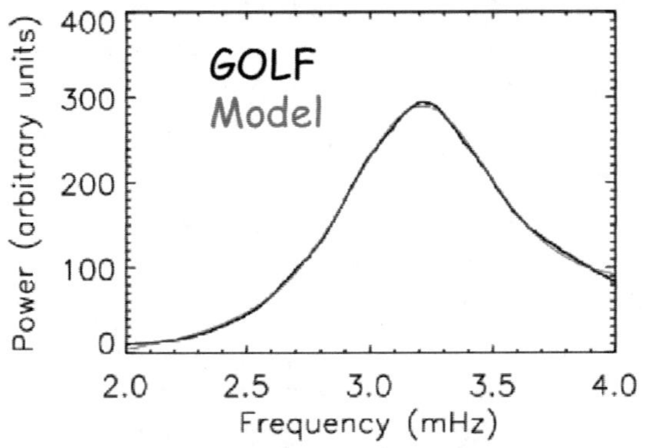

Figure 1. Smoothed oscillation spectrum , over 3.$\Delta\nu$, from GOLF data (in black) fitted by a Gaussian-like model (in red).

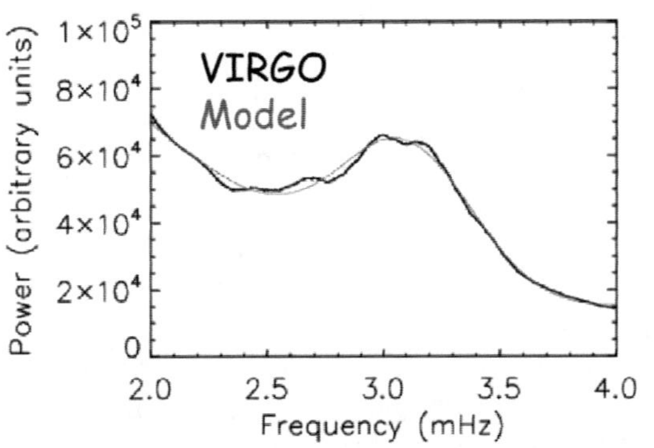

Figure 2. Smoothed oscillation spectrum, over 3.$\Delta\nu$, from VIRGO data (in black) fitted by a Gaussian-like model (in green).

3. Results

The variation of $\nu_{\rm max}$ with time shows a correlation with the number of sunspots for GOLF data but not for VIRGO data (figure 3). We see also on figure 3 that the $\nu_{\rm max}$ obtained from GOLF data is systematically larger than the one obtained from VIRGO data. This still needs to be investigated and is certainly related to the different shape of the background spectrum and its possible variation along the cycle.

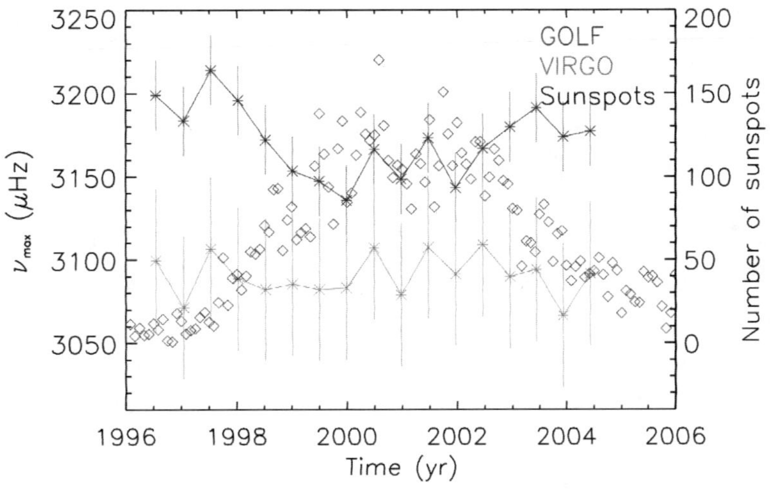

Figure 3. ν_{max} computed from GOLF data (blue) and VIRGO data (in red) and number of sunspots (http://solarscience.msfc.nasa.gov/SunspotCycle.shtml) as function of time. Error bars on ν_{max} are derived from Monte-Carlo simulations : their values are 21 Hz for GOLF and 43 Hz for VIRGO.

ν_{max} varies from 3135 to 3214 mHz for GOLF data along the solar cycle. If we compute the

Journal of Physics: Conference Series **440** (2013) 012031 doi:10.1088/1742-6596/440/1/012031

radius and the mass of a given star using these two values with Eq. (1) and (2) (e.g. Chaplin et al. 2011), this implies a difference in the stellar radius by 4 % and in the stellar mass by 12 %.

$$\left(\frac{R}{R_\odot}\right) = \left(\frac{\nu_{\rm max}}{\nu_{\rm max,\odot}}\right)\left(\frac{\Delta\nu}{\Delta\nu_\odot}\right)^{-2}\left(\frac{T_{\rm eff}}{T_{\rm eff,\odot}}\right)^{0.5} \tag{1}$$

$$\left(\frac{M}{M_\odot}\right) = \left(\frac{\nu_{\rm max}}{\nu_{\rm max,\odot}}\right)^{3}\left(\frac{\Delta\nu}{\Delta\nu_\odot}\right)^{-4}\left(\frac{T_{\rm eff}}{T_{\rm eff,\odot}}\right)^{1.5} \tag{2}$$

For VIRGO data, $\nu_{\rm max}$ varies and from 3066 to 3110 mHz implying a difference in the stellar radius by 4.3 % and in the stellar mass by 13 %.

If we consider two different kind of solar data, intensity (VIRGO) and velocity (GOLF) data, the difference in the stellar radius is about 7% and about 22% in the stellar mass.

4. Conclusions

In this work, we have found different values of $\nu_{\rm max}$ depending on the way the oscillations are observed (intensity vs velocity observations) and on the period considered because of the effect of the solar activity. This kind of study is important to be able to estimate the accuracy of stellar paramaters derived from scaling relations.

In a near future, we will investigate other methods to determine $\nu_{\rm max}$ (e.g. based on individual frequencies) as well as more sophisticated model to better reproduce the lower and higher frequency part of the excess power due to the oscillations in the power spectrum.

References

[1] Belkacem K., 2012, SF2A-2012: Proceedings of the Annual meeting of the French Society of Astronomy and Astrophysics. Eds.: S. Boissier, P. de Laverny, N. Nardetto, R. Samadi, D. Valls-Gabaud and H. Wozniak, pp.173-188

[2] Belkacem K. Goupil M.J., Dupret M.A., 2011 *A&A* **530**, 142

[3] Chaplin W.J., Kjeldsen H., Christensen-Dalsgaard J., et al. 2011 *Science* **332**, 213

[4] Kallinger Th., Weiss W., Barban C., et al., 2010 *A&A* **509**, 77

Temporal variations in solar magnetic bright points intensity and plasma parameters

J Jurčák[1], D Utz[2,3], L R Bellot Rubio[2]

[1] Astronomical Institute of the Academy of Sciences, Fričova 298, 251 65, Ondřejov, Czech Republic

[2] Instituto de Astrofísica de Andalucía (CSIC), Apdo. de Correos 3004, E-18080 Granada, Spain

[3] IGAM/Institute of Physics, University of Graz, Universitätsplatz 5, A-8010 Graz, Austria

E-mail: jurcak@asu.cas.cz

Abstract. Magnetic bright points are one of the finest magnetic structures observed in the solar atmosphere. They possibly represent single flux tubes in quiet Sun regions. Their formation is described by the convective collapse model, while the decay phase of these structures is not well characterized yet. We attempt to follow the evolution of a few selected examples of MBPs and to study their changes in brightness and also the variations of plasma parameters during their lifetime. We use data from the Hinode satellite and the Sunrise mission. The G-band observations taken with a cadence of 30 seconds by the Hinode Solar Optical Telescope (SOT) show very fast changes of the maximum intensity of these structures. The complementary spectropolarimetric data, which are used to estimate the plasma parameters, were taken with a cadence of approximately two minutes. The variations of plasma parameters cannot be matched one to one to the changes in intensity due to the different temporal resolution. However, the slow changes of intensity with large amplitude are matched with variations of magnetic field strength and line-of-sight (LOS) velocity. The Sunrise/IMaX data have a temporal resolution of 32 seconds and show fast variations in the line wing intensity. These variations are associated with changes in the magnetic field strength and LOS velocity.

1. Introduction

Magnetic bright points (MBPs) are structures with field strengths exceeding the equipartition field strength in the photosphere. The formation mechanism called convective collapse (superadiabatic effect) was suggested first by Parker [1] and further studied by e.g. Spruit [2]. According to theory, the intensification of magnetic field takes place simultaneously with fast downflows. This behaviour was observationally confirmed during the formation of flux tubes with strong magnetic fields by e.g. Bellot Rubio [3] and Nagata [4].

The decay phase of MBPs is not yet understood well. Bellot Rubio [3] found that the convective collapse does not lead to a stable kilogauss magnetic element, but the magnetic field decreases again below equipartition field strength in the late phase when upflows prevail in the flux tube. This is in agreement with the study by Takeuchi [5] who found that flux tubes resulting from the collapse of weak magnetic fields in large structures are destroyed by upward-moving shocks. However, there are some discrepancies between the maximum field strengths found in observations by Bellot Rubio (600 G) and reached in simulations by Takeuchi (over 1000 G).

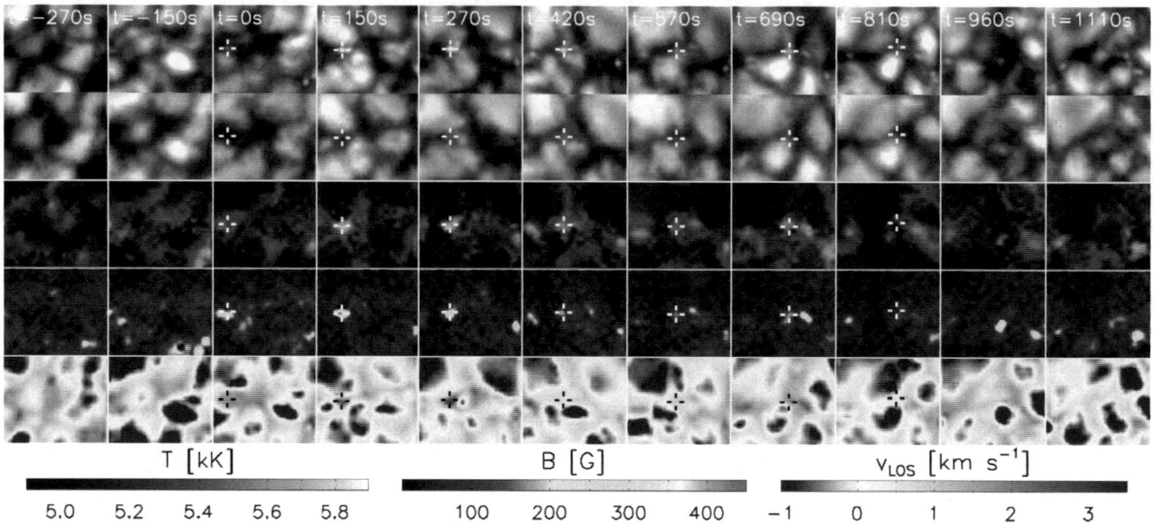

Figure 1. Maps of G-band intensities and plasma parameters corresponding to the MBP observed by the Hinode/SOT. From top to bottom: G-band filtergram images, SP continuum intensity, temperature, magnetic field strength, and line-of-sight velocity. The maps of plasma parameters are displayed at $\log(\tau) = -1$. Seconds shown in the G-band filtergram images correspond to the time between the first detection of the MBP and the given frame. The first two and last two columns show the situation before and after the MBP could be identified. The white (black) cross-hair mark the position of maximum G-band intensity of the MBP.

We follow the evolution of several MPBs identified in the datasets obtained by the Hinode satellite and the balloon-borne Sunrise observatory. The first results presented here show how the intensity changes correspond to the changes of plasma parameters during the lifetime of the studied magnetic structures.

2. Observations and data analysis

We use data from the Hinode Solar Optical Telescope [6, 7]. The data were taken on January 20, 2007 at 8 UT. From the broad-band filtergram imager we have G-band images taken with a cadence of 30 seconds and a pixel sampling of $0.11''$. The diffraction-limited spatial resolution due to the 0.5 m primary telescope aperture is $0.22''$. From the Hinode spectropolarimeter (SP) we have full Stokes profiles of the Fe I lines around 630.2 nm. Scans of a narrow solar surface region, comprising 25 slit positions, were taken with a cadence of approximately two minutes. The resulting spatial resolution of the SP data is $0.32''$. The data were calibrated with standard routines available within the Hinode SolarSoft package.

We also use data taken by the Sunrise/IMaX instrument [8] on June 9, 2009 at 2 UT. IMaX is an imaging spectropolarimeter based on Fabry-Perot etalon. It scanned all Stokes profiles of the Fe I line at 525.02 nm at four wavelength positions within the line and one in the continuum. The pixel sampling of this instrument corresponds to $0.055''$, which results into diffraction limited spatial resolution of $0.16''$. The IMaX data have been corrected for instrumental effects by performing dark-current subtraction, flat-field correction, and cross-talk removal.

The MBPs were identified manually in the Hinode G-band images and automatically in the Sunrise/IMaX data using the intensities in the blue wing of the scanned line (at 524.98 nm) and the code developed by Utz [9]. The plasma parameters were computed from the observed Stokes

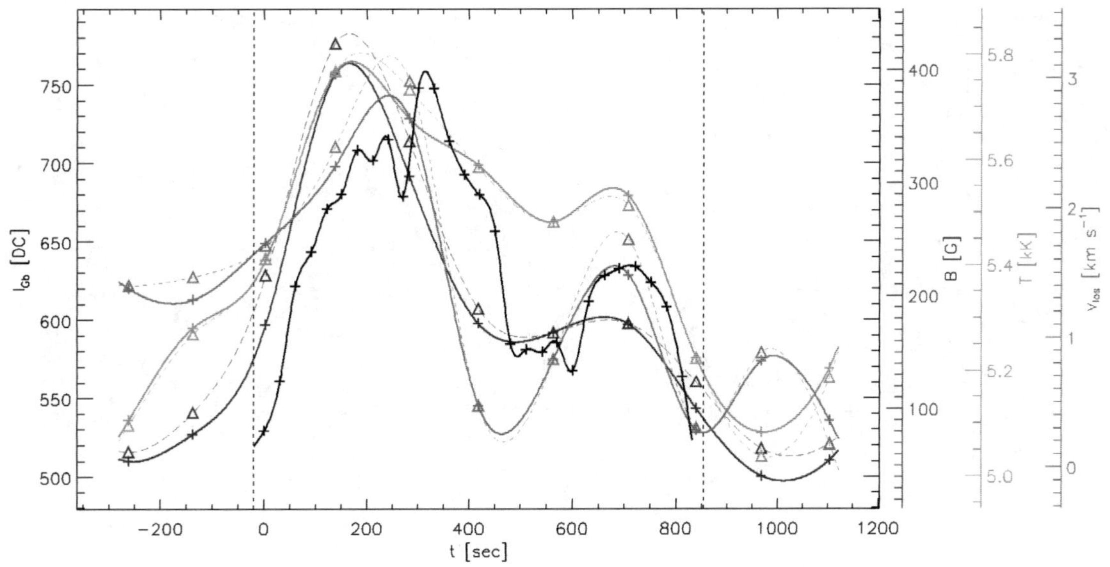

Figure 2. Temporal evolution of maximum G-band intensity (black + symbols) of the MBP shown in Fig. 1. The values of magnetic field strength (blue + symbols), temperature (green + symbols), and LOS velocity (red + symbols) are taken at the same position as the maximum G-band intensity (marked in Fig. 1) at $\log(\tau) = -1$. The Δ symbols correspond to the plasma parameters at the position of the strongest magnetic field strength in the MBP region. Spline functions are used to connect the + and Δ symbols. The dashed lines mark the lifetime of the MBP.

profiles using the inversion code SIR [10]. To analyze the Hinode SP data, we assumed plasma parameters changing linearly with height except for temperature, which has a more complex stratification. In the case of the Sunrise/IMaX data, we used plasma parameters constant with height (except for temperature), as the number of wavelength points does not allow a more complex atmospheric model.

3. Results

In Fig. 1, we show one of the MBPs manually identified in the Hinode dataset. In this case, the MBP lasted for approximately 14 minutes as it was detected on 28 consecutive G-band filtergrams. The first two columns in Fig. 1 show the G-band intensity images and maps based on SP data before the formation of the MBP, while the last two columns illustrate the situation after its disappearance. The maximum brightness of the MBP was reached 5 minutes after its formation (270 sec), then it became fainter (570 sec), followed by another brightness peak around 690 sec. As can be seen in the second row of Fig. 1, the MBP cannot be identified in the continuum map obtained from the SP scans. This is probably due to insufficient spatial resolution and lower contrast of MBPs in continuum intensities. The MBP is however visible as a bright structure in the temperature maps at $\log(\tau) = -1$, which are proportional to the intensity of the line wings. In the magnetic field strength maps (fourth row), an intensification of the magnetic field can be clearly seen around the MBP during the initial phase. Also the downflow velocity (LOS velocity maps are shown in the fifth row) is increasing during this phase. This is in agreement with previous studies [3, 4] and the convective collapse scenario.

In Fig. 2, we show the evolution of MBP maximum G-band intensity and the plasma

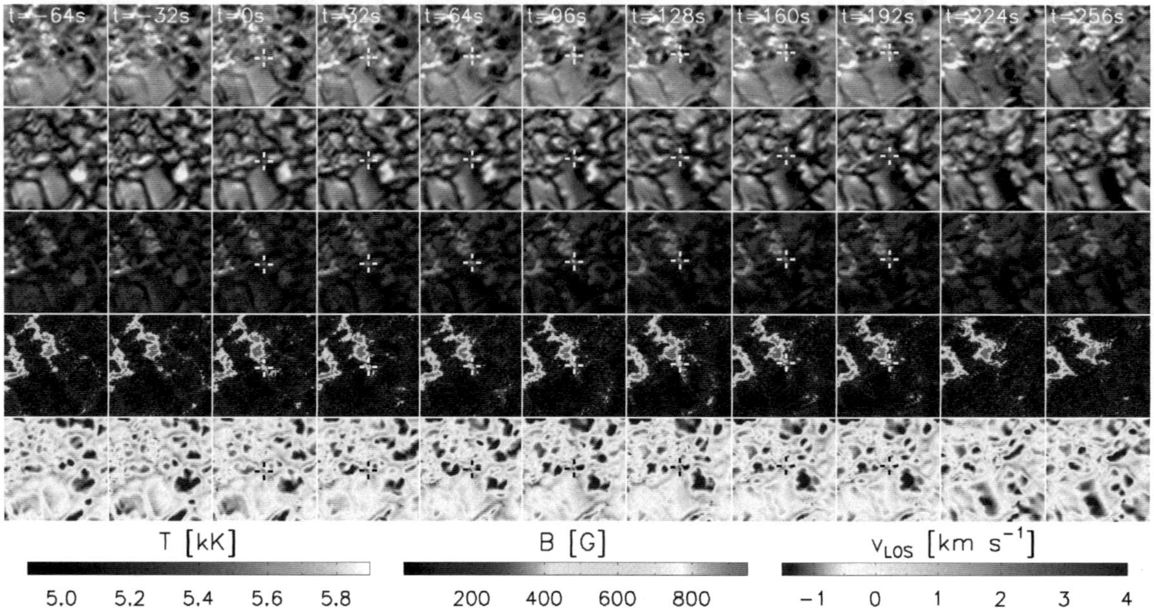

Figure 3. Analogous to Fig. 1 but showing the MBP automatically identified in the IMaX data. In the top row, the line-wing intensities are shown. The white (black) cross-hair mark the position of barycentre of the MBP.

parameters. As the pixel with the maximum G-band intensity is not necessarily co-spatial with the pixel where the maximum field strength can be found, we display the evolution of plasma parameters for both of these pixels. It is clear that the G-band intensity is rapidly changing and that the temporal resolution of SP measurements is insufficient to follow these fast fluctuations. However, we selected this MBP as there is a second peak of G-band intensity, which can be temporarily resolved also in the SP data. It is clear that each of these MBP intensity brightenings is accompanied by an intensification of the magnetic field strength and a higher downflow velocity.

In Fig. 3, we show one of the MBPs automatically identified in the Sunrise/IMaX data. The intensity observed in the blue wing of the scanned line was used for the identification, as the contrast of the structure is higher than in the continuum wavelength. The first two and the last two time steps show the situation before and after the algorithm identified the MBP. Due to significantly higher spatial resolution, the MBP can be seen also in continuum maps displayed in second row of Fig. 3. Compared to the MBP selected from the Hinode dataset, this MBP lasts only for five minutes and has only one main peak in brightness. Similarly to the results based on Hinode SP, the field strength is increasing rapidly in the initial phase of the MBP formation and the fastest downflows are observed during this phase.

In Fig. 4, we show the evolution of plasma parameters and intensities of the MBP shown in Fig. 3. Again one can observe fast changes of intensity, now matching with fast changes of plasma parameters. In most cases, local maximum in intensity is accompanied by local maximum of magnetic field strength and LOS velocity. Contrary to the results based on Hinode SP, the maximum values of magnetic field strength and LOS velocity at the MBP barycentre (Δ symbols and dashed lines in Fig. 4) are not co-temporal with the maximum of the line-wing intensity. The peak value of intensity corresponds only to the peak value of LOS velocity in the 5×5 pixels

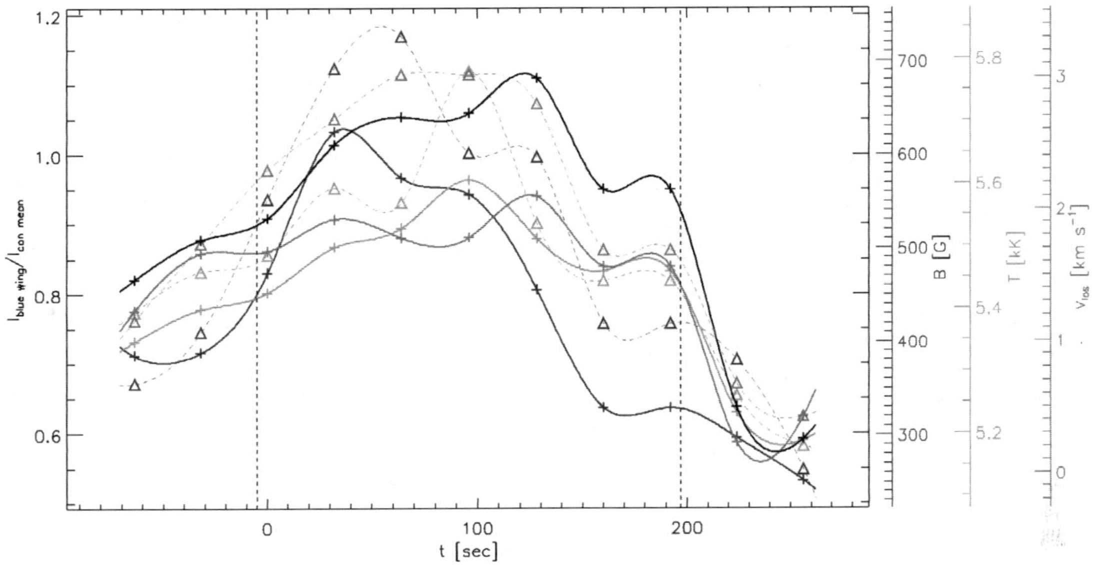

Figure 4. Analogous to Fig. 2 but showing the temporal evolution of the maximum line-wing intensity (black + symbols) of the MBP shown in Fig. 3. The + symbols represent averaged values from 5×5 pixels region around the barycentre of the MBP. The Δ symbols are the values of plasma parameters at the barycentre.

region around the barycentre of the MBP (+ symbols and solid lines in Fig. 4).

4. Discussion

The main problem arising in our analysis comes from the uncertainties caused by the inversion. We have satisfactory spectral resolution from the Hinode SP, but these data have still insufficient spatial resolution. As we used only an one-component atmospheric model with a fixed amount of stray-light, it results into underestimation of magnetic field strength, i.e., a maximum of only 400 G in the presented MBP. Also, the supposed atmospheric model is not complex enough to fit the highly asymmetric Stokes V profiles observed in the MBPs. To fit such profiles, it is necessary to use complex atmospheric models like Bellot Rubio [3] did.

Another problem is the temporal resolution of the Hinode SP measurements. We have to select a long lasting MBP with multiple brightenings to be able to compare it to the evolution of plasma parameters. In most cases, the MBPs have shorter lifetimes (comparable to the example selected from the Sunrise/IMaX dataset) and better temporal resolution is necessary.

The modest spectral resolution of the Sunrise/IMaX data and the fact that only one line is scanned does not allow for complex atmospheric models even for these data. However, the spatial resolution is high enough to provide more reliable values of magnetic field strength than those obtained from the simplified inversions of Hinode SP data.

5. Conclusions

In agreement with [3, 4], we confirmed observationally the convective collapse as the mechanism responsible for formation of MBPs. In the presented examples, the magnetic field strength does not reach the values predicted by the theory of the convective collapse, but it might be due to an oversimplified inversion setup. However, the resulting flux tube is not stable as the theory claims and dissolves rapidly, i.e., the convective collapse of the flux tube can be possibly stopped by some mechanism even before a stable kilogauss field concentration is reached.

In the presented MBP observed by Hinode, there is a relation between G-band intensity and magnetic field strength, i.e., the brighter the structure is, the stronger is the magnetic field. This does not seem to be true if we compare different MBPs. Some of the other MBPs identified in the Hinode data have higher G-band intensities than the presented one, but lower magnetic field strengths.

The Sunrise/IMaX data cannot be used to derive more complex atmospheric models as there is not enough information contained in the four measured wavelength points within the spectral line. The Hinode SP data can provide us with more accurate atmospheric models, if specially designed inversion codes are used, e.g. following the analysis of Bellot Rubio [3]. However, the insufficient spatial resolution must be kept in mind.

Acknowledgments

The support from GA CR P209/12/0287 and RVO:67985815 is gratefully acknowledged. D. U. wishes to acknowledge the support given by the Austrian Science Fund (FWF) project: J3176. This work has been funded by the Spanish MICINN through project AYA2011-29833-C06-04, including a percentage from European FEDER funds. J.J and D.U. want to thank ÖAD and MŠMT for finances provided in frame of the project MEB 061109. Hinode is a Japanese mission developed and launched by ISAS/JAXA, with NAOJ as domestic partner and NASA and STFC (UK) as international partners. It is operated by these agencies in cooperation with ESA and NSC (Norway). The German contribution to Sunrise is funded by the Bundesministerium für Wirtschaft und Technologie through Deutsches Zentrum für Luft- und Raumfahrt e.V. (DLR), Grant No. 50 OU 0401, and by the Innovationsfond of the President of the Max Planck Society (MPG). The Spanish contribution has been funded by the Spanish MICINN under projects ESP2006- 13030-C06 and AYA2009-14105-C06 (including European FEDER funds). The HAO contribution was partly funded through NASA grant NNX08AH38G.

References

[1] Parker E N 1978 *ApJ* **221** 368–377
[2] Spruit H C 1979 *Sol. Phys.* **61** 363–378
[3] Bellot Rubio L R, Rodríguez Hidalgo I, Collados M, Khomenko E and Ruiz Cobo B 2001 *ApJ* **560** 1010–1019
[4] Nagata S e a 2008 *ApJ* **677** L145–L147
[5] Takeuchi A 1999 *ApJ* **522** 518–523
[6] Kosugi T e a 2007 *Sol. Phys.* **243** 3–17
[7] Tsuneta S e a 2008 *Sol. Phys.* **249** 167–196 (*Preprint* `arXiv:0711.1715`)
[8] Barthol P e a 2011 *Sol. Phys.* **268** 1–34 (*Preprint* `1009.2689`)
[9] Utz D, Hanslmeier A, Möstl C, Muller R, Veronig A and Muthsam H 2009 *A&A* **498** 289–293
[10] Ruiz Cobo B and del Toro Iniesta J C 1992 *ApJ* **398** 375–385

Journal of Physics: Conference Series **440** (2013) 012033 doi:10.1088/1742-6596/440/1/012033

The Murchison Widefield Array: solar science with the low frequency SKA Precursor

S.J. Tingay[1,13], D. Oberoi[2], I. Cairns[3], A. Donea[4], R. Duffin[1], W. Arcus[1], G. Bernardi[5], J.D. Bowman[6], F. Briggs[7,12], J.D. Bunton[8], R.J. Cappallo[9], B.E. Corey[9], A. Deshpande[1]0, L. deSouza[8], D. Emrich[1], B.M. Gaensler[11,12], Goeke, R.[13], L.J. Greenhill[5], B.J. Hazelton[14], D. Herne[1], J.N. Hewitt[13], M. Johnston-Hollitt[15], D.L. Kaplan[16], J.C. Kasper[5], J.A. Kennewell[1], B.B. Kincaid[9], R. Koenig[8], E. Kratzenberg[9], C.J. Lonsdale[9], M.J. Lynch[1], S.R. McWhirter[9], D.A. Mitchell[12,17], M.F. Morales[14], E. Morgan[13], S.M. Ord[1], J. Pathikulangara[8], T. Prabu[10], R.A Remillard[13], A.E.E. Rogers[9], A. Roshi[1]0, J.E. Salah[9], R.J. Sault[17], N. Udaya-Shankar[10], K.S. Srivani[10], J. Stevens[8], R. Subrahmanyan[10,12], M. Waterson[1], R.B. Wayth[1,12], R.L. Webster[17,12], A.R. Whitney[9], A. Williams[18], C.L. Williams[13] and J.S.B. Wyithe[17,12]

[1] ICRAR - Curtin University, Perth, Australia
[2] National Centre for Radio Astrophysics, Pune, India
[3] The University of Sydney, Sydney, Australia
[4] Monash University, Melbourne, Australia
[5] Harvard-Smithsonian Center for Astrophysics, Cambridge, MA, USA
[6] Arizona State University, Tempe, AZ, USA
[7] The Australian National University, Canberra, Australia
[8] CSIRO Astronomy and Space Science, Australia
[9] MIT Haystack Observatory, Westford, MA, USA
[10] Raman Research Institute, Bangalore, India
[11] Sydney Institute for Astronomy, The University of Sydney, Sydney, Australia
[12] ARC Centre of Excellence for All-sky Astrophysics (CAASTRO)
[13] MIT Kavli Institute for Astrophysics and Space Research, Cambridge, MA, USA
[14] University of Washington, Seattle, USA
[15] School of Chemical and Physical Sciences, Victoria University of Wellington, New Zealand
[16] University of Wisconsin–Milwaukee, Milwaukee, WI, USA
[17] The University of Melbourne, Melbourne, Australia
[18] ICRAR - University of Western Australia, Perth, Australia

E-mail: `s.tingay@curtin.edu.au`

Abstract. The Murchison Widefield Array is a low frequency (80 - 300 MHz) SKA Precursor, comprising 128 aperture array elements (known as tiles) distributed over an area of 3 km diameter. The MWA is located at the extraordinarily radio quiet Murchison Radioastronomy Observatory in the mid-west of Western Australia, the selected home for the Phase 1 and Phase 2 SKA low frequency arrays. The MWA science goals include: 1) detection of fluctuations in the brightness temperature of the diffuse redshifted 21 cm line of neutral hydrogen from the epoch of reionisation; 2) studies of Galactic and extragalactic processes based on deep, confusion-limited surveys of the full sky visible to the array; 3) time domain astrophysics through exploration of the variable radio sky; and 4) solar imaging and characterisation of the heliosphere and ionosphere

Journal of Physics: Conference Series **440** (2013) 012033 doi:10.1088/1742-6596/440/1/012033

via propagation effects on background radio source emission. This paper concentrates on the capabilities of the MWA for solar science and summarises some of the solar science results to date, in advance of the initial operation of the final instrument in 2013.

1. Introduction

The Murchison Widefield Array (MWA) is the only Square Kilometre Array (SKA: [2]) Precursor telescope at low radio frequencies. An SKA Precursor is a recognised SKA technology demonstrator located at one of the two sites that will host the SKA, the Murchison Radio-astronomy Observatory (MRO) in the Murchison region of Western Australia and the Karoo region of South Africa's Northern Cape. The MWA is sited at the MRO, along with a second SKA Precursor, the Australian SKA Pathfinder (ASKAP: [4]; [5]). The MeerKAT SKA Precursor will be located at the South African site[1]. The MWA is sited at the MRO due to the extremely low levels of human-made radio frequency intereference in this area of Western Australia, particularly within the FM band (87 - 108 MHz) which is encompassed by the MWA at the low end of its operating frequency range.

The technical capabilities of the MWA are described in detail by [8] and the full science case for the MWA is described by [1]. Broadly, the MWA science goals include: 1) detection of fluctuations in the brightness temperature of the diffuse redshifted 21 cm line of neutral hydrogen from the epoch of reionisation; 2) studies of Galactic and extragalactic processes based on deep, confusion-limited surveys of the full sky visible to the array; 3) time domain astrophysics through exploration of the variable radio sky; and 4) solar imaging and characterisation of the heliosphere and ionosphere via propagation effects on background radio source emission.

The MWA has reached the point of practical completion and will be available to the global user community from mid-2013, based on an "Open Skies" policy (http://www.mwatelescope.org). Users will be able to propose observational programs with the MWA in the first half of 2013, including for solar science. Considerable data have been obtained in advance of completion of the full instrument. A prototype array was operated for approximately 2 years, from 2009 to 2011, and during construction of the full array, science commissioning observations were undertaken.

This paper will briefly describe the capabilities of the MWA for solar physics and summarise some of the solar results thus far obtained. The results to date have been obtained with arrays four times less sensitive than the final array that will be available in 2013, and generally with lower angular resolution and image fidelity than will be available with the final array.

2. History of low frequency solar imaging in Australia

Low radio frequency imaging of the Sun was pioneered in Australia under the leadership of Paul Wild [3]. Wild and others imaged the Sun at frequencies similar to the MWA frequencies for the first time, revealing the nature of type II and type III bursts (these types were defined by the same group based on high quality dynamic spectra) and monitoring their evolution as a function of time and frequency [9]. This work used the Culgoora solar heliograph, built by Wild and his team in the late 1960s.

Since the closure of Culgoora in 1984, there has been no low frequency solar imaging capability in Australia. The MWA will therefore renew this capability in Australia using modern instrumentation on a radio quiet site.

[1] http://public.ska.ac.za/meerkat

3. Results from the MWA prototype and science commissioning

Important capabilities of the MWA for solar physics are the time and frequency resolution of imaging. The MWA can produce images at a rate of two per second, for each of the four Stokes parameters and for each frequency channel across the bandwidth. The bandwidth is 30.72 MHz, tunable in the range 80 to 300 MHz, with the 30.72 MHz broken up into a maximum of 3072 channels of 10 kHz each. Thus, in one minute it is possible for the MWA to produce almost 1.5 million images (number of frequency channels × number of Stokes parameters × number of time steps), allowing a comprehensive exploration of the evolution of the radio emission as function of both frequency and time.

The angular resolution of the MWA at these low frequencies is of order 1 arcminute, likely within a factor of a few of the expected scattering size due to the intervening plasma, at the top end of the MWA frequency range. Approximately 1000 resolution elements will cover the disk of the Sun, at the top end of the MWA frequency range (fewer resolution elements at lower frequency).

In Novemnber 2012 the MWA reached a major milestone in practical completion of the full instrument as described by [8] and in early 2013 the full capabilities of the MWA will be exercised for the first time, leading to commencement of the first science operations phase from mid-2013. During the period 2009 - 2011, a 32 tile prototype array was operated and used as a science and engineering testbed. This was decommissioned in early 2012, to make way for the construction of the final instrument. In the final stages of construction, from approximately September 2012, science commissioning of the final instrument commenced, overlapping with construction; 32 tile subsets of the final instrument were commissioned and tested as they were completed. The 32 tile sub-arrays ranged from configurations with only short baselines (low angular resolution) to configurations with close to the maximum length baselines (high angular resolution).

During both the prototype phase and science commissioning phase, scientific observations were performed across all MWA science themes. In the prototype phase, [7] performed the first demonstration of the spectroscopic imaging capabilities of the MWA (albeit with relatively low angular resolution), revealing a rich variety of phenomena in the observed frequency range of 171 - 202 MHz, consisting of many short-lived, narrow-band, non-thermal emission features. While the [7] images are of low angular resolution, they show a number of these events to be marginally resolved. A somewhat surprising feature of these first serious solar observations with the MWA was the rich variety of weak transient events, showing that to effectively study the Sun at these frequencies a surperb radio quiet location is required. Even to study the Sun, one of the brightest radio objects in the sky, very low levels of human-made interference appear critical to success.

Other MWA prototype observations of Type III flares from 2011 are currently being analysed in conjunction with SDO and RHESSI data (Cairns et al. 2013, in preparation). Even the prototype phase of the MWA development was scientifically productive in terms of solar science.

In the science commissioning phase of the final MWA instrument, since approximately September 2012, almost 100 TB of solar imaging data have been collected! This vast store of data covers observations with five different 32 tile sub-arrays, generally during quiet periods for the Sun. The most effective sub-array was the so-called "gamma" array, since the configuration of 32 tiles used included long baselines and therefore produced high angular resolution images (few arcminute angular resolution). An example is shown in Figure 1, from data obtained on October 29, 2012 (03:35:44 UT). This image represents a single 1 second integration and a 40 kHz channel at a frequency of approximately 150 MHz in a single linear polarisation. The disk of the Sun is well-resolved. The Sun on this day was rated as quiet. However, even in this state weak transient events are clearly seen when a series of images are made over a period of several minutes (Tingay et al. 2013, in preparation).

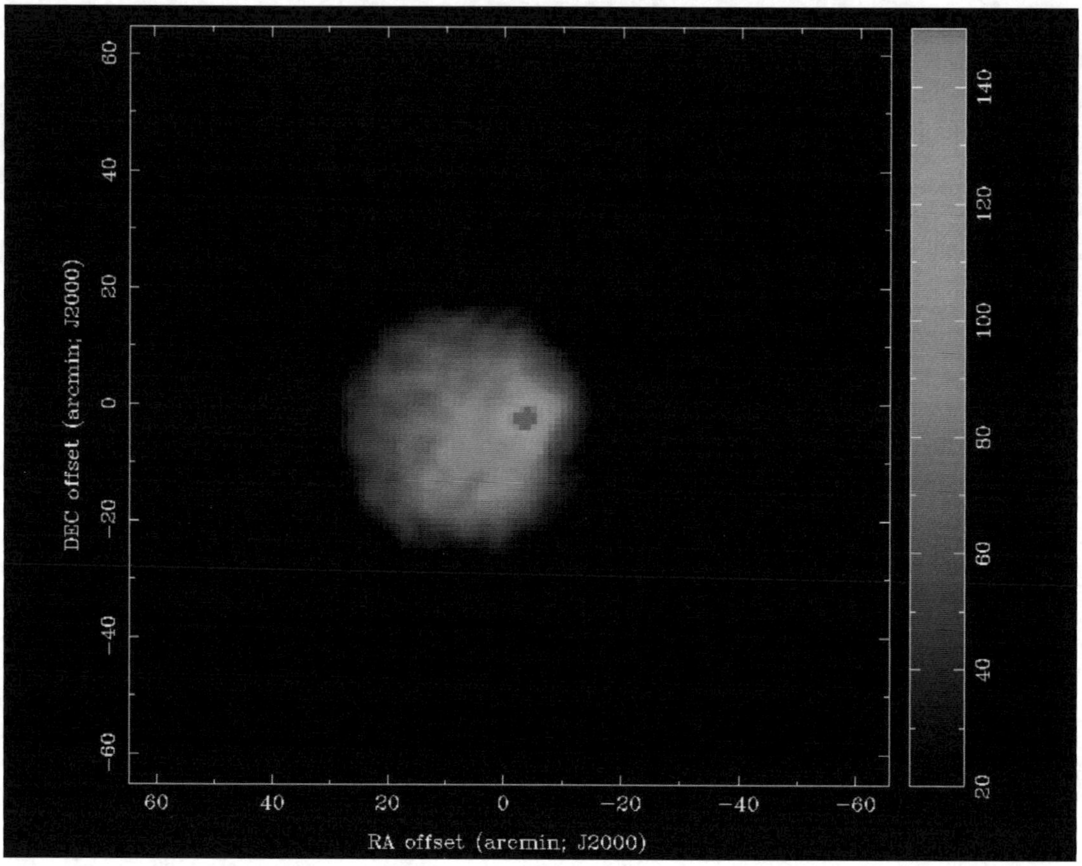

Figure 1. An image from the science commissioning phase of the MWA. Data obtained as described in the text. Normal astronomical units are used, with north at top and east to the left. The synthesised beam has a full width at half maximum of approximately 4 arcminutes and is close to circular.

4. Future prospects

In addition to the capabilities of the final instrument described above, the MWA will have the ability to record the raw voltage streams from each tile. This capability makes it possible to produce dynamic spectra and images at very high time and frequency resolution, sub-millisecond and sub-10 kHz (but not at maximum time and frequency resolution simultaneously). Such high resolution imaging will open up the ability to study the temporal and frequency sub-structure of both strong flares and the weak features seen by [7].

The MWA data will be of most use if they can be combined with data from other solar observation platforms in other wavelength ranges. Part of the motivation for attending this conference and providing an update on the capabilities of the MWA was to make contact with other instrument teams and to start to coordinate MWA solar observations with multi-wavelength campaigns.

As described in [1], the MWA could play a significant role in tracking large Coronal Mass Ejections and making early predictions regarding collisions with the Earth. It is thus timely that the MWA will come into full operation in the first half of 2013, presumably the year of maximum activity during Cycle 24.

Acknowledgments
This scientific work makes use of the Murchison Radio-astronomy Observatory. We acknowledge the Wajarri Yamatji people as the traditional owners of the Observatory site. Support for the MWA comes from the U.S. National Science Foundation (grants AST-0457585, PHY-0835713, CAREER-0847753, and AST-0908884), the Australian Research Council (LIEF grants LE0775621 and LE0882938), the U.S. Air Force Office of Scientic Research (grant FA9550-0510247), and the Centre for All-sky Astrophysics (an Australian Research Council Centre of Excellence funded by grant CE110001020). Support is also provided by the Smithsonian Astrophysical Observatory, the MIT School of Science, the Raman Research Institute, the Australian National University, and the Victoria University of Wellington (via grant MED-E1799 from the New Zealand Ministry of Economic Development and an IBM Shared University Research Grant). The Australian Federal government provides additional support via the National Collaborative Research Infrastructure Strategy, Education Investment Fund, and the Australia India Strategic Research Fund, and Astronomy Australia Limited, under contract to Curtin University. We acknowledge the iVEC Petabyte Data Store, the Initiative in Innovative Computing and the CUDA Center for Excellence sponsored by NVIDIA at Harvard University, and the International Centre for Radio Astronomy Research (ICRAR), a Joint Venture of Curtin University and The University of Western Australia, funded by the Western Australian State government.

References

[1] Bowman, J. et al. 2012, PASA submitted (arXiv:212.5151)
[2] Dewdney, P.. et al. 2010, SKA Memo #130, "SKA Phase 1: Prelimninary System Description", www.skatelescope.org/pages/page_memos.htm
[3] Ekers, R.D. & Frater, R.H. 2012, Biogr. Mems Fell R. Soc. 58, in press
[4] Johnston, S. et al. 2008, Ex.A., 22, 151
[5] Johnston, S. et al. 2007, PASA, 24, 174
[6] Kundu, M.R. 1965, "Solar Radio Astronomy", Wiley (New York)
[7] Oberoi, D. et al. 2011, ApJ, 728, L27
[8] Tingay, S.J. et al. 2012, PASA accepted (arXiv:1206.6945)
[9] Wild, J.P., Sheridan, K.V. & Kai, K. 1968, Nature, 218, 536

Eclipse on the Coral Sea: Cycle 24 Ascending (GONG 2012, LWS/SDO-5, and SOHO 27) IOP Publishing
Journal of Physics: Conference Series **440** (2013) 012034 doi:10.1088/1742-6596/440/1/012034

Global aspects of solar flares

Don Melrose
SIfA, School of Physics, The University of Sydney, NSW 2006, Australia
E-mail: melrose@physics.usyd.edu.au

Abstract. Existing global models for flares do not include an essential ingredient in the electrodynamics: the inductive electric field due to the time-varying magnetic field. How the large ($\approx 10^{10}$ V) electromotive force and the current it drives can be included in a model is discussed. Alfvén waves play an important role in transporting energy and potential, and in redistributing current on a global scale.

1. Introduction

Global aspects of solar flares were emphasized in flare models from the 1940s to the 1980s, as is evident from Hudson's solar flare cartoon archive.[1] Most of the flare models that emerged are variants of either the CSHKP model [1, 2, 3, 4] or of a circuit model [5, 6, 7]. In a CSHKP model, the magnetic energy release occurs through magnetic reconnection in a current sheet along the axis of a coronal streamer. In a circuit model, the energy release is attributed to current dissipation due to the turning on of an effective resistance in a current-carrying coronal loop. It is widely accepted that a flare involves release of magnetic energy stored in the corona prior to a flare. Some flares involve the eruption of a magnetic arcade, due to internal or external reconnection following post-emergence shearing or twisting, resulting in the 'two-ribbon' photospheric emission pattern observed from either side of a magnetic polarity-inversion line. In other (non-eruptive or compact) flares, the energy release is driven by newly emerging magnetic flux impinging on and reconnecting with existing magnetic flux, and the energy goes primarily into energetic electrons, rather than into the kinetic energy of a coronal mass ejection. The magnetic free energy is associated with coronal currents of order 10^{11} A that are observed in vector magnetogram data. The currents are assumed to flow along coronal loops and close deep inside the Sun. CSHKP and circuit models incorporate neither the effect of newly emerging magnetic flux nor the boundary condition that the current at each footpoint does not change significantly on the timescale of a flare. These deficiencies partly motivated the development of quadrupolar models [8, 9, 10, 11] in which magnetic reconnection allows transfer of flux and current between two pairs of bipolar footpoints. Figure 1 illustrates such a model. An essential feature of the physics that is not included in any existing model is an inductive electric field. From Faraday's equation, the 'potential' (the line integral of the electric field around any closed path) is equal to minus the rate of change of the enclosed magnetic flux, and this is very large in a flare, $\Phi = 10^9$–10^{10} V. In this paper I discuss how Φ can be included in a global model [12, 13], and I indicate how including it might help resolve two long-standing related problems.

[1] see http://solarmuri.ssl.berkeley.edu/~hhudson/cartoons/

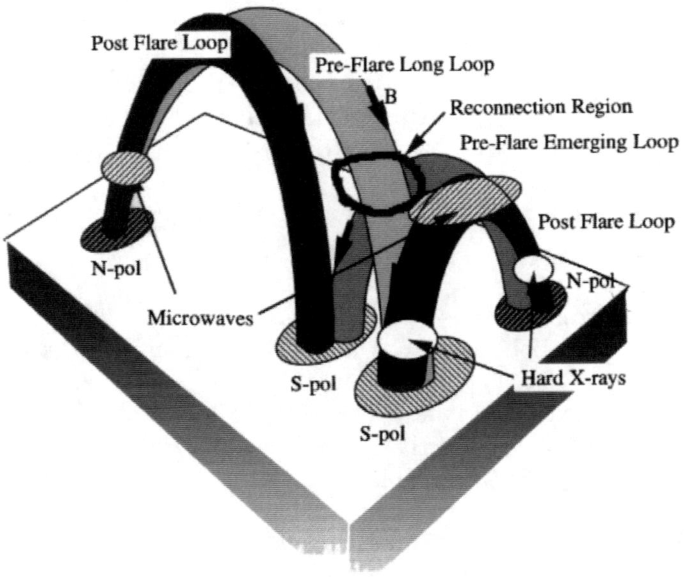

Figure 1. A cartoon of a quadrupolar model [10]. The pre-flare loops reconnect to form the post-flare loops, and the generator region corresponds to the indicated Reconnection Region.

The two problems are 'bulk energization' of electrons and the 'number problem'. Most of the nonthermal energy in a non-eruptive flare goes into bulk energization of the $\varepsilon = 10$–$20\,\mathrm{keV}$ electrons that produce hard X-ray bursts and type III radio bursts. The acceleration requires a parallel electric field [14], but how this develops is not understood; moreover, there is no obvious explanation as to why electrons are not accelerated to a much higher energy, of order $e\Phi$. Two aspects of the number problem are that the rate accelerated electrons precipitate, $\dot{N} > 10^{36}\,\mathrm{s}^{-1}$, implies a current, $e\dot{N}$, that greatly exceeds the maximum possible current that can exist in the corona [15], and that the total number of electrons that are accelerated during a flare greatly exceeds the number in the flux loop prior to the flare [16, 17]. These problems are related in that the power released, $I\Phi$, may be written as $\dot{N}\varepsilon$, with $e\dot{N} = MI$ and $e\Phi = M\varepsilon$, where the multiplicity factor is of order $M = 10^6$. A successful model needs to explain this multiplicity.

2. Parameters for a flare

A global model involves bulk parameters, and characteristic values for these parameters are needed. The power and energy released in flares cover several orders of magnitude. Here I assume a large flare with a power of $I\Phi = 10^{21}\,\mathrm{W}$ and a flare time $\tau = 10^3\,\mathrm{s}$, resulting in the release of $10^{24}\,\mathrm{J}$ of magnetic energy. The magnetic energy stored in a single magnetic loop is $\frac{1}{2}LI^2$, and assuming two loops one has $LI^2 = 10^{24}\,\mathrm{J}$. The current is assumed to be $I = 10^{11}\,\mathrm{A}$, from vector magnetogram data. The inductance depends primarily on the length of the loop, and a change in the stored energy requires a reduction in the length of the loop [9]. The resulting change in the inductance is assumed to be $\Delta L = 10^2\,\mathrm{H}$. The rate of change in the magnetic flux, LI, then implies $\Phi = \Delta L\, I/\tau = 10^{10}\,\mathrm{V}$.

The fact that such a large value of Φ must be present during a flare was already in the literature in the 1930s [18], but it is not included explicitly in CSHKP or quadrupolar models, and it is attributed incorrectly to a photospheric dynamo in most circuit models. This 'potential' is more appropriately interpreted as an electromotive force (EMF), which drives a current around a circuit. The current that this EMF drives must be 'new' in the sense that it did not exist prior to the onset of the flare. The currents flowing through the footpoints do not have time to change during a flare, so this 'new' current must involve redirection of the pre-flare currents. The path of this 'new' current must form a closed loop that encloses the changing magnetic flux, and includes a cross-field portion in the region where the reconnection occurs [13]. The flare is driven by a 'generator', where both I and Φ are across field lines in the corona. There is no obvious generator in a CSHKP model, and the generator is incorrectly identified with a photospheric dynamo in a circuit model. The generator in a quadrupolar model is indicated in Figure 1.

3. Analogy with substorms

The concept of a generator region is familiar in the context of geomagnetic substorms. Our understanding of substorms and flares has complementary aspects, such that the analogy between them is useful in identifying generic properties of magnetic explosions [12]. A specific property that is obvious for substorms is that the generator region is widely separated from the acceleration (of auroral electrons) region. This requires that the magnetic energy released in the generator propagate over a long distance before it dissipates through acceleration of electrons. Fletcher and Hudson [19] applied this to a model for flares, illustrated in Figure 2, in which they emphasized energy transport by Alfvén waves. Haerendel [20, 21] pursued the analogy between flares and substorms in further detail.

A specific model for energy transport between the generator and the acceleration region in substorms [22, 23] involves large-amplitude Alfvén waves. The electric field in the wave may be described in terms of a wave potential. The waves also transport potential along field lines, mapping the cross-field Φ at the generator onto the top of the acceleration region. There is also a parallel current, $J_\parallel$, associated with the wave. Alfvén waves have no net parallel current associated with them, that is, the integral of (the wave-associated) $J_\parallel$ over area perpendicular to the field lines must vanish. In the simplest case there is one area with current flowing up and a neighboring area with an equal current flowing down. These up and down currents close across field lines at the generator and across field lines in the ionosphere or photosphere. The Alfvén wave effectively sets up a loop current, with up and down field-aligned currents closing across field lines at boundaries. This closed loop pattern is well recognized in the context of substorms [24], where it is called a current wedge. Essentially the same effect had been recognized much earlier in a laboratory context [25]: a current driven across field lines in the body of the plasma tends to close through field-aligned currents to a boundary where Ohm's law allows closure. In the solar context, this conducting boundary can be identified as the photosphere. The cross-field (Pedersen) conductivity of the photosphere [26, 27] is large enough to allow such closure. The concept of a current wedge has already been invoked in at least one solar model [28, 29].

A further relevant idea, attributed to Gurnett [30], is that the equi-potential surfaces (above the acceleration region) must close across field lines somewhere above the highly conducting boundary. Closure of the equi-potential surfaces implies a parallel electric field, and this field is assumed to accelerate electrons, drawing energy from the Alfvén waves and causing them to damp. In other words, a postulated loss of wave energy to accelerated electrons implies that the waves damp, so that the wave potential decreases with decreasing height, with the resulting gradient of this potential implying the $E_\parallel$ that transfers the wave energy to electrons. Although a full understanding of the dissipation/acceleration process requires a specific mechanism for setting up $E_\parallel$ [31], this idea allows one to infer that the global physics requires such an $E_\parallel$ must be develop.

4. Discussion and Conclusions

Existing solar flare models omit an essential part of the basic electrodynamics: the inductive electric field and associated EMF. Release of magnetic energy implies that the magnetic flux in

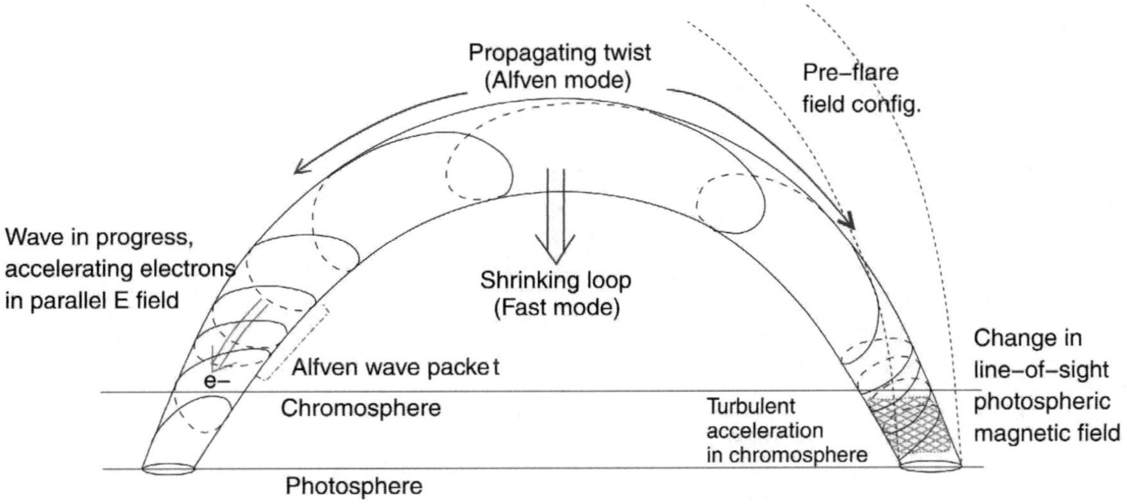

Figure 2. A cartoon showing a model due to Fletcher and Hudson [19] for energy release in a flare in which energy is transported from an energy release site (effectively at the top of the loop), via Alfvén waves to an acceleration region in the chromosphere.

the corona is decreasing, and Faraday's law implies an EMF (of order 10^{10} V) around any path that encloses this changing flux. This EMF drives a flare-related current that allows the coronal current distribution to change as the magnetic structure changes. The EMF plays a central role in any magnetic explosion, and an acceptable global model for flares must take it into account explicitly.

Alfvén waves transport energy and potential along field lines, and redistribute current across field lines. They link the widely separated generator and acceleration regions. Their damping in the acceleration region is due to the development of $E_\parallel \neq 0$ that accelerates the electrons.

A multiplicity M is needed to relate Φ to the potential Φ/M that observations imply is available to accelerate the electrons. The only plausible explanation for M was proposed by Holman [14]: the current flows back and forth M times between the generator and the acceleration region, with the effective EMF driving each pair of up and down flows being Φ/M.

4.1. Acknowledgments
I thank Mike Wheatland for helpful comments.

5. References
[1] Carmichael H 1964, in *AAS-NASA Symposium on Solar Flares* ed W N Hess (NASA SP-50), 451
[2] Sturrock P A 1966, *Nature*, **211**, 695
[3] Hirayama T 1974, *Solar Phys.*, **34**, 323
[4] Kopp R A and Pneuman G W 1976, *Solar Phys.*, **50**, 85
[5] Alfvén H and Carlqvist P 1967, *Solar Phys.*, **1**, 220
[6] Colgate S A 1978, *Astrophys. J.*, **221**, 1068
[7] Spicer D S 1982, *Space Sci. Rev.*, **31**, 351
[8] Uchida Y 1980, in *Solar Flares* ed P A Sturrock (Colorado: Colorado Assoc. Univ. Press), p 67
[9] Melrose D B 1997, *Astrophys. J.*, **486**, 521
[10] Nishio, Yaji K, Kosugi T, Nakajimi H and Sakurai T 1997, *Astrophys. J.*, **489**, 976
[11] Hirose S, Uchida Y, Uemura S, Yamaguchi T and Cable S A 2001, *Astrophys. J.*, **551**, 586

[12] Melrose, D B 2012a, *Astrophys. J.*, **749** 58M
[13] Melrose, D B 2012b, *Astrophys. J.*, **749** 59M
[14] Holman G D 1985, *Astrophys. J.*, **293**, 584
[15] Spicer D S and Sudan R N 1984, *Astrophys. J.*, **280**, 448
[16] Hoyng P, Brown J C and van Beek H F 1976, *Solar Phys.*, **48**, 197
[17] Emslie A G and Hénoux J C 1995, *Astrophys. J.*, **446**, 371
[18] Swann W F G 1933, *Phys. Rev.*, **43**, 217
[19] Fletcher L and Hudson H S 2008, *Astrophys. J.*, **675**, 1645
[20] Haerendel G 1994, *Astrophys. J. Suppl.*, **90**, 765
[21] Haerendel G 2012, *Astrophys. J.*, **749**, 166
[22] Vogt J and Haerendel G 1998, *Geophys. Res. Lett.*, **25**, 277
[23] Paschmann G, Haaland S and Treumann R 2002, *Space Sci. Rev.*, **103**, 1
[24] McPherran R L, Russel C T and Aubry M P 1978, *J. Geophys. Res.*, **78**, 3131
[25] Simon A 1955, *Phys. Rev.* **98**, 317
[26] Hénoux, J C and Somov, B V 1987, *Astron. Astrophys.*, **185**, 306
[27] Melrose, D B and Khan, J I 1989, *Astron. Astrophys.*, **219**, 308
[28] Ugai, M 2007, *Phys. Plasmas*, **14**, 102904
[29] Ugai, M 2008, *Phys. Plasmas*, **15**, 032902
[30] Alfvén H 1977, *Rev. Geophys. Space Phys..*, **15**, 271
[31] Bian N H, Kontar E P and Brown J C 2010, *Astron. Astrophys.*, **519**, 114

Eclipse on the Coral Sea: Cycle 24 Ascending (GONG 2012, LWS/SDO-5, and SOHO 27) IOP Publishing
Journal of Physics: Conference Series **440** (2013) 012035 doi:10.1088/1742-6596/440/1/012035

Puzzling nature of the fine structure of quiescent prominences and filaments

Stanislav Gunár, and Petr Heinzel
Astronomical Institute, Academy of Sciences of Czech Rep., 25165 Ondřejov, Czech Republic
E-mail: stanislav.gunar@asu.cas.cz

Ulrich Anzer
Max-Planck-Institut für Astrophysik, 85740 Garching bei Mü nchen, Germany

Duncan H. Mackay
School of Mathematics and Statistics, University of St Andrews, St Andrews, KY16 9SS, UK

Abstract. Even after more than 160 years of observations and modelling of solar prominences their true nature contains many open questions. In this work we argue that current 2D prominence fine structure models can help us to understand the puzzling connection between quasi-vertical fine structures often seen in quiescent prominences observed on the solar limb and horizontally aligned dark fibrils representing the fine structures of prominences observed in absorption against the solar disk (filaments).

1. Introduction

Some of the most striking features visible during total solar eclipses, of course besides the beautiful renderings of the coronal structures, are solar prominences. Indeed, these were first observed during total solar eclipses and studied since. In the last decades prominences were observed by powerful instruments and extensively modeled but their nature still puzzles solar physicists, even after more than 160 years. One of the most puzzling problems in current prominence physics is the presence of quasi-vertical fine-structure threads often seen in quiescent prominences observed in emission above the solar limb (see e.g. [1] and their Fig. 1). How can such vertically aligned structures be embedded in the predominantly horizontal magnetic field [2]? How can they be related to fine densely-packed horizontal dark fibrils usually observed in filaments in absorption against the solar disk (see e.g. [3] and Figs. 1 and 2 therein)? Are these the same structures?

Our contribution to answering the above questions is based on a detailed 2D modelling of the prominence fine structures using either gravity-induced Kippenhahn-Schlüter type (KS-type) models of [4] or newly developed non-linear force-free (NLFF) magnetic dip models of [5]. The 2D KS-type models represent the magnetic dips in initially horizontal field caused entirely by the weight of the prominence mass. Such dipped regions could propagate vertically and might lead to formation of quasi-vertical structures. Gunár et al. (in [6], [7], [8], [1], and [9]) showed that these models can produce synthetic hydrogen spectra in very good agreement with optical

and UV observations, especially in case of multi-thread configurations. On the other hand, the NLFF magnetic dips result from large-scale 3D NLFF modelling of the whole prominence magnetic field [10]. They can be then filled with the realistic prominence plasma for example by the technique described in [5].
We demonstrate here that such prominence fine structure models, either gravity-induced or NLFF, can be also consistent with observations of fine structures of both prominences and filaments.

2. Prominence fine structure models

Both types of prominence fine structure models used in this work assume 2D vertically infinite geometry with all quantities uniform along the vertical axis and varying within the horizontal x-y plane (see Fig. 1). Such an approach is valid for modelling of quasi-vertical prominence fine structures observed above the solar limb, because the dominant incident radiation from the disk penetrates the prominence body along the horizontal plane. In this plane we solve, in both cases, the 2D multi-level non-LTE (i.e. departures from Local Thermodynamic Equilibrium) radiative transfer problem to obtain the synthetic hydrogen spectra, including the Hα spectral line.
In order to model fine structures of filaments observed against the solar disk we have to artificially cut our 2D vertical threads at an arbitrary height and use resulting 3D structures to compute the radiation output along the line of sight pointing vertically. To this end we use the plasma properties and source function variation obtained by 2D radiative transfer modelling in the x-y plane distributed uniformly in the vertical direction. While this approach does not conserve the magneto-hydrostatic (MHS) equilibrium of the KS-type models and is not suitable to model the spectral lines with significant optical thickness, such as Lyman lines, it is useful for obtaining synthetic Hα line profiles. Thus we can produce synthetic narrow-band observations in the Hα centre for both cases, prominences on the limb and filaments against the disk.

The above idea of projection of 2D vertical threads produced by KS-type models against the disk was presented by [11], who computed the Hα line contrast of dark fibrils. Here we use a multi-thread model consisting of identical 2D prominence fine-structure threads, without any mutual radiative interaction. We arrange this set of threads in a way that qualitatively corresponds to the observed prominence/filament structure, however, we do not attempt to reproduce any particular prominence or filament. Individual threads are arranged so that when viewed from the top they resemble the main body of a filament with dark fibrils aligned along a sheared magnetic field. Each thread is shifted by a randomly selected distance with respect to its central position along the spine of the filament (see Fig. 2). The width of individual threads is 1000 km and the gap between threads is also 1000 km wide. These dimension can be too large in comparison with the observed widths of the prominence/filament fine structures that may be as small as 100

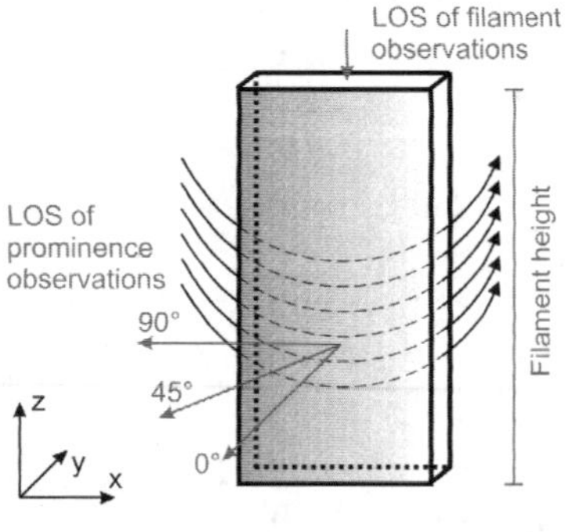

Figure 1. Scheme of a 2D vertically infinite fine structure thread with indicated lines of sight for prominence and filament observations. Drawn magnetic field lines approximate the shape of the magnetic dips. Note that we assume $B_y = 0$

km [3]. However, with this multi-thread set-up (Fig. 2) we aim to demonstrate that our models can consistently simulate the prominence and filament fine structure observations. For that we use the 2D model (MODEL1) from [8] that produces synthetic hydrogen Lyman spectra in very good agreement with SOHO/SUMER [12] observations. The width of the thread (1000 km) is an arbitrarily chosen input parameter of the model, and the length of the thread (approximately 28000 km) is determined by the MHS equilibrium computations. We have artificially cut the 2D vertical threads of the model to obtain height of 10000 km. Figure 2 shows, in the bottom panel, the synthetic Hα line centre image of fine-structure threads as dark fibrils in projection against the solar disk. The panel on the top shows the same multi-thread set-up as seen with the LOS along the y-axis (see Fig. 1). This produces a synthetic Hα line centre image of the vertical prominence fine structures seen above the solar limb. Here we see relatively narrow vertical structures even though the LOS is perpendicular to the longer of the thread dimensions. This is due to the fact, that the most intense (and thus easily observable) Hα line emission is produced in the relatively confined region in the middle of each thread. On the other hand, an observable absorption producing the dark fibrils in the projection against the solar disk occurs in more extended volume, which produces significantly elongated fibrils. Their length depends on the assumed height of the threads (i.e. vertically aligned dips), however, [11] showed that this dependence is not very strong (see also Fig. 3, bottom).

In the next step we employ the NLFF magnetic dip model (the DEEP_DIP from [5]) not in a multi-thread configuration but to demonstrate the properties of the synthetic Hα line centre observations of a single fine structure. This model, produced by insertion of the prominence plasma into dipped magnetic field created by large-scale 3D NLFF modelling of the whole prominence magnetic field [10], can be restricted in height naturally, without the loss of the plasma hydrostatic equilibrium.
The top panel of Fig. 3 shows, as a synthetic Hα line centre prominence image, a single fine structure viewed with three different orientations of the LOS, all lying in x-y plane (Fig. 1).

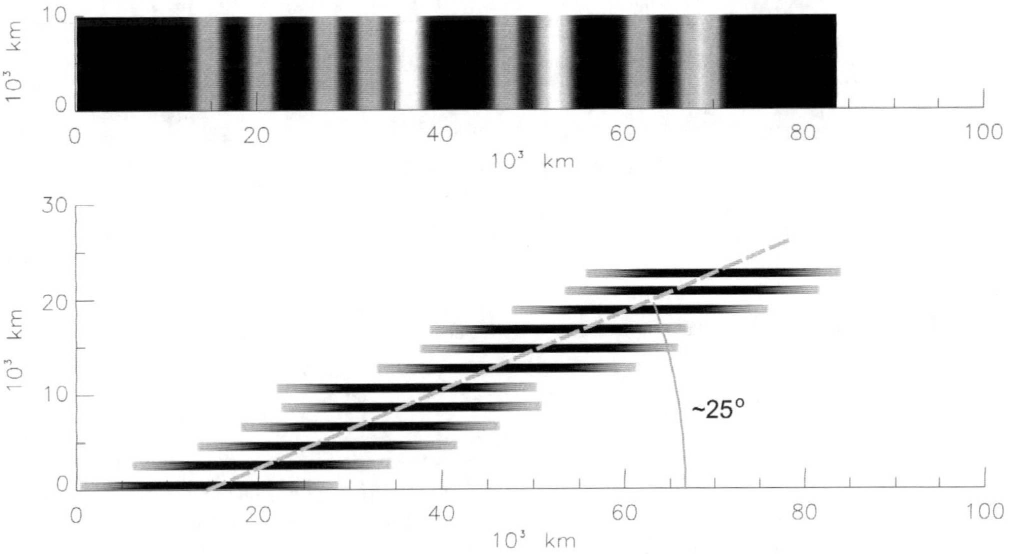

Figure 2. Multi-thread configuration of the 2D KS-type model. The top panel shows the synthetic on-the-limb image in the Hα line centre. The bottom panel shows the Hα line-centre projection against the disk. The dashed line represent the filament spine.

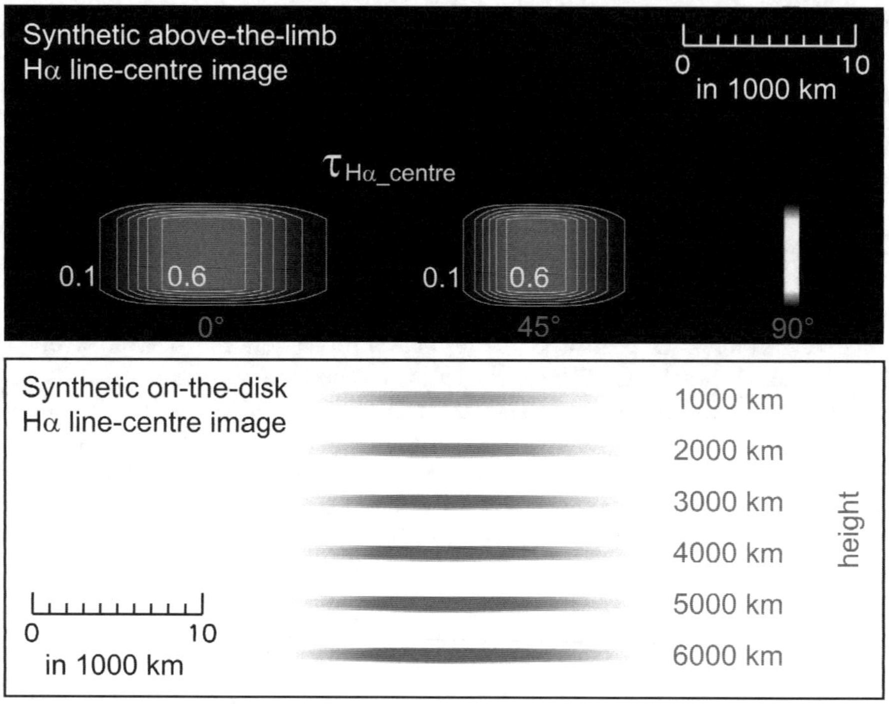

Figure 3. NLFF magnetic dip model observed as a single fine structure. The top panel shows the synthetic on-the-limb image in the Hα line centre. Over-plotted contours represent the optical thickness in the Hα centre. The bottom panel shows the Hα line-centre projection against the disk in dependence on the height of the fine structure.

Over-plotted contours represent the optical thickness in the Hα centre. The maximum $\tau_{\mathrm{H}\alpha\_\mathrm{centre}}$ in case of LOS along y-axis (0°) and with 45° inclination is well bellow unity. This would, in case of randomly arranged multiple fine structures along the given LOS, lead to occurrence of smaller-scale regions with highest intensities where emission from more fine structures would add up. Note that the synthetic intensities are plotted in the same intensity scale for all three LOS orientations. The bottom panel shows the synthetic Hα line centre images of the same NLFF magnetic dip model against the solar disk. Here we show the geometrical extension of such produced dark fibrils in dependence on the height of the fine structure.

3. Conclusions

In this work we show that current 2D prominence fine structure models, either gravity-induced or NLFF, can be adapted to consistently produce synthetic narrow-band images (e.g. in the Hα line centre) of prominence and filament fine structures resembling the real observations. However, the limitations of currently used 2D geometry don't allow us to look for models consisting of multiple fine structures that will be able to explain all observed characteristics of prominences. To this end we will have to concentrate on development of truly 3D fine-structure models with well described structure of supporting magnetic field and with detailed treatment of the 3D radiative transfer. Such models will for the first time allow us to consistently visualize the fine structures of prominences observed on the limb and against the disk. This will help us to understand the true nature of these fascinating solar features.

Acknowledgments

S.G. and P.H. acknowledge the support from grant P209/12/0906 of the Grant Agency of the Czech Republic. P.H. acknowledges the support from grant P209/10/1680 of the Grant Agency of the Czech Republic. S.G. and P.H. acknowledge the support from the MPA Garching; U.A. thanks for support from the Ondřejov Observatory. S.G. acknowledges the support from St Andrews University. Work of S.G. and P.H. was supported by the project RVO: 67985815. DHM acknowledges financial support from the STFC and the Leverhulme Trust. This research has also received funding from the European Commission's Seventh Framework Programme (FP7/2007-2013) under the grant agreement SWIFF (project n 263340, www.swiff.eu).

References

[1] Berlicki A, Gunar S, Heinzel P, Schmieder B and Schwartz P 2011 *Astron. Astrophys.* **530** A143+

[2] Leroy J L 1989 *Dynamics and Structure of Quiescent Solar Prominences* (*Astrophysics and Space Science Library* vol 150) ed Priest E R pp 77–113

[3] Lin Y, Engvold O, Rouppe van der Voort L, Wiik J E and Berger T E 2005 *Solar Phys.* **226** 239–254

[4] Heinzel P and Anzer U 2001 *Astron. Astrophys.* **375** 1082–1090

[5] Gunár S, Mackay D, Anzer U and Heinzel P 2013 *Astron. Astrophys.(in press)*

[6] Gunár S, Heinzel P, Schmieder B, Schwartz P and Anzer U 2007b *Astron. Astrophys.* **472** 929–936

[7] Gunár S, Heinzel P, Anzer U and Schmieder B 2008 *Astron. Astrophys.* **490** 307–313

[8] Gunár S, Schwartz P, Schmieder B, Heinzel P and Anzer U 2010 *Astron. Astrophys.* **514** A43+

[9] Gunár S, Mein P, Schmieder B, Heinzel P and Mein N 2012 *Astron. Astrophys.* **543** A93

[10] Mackay D H and van Ballegooijen A A 2009 *Solar Phys.* **260** 321–346

[11] Heinzel P and Anzer U 2006 *Astrophys. J. Lett.* **643** L65–L68

[12] Wilhelm K, Curdt W, Marsch E, Schühle U, Lemaire P, Gabriel A, Vial J C, Grewing M, Huber M C E, Jordan S D, Poland A I, Thomas R J, Kühne M, Timothy J G, Hassler D M and Siegmund O H W 1995 *Solar Phys.* **162** 189–231

Eclipse on the Coral Sea: Cycle 24 Ascending (GONG 2012, LWS/SDO-5, and SOHO 27) IOP Publishing
Journal of Physics: Conference Series **440** (2013) 012036 doi:10.1088/1742-6596/440/1/012036

Making global map of the solar surface B_r from the HMI vector magnetic field observations

K Hayashi[1], Y Liu[1], X Sun[1], J T Hoeksema[1], R Centeno[2], G Barnes[3], K D Leka[3]

[1]W.W. Hansen Experimental Physics Laboratory, Stanford Univertsity, 452 Lomita Mall, Stanford, CA 94305, USA
[2]HAO
[3]NWRA

E-mail: keiji@sun.stanford.edu

Abstract. The Helioseismic Magnetic Imager (HMI) has made full-disk vector magnetic field measurements of the Sun with cadence of 12 minutes. The three-component solar surface magnetic field vector data are from the HMI observations with the data process pipeline modules, VFISV (Very Fast Inversion of the Stokes Vector, Borrero et al., 2011) for Milne-Eddington inversion and the minimum-energy disambiguation algorithm (Metcalf 1994, Leka et al, 2009). The models of the global corona and solar wind, such as the PFSS (potential-field source-surface) model and the MHD simulations, often use the maps of solar surface magnetic field, especially the radial component (B_r) as the boundary condition. The HMI observation can provide new B_r data for these model. Because of weak magnetic signals at the quiet regions of the Sun, the limb darkening, and geometric effects near solar poles, we need to apply an assumption to make a whole-surface map. In this paper, we tested two assumptions for determining B_r at weak-field regions. The coronal structures calculated by the PFSS model with the vector-based B_r are compared with those with the magnetogram-based B_r and the corona observed by the SDO/AIA (Atmospheric Imaging Assembly). In the tested period, CR 2098, the vector-based B_r map gives better agreements than the line-of-sight magnetogram data, though we need further investigation for evaluation.

1. Introduction

For realistically modeling the global solar corona, the models, such as the potential-field source-surface (PFSS) model [6] and MHD simulations, use the solar surface magnetic field data as the boundary conditions to specify the time of interest. Models, in general, use the magnetic field component normal to the boundary (the radial component B_r in the global model constructed in the spherical coordinates system), and the B_r synoptic maps are needed for the global coronal models. Earlier, with the line-of-sight (LoS) magnetogram observations, the radial method [10] is widely applied to determine the solar surface B_r from the observed LoS component B_l, by assuming the field is radial on the photosphere, expressed as a relation, $B_r = B_l/\sin\rho$, where ρ is the arc-distance from the disk center.

The HMI vector magnetic field measurement [7, 8], in principal, can provide the radial component on the solar photosphere straightforwardly. The final data of the HMI vector magnetic field are, in brief, made through the VFISV Milne-Eddington inversion for the Stokes vector [1] and the minimum energy method [3,5] for disambiguating the 180-degree ambiguity

Journal of Physics: Conference Series **440** (2013) 012036 doi:10.1088/1742-6596/440/1/012036

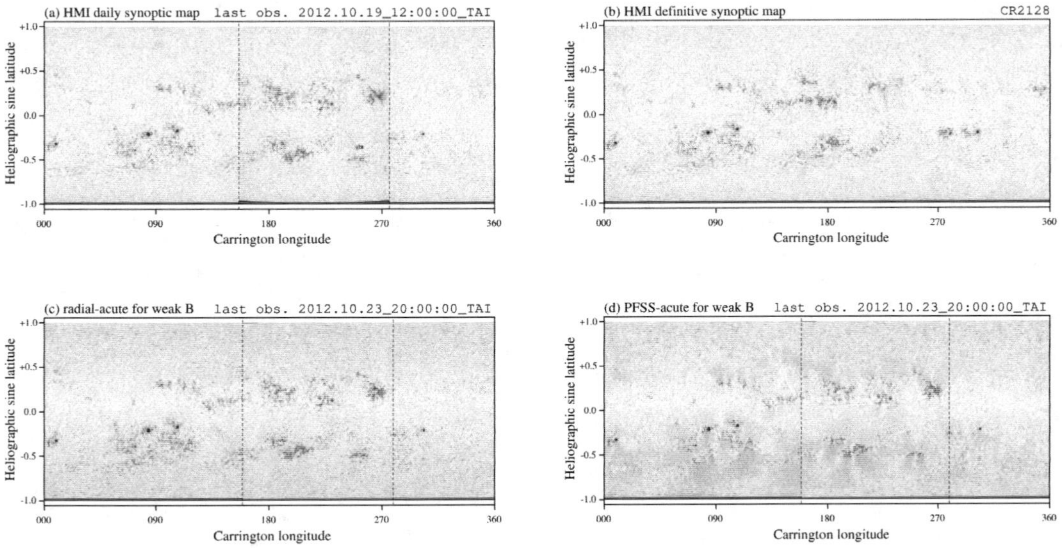

Figure 1. Synoptic maps of B_r, for October 2012. Top row: (a) daily-updated synoptic map of the magnetogram observation and the radial method, and (b) a standard synoptic map of the magnetogram observation for a Carrington rotation (CR) 2128. Bottom row: Synoptic map of vector-based B_r with (c) radial-acute assumption, and (d) PFSS-acute assumption. Vertical dashed lines in plots (a), (c), and (d) approximate the longitudes 60 degrees east and west from the central meridian viewed from Earth, on 2012 October 19. In plot (a) regions between the two dashed lines are the part updated. We extended the sampling window about 2 days to make vector-based B_r maps, (c) and (d).

in the azimuth angle. For making a whole-surface map, the disambiguated vector data are projected onto the coordinate of the synoptic or synchronic frame.

For strong-field regions such as the sunspots, the disambiguated vector data are generated regularly. However, an assumption is needed to determine B_r at weak-field regions, for which the disambiguation algorithm sometimes cannot determine the minimum energy state because of the weak magnetic signal. In this paper, we apply two assumptions, the potential-field acute model and the radial-acute mode, for determining the vector of the magnetic field, then apply the B_r maps to PFSS model, and compare the results with those with the HMI magnetogram [4] and the solar corona observed by the AIA.

2. B_r Maps

For determining one B_r value at weak field regions out of two candidate answers with the 180-degree ambiguity, we tested two assumptions. The first one is the radial-acute model where a candidate that has a larger absolute value of the radial component, $|B_r|$, is selected. The second one is the potential-field acute model in which the vector of PFSS field on the surface, $\vec{B}_{\rm pot}$, is first calculated with the LoS component then the candidate that has greater (positive) dot product with the potential field, $\vec{B} \cdot \vec{B}_{\rm pot}$, is chosen. In Figure 1, the global B_r maps obtained in these ways are shown together with the maps from the LoS magnetogram observations. It is easily noticed that the distribution of B_r derived with the PFSS-acute model is different from the other three. We made a tentatively conclusion that the radial-acute model (c) is more reasonable than the PFSS-acute model (d), after preliminary test similar to that in Section 3 and others. A supplemental reason was that the polarity inversion lines, dividing the positive

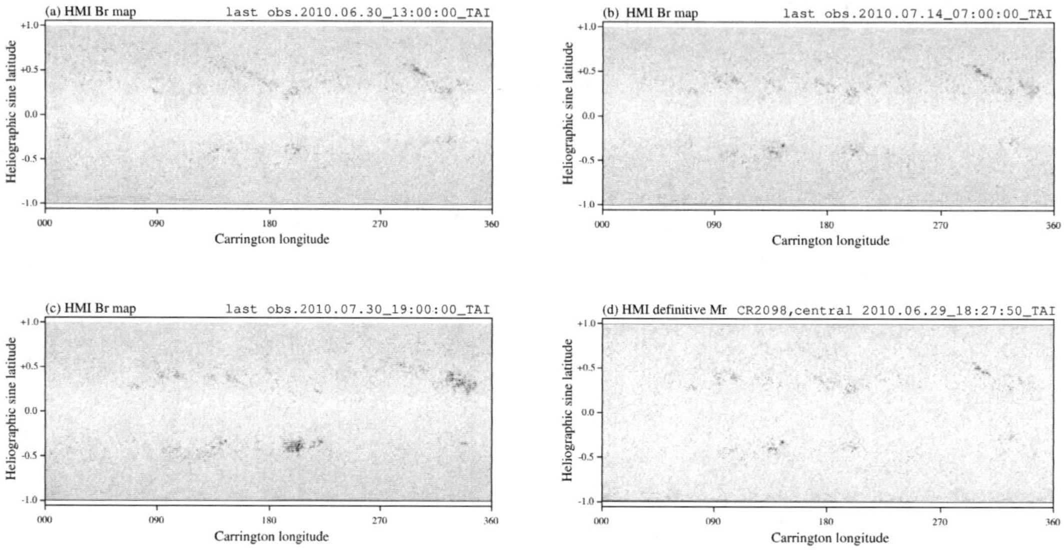

Figure 2. The four synoptic maps of B_r, around CR 2098, used for the PFSS calculations shown in Figure 3. The vector-based B_r maps, with the last full-disk observation data at (a) 2010/06/30, (b) 2010/07/14, and (c) 2010/07/30, are shown. The end time of the 27.3-day window are chosen so that the times of comparison in Figure 3 will be well near the center. The synoptic map of B_r, with the radial method applied to the LoS magnetogram data, for CR 2098 is shown in plot (d).

and negative B_r, of the map (d) did not coincide with the filaments observed in the period.

3. Comparisons

We use the B_r maps in the PFSS calculation and compare the bases of the open-field lines with the low-intensity regions of AIA 193. We chose a period, CR 2098, because the solar corona had the coronal holes (CHs) of various types of shape during this period.

As the input to the PFSS, we use the maps of the radial-acute vector B_r and those of the magnetogram-based B_r. Figure 2 shows the data. An additional choice, whether or not to adjust values of B_r to make the whole-surface integration, $\int B_r dS$ will be zero, is also critically important in the PFSS model and other coronal models. In this study, the adjustment is made by offsetting the B_r on whole surface uniformly. We run four PFSS calculations with iterative direct Laplace solver, determine the base of the open-field region by tracing the field lines, and compare with the AIA data at each instant.

Figure 3 is the comparison plots: The AIA 193 images are in the leftmost column, and the other four columns are for the base of the open field lines in the potential field viewed from the position of the Earth at instants of the AIA observation. The rectangles, circles, diamond and triangle boxes emphasize the differences of the base of the open field lines among the four calculations.

Here we briefly summarize comparisons: (1) In the parts marked with rectangles at the topmost row, we can see that the north-south (trans-equatorial) stretched coronal hole is better reproduced with the vector-based B_r, with or without B_r offset, than those with the magnetogram data. (2) In the parts marked with the circles, the stretched narrow coronal holes are well reproduced with the vector-based B_r with the offset, while the magnetogram-based B_r maps without offset gives best agreements. (3) The northern polar coronal hole marked with

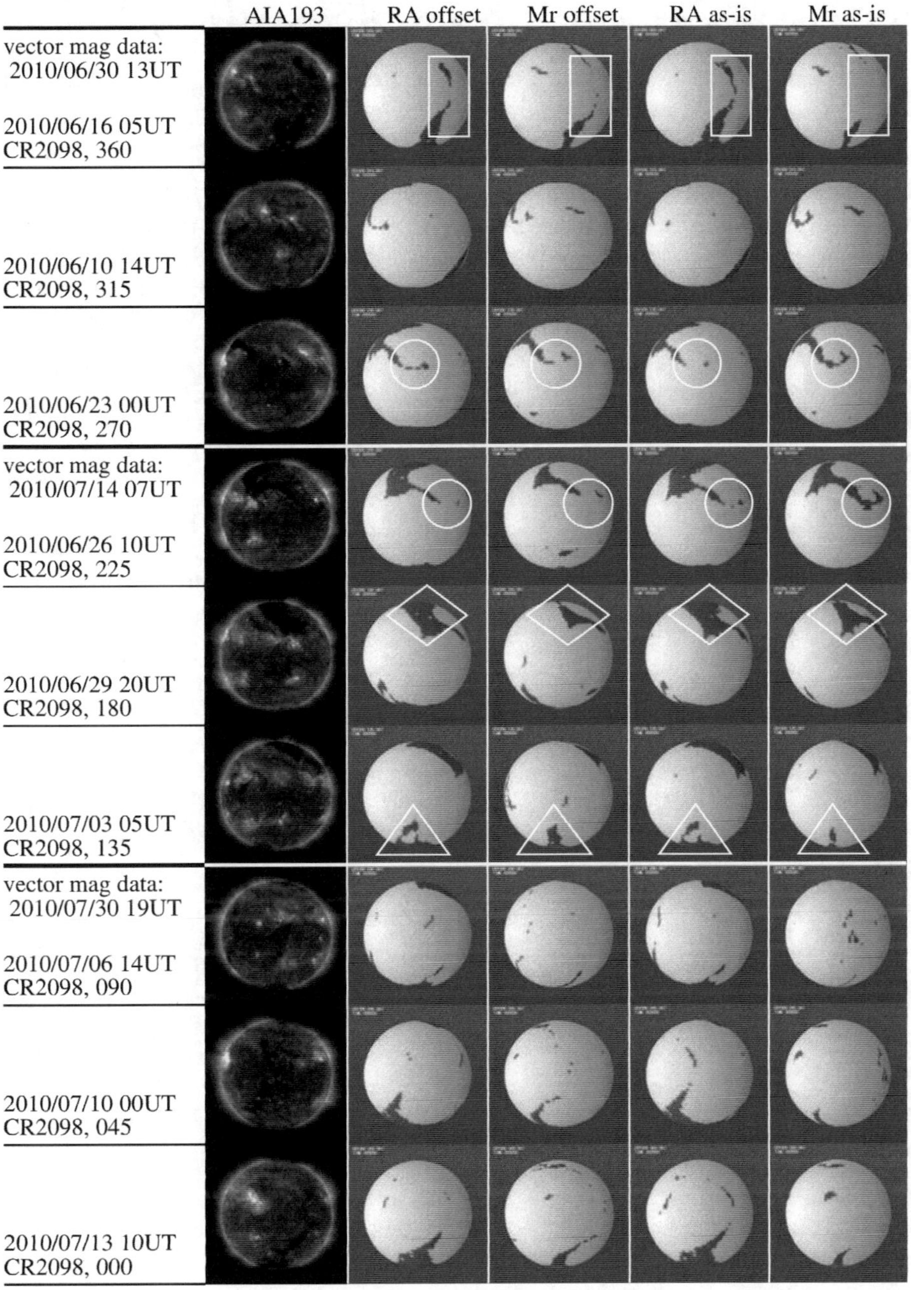

Figure 3. Comparison of the base of the PFSS open-field lines with the AIA 193 observations, at the Carrington longitudes 45 degrees apart, over a period of CR 2098. Time of the AIA observations and the point of view of open-field base of the PFSS model (darker grays) are denoted at the left. From the left, the AIA observation, the base of the open field derived from the PFSS model using the "monopole-offset" radial-acute vector B_r, the offset magnetogram-based B_r with the radial method, and the radial-acute B_r without the offset correction, and the magnetogram-based B_r without the offset correction. For each three of nine rows, one vector-based synoptic map is made from dataset with a different 27.3-day window of the HMI observations.

Journal of Physics: Conference Series **440** (2013) 012036 doi:10.1088/1742-6596/440/1/012036

diamonds is best reproduced with the vector-based B_r with the offset. (4) The open field region at the southern polar region marked with the triangles is not well seen in the AIA image data. Probably this is because the Earth was at about 4 degrees north of the solar equatorial plane.

4. Summary

We test two types of the assumptions to solve the 180-degree ambiguities at the weak field regions, and select one (the radial-acute) to compare with the standard LoS magnetogram and the AIA data. The vector-based B_r map with the offset gives overall better agreements with the AIA images than the magnetogram-based one; however, the advantage of vector-based B_r over the B_r maps from the magnetogram with the radial method is not sufficiently outstanding at this time.

As seen in the differences between the second and forth column, and between the third and fifth (rightmost) column of Figure 3, the treatment of the surplus of the magnetic field is critically important. The apparent "monopole" (the surface integration of B_r) may come from the fact that the synoptic maps are made by collecting solar-disk observation data at various instants over 27.3 days. In addition, if a synoptic map has one of two active-region polarities (i.e., the boundary line of longitude happens to split an active region), then the unbalance of B_r will be substantial [2]. For better determining the magnetic field in the polar regions, the polar field correction [9] is a method to apply. We continue to examine these factors to make "best" global map of B_r.

References

[1] Borrero J M, Tomczyk S, Kubo M, Socas-Navarro H, et al. 2011 *Solar Physics* **273** 267
[2] Hayashi K, Liu Y, Sun X and Turmon M J 2012 *AAS 43rd meeting* 207.16, June 2012, Anchorage, AK USA
[3] Leka K D, Barnes G, Crouch A D, Metcalf T R, et al. 2009 *Solar Physics* **260** 83
[4] Liu Y, Hoeksema J T, Scherrer P H, Schou J et al. 2012 *Solar Physics* **279** 295
[5] Metcalf T R 1994 *Solar Physics* **155** 235
[6] Schatten K H, Wilcox J M and Ness N F 1969 *Solar Physics* **6** 442
[7] Scherrer P H, Schou J, Bush R I, Kosovichev A G et al. 2012 *Solar Physics* **275** 207
[8] Schou J, Scherrer P H, Bush R I, Wachter R et al. 2012 *Solar Physics* **229** 207
[9] Sun X, Liu Y, Hoeksema J T, Hayashi K and Zhao X 2011 *Solar Physics* **270** 9
[10] Wang Y M and Sheeley Jr. N R 1992 *ApJ* **392** 392

Journal of Physics: Conference Series **440** (2013) 012037 doi:10.1088/1742-6596/440/1/012037

The state of nonlinear force-free magnetic field extrapolation

M. S. Wheatland and S. A. Gilchrist

Sydney Institute for Astronomy, School of Physics, The University of Sydney, NSW 2006, Australia

E-mail: m.wheatland@physics.usyd.edu.au

Abstract. Magnetic field extrapolation is the construction of a model solution for the coronal magnetic field in active regions from magnetic boundary data originating close to the Sun's surface. The nonlinear force-free model (in which the electric current density is parallel to the magnetic field) is often adopted to describe the coronal field. The solution of the nonlinear force-free equations is a challenging computational task, and the application of codes to available boundary data has revealed a number of significant problems with nonlinear force-free extrapolation. This paper summarises the present status of coronal field extrapolation, and describes some recent developments.

1. Background

The structure glimpsed in the solar corona during a total eclipse (see Figure 1) is determined by magnetic fields at the Sun's surface. The energy associated with strong fields around sunspots drives large-scale solar activity, namely flares and coronal mass ejections, which produce space weather storms in our local space environment, posing risks to critical communications infrastructure [1].

Solar activity and space weather motivate the modeling of active region magnetic fields. Ideally we would like to use modeling to address fundamental questions concerning the magnetic energy release process underlying solar activity, as well as to improve the accuracy of space weather forecasts.

1.1. Vector magnetogram data

Vector magnetogram data provide a basis for active region magnetic field modeling. The data are derived from determinations of Stokes polarisation profiles for magnetically sensitive lines formed close to the photosphere [2], for fields of view encompassing active regions. Magnetic field values are obtained from the Stokes profiles via "Stokes inversion," a process involving modeling radiative transport of polarised radiation through the magnetised solar atmosphere. It is important to recognise that field values obtained in this way are inferences, not direct measurements or observations, and as such are subject to substantial a priori uncertainty. The field values are highly method- and model-dependent. Additionally, the intrinsic 180° ambiguity in the direction of the component of the field perpendicular to the line of sight must be resolved to obtain unique magnetic field values [3, 4, 5].

Figure 1. The Sun at totality during the eclipse of 14 November 2012, photographed from Palm Cove beach by the author.

The resulting vector magnetogram data consist of a photospheric map of magnetic field values $\mathbf{B} = (B_x, B_y, B_z)$ over a region at the photosphere, where x, y, and z are local heliocentric coordinates, with z denoting the local radial direction. It is common to neglect the curvature of the photosphere for data on an active region scale, and to refer to z as the vertical direction.

Two space-based vector magnetographs – the Solar Optical Telescope/Spectro-Polarimeter (SOT/SP) on the Hinode satellite [7], and the Helioseismic & Magnetic Imager on the Solar Dynamics Observatory (SDO/HMI) – are now providing data with unprecedented quality, resolution, and cadence [8].

In principle, the data provide boundary conditions for coronal field modeling. Assuming the field is static, a specific coronal field model may be solved as a boundary value problem, a process often called "coronal field reconstruction." The vertical current density J_z at the photosphere provides an important boundary value for coronal field models, which determines the magnetic energy of a solution to the model (the minimum energy magnetostatic field with a prescribed normal component of the field in the boundary is the potential or current-free field, for which $J_z = 0$ in the boundary [6]). The vertical electric current density at the photosphere may be estimated from vector magnetogram data using

$$\mu_0 J_z|_{z=0} = \left[\frac{\partial B_y}{\partial x} - \frac{\partial B_x}{\partial y}\right]_{z=0}. \qquad (1)$$

1.2. The nonlinear force-free model

The coronal magnetic field is often modeled as a nonlinear force-free field [9], described by

$$\mathbf{J} \times \mathbf{B} = 0 \quad \text{and} \quad \nabla \cdot \mathbf{B} = 0. \qquad (2)$$

The model describes a static coronal field in which the electric current density $\mathbf{J} = \mu_0^{-1}\nabla \times \mathbf{B}$ is everywhere parallel to $\mathbf{B}$. Equations (2) may be rewritten in the form

$$\mathbf{B} \cdot \nabla\alpha = 0 \quad \text{and} \quad \nabla \times \mathbf{B} = \alpha\mathbf{B} \tag{3}$$

by introducing the force-free parameter α, defined by $\mathbf{J} = \alpha\mathbf{B}/\mu_0$. Physically α represents the constant ratio of the electric current density to the magnetic field along a field line. Equations (3) are a coupled system of four nonlinear PDEs for the four dependent variables (B_x, B_y, B_z, and α), which may be solved subject to suitable boundary conditions.

The boundary value problem in a half space ($z > 0$) requires specification of the normal component of the field B_z over the boundary $z = 0$, together with the specification of α over one polarity of the field in the boundary (i.e. the values of α must be given over $z = 0$ where $B_z > 0$, or the values of α must be given over $z = 0$ where $B_z < 0$) [10]. It is necessary only to specify α over one polarity because α is a constant along field lines ($\mathbf{B} \cdot \nabla\alpha = 0$), so values at one polarity are "mapped" to the other polarity by the field lines of the solution. We refer to the polarities defined by $B_z > 0$ and $B_z < 0$ as P and N respectively.

A variety of methods of solution of Equations (3) have been developed [9]. All methods are iterative, and work by solving a system of linear equations, such that an iterative sequence of solutions converges to a solution to the nonlinear Equations (3). Popular methods include Grad-Rubin iteration [10, 11], optimization [12, 13], and the magnetofrictional method [14, 15].

Some force-free methods (e.g. optimization) use the vector field values $\mathbf{B} = (B_x, B_y, B_z)$ at $z = 0$ over both polarities (P and N) as boundary conditions. This is formally an over-prescription, but should lead to no error provided the boundary conditions are consistent with a solution to the force-free model.

The Grad-Rubin scheme requires solution, at at iteration k, of the linear system

$$\mathbf{B}^{[k-1]} \cdot \nabla\alpha^{[k]} = 0 \quad \text{and} \quad \nabla \times \mathbf{B}^{[k]} = \alpha^{[k]}\mathbf{B}^{[k-1]}. \tag{4}$$

The boundary conditions are imposed on $B_z^{[k]}$ at $z = 0$, and on $\alpha^{[k]}$ at $z = 0$ over either P or N, as explained above. The iteration sequence is typically started with a potential (current-free) field for $\mathbf{B}^{[0]}$. A fixed point of the iteration scheme defined by Equations (4) is clearly a solution to the nonlinear force-free equations.

2. The inconsistency problem

2.1. *The problem*

A sequence of yearly workshops was organised by Lockheed Martin for the years 2005 to 2009, directed at improving the state of nonlinear force-free modeling. The workshops revealed that force-free methods work for test cases but fail, in specific ways, when applied to vector magnetogram data [16, 18, 17, 19]. A basic problem is that the boundary data are inconsistent with the model. Vector magnetogram data provide two sets of boundary conditions for the force-free model, since values of $\alpha = \mu_0 J_z/B_z$ may be determined for both the P and N polarities over the magnetogram field of view. This admits the possibility of two different solutions (the P and N solutions) to the model, depending on the choice of polarity for the boundary conditions. The workshops showed that two distinctly different solutions were obtained, for the Hinode SOT/SP vector magnetograms used in the studies. The field line configurations for the P and N solutions were found to be significantly different, and moreover the magnetic energies of the two solutions were substantially different [17, 19]. Vector magnetogram data typically fail to meet necessary conditions for the existence of a force-free solution [20], which may be attributed to non-magnetic forces at the photospheric level of the atmosphere. The solar atmosphere is thought to become force-free $\approx 400\,\text{km}$ above the photosphere [21].

Figure 2 illustrates the inconsistency problem. The left hand diagram shows field lines for a P solution, and the right hand diagram shows field lines for an N solution, for boundary data which are inconsistent with the force-free model. The field line configurations are distinctly different:

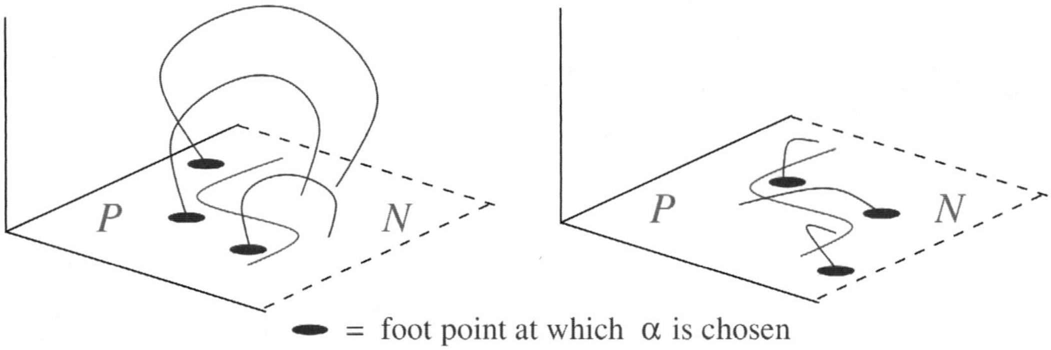

Figure 2. A schematic illustration of the inconsistency problem in nonlinear force-free modeling. If the boundary data are inconsistent with the force-free model, the P and N solutions (left and right diagrams, respectively) will be different. The field lines for the two solutions are shown by the loops.

the P solution has tall, nearly current-free loops over the neutral line separating the polarities, and the N solution has low current-carrying loops which are strongly sheared along the neutral line. This situation may occur if the α values are systematically larger in the N polarity than the P polarity. (Note that an extreme case of inconsistency is shown, for illustration. Actual solar data are unlikely to show systematic discrepancies.)

We would like to estimate the free magnetic energy of the extrapolated solutions, i.e. the energy $E_f = E - E_0$, where E is the total magnetic energy, and E_0 is the energy of the potential field configuration with the same boundary conditions on B_z. However, in general the P and N solutions obtained from extrapolating vector magnetogram data have different magnetic energies. Typically $E_f \ll E_0$, so the free magnetic energy of the P and N solutions may differ by a large factor, as illustrated by an example presented in Section 3.

This discussion applies to methods of solution of the model using values of α over one polarity as boundary conditions. Some methods of solution of the nonlinear force-free model use values of $\mathbf{B}$ over both P and N as BCs. In that case the inconsistency of the boundary data with the model leads to "solutions" with $\mathbf{J} \times \mathbf{B} \neq 0$ and/or $\nabla \cdot \mathbf{B} \neq 0$ somewhere in the coronal volume.

2.2. Practical methods of resolution

The "preprocessing" procedure [22] provides one method of improving (but not resolving) the inconsistency problem. The vector field components in the boundary are modified to meet certain conditions necessary for the force-free model. However, the conditions imposed are not sufficient to ensure a solution to the model, and in general preprocessed boundary data remain inconsistent with the model. The procedure is found to improve results in some cases [19].

A solution consistent with the model is obtained by the "self-consistency procedure" [23, 24], which works as follows. The vector magnetogram data are used to calculate P and N solutions. The two solutions are then used to identify a single force-free solution which is in some sense "closest" to the P and N solutions, and in particular has a very similar mapping of foot points between polarities. The method treats the P and N solutions as equally valid, except that uncertainties in boundary values of α are taken into account (preference is given to boundary data with small uncertainties). The self-consistent solution has the same boundary conditions on B_z as the vector magnetogram, but modified BCs on α.

To give a more detailed account, there are three steps in the self-consistency procedure. Step 1. is the calculation of the P and N solutions, from (unpreprocessed) vector magnetogram data.

The Grad-Rubin scheme of Equation (4) is used [11]. Step 2. is an adjustment of the boundary values, taking uncertainties in α values into account, as follows. Each solution defines a mapping of α values from one polarity to the other, since α is a constant along field lines in a force-free solution (the P solution maps boundary values of α from the P to the N polarity, and the N solution maps boundary values of α from N to P). The mappings define two complete sets of boundary values of α, which we denote by α_P and α_N. (The α_P data consist of the magnetogram values over the P polarity, and mapped values over the N polarity, and the α_N data consist of the magnetogram values over the N polarity, and mapped values over the P polarity.) There are also two complete sets of corresponding uncertainties, σ_P and σ_N, since the uncertainties map also along field lines. Together there are two possible sets of boundary values and uncertainties:

$$\alpha_P \pm \sigma_P \quad \text{and} \quad \alpha_N \pm \sigma_N \quad \text{at } z = 0. \tag{5}$$

The two sets of values are regarded as two (imperfect) determinations of the true values. A Bayesian estimate of the true values and associated uncertainties, assuming Gaussian-distributed errors, is [23]:

$$\alpha_{\text{est}} = \frac{\alpha_P/\sigma_P^2 + \alpha_N/\sigma_N^2}{1/\sigma_P^2 + 1/\sigma_N^2} \quad \text{and} \quad \sigma_{\text{est}} = \frac{1}{\sqrt{1/\sigma_P^2 + 1/\sigma_N^2}}. \tag{6}$$

The values α_{est} will in general still be inconsistent with the force-free model, but they are expected to be closer to consistency. Step 3. is the process of iteration. Steps 1. and 2. are repeated, with the vector magnetogram-derived values of α replaced by the new estimates $\alpha_{\text{est}} \pm \sigma_{\text{est}}$. (The first step in the next iteration is the calculation of P and N solutions using the magnetogram boundary values of B_z together with the new boundary values α_{est} for the force-free parameter.)

In practice the self-consistency procedure is found to arrive at a self-consistent solution within about a dozen "cycles" of the three steps (the iterations of the steps are termed cycles to distinguish them from the iterations of the Grad-Rubin procedure used to construct the P and N solutions). A practical convergence criterion for the slef-consistency cycles is that the P and N solution energies agree to $\lesssim 1\%$ for consecutive cycles.

Other approaches to the inconsistency problem have also been presented, in particular an "optimization Grad-Rubin method" [25], and an optimization method incorporating measurement errors [26].

3. A modeling example

To illustrate the application of the self-consistent modeling procedure, we consider the treatment of active region 11029, a small but highly-flare productive region which emerged on the disk on 21-22 Oct 2009. The region produced 73 GOES events above B class during its transit of the disk, briefly interrupting the extended minimum at the end of cycle 23. Because it flared in relative isolation, active region 11029 provided a good subject for a study of the statistics of flare occurrence in an individual active region [27]. The size distribution of soft X-ray events in the region appeared to show a relative deficit of large events – a departure from the well known power-law flare frequency-size distribution. In a follow-up study the region was modeled using the self-consistency procedure, to attempt to determine the region's free energy [28].

Figure 3 shows the data and the results of the nonlinear force-free modeling of the region. The upper left panel illustrates the boundary conditions, which use Hinode SOT/SP vector field determinations together with Solar and Heliospheric Observer/Michelson Doppler Interferometer (SoHO/MDI) line-of-sight magnetic field data [29]. The upper sub-panel shows B_z and the lower sub-panel shows J_z. Uncertainties are assigned to the data based on uncertainties from the Stokes inversion procedure. The upper right panel in Figure 3 demonstrates the convergence of the self-consistency procedure. The energy of the P and N solutions calculated at each self-consistency

Journal of Physics: Conference Series **440** (2013) 012037 doi:10.1088/1742-6596/440/1/012037

cycle is plotted versus the cycle number (the P energies are shown by plus signs and a dashed line, and the N energies are shown by diamonds and a solid line) with the energy given in units of the potential-field energy E_0. Initially the P and N solutions have very different energies, illustrating the inconsistency, but after 10 cycles the energies are very similar, as expected for a self-consistent solution (i.e. one for which the P and N solutions match). The lower panel shows the magnetic field configurations after 10 self-consistency cycles. The view looks down on the computational domain. The boundary values of B_z are shown by the color image, with regions in the P polarity colored blue, and regions in the N polarity colored red. The corresponding field lines for the P and N solutions are shown by the blue and red curves respectively. There is a good correspondence between the two solutions, demonstrating the achievement of a self-consistent solution. (Some discrepancy is seen for longer field lines.)

The calculation was performed on a $440 \times 300 \times 200$ grid with 20 Grad-Rubin iterations per self-consistency cycle. The code is a parallel implementation for shared memory multi-processor computers.

The modeling of active region 11029 allows estimation of the magnetic free energy from the extrapolated solutions. The energy of the self-consistent solution illustrated in Figure 3 is $E/E_0 = 1.04$ (see the top right panel of Figure 3), i.e. the free magnetic energy is $\approx 4\%$ higher than the potential energy. In absolute terms the potential field energy is $E_0 = 1.7 \times 10^{33}$ erg (a relatively large value) and the free energy is $E_f = E - E_0 = 6 \times 10^{31}$ erg. Varying the details of the modeling leads to slightly different self-consistent solutions, with slightly different energies. Based on a sequence of self-consistent calculations the free energy is estimated to lie in the range $E_f = 6 \times 10^{31} - 8 \times 10^{31}$ erg [28]. This example shows that reliable estimation of the magnetic free energy is difficult. Other recent detailed modeling studies confirm this [30].

A self-consistent solution is obtained for active region 11029, but it is unclear whether the model solution provides a close match to the actual coronal magnetic field at the time of the observations. The criteria for validating the modeling in this study are consistency and robustness of the solutions. However, comparison with independent data is needed to confirm the accuracy of the modeling. It is difficult to validate magnetic field extrapolation. Extrapolated field lines may be compared, in projection, with coronal loops in short wavelength images of the corona, but this test is qualitative. In some cases, the three-dimensional trajectories of model field lines have been compared with stereoscopically determined coronal magnetic loop trajectories [19]. It is fair to say that the validation of magnetic field extrapolation remains an outstanding problem.

4. Recent developments

4.1. Extrapolation in spherical geometry

The Helioseismic and Magnetic Imager on the Solar Dynamics Observatory is providing full disk vector magnetogram data [8]. In principle this allows extrapolation of coronal magnetic fields from full disk data, for a given magnetic field model in spherical geometry.

Spherical force-free codes have been developed, for example a generalisation of the optimization method [31, 32], and a new force-free electrodynamics method [33].

Here we provide a brief report on a new implementation of a Grad-Rubin solution to the force-free equations in spherical geometry, which is similar in approach to the existing cartesian code [11]. The method achieves an iterative solution of Equations (3) in spherical polar coordinates, representing the field using a vector potential to ensure $\nabla \cdot \mathbf{B} = 0$. The current updating step [the first of the first of Equations (3)] is achieved via field line tracing, and the field updating step [the second of Equations (3)] uses a Fourier solution of the vector Poisson equation for the vector potential. The boundary conditions are the specification of B_r at the solar surface $r = R_\odot$, the specification of α over one polarity of the radial field at $r = R_\odot$, and the requirement that $\mathbf{B} \to 0$ as $r \to \infty$.

Journal of Physics: Conference Series **440** (2013) 012037 doi:10.1088/1742-6596/440/1/012037

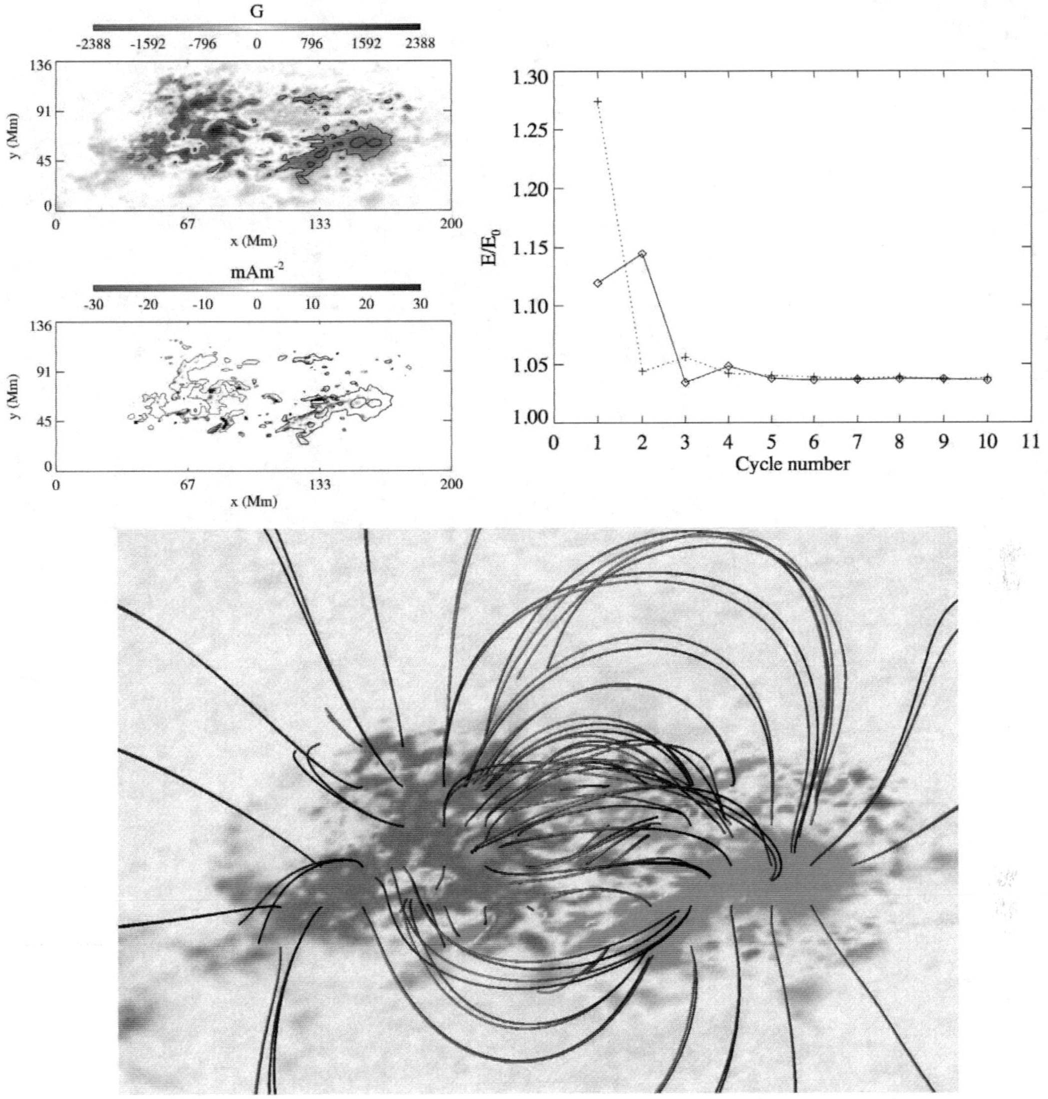

Figure 3. Self-consistent nonlinear force-free modeling of active region 11029 [28]. The upper left panel (with two sub-panels) shows the boundary data, the upper right panel illustrates the achievement of a self-consistent solution, and the lower panel shows the final magnetic field configuration. (See Section 3 for details.)

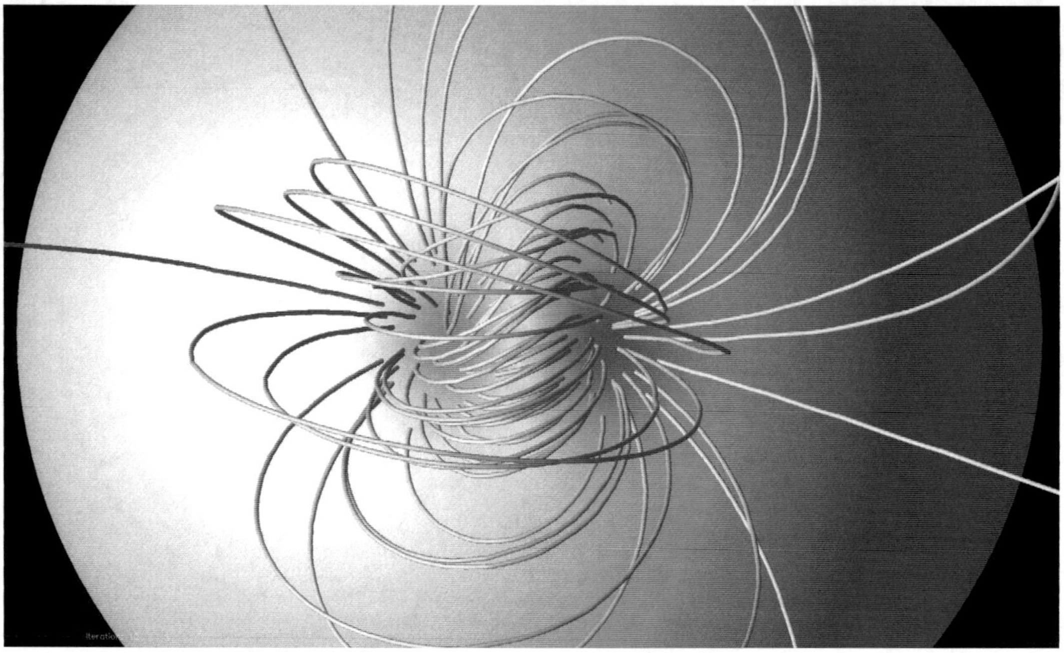

Figure 4. A test of a new full-Sun nonlinear force-free magnetic field calculation code, for a localised current-carrying bipole field configuration (see text).

Figure 4 shows a test calculation using the new code, for a localised current-carrying bipole field configuration. The radial field at the Sun's surface is shown by the color on the sphere (blue for a positive polarity radial field, and red for negative polarity). The boundary field consists of two Gaussian spots at the points $(R_\odot, \phi_1, \theta_1)$ and $(R_\odot, \phi_2, \theta_2)$, where ϕ and θ denote the usual azimuthal and polar angles in spherical polar coordinates. More precisely the radial magnetic field over the model photosphere is

$$B_r(\theta, \phi) = B_0 \left(e^{-s_1^2/\sigma^2} - e^{-s_2^2/\sigma^2} \right), \tag{7}$$

where B_0 is a normalisation constant, and where σ sets the widths of the spots. The variables $s_1 = s_1(\theta, \phi)$ and $s_2 = s_2(\theta, \phi)$ are arc lengths along great circles, given by

$$s_1 = R_\odot \cos^{-1}[\sin\phi_1 \sin\phi + \cos\phi \cos\phi_1 \cos(\theta - \theta_1)], \tag{8}$$

and

$$s_2 = R_\odot \cos^{-1}[\sin\phi_2 \sin\phi + \cos\phi \cos\phi_2 \cos(\theta - \theta_2)]. \tag{9}$$

The boundary condition on electric current is $\alpha = \alpha_0$ where $B_r > 0.9B_0$, and $\alpha = 0$ otherwise, which introduces current at the centre of the positive polarity bipolar spot. (The P solution is calculated in this test case.) The white curves in Figure 4 show the field lines of the field obtained after convergence of a sequence of Grad-Rubin iterations. A force-free twisted bipolar structure is obtained.

4.2. New nonlinear force-free modeling workshops

As mentioned in Section 2, a series of nonlinear force-free modeling workshops was held 2005-2009, organised by Lockheed-Martin personnel, to test and compare extrapolation methods. The workshops have been relatively influential: the four publications from the workshops [16, 18, 17, 19] have more than 400 citations, and techniques from the papers, e.g. the use of metrics to compare vector field calculations, have become standard.

A new series of workshops will start in 2013, funded by an International Space Sciences Institute (ISSI) International Teams in Space and Earth Sciences grant awarded in 2012 to Marc De Rosa and Mike Wheatland ("Nonlinear Force-Free Modeling of the Solar Corona: Towards a New Generation of Methods"). The workshops will permit testing of recent developments in extrapolation, using Hinode and SDO/HMI data, and the results will be published in Space Science Reviews. The first meeting is scheduled for late 2013.

5. Summary and conclusions

This paper presents an overview of the state of coronal magnetic field extrapolation. The modeling is motivated by a need to understand solar activity, and the potential for improved space weather prediction.

The data used are from vector magnetograms, which are determinations of all three components of the photospheric magnetic field over areas on the Sun including active regions. The model described here is the nonlinear force-free model.

The nonlinear force-free model is popular, but the application of the model to the data presents a variety of challenges. In particular the solar data are inconsistent with the model, leading to unreliable results, as highlighted in a series of nonlinear force-free workshops held during 2005-2009.

Approaches to the inconsistency problem are here discussed, with a focus on the "self-consistency procedure." The method is illustrated in application to active region 11029, a small but flare-productive region observed in late 2009. The modeling of active region 11029 emphasizes the difficulties of estimating the free magnetic energy based on extrapolated solutions, and of validating the solutions.

Two recent developments are also discussed: a new spherical extrapolation method, suitable for application to full-disk vector magnetic field data, and a new series of International Space Science Institute (ISSI) workshops on nonlinear force-free modeling due to start in 2013.

There has been progress in recent years in coronal magnetic field extrapolation, but the state of the art remains unsatisfactory. There are problems associated with the reliability of the data, the ability of the codes to solve the model, the inconsistency of the available data with the model, and we lack ways to confirm results. At this stage extrapolation techniques are not standard packages ready to be included in data processing pipelines: caveat emptor! It is hoped that additional work and new developments will produce reliable methods of extrapolation, for general application.

Acknowledgments

Mike Wheatland thanks the organisers of the meeting. S.A. Gilchrist acknowledges the support of an Australian Postgraduate Research Award. We thank an anonymous referee for a careful reading of the paper.

References

[1] Committee On The Societal and Economic Impacts Of Severe Space Weather Events 2008 *Severe Space Weather Events: Understanding Societal and Economic Impacts Workshop Report, Space Studies Board ad hoc Committee: A Workshop* (Washington: The National Academies Press) DC ISBN: 978-0-309-12769-1
[2] Del Toro Iniesta J C 2003, *Introduction to Spectropolarimetry* (Cambridge: Cambridge University Press) UK ISBN: 0521818273
[3] Metcalf T R 1994 *Sol. Phys.* 155 235
[4] Metcalf T R et al. 2006 *Sol. Phys.* **237** 267
[5] Leka K D Barnes G Crouch A D Metcalf T R Gary G A Jing J Liu Y 2009 *Sol. Phys.* **260** 83

[6] Sakurai T 1979, *Proc. Astronom. Soc. Jap.* **31** 209
[7] Tsuneta S et al. 2008 *Sol. Phys.* **249** 167
[8] Schou J et al. 2012 *Sol. Phys.* **275** 229
[9] Wiegelmann T and Sakurai, T 2012 *Living Rev. Sol. Phys.* 9 5
[10] Grad H and Rubin H 1958 *Proc. 2nd Int. Conf. on Peaceful Uses of Atomic Energy* vol 31 (Geneva: UN) 190
[11] Wheatland M S 2007 *Sol. Phys.* **245** 251
[12] Wheatland M S Sturrock P A Roumeliotis G 2000 *Astrophys. J.* **540** 1150
[13] Wiegelmann T 2008 *J. Geophys. Res. (Space Phys.)* **113** 3
[14] Chodura R and Schlueter A 1981 *J. Computational Phys.* **41** 68
[15] Valori G Kliem B Keppens R 2005 *Astron. Astrophys.* **433** 335
[16] Schrijver C J De Rosa M L Metcalf T R Liu Y McTiernan J Régnier S Wheatland M S Wiegelmann T *Sol. Phys.* **235** 161
[17] Schrijver C J et al. 2008 *Astrophys. J.* **675** 1637
[18] Metcalf T R et al. 2008 *Sol. Phys.* **247** 269
[19] De Rosa M L et al. 2009 *Astrophys. J.* **696** 1780
[20] Molodenskii M M 1969 *Sov. Astron.* **12** 585
[21] Metcalf T R Jiao L McClymont A N Canfield R C Uitenbroek H 1995 *Astrophys. J.* **439** 474
[22] Wiegelmann T Inhester B Sakurai T 2006 *Sol. Phys.* **233** 215
[23] Wheatland M S Régnier S 2009 *Astrophys. J.* **700** L88
[24] Wheatland M S Leka K D 2011 *Astrophys. J.* **728** 112
[25] Amari T Aly J J 2010, *Astron. Astrophys.* **522** A52
[26] Wiegelmann T Inhester B 2010 *Astron. Astrophys.* **516** A107
[27] Wheatland M S 2010 *Astrophys. J.* **710** 1324
[28] Gilchrist S A Wheatland M S Leka, K D 2012 *Sol. Phys.* **276** 133
[29] Scherrer P H et al. 1995 *Sol. Phys.* **162** 129
[30] Wiegelmann T Thalmann J K Inhester B et al. 2012 *Sol. Phys.* **281** 37
[31] Wiegelmann T Neukirch T Ruan P Inhester B 2007 *Astron. Astrophys.* **475** 701
[32] Tadesse T Wiegelmann T Inhester B 2009 *Astron. Astrophys.* **508** 421
[33] Contopoulos I 2012 *Sol. Phys.* **250** DOI 10.1007/s11207-012-0154-y

Eclipse on the Coral Sea: Cycle 24 Ascending (GONG 2012, LWS/SDO-5, and SOHO 27) IOP Publishing
Journal of Physics: Conference Series **440** (2013) 012038 doi:10.1088/1742-6596/440/1/012038

How do the magnetic field strengths and intensities of sunspots vary over the solar cycle?

A.A. Norton[1], E.H. Jones[2] and Y. Liu[1]

[1] HEPL Solar Physics, Stanford University, CA 94305-4085 USA
[2] University of Southern Queensland, Toowoomba, QLD, 4350 Australia

E-mail: aanorton@stanford.edu

Abstract. Many efforts have been made to determine if sunspot umbrae continuum intensities and magnetic field strengths are different at sunspot maximum than at sunspot minimum. The results are inconsistent, probably due to differences in sample size and analysis methodology. However, five out of six studies reviewed in this paper agree that sunspots are darker and stronger at sunspot maximum than later in the same cycle, i.e. sunspots brighten during the declining phase of the sunspot cycle. The trend during the rising phase is not agreed upon. Better statistics during the rising phase is crucial to determine if umbrae exhibit a cyclical or linear brightness trend over the cycle. We further this work by analyzing the intensities of 179 sunspots observed with the Helioseismic Magnetic Imager (HMI) for the rising phase of Sunspot Cycle 24. We find no significant trend in the brightness of sunspot umbrae in HMI data during Carrington Rotations 2097-2129 in either hemisphere. Future studies should place limits on sunspots included in the data sample, i.e. use only the leading sunspot in a bipolar active region after most of the flux has emerged but prior to sunspot decay, hopefully separating the effects of surface conditions from those of the interior where the magnetic flux is generated.

1. Introduction

The strength of a sunspot cycle can be defined by the number and area of sunspots generated during the cycle. Although the number of sunspots differ greatly from one cycle to the next, indicating an inherent variability in the process that generates sunspots, it has been difficult to identify trends in the average properties of individual sunspots during a sunspot cycle or between successive sunspot cycles. For example, a strong cycle may produce sunspots that are on average stronger, darker and larger than those from a weaker cycle. Within a cycle, we might also expect stronger, larger sunspots to appear more frequently at maximum than during the rising or declining phase of the cycle.

Sunspot umbrae, in particular the dark nuclei that evolve the least during sunspot formation and dissipation [1] may prove to be a diagnostic of the dynamo mechanism. This should be true if the flux formation and eruption process determines the peak magnetic field strength of sunspots. One of the reasons the intensity of sunspots is interesting is that the brightness ratios (BR), the ratio of the continuum intensity in the darkest part of umbra to the average surrounding continuum intensity of the photosphere, is a very good proxy for magnetic field strength.

$$BR = \frac{minimum(I_c\ umbra)}{average(I_c\ photosphere)} \quad (1)$$

Journal of Physics: Conference Series **440** (2013) 012038 doi:10.1088/1742-6596/440/1/012038

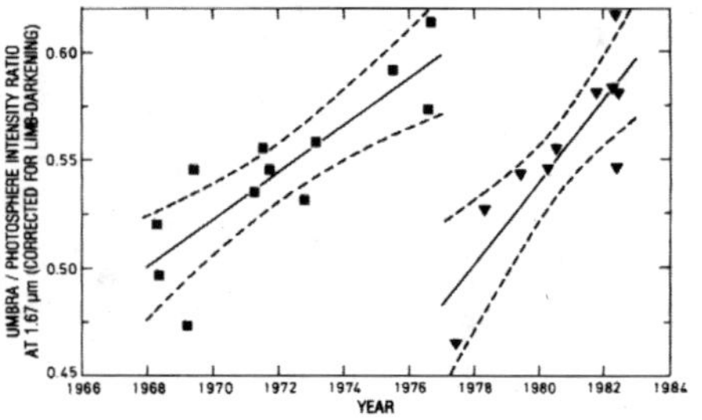

Figure 1. LINEAR FIT: BR of 22 sunspots observed in Cycle 20 & 21 by Maltby *et al* [6].

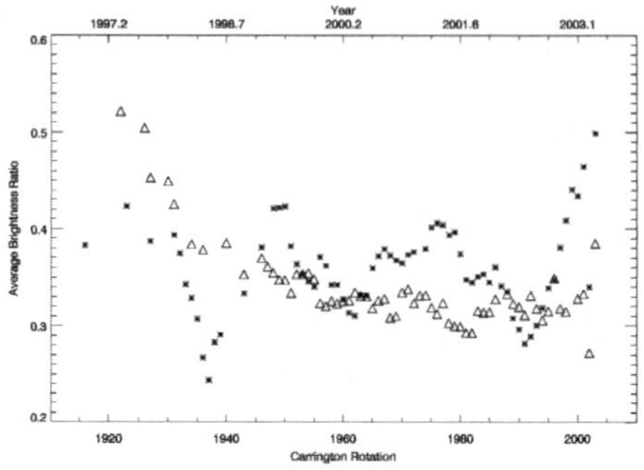

Figure 2. CYCLICAL: Norton & Gilman plot binned BRs of 650 sunspots observed with MDI separated by hemisphere for Cycle 23 [7].

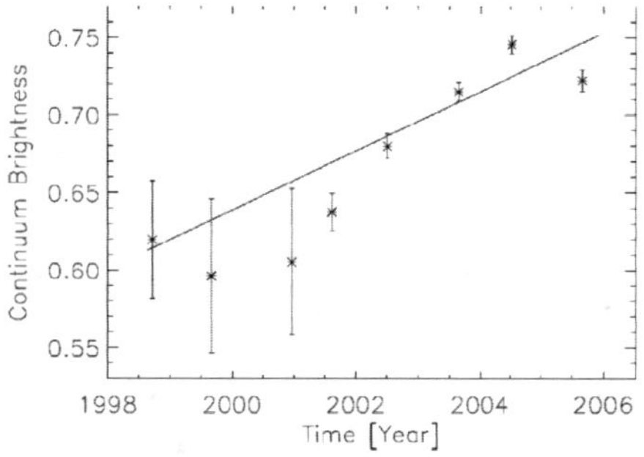

Figure 3. LINEAR FIT: Average annual BRs of over 900 umbrae observed by Penn & Livingston [11] for Cycle 23 with the Fe I 1564.8 nm line at NSO McMath Pierce.

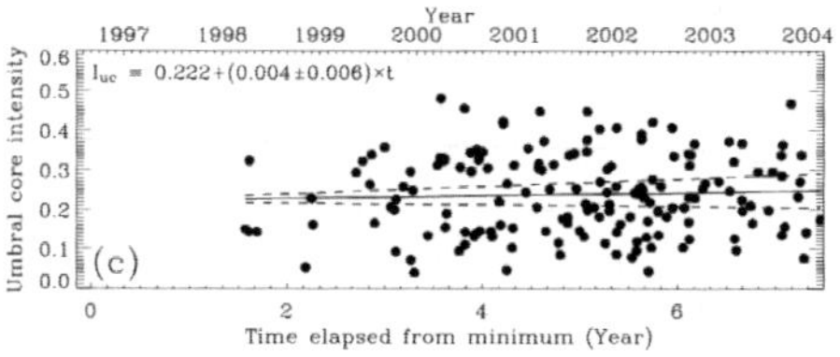

Figure 4. NO TREND: Mathew *et al* [13] studied 160 sunspots from 1998−2004 in Cycle 23 using MDI data.

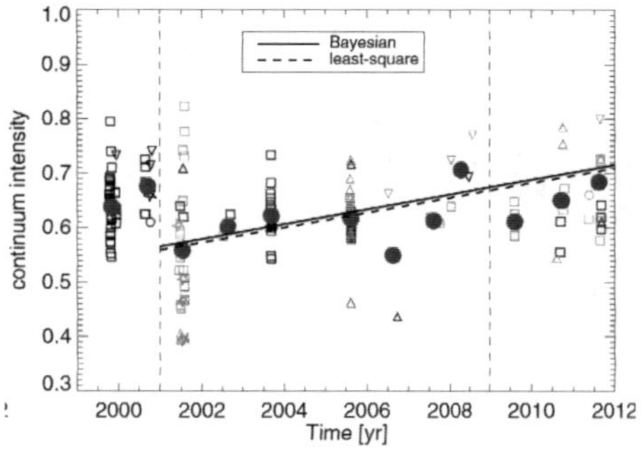

Figure 5. LINEAR FIT: Rezaei *et al*[14] plot the binned BRs from 183 sunspots observed during 1999−2011 (Cycle 23−24) with TIP.

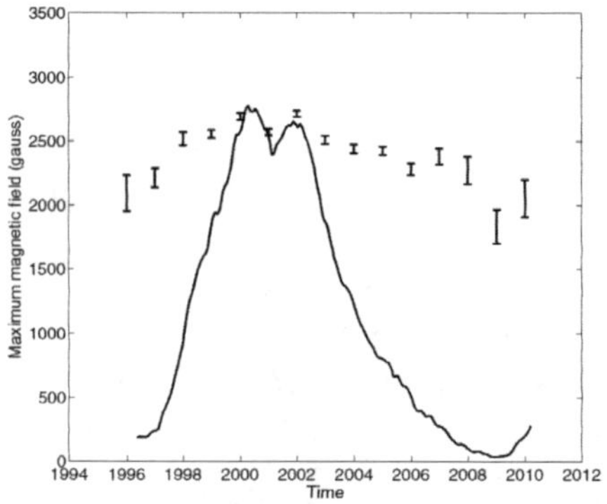

Figure 6. CYCLICAL: Maximum magnetic fields in Cycle 23 from over 30000 umbrae in MDI data as studied by Watson *et al* [17].

The BR is free from many problems experienced in the inversion of Stokes profiles to determine field strength and therefore is useful in inferring field strengths over several solar cycles using

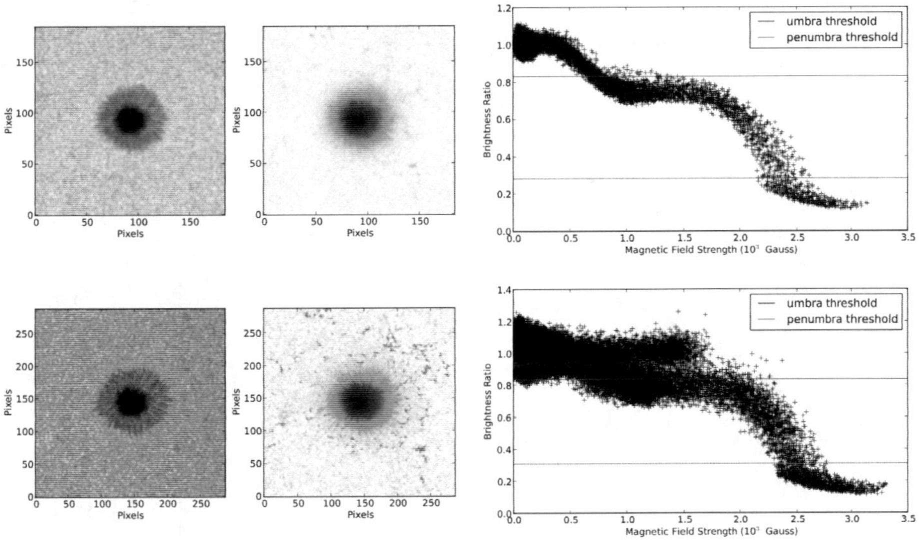

Figure 7. Continuum image (left), Magnetogram (center), and Brightness Ratio vs. Magnetic Field Strength (right) for Active Region 11084, as seen by SDO HMI (top) and Hinode SOT-SP (bottom).

different telescopes with varied spectral lines. The BR does not experience saturation effects due to large wavelength shift from Zeeman splitting or spacecraft velocity [2]. Martínez Pillet & Vázquez [3] studied the distribution of continuum intensity and magnetic field in eight umbrae and demonstrated that the smaller continuum intensities corresponded to larger field strengths. This relationship can be understood in terms of the sunspot response to a horizontal force balance. Stronger umbral magnetic fields (higher magnetic pressure) need less plasma (lower gas pressure) to balance the external forces. The local relationship between temperature and magnetic field in sunspot umbrae dictates that the darker, cooler umbrae contain stronger magnetic fields.

Sunspot cycles are not symmetric in time [4-5]; the cycle rises quickly from sunspot minimum to maximum then declines slowly from maximum to minimum. The Solar International Data Centre (SIDC) monthly smoothed sunpot numbers show Cycle 23 spent 47 months rising (May 1996 - Apr 2000) and 97 months declining (Apr 2000 - Dec 2008). Cycle 22 spent 34 months rising (Sep 1986 - July 1989) and 82 months declining (July 1989 - May 1996). The declining phase is approximately twice as long as the rising phase. Consequently, any trend in sunspot BR field strength is poorly sampled in the rising phase as compared to the declining phase. Perhaps this fact contributed to the difficulty in determining trends in the rising phase of the cycle. A brief literature review of intensity trends observed in sunspots follows.

2. Survey of Six Previous Studies on Umbral Brightness Trends

Maltby *et al.* [6] observed twenty-two sunspots from 1968−1983 including data from sunspot cycles 20−21, see Figure 1. They limited the data sample to large sunspots with the criteria that umbrae radii be $>5''$, and concluded that sunspots become linearly brighter and hotter, corresponding to weaker magnetic fields, throughout an individual sunspot cycle. Mid to late cycle sunspots were modeled as 190 K and 380 Kelvin hotter than early cycle sunspots, respectively. (Early, middle and late sunspots were defined as 0.1, 0.5 and 0.9 percentage of time into the cycle.) A strength of this analysis was that multiple wavelengths were used

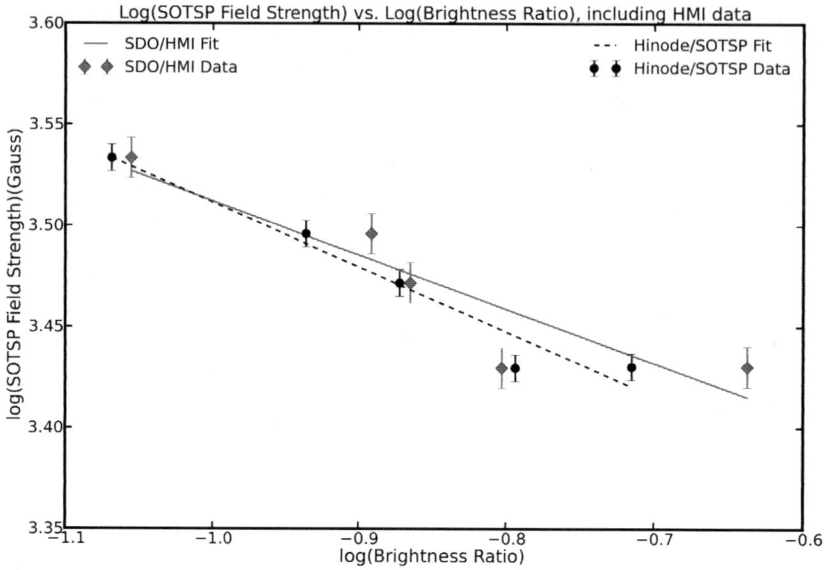

Figure 8. A log plot of SOT-SP field strength versus SOT-SP and HMI brightness ratio, with best-fit lines for each instrument showing extremely good correlation between the two. Error bars represent the standard deviation between the observed magnetic field strength and the value expected from the best-fit of the data. The standard deviation values are 62 and 41 Gauss for HMI and SOT, respectively.

and that the model umbral atmospheres developed from this work are the standard Maltby-E, Maltby-M and Maltby-L models. A weakness is the very small sample size: twenty-two sunspots over sixteen years.

Norton & Gilman [7] studied the sunspot cycle trends from 1998−2003 (Cycle 23) using over 650 sunspots observed with Michelson Doppler Imager (MDI) [8], see Figure 2. MDI BRs were compared to the BRs and field strengths obtained with the Advanced Stokes Polarimeter [9,10] for the same sunspots. Data were separated into N and S hemispheres. A decrease (not linear) in continuum intensity brightness was seen until sunspot maximum in the North, then sunspots began to brighten again. The late declining phase of Cycle 23 was not sampled as their data ended in 2003. A maximum difference of 650 K was found in the N hemisphere. The S hemisphere was disorganized. A weakness of this study was that no scattered light correction was applied.

Penn and Livingston [11] observed over 900 sunspot umbrae from 1998−2005 (Cycle 23) using the Fe I 1564.8 nm line at NSO McMath Pierce Telescope. They find that the maximum sunspot magnetic fields decreased at about 52 Gauss yr^{-1} and that the umbrae became linearly brighter with the normalized umbral intensity changing from 0.60 to 0.75 (corresponding to a rise of 550 Kelvin), see Figure 3. It could be argued that the trend is not linear in as much as it is cyclical with the brightness of spots decreasing until sunspot maximum and then increasing until the minimum that follows. As the rising phase of the sunspot cycle is shorter than the declining phase, the rising phase is not as well sampled. It is the trend of sunspots in the rising phase of the cycle which is controversial. A later study [12] includes data from Cycle 24 and reports the increasing trend in umbrae brightness continuing into Cycle 24 with sunspots in general being brighter and weaker in Cycle 24 as compared to sunspots early in Cycle 23. This results caused

Table 1. Active regions with associated observation dates, HMI and SOT-SP brightness ratios, magnetic field strengths, and μ values.

AR No.	Date	BR_{HMI}	BR_{SOT}	$B_{HMI}(Gauss)$	$B_{SOT}(Gauss)$	μ
11084	20100702	0.13	0.12	2910	3130	0.93
11064	20110301	0.16	0.16	2640	2690	0.71
11066	20110310	0.14	0.13	2680	2960	0.95
11076	20110326	0.23	0.19	2440	2690	0.78
11339	20111106	0.09	0.09	3440	3420	0.82

speculation that sunspots may disappear soon if this trend continues and the Sun enters another grand minimum.

Mathew *et al.* [13] studied 160 sunspots from 1998−2004 (Cycle 23) using MDI data. They reported no trend in umbral brightness, see Figure 4. They limited their data to sunspots with radii between 5−15″ and they corrected for stray light. They found a significant decrease in core brightness with increasing umbral radii. The slight decrease in umbral core intensity in the N hemisphere was attributed to a trend in sunspot radius in time during the cycle.

Rezaei, Beck and Schmidt [14] studied 183 sunspots from 1999−2011 (Cycle 23−24) with Tenerife Infrared Polarimeter [15,16]. They found a trend that sunspots in the late stage of cycle 23 are weaker than those at the start of the cycle, see Figure 5. The decrease in the field strength with time is about 94 Gauss yr^{-1}. In the same time interval, the continuum intensity of the umbra increases linearly with a rate of 1.3% of I_c yr^{-1} from maximum to the following minimum, while the umbral area does not show any trend. Sunspots in the new Cycle 24 show higher field strengths and lower continuum intensities than those at the end of cycle 23, interrupting the trend, and conflicting with [12].

Watson, Fletcher & Marshall [17] analyze over 30000 sunspot observations using MDI and find a cyclical trend that follows sunspot number (not linear) with sunspots having stronger maximum magnetic fields at sunspot maximum and weaker fields at sunspot minimum, see Figure 6. They focus on distribution of the sunspots, area and yearly average of maximum umbral strength. Unfortunately, they do not report on intensity or BR changes.

Why may there be a trend in umbral brightness during a sunspot cycle? Yoshimura [18] postulated that “the depths of the roots of the flux ropes decreases in time.” Flux ropes rising from greater depths early in the cycle may be darker than those rising from shallower depths later. Alternatively, the surface or near-surface processes in which emerging flux coalesces into sunspot umbrae could vary with the solar cycle and emergent latitudes. For example, rates of mass removal from entrained U-loops may vary [19]. There could be a stronger convergence flow at depths below sunspots in the mid-latitudes of the sunspot belts resulting in the coalescence of more magnetic flux in the middle of the cycle. Another possibility is that there could simply be more flux or stronger flux emerging at sunspot maximum.

3. Sunspot Brightness of Cycle 24 Rising Phase Observed by HMI

We wish to contribute to the discussion of umbral brightness trend by studying the sunspots of Cycle 24 during the rising phase using HMI data. HMI is a filter-based instrument observing Stokes I, Q, U and V at six wavelength positions across the 6173 Å Fe-I line with 76 mÅ filter space sampling [20,21]. The HMI pipeline uses a Milne-Eddington inversion code named the Very Fast Inversion of the Stokes Vector (VFISV) [22,23] to detexcrmine the solar atmospheric parameters based on Stokes observations.

Journal of Physics: Conference Series **440** (2013) 012038 doi:10.1088/1742-6596/440/1/012038

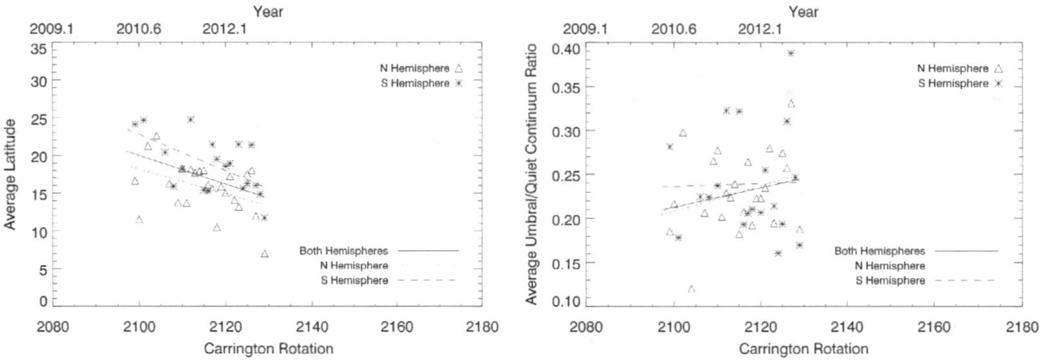

Figure 9. (Left) Cycle 24 sunspot latitude (left) and BR (right) as a function of time as observed by HMI. The data are separated into N and S hemispheric data and average of each CR are plotted. There is no smoothing. Some CRs did not have any sunspots and hence do not have a data point. The N and S hemispheric and total-Sun data are fit with a linear regression and overplotted as lines. The linear fits to the latitude data (left) have greater than a 3σ confidence level. The linear fits to the BR data (right) are not significant at even a 2σ level.

Norton et al. [7] used the Advanced Stokes Polarimeter (ASP) to verify that MDI BR data could be used to infer field strengths. In a similar manner, we use the Hinode Solar Optical Telescope Spectropolarimeter (SOT-SP), which uses a slit and a diffraction grating to sample Stokes I, Q, U and V with a 21.5 mÅ spectral resolution in the 6301.5 and 6302.5 Å Fe-I line pair, to verify that HMI BR data can be used to infer field strengths. We chose the active region NOAA 11804 to begin our analysis, as it was observed by Hinode SOT-SP and HMI on 2 Jul 2010 on 5:12:00 UT. HMI continuum and magnetic field strength products of VSIFV were obtained from Joint Science Operations Center (JSOC) at Stanford University for this date and time, while Hinode SOT-SP continuum and level2 MERLIN [24,25] magnetic field strength products were obtained from the Community Spectro-polarimetric Analysis Center (CSAC) hosted by HAO NCAR.

In Figure 7, the Active Region 11084 continuum intensity and field strength are shown as context plots as well as the scatter plot of intensity versus magnetic field strength as observed by Hinode SOT-SP and HMI respectively. Due to the difference in resolution between SOT-SP and HMI, data from SOT-SP was binned 2×2 and then interpolated to have the same resolution as HMI of $0.5''$. Much of the difference between Hinode SOT-SP results and HMI results are due to the fact that the HMI VFISV inversion code assumes the filling fraction of the pixel is one, meaning that all the plasma sampled within the pixel is magnetized and contributing to the Stokes polarization. No scattered light corrections are applied at this time since the appropriate point-spread function is still being developed.

To determine how well Hinode SOT-SP and HMI correlate, we calculate the BRs and peak magnetic field strengths of several simultaneously observed active regions. Thresholding in continuum intensity was used to determine the umbra in each region. Then the darkest pixel in the umbral continuum intensity was selected and the associated field strength recorded. Table 1 lists each active region, observation date, BR, peak magnetic field strength, and μ for SOT-SP and HMI. The uncertainties of the BR values are small. For example, active region 11804 was observed over a seven hour period with BRs determined at five equal time intervals, providing a mean BR of 0.128 ± 0.004, where the uncertainty is the standard deviation of the five measurements.

Figure 8 is a log plot of the umbral peak magnetic field strengths versus BRs for 5 sunspots simultaneously observed in 2010-2012, with linear fits to HMI and SOT-SP BR data. In this plot, all field strengths are determined by Hinode SOT. We do not use HMI magnetic field data because the HMI algorithm can become saturated in the strongest part of the umbrae depending upon the spacecraft velocity [2]. Error bars are determined using the standard deviation between the observed magnetic field strength and the value expected from the best linear fit to the data. The standard deviation values are 62 and 41 Gauss for HMI and SOT, respectively. This is higher than the error determined for the BR of 11804 observed over the seven hour time period as reported above; the BR of 0.128 ± 0.004 is equivalent to a field strength of 3045 ± 25 Gauss when using the fit to the HMI BR to determine field strength. The large range of μ values for data shown in Figure 8, $0.71 - 0.95$, is not ideal, but was necessary to compare simultaneous data from SOT-SP and HMI.

For our analysis of BR trends observed with HMI, we use a continuum intensity Carrington rotation map from HMI that samples sunspots as they cross the central meridian. Thirty-three Carrington rotation maps are sampled, from CR 2097 - 2129. As most sunspots appear within 30° of the equator, BR data are in general determined from $cos(\theta) = 0.87$ or greater. Specifically, 90% of the sunspots have $\mu \geq 0.90$ with only 11 sunspots having $\mu < 0.90$. The HMI synoptic continuum intensity data is determined from observations taken every 720 seconds and a fifth order polynomial in $log(\mu)$ is used to remove limb-darkening. These maps are normalized so that faculae have values above 1.0 and the entire range of the data is from $0.3 - 1.2$.

For Figure 9, Carrington rotation synoptic intensity maps from HMI are used, corrected for limb darkening and flat fielded, but stray light has not been removed yet. Sunspots are hand-selected and include 179 sunspots in total with 101 (78) in the N (S) hemisphere, respectively. Their latitude, BR and peak field strength is recorded. The average latitudes and BRs of umbrae are plotted as a function of time for each hemisphere. We assume that Cycle 24 began in 2009 in CR 2083 and have adjusted the origin of the plot accordingly. We calculate a mean for all data in each CR. The left panel of Figure 9 shows the well-documented migration of the sunspot latitudes toward the equator. The right panel of Figure 9 plots the average BR for each CR that sunspots were present in the separate hemispheres. The linear fit to the BR data was not signifcant at the 2σ level for either the N and S hemisphere or the total-Sun average. We argue that there is always a need to separate data into N and S hemispheres in order to not blur mechanics that are operating slightly out-of-phase with each other.

4. Conclusions

A consensus is forming that sunspots are darker and stronger at sunspot maximum than at sunspot minimum. The observed intensity difference could represent a change of 300-600 Kelvin in average umbral temperatures. There are conflicting results, though, for the overall trend during the cycle since the trend during the rising phase is not agreed upon. The underlying question is whether there is

- a linear increase in brightness from the start of the cycle until the end,
- a cyclical change where brightness decreases in the rising phase until maximum, then increases again until the following minimum, or
- no significant change whatsoever.

The rising phase of the sunspot cycle is not as well-sampled as the declining phase; it is shorter and produces less sunspots. If the brightness of umbrae decrease during the rising phase, then the trend will be cyclical, as reported by [7, 17] and seen in Figures 2 and 6. Alternatively, if the umbrae in the rising phase increase in brightness on average, then the results of [6, 11, 14] will be confirmed as seen in Figures 1, 3 and 5. Another option is that after careful correction of stray light, no trend may be present as reported by [13] and seen in Figure 4. Understanding the behavior of umbral brightness of field strengths during the rising phase of the sunspot cycle is key.

We find no significant intensity trend in either hemisphere in the HMI BR data for the rising phase of Cycle 24, as can be seen in the right panel of Figure 9, thereby confirming the results of [13]. If a linear fit is forced to the data, the total Sun and N hemisphere show an increasing brightness trend but the p-value for the linear regression is 0.19 and 0.20 respectively, much higher than the p=0.05 expected for a 3σ detection. Therefore, we must conclude that no trend is detected for BRs in HMI data. We need, however, to carefully apply a stray light correction to HMI data and re-evaluate the umbral brightness trends.

We stress that it is very important to separate the hemispheric data since the sunspot cycles may be out of phase and have slightly different amplitudes in the two hemispheres. Future studies should place limits on sunspots included in the data sample, i.e. separate the leading and following sunspot populations in bipolar active regions after most of the flux has emerged but prior to sunspot decay, hopefully separating the effects of surface conditions from those of the interior where the magnetic flux is generated. The leading and following sunspot asymmetry may be worth studying as a function of solar cycle to ascertain changes in the flux emergence process. Future studies could also determine if similar umbral trends are observed in BR, mean umbral strength, or parameters fit to the umbral distribution of strengths. The relationship between umbral radius (size) and BR is controversial. We argue that the mechanism which determines the maximum field strength of an umbra may be the same mechanism that determines the umbral size. Therefore, even if a clear relationship between radius and maximum umbral field strength were established, the trend during a sunspot cycle would still be interesting.

Comparing Cycle 23 to Cycle 24, we find the peak umbral field strength based on BRs was 2660 (MDI) [7] and 2690 Gauss (HMI) at a time of forty CRs (3 years) into each cycle. The average umbral latitude at a time of 40 CR into the cycle is 22° and 17° for Cycle 23 and 24 respectively. We do not observe a decrease in the peak field strengths of sunspots between Cycle 23 and 24.

References

[1] Socas-Navarro, H, 2003 *ASP Conf. Ser.* **286** 267
[2] Liu, Y, Norton, A and Scherrer P 2007 *Solar Phys.* **241** 185
[3] Martínez Pillet, V and Vázquez, M 1993 *Astron. & Astroph.* **270** 494
[4] Waldmeier, M 1935 *Astron. Mitt. der Eidgenössischen Sternwarte Zürich* **14** 105
[5] Hathaway, D, Wilson, R, Reichmann, E 1994 *Solar Phys.* **151** 177
[6] Maltby, P, Avrett, H, Carlsson, M, Kjeldseth-Moe, O, Kurucz, R L and Loeser, R 1986 *ApJ* **306** 284
[7] Norton A A and Gilman P A 2004 *ApJ* **603** 348
[8] Scherrer P H *et al* 1995 *Solar Phys.* **162** 129
[9] Elmore, D F *et al* 1992 *SPIE Proceedings* **1746** 22
[10] Lites, B W, Elmore, D F, Seagraves, P and Skumanich, A 1993 *ApJ* **418** 928
[11] Penn, M J and Livingston, W 2006 *ApJ* **649** L45
[12] Livingston, W, Penn, M J, and Svalgaard, L 2012 *ApJL* **757** L8
[13] Mathew, S K, Martinez Pillet, V, Solanki, S K and Krivova, N A, 2007 *Astron. & Astroph.* **465** 291
[14] Rezaei, R, Beck, C and Schmidt, W 2012 *Astron. & Astroph.* **541** A60
[15] Martínez Pillet, V, Collados, M, Sánchez Almeida, J, *et al* 1999 *ASP Conf. Ser.* **183** 264
[16] Collados, M, Lagg, A, Díaz García, J J, *et al* 2007 *ASP Conf Ser.* **368** 611
[17] Watson, F T, Fletcher, L and Marshall, S 2011 *Astron. & Astroph.* **533** A14
[18] Yoshimura H 1983 *Solar Phys.* **87** 251
[19] Cheung, M C M, Rempel, M, Title, A M, and Schüssler, M 2010 *ApJ* **720** 233
[20] Scherrer, P H, *et al* 2012 *Solar Phys.* **275** 207
[21] Norton, A A, Pietarila Graham, J.D., Ulrich, *et al* 2006 *Solar Phys.* **239** 69
[22] Borrero, J M, Tomczyk, S, Kubo, M., *et al* 2011 *Solar Phys.* **273** 267
[23] Borrero, J M, Tomczyk, S, Norton, A A, *et al* 2007 *Solar Phys.* **240** 177
[24] Lites, B W, *et al* 2007 *ASP Conf Ser.* **369** 55
[25] Ichimoto, K, *et al* 2008 *Solar Phys.* **249** 233

Journal of Physics: Conference Series **440** (2013) 012039 doi:10.1088/1742-6596/440/1/012039

The Next Generation of Canadian Solar Flux Monitoring

Kenneth F Tapping and Donald C Morton

[1]National Research Council, P. O. Box 248, Penticton, BC, V2A 6J9, Canada

[1]National Research Council, 5071 West Saanich Dr., Victoria, BC, V9E 2E7, Canada

Email: ken.tapping@nrc-cnrc.gc.ca, don.morton@nrc-cnrc.gc.ca

Abstract. The 10.7 cm solar radio flux (F10.7), provided by the National Research Council of Canada since 1947, is widely used as an index of solar activity and as a proxy for other solar quantities that are harder to measure. Over recent years needs have arisen that are difficult to meet with solar flux measurements at a single wavelength. F10.7 comprises contributions from multiple emission mechanisms. To separate these, multi-wavelength measurements are needed. A new instrument is under construction that will measure fluxes precisely in six bands at 2.8, 3.6, 6.0, 10.7, 18 and 21 cm.

1. History and Present Status of the Radio Flux Monitoring Programme

At the end of World War II, engineers in various countries, including Australia and Canada, used ex-military radar equipment to make radio telescopes. In Canada the first radio astronomical experiments were carried out by a team headed by Arthur Covington, of the National Research Council of Canada. The instrument was derived from a radar system operating at 10.7 cm wavelength [1] with a "dish" antenna 1.2 m (4 ft) in diameter and was located in the Ottawa area (longitude W75°). Its sensitivity was poor so the only radio source that could be detected and usefully measured was the Sun.

However, carefully calibrated measurements made over many days showed the solar radio emission to vary from day to day, reflecting changes in activity inferred by other means such as sunspot number counts. This led to the simultaneous discovery by Covington [2] (at 10.7 cm wavelength) and Pawsey and Yabsley [3] (at 21 cm), in Australia, of the *Slowly-Varying* or *S-Component* of solar radio emission. In 1947 Covington, using a new instrument with a 1.8 m (6-ft) dish, began a programme of systematic daily measurements of what became known as the 10.7 cm solar radio flux, or simply F10.7, which continues today, as plotted in Figure 1. Two noteworthy features are the weakness of the present Cycle 24 and the last minimum 4 % lower than the previous ones.

The Ottawa location suffered from increasing interference. Consequently, when the Algonquin Radio Observatory (ARO) was opened in 1962 at a very radio-quiet location in a Provincial Park (longitude W78°) some 250 km from Ottawa, Covington moved the monitor there. This site offered the opportunity for additional solar radio facilities.

While operating in the Ottawa area, experiments had been carried out with multi-antenna arrays in order to make one-dimensional scans of the solar disc [4]. A larger form of this interferometer with 32 dishes each 3 m in diameter was installed at ARO on a 215 m baseline and used to make daily east-west scans of the solar disc at 10.7 cm (2800 MHz) with a resolution a little less than 2 arcmin on the meridian. These measurements complemented Paul Wild's radioheliograph at Culgoora [5, 6], Australia, which made solar images at 43, 80, 160 and 327 MHz, showing burst sources in the solar corona.

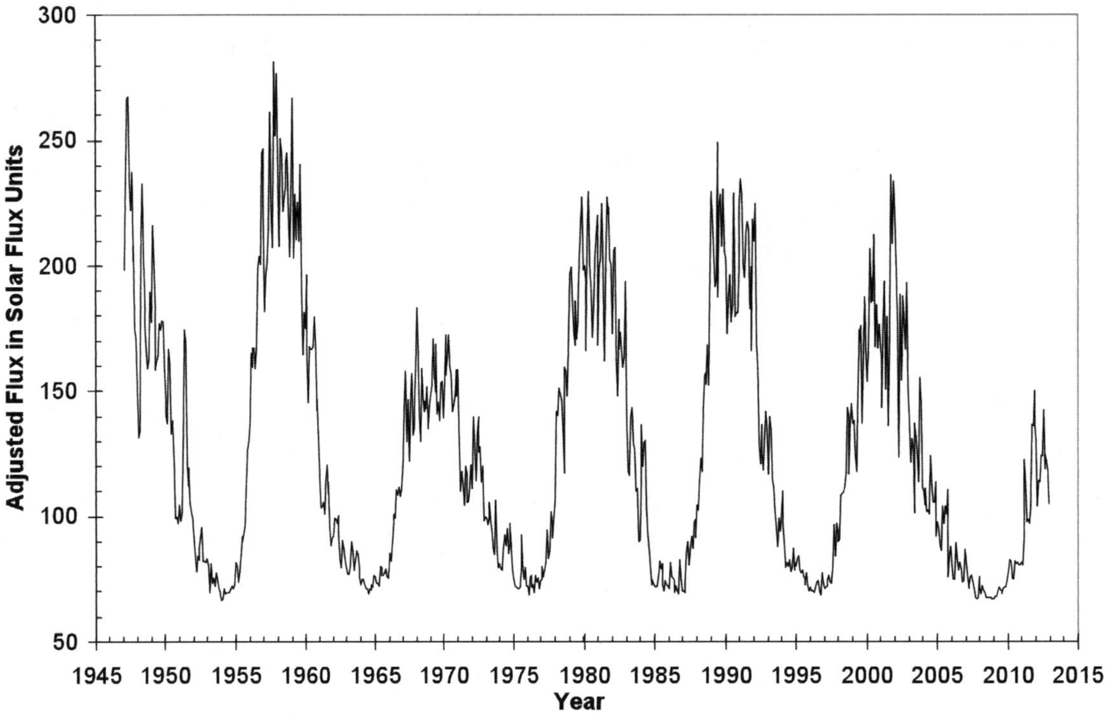

Figure 1. F10.7 monthly averages to the end of 2012 adjusted to the mean Earth-Sun distance. A solar flux unit (sfu) = 10^4 Jansky = 10^{-22} watt m^{-2} Hz^{-1}.

In the late 1980's serious budget constraints forced the National Research Council to close the Algonquin Radio Observatory. Fortunately, the other Canadian national radio facility, the Dominion Radio Astrophysical Observatory (DRAO), located near Penticton, British Columbia (W120°), could be kept open. Moreover, there was already a solar radio observatory on-site, operating an old, backup flux monitor installed to increase time coverage. So in 1990 the Solar Radio Monitoring Programme was relocated there. The old solar instrument was removed, and a new one developed and installed. The new flux monitor was operated for several months in parallel with the ARO instrument to establish continuity of calibration and to evaluate the new instrument. The ARO instrument was then shut down and moved to DRAO, where both now are operated in parallel, to increase reliability. The radio interferometer was shut down because it was too large to move or replace.

Initially, Covington calibrated the flux measurements absolutely using a black body source that could be attached to the antenna. However, later he implemented a horn calibration standard. Simple, gently tapering pyramidal horns have the highly desirable quality that their gains (effective collecting areas) can be calculated accurately from their physical dimensions. This horn calibration standard has been refurbished and installed at DRAO, providing calibration consistency over generations of flux monitors. Several horn comparison measurements are made each northern summer when the sun has its highest altitude.

The 10.7 cm solar radio flux (F10.7) correlates with other indices of solar activity, such as the subjective, but remarkably robust sunspot number index. (See for example www.ips.gov.au.Educational/2/2/5). To a large degree these correlations are due to the overriding power of the variation of magnetic activity over the solar cycle as a modulator of solar processes. Comparisons between these indices afford a subtle means to detect changes in the solar activity cycle [7].

When Arthur Covington retired in 1978, other members of his group and their successors kept the programme running. The observations are ongoing, and the F10.7 database, extending back to 1947, is a record of solar activity of length second only to the sunspot number. The F10.7 measurements are available from the Natural Resources Canada site www.spaceweather.gc.ca/datadonnee/sol flux/sx-4-eng.ph and the United States NOAA site www.swpc.noaa.gov//SolarCycle/f10.gif.

2. The Next Generation Solar Flux Monitor

The 10.7 cm solar flux is an excellent index of solar activity, but there are user needs that could be better met through an augmentation of this data service. First there are requirements for accurate flux values at other wavelengths that cannot be usefully derived from F10.7 values. Secondly, a major application of F10.7 is as a proxy for other important solar emissions, such as the extreme-ultraviolet (EUV) and ultraviolet (UV) fluxes. These proxies are used in modelling ionospheric processes, forecasting communication conditions, and estimating the temperature structure of the earth's atmosphere. As the applications become more demanding, there is a need to improve the proxies, or at least to better understand them.

However F10.7 comprises contributions from multiple emission mechanisms that relate differently to the proxied quantities. Emissions from flares are in principle identifiable, but in their absence F10.7 is dominated by two thermal contributions. These are free-free emission originating where the ambient magnetic fields are too weak for the electron gyrofrequency to exceed about a third of the observing frequency, and gyroresonance emission where the magnetic fields are stronger than that value. See Kundu [8] for a description of these processes. When present, thermal gyroresonance produces greater optical thicknesses and therefore brighter emission than thermal free-free emission. Therefore, at 10.7 cm wavelength, sunspots appear very bright, whereas at optical and shorter wavelengths they appear dark.

If the variations in EUV and UV fluxes are entirely due to changes in the brightness of active-region structures, proxies derived more completely from the free-free emission contribution should be more accurate. However, if the EUV fluxes are also modulated by the energy transfer through the convection zone, then a more complete indicator of magnetic activity such a mix of the two contributions, either the original F10.7 value or in some other proportion might be better. It would be of great interest to see how the relative contributions by these mechanisms to the S-component vary over the solar cycle.

The spectra of thermal free-free emission and gyroresonance are very different. The free-free emission shows a monotonic decrease with increasing wavelength, whereas thermal gyroresonance from solar active regions has a broad peak due to the superposition of many resonances, but drops to low values both at short wavelengths due to rapidly falling optical thickness and at long wavelengths where it is blocked by the overlying optically thick free-free emission. Tapping and DeTracey [9] showed that the spectrum of solar free-free emission is relatively insensitive to different coronal density models but sensitive to the density at the base of the corona. This suggests that a fit to identifiably free-free contributions to the S-Component spectrum would remove most of the free-free contribution, leaving that due to gyroresonance. In Figure 2, the green curve is a modelled spectrum of the quiet sun emission, assumed to be totally free-free. The black curve represents a spectrum of a more active Sun, assembled from observations. By increasing the density at the base of the corona the free-free curve can be made to fit the observed spectrum at the long and short wavelengths, where the emission may be assumed to be completely free-free (blue curve). By subtracting that spectrum from the total, we obtain the contribution due to gyroresonance (the red curve). This procedure would not be practical for individual active regions, but should work with the integrated emission from all sources present on the solar disc.

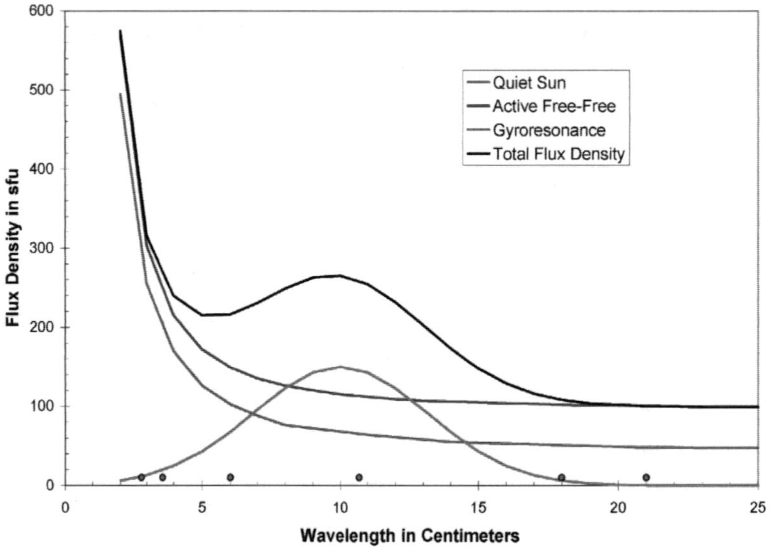

Figure 2. Schematic spectra showing total flux (black), free-free quiet sun emission (green), this emission scaled to fit the total flux at long wavelengths (blue), and the gyroresonance contribution (red).

In practice, resolving the components will require precise flux density measurements at several wavelengths made in exactly the same manner, at the same time and, as much as possible, using the same hardware. In addition to meeting user needs for flux values at additional wavelengths, this spectral decomposition was the rationale for the development of the multifrequency "Next Generation Solar Flux Monitor" (NGSFM) currently under construction at DRAO.

The specifications for the system are as follows:

1) Solar fluxes will be measured at 6 wavelengths: 21cm (1415, 20 MHz), 18 cm (1660, 10 MHz), 10.7 cm (2800, 100 MHz), 6.0 cm (4995, 50 MHz), 3.6 cm (8.3 GHz, 50 MHz), 2.8 cm (10.6 GHz, 100 MHz), where the figures in the parentheses are respectively the frequencies corresponding to the observing wavelengths and the receiver bandwidths. Once the NGSFM is in operation and its performance has been evaluated, some wavelengths might be changed and longer wavelengths added to more effectively sample the spectrum.

2) All frequencies will be measured at 1000 samples/sec to provide precise observations of solar bursts and other transient events.

3) Duplicate equipment will use orthogonal linear polarizations from the antenna feed to achieve redundancy in the observations.

4) Seven different calibration noise levels will be used on each channel in order to monitor linearity and to increase the chance of a calibration level being close to the solar noise level being measured.

5) To measure the total flux from all the emissions from the solar disc, the antenna beamwidth must be at least 3.5 degrees (for a Gaussian beamshape) over the whole operating wavelength range. To do this the feed will progressively under-illuminate the dish with decreasing observing wavelength. The antenna is a 3.7 m (12-ft) dish that was originally used for geophysical VLBI observations by Natural Resources Canada.

To better observe the spectral structure in solar bursts and other emissions, a team headed by Professor David Thompson (Queen's University, Kingston, Ontario) will provide a radio spectrometer covering the 30 - 2.5 cm wavelength range. This also could provide fill-in information that can be used with the flux measurements to construct the whole spectrum.

Each of the orthogonally polarized broadband signals from the focus will be individually amplified without any filtering or bandwidth limiting and fed to two identical receiver modules mounted on the back of the dish. In each module the signal will be split into eight identical channels. Six of them will be fed to the receiver modules, one for each observing wavelength, one of the remaining two channels will be fed to the Queen's University spectrometer; the other will be a test output.

Calibration of an under-illuminated paraboloidal antenna is particularly difficult, so this will be done using an array of horn antennas on a common mount, making absolute flux determinations at all six wavelengths. It is anticipated that the NGSFM will start operation in 2013 with calibrated flux values being available about a year later.

Ground-based solar monitoring has the advantages over monitoring from space-based platforms that they are comparatively inexpensive, very flexible and easy to keep running for a long time. However, they have the disadvantage that continuous, 24-hour coverage is impossible from any single location. Consequently we are actively collaborating with the Royal Observatory of Belgium and communicating with additional ones, with the objective of having several similar or identical instruments distributed in longitude around the world.

Acknowledgments
The development of the next generation flux monitor as well as the current monitoring programme are managed and operated by the National Research Council of Canada with the participation of Natural Resources Canada and the Canadian Space Agency.

References
[1] Covington A E 1952 NRC 10.7-centimeter radio telescope and radiometer *NRC Report* **ERA-216**, National Research Council of Canada
[2] Covington A E 1948 Solar radio noise observations at 10.7 cm *Proc IRE* **36** 454
[3] Pawsey J L and Yabsley D E 1949 Solar radio-frequency radiation of thermal origin *Australian J. Scientific Research - Series A* **2** 198
[4] Covington A E and Broten N W 1957 An interferometer for radio astronomy with a single-lobed antenna pattern *Transactions IRE Antennas and Propogation* **AP-5** 247
[5] Labrum N R 1972 The Culgoora solar radio observatory *Solar Physics* **27** 496
[6] Frater R H and Ekers R D 2012 John Paul Wild 1923 - 2008 *Historical Records of Australian Science* **23** 212
[7] Tapping K F and Valdes J J 2011 Did the sun change its behaviour during the decline of cycle 23 and into cycle 24? *Solar Physics* **272** 337
[8] Kundu M R 1965 *Solar Radio Astronomy* (New York: Wiley-Interscience) Chapters 5 and 6
[9] Tapping K F and DeTracey B 1990 The origin of the 10.7 flux *Solar Physics* **127** 32

Journal of Physics: Conference Series **440** (2013) 012040 doi:10.1088/1742-6596/440/1/012040

The Sun-as-a-star observations: GOLF & VIRGO on SoHO, and BiSON network

R. A. García[1], G.R. Davies[1,2], A. Jiménez[3,4], J. Ballot[5,6], S. Mathur[7,8], D. Salabert[9], W. J. Chaplin[2], Y. Elsworth[2], C. Régulo[3,4], S. Turck-Chièze[1]

[1] Laboratoire AIM, CEA/DSM-CNRS-Université Paris Diderot; CEA, IRFU, SAp, F-91191, Gif-sur-Yvette, France
[2] School of Physics and Astronomy, University of Birmingham, Edgbaston, Birmingham, B15 2TT, UK
[3] Instituto de Astrofísica de Canarias, E-38200 La Laguna, Tenerife, Spain
[4] Dept. de Astrofísica, Universidad de La Laguna, E-38206 La Laguna, Tenerife, Spain
[5] CNRS, Institut de Recherche en Astrophysique et Planétologie, 14 avenue Edouard Belin, 31400 Toulouse, France
[6] Université de Toulouse, UPS-OMP, IRAP, 31400 Toulouse, France
[7] High Altitude Observatory, 3080 Center Green Drive, Boulder, CO, 80302 USA
[8] Space Science Institute, 4750 Walnut Street, Suite 205, Boulder, Colorado 80301 USA
[9] Laboratoire Lagrange, UMR7293, Université de Nice Sophia-Antipolis, CNRS, Observatoire de la Côte d'Azur, Bd. de l'Observatoire, 06304 Nice, France

E-mail: rgarcia@cea.fr, davies@bison.ph.bham.ac.uk, ajm@iac.es, Jerome.ballot@irap.omp.eu, smathur@SpaceScience.org, salabert@oca.eu, wjc@bison.ph.bham.ac.uk, ype@bison.ph.bham.ac.uk, crr@iac.es, sylvaine.turck-chieze@cea.fr

Abstract.
Sun-as-a-star observations are very important for the study of the conditions within the Sun and in particular for the deep interior where higher degree modes do not penetrate. They are also of significance in this era of dramatic advances in stellar asteroseismology as they are comparable to those measured in other stars by asteroseismic missions such as CoRoT, *Kepler*, and MOST. More than 17 years of continuous measurements of SoHO and more than 30 years of BiSON observations provide very long data sets of uninterrupted helioseismic observations. In this work, we discuss the present status of all these facilities that continue to provide state-of-the-art measurements and invaluable data to improve our knowledge of the deepest layers of the Sun and its structural changes during the activity cycle.

1. Introduction

Helioseismology entered in a new era in the mid nineties with the deployment of the ground-based networks around the world such as the Birmingham Solar Oscillation Network (BiSON, [30]), the Global Oscillation Network Group (GONG, [2]), and the launch of the Solar and Heliospheric Observatory (SoHO) mission [3]. All these facilities allowed a near continuous monitoring of the Sun, drastically improving the quality of the datasets available.

By measuring the Sun as a star, i.e., without spatial resolution, we can access low-degree modes. These modes are particularly interesting because they reach the deepest regions of the

Sun providing us with invaluable information of its structure (e.g. [4, 5, 6, 7]), and dynamics (e.g.[8, 9, 10]). They have also been used very recently as new indicators of activity of the sub-surface layers [11, 12]. Both, SoHO and BiSON, will be in operation until, at least, 2015 ensuring the continuity of the activity-cycle studies as well as progressing in the detection of low-degree low-order p modes and hopefully g modes.

2. SoHO Instruments: GOLF and VIRGO/SPM

The ESA/NASA SoHO mission was launched on December 2nd 1995. As part of the scientific payload, it contains three helioseismic instruments, two of them dedicated to study the Sun as a star: the Global Oscillations at Low Frequency (GOLF, [13]) and the Variability of solar IRradiance and Gravity Oscillations (VIRGO, [14]). The overall duty cycle of these instruments is above 95% for the last ~17 years. Most of the missing data correspond to the period July - October 2008 due to the loss and recovery of the spacecraft.

GOLF is a resonant scattering spectrophotometer measuring the Doppler wavelength shift - integrated over the solar surface - of the D1 and D2 Fraunhofer sodium lines at 589.6 and 589.0 nm respectively [15]). The instrument was also designed to measure the solar mean magnetic field of the Sun [16]. Unfortunately, due to occasional malfunctions of the polarization mechanisms switching between both wings of the Na profile and between the two sigma components of the solar light, the instrument was operated in a single-wing configuration in early 1996 [17]. The low instrumental noise, the fast sampling rate, and the high stability of the platform orbiting around the Lagrangian L1 point allows GOLF to measure: a) the oscillation properties of the Sun at very low frequencies, allowing the first detection of the asymptotic properties of dipole g modes [18, 19] and the dynamics of the radiative interior [20, 21]; b) the low-order p modes at around 1 mHz [22, 23]; c) the pseudomodes region above the cut-off frequency [24]. In Fig. 1 the low-frequency range between 500 and 1500 μHz is shown. Some non previously reported p modes seem to rise from the background, e.g. the l=0, n=7 at around 1118 μHz.

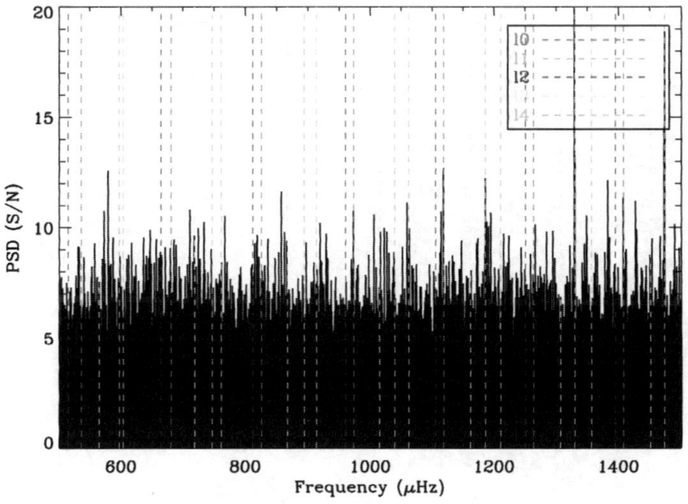

Figure 1. GOLF Power Spectrum Density in units of signal-to-noise ratio showing the low p-mode frequency region (500 to 1500 μHz) computed from continuous 6000 days velocity time series starting on April, 11 1996.

VIRGO is composed of three type of instruments: two active-cavity radiometers, three Sun spectrophotomenters (SPM) at 402 nm (Blue), 500 nm (Green), and 862 nm (Red), and one Low resolution imager (LOI). Although the solar convective background is higher at lower frequencies compared to velocity measurements (see Fig 2), VIRGO/SPM data have been very useful to study solar oscillation properties from ~ 2 mHz up to the cut-off frequency, including also the pseudo-modes region [25, 26] (see Fig.3). A comparison of the p-mode properties measured in velocity and intensity can be found in [27].

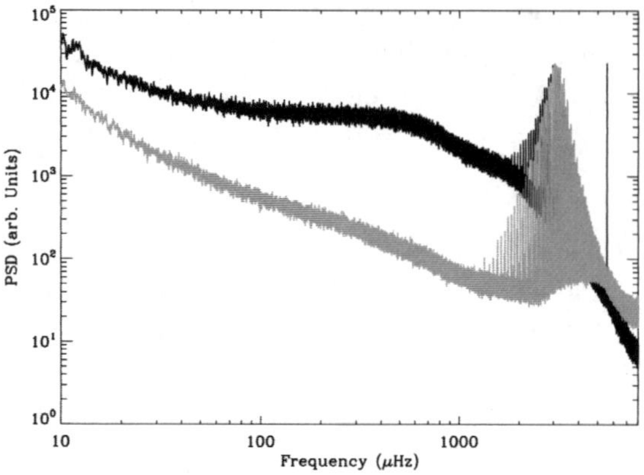

Figure 2. Power Spectrum Density (PSD) of GOLF (magenta) and VIRGO/SPM green channel (black). Both PSDs have been normalized by the maximum of the p-mode hump and smoothed by a boxcar of 100 bins. The spike at 5555 μHz is due to the period of the calibration reference used by the VIRGO/SPM Data Acquisition System.

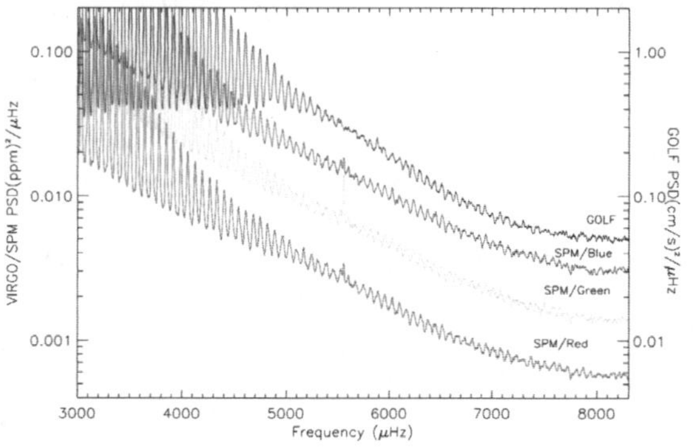

Figure 3. High-frequency region of the Power Spectrum Density of GOLF and the three VIRGO/SPM is shown. The pseudo-mode region is clearly visible above the cut-off frequency.

3. The BiSON network

The Birmingham Solar Oscillation Network (BiSON) is a 6-station network capable of observing the Sun 24h a day (see Fig. 4). Although many technical improvements in design have been implemented in the newer stations, the fundamental principles of the Resonant Scattering Spectrometer are unchanged [28].

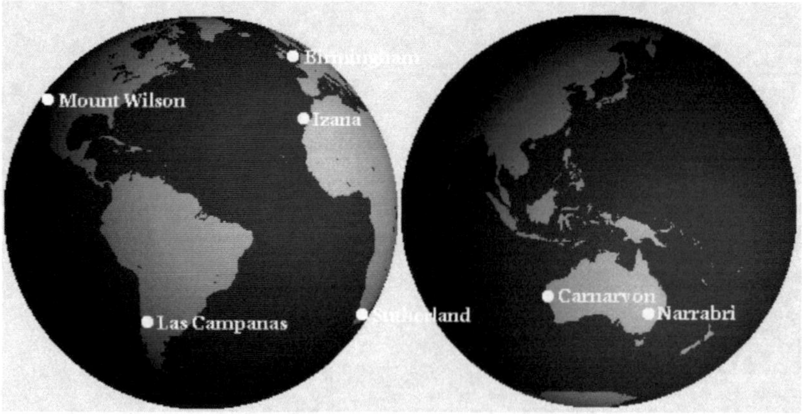

Figure 4. Distribution of the BiSON sites all over the world.

BiSON instruments measure the Doppler-velocity shifts in the K Fraunhofer line at 770 nm. Near-continuous observations have been taken since July 31st, 1978 [29], with year-by-year duty cycles regularly exceeding 80% since the network was completely deployed [30]. BiSON is a unique tool for the study of the oscillation frequencies [31] and their changes that accompany the solar cycle because it has been collecting data for the past 30 yrs. BiSON observations cover most of cycle 21, the complete cycles 22 and 23, also cycle 24 to present [32]. An example of the BiSON spectrum is shown in Fig. 5.

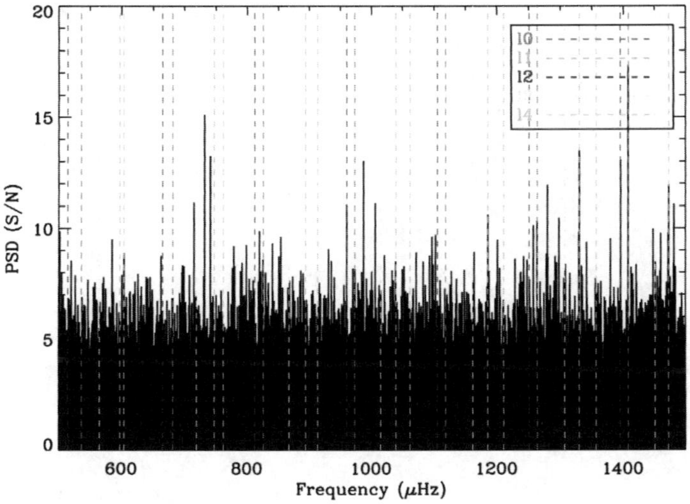

Figure 5. BiSON Power Spectrum Density expressed in units of signal-to-noise ratio in the low-frequency regime (500 to 1500 μHz). It corresponds to a velocity time series of 2800 days starting January 1, 1995 with an overall duty cycle of 71%.

4. Conclusions

The BiSON network and the GOLF and VIRGO/SPM instruments on board SoHO are still running facilities with ensured funding until, at least, 2015 and 2016 respectively. Their overall performances are still good, although in the case of SoHO, the instruments have followed the expected degradation compared to the original performances after 17 years in space. These observations provide the required continuity on the helioseismic observations, while monitoring long-term variability of the properties of the global acoustic modes. These observations are also very important in the asteroseismic context because the Sun is often used as a reference, and only low-degree modes are available for other stars.

Acknowledgments

SoHO is a space mission of international cooperation between ESA and NASA. R.A.G., S.T-C., and D.S. thank the support from CNES. This work was supported in part by the NASA grant NNX09AE59G, by the White Dwarf Research Corporation through the Pale Blue Dot project, and by the grant AYA2010-17803 from the Spanish National Research Plan. NCAR is supported by the National Science Foundation. BiSON is funded by the UK Science Technology and Facilities Council (STFC).

References

[1] Chaplin W J, Elsworth Y, Howe R et al. 1996 *Solar Phys.* **168** 1
[2] Harvey J W, Hill F, Hubbard R et al. 1996 *Science* **272** 1284
[3] Domingo V, Fleck B, Poland A I 1995 *Solar Phys.* **162** 1
[4] Turck-Chièze S, Couvidat S, Kosovichev A G et al. 2001 *Astrophys. J.* **555** L69
[5] Mathur S, Turck-Chize S, Couvidat S, García R A 2007 *Astrophys. J.* **668** 594
[6] Basu S, Chaplin W J, Elsworth Y, New R, Serenelli A M 2009 *Astrophys. J.* **699** 1403
[7] Turck-Chièze S, Lopes I 2012 *RAA* **112** 1107
[8] Elsworth Y, Howe R, Isaak G R et al. 1995 *Nature* **376** 669
[9] Turck-Chièze S, Palacios A, Marquès J et al. 2010 *Astroph. J* **715** 1539
[10] Turck-Chiéze S, Couvidat S 2011 *Report on Progress in Physics* **74** 086901
[11] Salabert D, García R A, Pallé P L, Jiménez-Reyes S J 2009 *Astron. & Astroph.* **504** L1
[12] Simoniello R, Jain K, Tripathy S C et al. 2013 *Astrophys. J.* In Press (arXiv:1210.6796)
[13] Gabriel A H, Grec G, Charra J et al. 1995 *Solar Phys.* **162** 61
[14] Frohlich C, Romero J, Roth H et al. 1995 *Solar Phys.* **162** 101
[15] García R A, Roca Cortés T, Régulo C 1998 *Astron. & Astroph. Supl. Ser.* **128** 389
[16] García R A, Boumier P, Charra J et al. 1999 *Astron. & Astroph.* **346** 626

[17] García R A, Turck-Chièze S, Boumier P et al. 2005 *Astron. & Astroph.* **442** 385
[18] García R A, Turck-Chièze S, Jiménez Reyes S J et al. 2007 *Science* **316** 1591
[19] García R A, Jiménez A, Mathur S et al. 2008 *Astron. N.* **329** 476
[20] García R A, Corbard T, Chaplin W J et al. 2004 *Solar Phys.* **220** 269
[21] Mathur S, Eff-Darwich A, García R A, Turck-Chièze S 2008 *Astron. & Astroph.* **484** 517
[22] Bertello L, Varadi F, Ulrich R K et al 2000 *Astrophys. J.* **537** 143
[23] García R A, Régulo C, Turck-Chièze S et al. 2001 *Solar Phys.* **200** 361
[24] García R A, Pallé P L, Turck-Chièze S et al. 1998 *Astrophys. J.* **504** L51
[25] Jiménez A, Jiménez-Reyes S J, García R A 2005 *Astrophys. J.* **623** 1215
[26] Karoff C, Kjeldsen H 2008 *Astrophys. J.* **678** L73
[27] Toutain T, Appourchaux T, Baudin F et al. 1997 *Solar Phys.* **175** 311
[28] Brookes J R, Isaak G R, van der Raay H B 1976 *Nature* **259** 92
[29] Claverie A, Isaak G R, McLeod C P et al. 1979 *Nature* **299** 704
[30] Chaplin W J, Elsworth Y, Howe R et al. 1996 *Solar Phys.* **168** 1
[31] Broomhall A-M, Chaplin, W J, Davies, G R et al. 2009 *MNRAS* **396** L100
[32] Basu S, Broomhall A M, Chaplin W J, Elsworth Y 2012 *Astrophys. J.* **758** 43

Journal of Physics: Conference Series **440** (2013) 012041 doi:10.1088/1742-6596/440/1/012041

Long-term synoptic observations of the Sun at the National Astronomical Observatory of Japan

Yoichiro Hanaoka

Solar Observatory, National Astronomical Observatory of Japan, 2-21-1 Osawa, Mitaka, Tokyo 181-8588, Japan

E-mail: hanaoka@solar.mtk.nao.ac.jp

Abstract. The National Astronomical Observatory of Japan started regular synoptic solar observations about 100 years ago. At the beginning, Ca K spectroheliograms and white-light photographs were taken, and various other types of observations have been added. These historical data have been digitized and are now open at our web site (http://solarwww.mtk.nao.ac.jp/en/solarobs.html). Currently we are operating high-resolution imaging observations in the Hα line, the green continuum, and the G-band. Besides various kinds of imaging observations, magnetic field measurements have been carried out for about 30 years. We recently started to conduct full-Sun spectropolarimetry observations in the lines of He 10830 Å/Si 10827 Å and Fe 15648 Å. These near infrared observations show the magnetic field evolutions in both the photosphere and the chromosphere. In this paper, we present our data, which are expected to contribute to studying the long-term change of the solar activity.

1. Introduction: NAOJ History of Solar Observation

Long-term solar variability is not only an important clue to study the stellar dynamo, but also it is a source of the change in the solar system's environment, including the earth's climate. To study the long-term variation of the solar magnetic activity, synoptic solar observations are essentially important. Sunspot records using telescopes date back to about 400 years ago, and the sunspot relative number has been used as a canonical index of the solar activity. However, the sunspot relative number shows only one aspect of the solar activity. Two-dimensional, full-disk information of the upper solar atmosphere and the solar magnetic field is more important, and both utilizing old data and ensuring to continue the current observation for long time are required to understand the long-term variability.

The Tokyo Astronomical Observatory, the predecessor of the National Astronomical Observatory of Japan, started the regular solar observation in the 1910's in Tokyo, Japan. Photographic and hand drawing observations were continued until the 1990's, and now the observations are carried out with electric devices. The accumulated analog data had been difficult to be distributed, and therefore digitization of the plates, films, and hand drawings was started about ten years ago. The project to digitize analog records was led by Prof T. Sakurai, and was carried out with some funds from the Japan Society for the Promotion of Science and Nagoya University, and recently it was completed.

About nine solar cycles are covered by various types of observations, and the following data, not only digitally acquired data but also digitized old analog data, are available online now.

- Sunspot relative numbers: 1929–
- Sunspot drawings: 1939–1998, digitized
- Full-disk white-light images: 1918–1998 digitized photographs, 1997– electronic data
- Full-disk Ca K images: 1917–1974 digitized photographs
- Full-disk Hα images: 1949–1964 digitized drawings, 1957–1992 digitized photographs, 1991– electronic, 2011– high-resolution with off-band images
- Full-disk magnetograms: 1993–2006 imaging polarimetry at Fe 6303 Å, longitudinal B and Doppler, 2010– spectropolarimetry, He 10830 Å/Si 10827 Å and Fe 15648 Å lines
- Full disk He 10830 Å: 1991–1998 spectroheliograms, electronic
- Active region Hα images: 1990–2007, electronic, partly with polarimetry
- Active region magnetograms: 1982–1995 spectropolarimetry at Okayama, 1992–2007 imaging polarimetry at Mitaka, Tokyo
- Coronal green-line images: 1978–1991 digitized photographs, 1997–2009 electronic
- Coronal green-line intensities: 1951–2009

In the following section, we will briefly intoduce the observations for various types of the targets to promote data use for synoptic solar physics. A review of scientific results from the old observations is found in [1].

2. Data

2.1. Full-disk Chromosphere

The full-disk Ca K observation was started in 1917 with a Grubb 30-cm sidelostat and a Toepfer double-prism spectroheliograph, and was continued until 1974. Basically one scan was carried out per day, and Ca K images were recorded on photographic plates (–1960) and 35mm films (1960–). In total, 8585 images were recorded on 8524 observing days. They show magnetic activity of the Sun before the magnetograph observations became available, as shown in Figure 1(a). Furthermore, Ca K activity is considered to be closely related with the UV irradiance of the Sun. The Ca K data during recent 100 years taken at some observatories (Kodaikanal from 1905, Mt. Wilson from 1915, and Arcetri from 1930) have been already digitized, and some attempts to reproduce the correct Ca K activity have been made (see [2], and references therein). Our data will contribute to more accurate estimation of the Ca K activity in the 20th century (Figure 1(b)).

Since 1949, the full-disk Hα observation was carried out with hand drawing using a spectrohelioscope. In 1957 photographic observation was started with a OPL Lyot filter with a 0.75 Å passband, and later on CCD imaging with a Halle Lyot filter succeeded it. Recently we started high-resolution imagings, with a 2k$\times$2k camera and a 0.25 Å passband Zeiss Lyot filter, and the obtained images are comparable to those taken by the Global High Resolution Hα Network (see http://swrl.njit.edu/ghn_web/). In addition, now we are taking images at several wavelengths around the Hα line (center, $\pm$0.5 Å, $\pm$0.8 Å, and +3.5 Å). Those data show flares and filament activities not only with brightness changes but also with Doppler velocities. Hα data give us three dimensional information of mass motions, and this fact makes that the Hα observations are unique and difficult to be substituted with simple imaging observations in other wavelengths.

2.2. Full-disk Photosphere

White-light photographic imaging observation was started in 1918, and now we are taking white-light images with a CCD camera. Sunspot relative number derivation was started in 1929 [3]. Sunspots were formerly counted on the hand-drawn sketches, but now the images taken by the

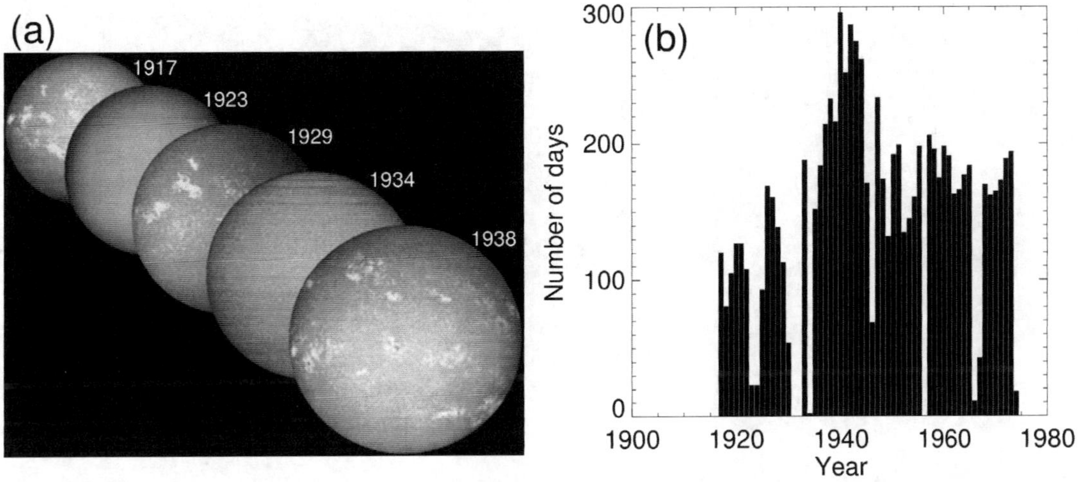

Figure 1. (a) Some samples of the Ca K spectroheliograms taken by NAOJ during the early 20th century. Solar activity cycle can be seen in these images. (b) Annual number of Ca K observing days. During 1917–1974, the Ca K spectroheliograms were taken on 8524 days.

CCD camera is used for automatic pick-up of sunspots [4]. Recently we are working to replace the current system with a new imaging system, which takes G-band (CH 4305 Å) images as well as green continuum images.

2.3. Magnetic Field

NAOJ is the only institute which continues regular solar magnetic field observations in Japan. In 1982, vector magnetic field measurements of active regions was started at the Okayama Observatory of NAOJ with a spectropolarimetry instrument [5]. In 1992, imaging polarimetry at Fe I 6303 Å line was started with the Solar Flare Telescope [6] at Mitaka, Tokyo. In 2002, polarimetry in the Hα line was added to study polarization signals in the chromosphere. These active region polarimetry was continued until 2007, and then, we have installed a full-disk near-infrared spectropolarimeter on this telescope, and from 2010 it has been under operation (see [7]; full description is now in preparation by Sakurai *et al.*). Observing wavelengths are as follows, and sample Stokes maps are shown in Figure 2.

- He I 10830 Å: This line shows the polarization signals in the chromosphere. In addition to the Zeeman polarizations in sunspots and plages, prominences show atomic and Hanlé polarizations, and they contain information of the magnetic field in prominences. A nearby line, Si I 10827 Å, which show the photospheric magnetic field, is observed simultaneously with the He line.
- Fe I 15648 Å: This line shows the photospheric magnetic field. This line has a large Landé factor, and with the long wavelength, it shows much larger Zeeman splitting than usual lines used in the solar polarimetry. Therefore, we can estimate the magnetic field strength down to a couple of hundred gauss directly from the separation of Zeeman split components, regardless of the filling factor of the magnetized area.

This infrared polarimeter is intended to obtain basic magnetic information of both the chromosphere and photosphere for long time. Such an instrument, which regularly measures full-disk, full-Stokes signals in He I 10830 Å and Fe I 15648 Å, is unique, and is expected to contribute to the solar dynamo study.

Journal of Physics: Conference Series **440** (2013) 012041 doi:10.1088/1742-6596/440/1/012041

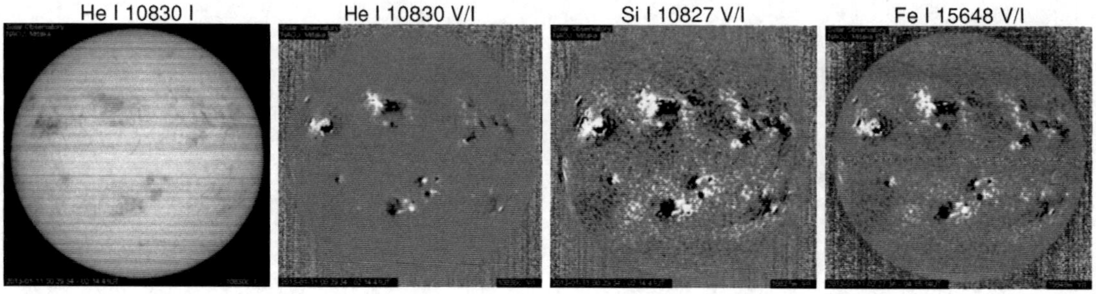

Figure 2. Sample Stokes I image of the He I 10830 Å line and Stokes I/V images of the He I 10830 Å, Si I 10827 Å, and Fe I 15648 Å lines taken with the infrared spectropolarimeter on 2013 Jan 11. The celestial north is to the top.

2.4. Corona: Coronagraphs and Solar Eclipses
The observation with a coronagraph was started in 1949 in the central mountain area in Japan, and until the shut-down in 2009 the Norikura Corona Observatory was operated by NAOJ with some coronagraph systems. The long-term coronal activity variation has been monitored for about 60 years. Latitudinal distribution of the green line (Fe XIV 5303 Å) brightness was also measured; basically it shows similar variation to that seen in sunspot butterfly diagrams, but coronal observations cover to the poles, where the sunspots do not appear. Such coronal observations show synoptic magnetic features before the space observations of the corona became available.

The expeditions for eclipse observations have been dispatched since the 19th century. The total eclipses still provide us unique chances to observe the corona, particularly in the visible wavelengths. In recent eclipses including the very eclipse on 2012 November 14, we collaborated with amateur astronomers to raise the probability to succeed in taking images of the corona under good weather conditions and to obtain data at more than one site [8].

3. Future Prospect
The solar and heliophysics community in Japan is proposing to establish a framework of the cooperation for the ground-base synoptic solar observations. There are various plans to construct or upgrade telescopes and establish data circulation networks. We will promote these projects collaboratively to continue the synoptic observations in the future.

References
[1] Sakurai T 1998 *ASP. Conf. Ser.* **140** *Synoptic Solar Physics* ed K S Balasubramaniam *et al.* (San Francisco: ASP) p 483
[2] Ermolli I, Solanki S K, Tlatov A G, Krivova N A, Ulrich R K and Singh J 2009 *Astrophys. J.* **698** 1000
[3] Natori T 1991 *Rep. Natl. Astron. Obs. Japan* **1** 71 (in Japanese)
[4] Imai H, Suematsu Y, Miyashita M and Kumagai K 1998 *Rep. Natl. Astron. Obs. Japan* **4** 7 (in Japanese)
[5] Makita M, Sakurai T, Shibasaki K and Koyano H 1993 *ASP. Conf. Ser.* **46** *The Magnetic and Velocity Fields of Solar Active Regions, IAU Colloq. 141* ed H Zirin *et al.* (San Francisco: ASP) p 180
[6] Sakurai T. *et al.* 1995 *Publ. Astron. Soc. Japan* **47** 81
[7] Hanaoka Y *et al.* 2011 *ASP. Conf. Ser.* **437** *Solar Polarization 6* ed J R Kuhn *et al.* (San Francisco: ASP) p 371
[8] Hanaoka Y, Kikuta Y, Nakazawa J, Ohnishi K and Shiota K 2012 *Solar Phys.* **279** 75

The sunspot cycle revisited

Nick Lomb

Sydney Observatory, PO Box K346, Haymarket NSW 2000, Australia

Email: nickl@phm.gov.au

Abstract. The set of sunspot numbers observed since the invention of the telescope is one of the most studied time series in astronomy and yet it is also one of the most complex. Fourteen frequencies are found in the yearly mean sunspot numbers from 1700 to 2011using the Lomb-Scargle periodogram and prewhitening. All of the frequencies corresponding to shorter term periods can be matched with simple algebraic combinations of the frequency of the main 11-year period and the frequencies of the longer term periods in the periodogram. This is exactly what can be expected from amplitude and phase modulation of an 11.12-year periodicity by longer term variations. Similar, though not identical, results are obtained after correcting the sunspot number series as proposed by Svalgaard. On looking separately at the amplitude and phase modulation a clear relationship is found between the two modulations although this relationship has broken down for the last four solar cycles. The phase modulation implies that there is a definite underlying period for the solar cycle. Such a clock mechanism does seem to be a possibility in models of the solar dynamo incorporating a conveyor-belt-like meridional circulation between high polar latitudes and the equator.

1. Introduction

Sunspots are an indication of solar activity. Records of sunspot numbers over hundreds of years can be studied to examine the variability of the Sun. These studies can give clues to the nature of the mechanism driving the solar cycle and provide constraints that must be met by any viable theory.

A quick look at a plot of the sunspot number time series indicates a quasi-regular variation of about 11 years that changes in amplitude and length. This 11-year period can largely disappear during great minima such as the Maunder Minimum, which persisted for much of the 17th century.

An obvious feature of the sunspot number time series, one that is so obvious that it is usually missed, is that the definition of the series constrains it to be always positive. Hence, and unlike most other time series with which scientists are familiar, the fluctuation in sunspot numbers is not about zero but about the mean value. Stated mathematically the series is not in the form

$$A(t)\sin(2\pi ft + P(t))$$

where A(t) and P(t) represent amplitude and phase modulation terms, respectively, but in the form

$$A(t)\,[1 + \sin(2\pi ft + P(t))] \qquad (1)$$

As was shown in a previous paper [1], to be referred to as paper 1, this peculiar, always positive, nature of the sunspot time series means that any modulation period must show in the spectrum both as a sidelobe to the main period and as a separate periodicity.

In paper 1 the yearly mean Wolf Sunspot Numbers [2] from 1700 to 1964 were studied by frequency analysis using least squares or LS frequency analysis [3], which has since then been widely used under the name Lomb-Scargle Periodogram [4]. In the paper a set of 14 statistically significant periodicities was found in the frequency spectrum. These periodicities could be described in terms of just three main periods. This was interpreted as indicating that the sunspot series results from the amplitude and phase modulation of the main 11-year period by two longer term periodicities. In addition, a correlation was found between the amplitude and phase modulation in the sense that the stronger the peak of a cycle the earlier it occurred.

The current paper aims to extend the analysis of paper 1 by including the most recent solar cycles and to check whether the sunspot numbers still show the same structure with more recent data. It also aims to better interpret the results on the basis of modern models of the solar dynamo. Additionally, the analysis is to be repeated after increasing the pre-1945 sunspot numbers by 21% following Svalgaard [5] who has found that such corrections are necessary to maintain the uniformity of the series.

2. Analysis

In this paper the yearly mean Wolf sunspot numbers are again used, but this time from 1700 to 2011 with the data taken from the Solar Influences Data Analysis Center (http://sidc.oma.be/sunspot-data/).

2.1. Frequency analysis

The frequency analysis was carried out with the Lomb-Scargle Periodogram using the Peranso period analysis software (http://www.peranso.com/). After each frequency was found the data was prewhitened and the calculation repeated to obtain the next frequency.

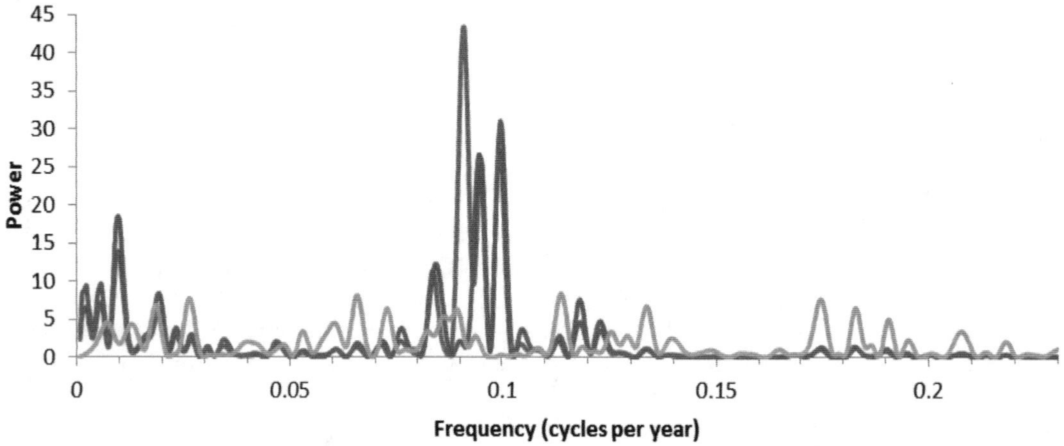

Figure 1. Periodograms of the yearly mean sunspot numbers, 1700-2011: the periodogram of the original data (blue), the periodogram after prewhitening by the main 11-year periodicity (red) and the periodogram after prewhitening by all 14 periods found in the data (green). Power on the y axis represents the reduction in the sum of squares due to each frequency with the sunspot numbers scaled to vary from +1 to -1.

The frequency analysis was stopped after finding 14 periods in the data. Figure 1 shows the frequency spectrum of the raw sunspot series together with the spectra after the removal of the main period and after the removal of all 14 periods. As can be seen from the figure, prewhitening is important as the mutual influence of adjacent frequency peaks can shift the position of the peaks and hence the determined frequency.

After the removal of all found periods a Monte-Carlo analysis indicated that the next remaining peak had a false alarm probability of only 0.045±.015. There are thus probably further statistically significant periodicities remaining in the data, however, they are unlikely to make much contribution to our understanding of its structure. Table 1 lists the periods/frequencies that were found together with their uncertainties. The Peranso program calculates uncertainties with a method due to Schwarzenberg-Czerny [6].

Table 1. Periodicities in the yearly mean sunspot numbers, 1700–2011.

Observed period (yr)	Observed frequency (cycles/yr)	Predicted frequency (cycles/yr)	Identification
11.00±0.06	0.0909±0.0005	0.0899	f_1
10.03±0.05	0.0997±0.0005		f_1+f_{01}
10.61±0.08	0.0942±0.0007	0.0944	f_1+2f_{02}
102.07±7.83	0.0098±0.0007		f_{01}
11.86±0.14	0.0843±0.0010	0.0833	f_1-3f_{02}
449.24±94.61	0.0022±0.0005		f_{02}
8.46±0.05	0.1182±0.0007	0.1193	f_1+3f_{01}
13.08±0.17	0.0764±0.0010	0.0743	f_1-f_{03}
8.12±0.08	0.1232±0.0012	0.1213	f_1+2f_{03}
63.80±5.02	0.0157±0.0012		f_{03}
9.40±0.09	0.1064±0.0010	0.1056	f_1+f_{03}
43.56±1.42	0.0230±0.0007		f_{04}
10.34±0.11	0.0967±0.0010	0.0966	f_1+3f_{02}
28.55±1.02	0.0350±0.0012		f_{05}

As in paper 1 the shorter periods can all be matched by simple algebraic combinations of the frequency of the main 11-year period and the frequencies of the longer term periods in the periodogram. This is exactly what can be expected when one periodicity of the form in equation 1 is amplitude and phase modulated by longer periods. The predicted value of f_1 was taken from the determined value of f_1+f_{01} as paper 1 showed that term was determined with smaller errors.

The analysis was repeated with the sunspot numbers modified following the scheme proposed by Svalgaard [5]. Results similar to those with the unmodified data are obtained though, as expected, with changes to the longer term periods. The longer term periods equivalent to f_{01} to f_{05} in table 1 become, 99.79, 67.32, 54.19, 170.49 and 44.31 years. The long trend period f_{02} from table 1 is now gone, but the period around 100 years remains with the modulation by this period obvious in a visual examination of a plot of the modified sunspot number data. Most interestingly, all five of these longer-term periods in the modified sunspot data are subharmonics of the main 11-year periodicity.

2.2. Amplitude and phase modulation

Various authors have looked at the modulation in phase, generally interpreted as variation in frequency, using highly sophisticated modern time series methods. For example, Kolláth and Oláh [7] follow the frequency variation of the 11-year period, its first harmonic and that of a longer period using the pseudo-Wigner distribution from sound processing. Here, instead, we looked at the modulation in more detail using the conceptually simple technique of fitting by least squares a sine wave with an 11-year period and a constant for each of the 28 minimum-to-minimum cycles since 1700. The original unmodified yearly mean Wolf sunspot numbers were used. The longer period modulation of the sunspot cycle can be seen in both the amplitudes and the phases as shown in figure 2. As expected from (1), there is a good fit between the constants and amplitudes. As well there is also a clear relation between the phases and the amplitudes. To explore this relationship a linear regression model was fitted between the phases and the amplitudes together with a time term to allow for differences in period from the assumed 11 years.

The regression coefficient for the period of -0.33±0.12° per sunspot yields a frequency of 0.0900 cycles per year or a period of 11.11 years. This value for the main solar cycle is in good agreement with the predicted value listed in table 1 that equates to 11.12 years.

As shown in figure 2, there is a clear relationship between the amplitude and the phase modulation, though with occasional deviations as in the late 18th century. Interestingly, the relationship seems to have completely broken down for the last four solar cycles. This relationship is connected to the Waldmeier effect [8], according to which there is an inverse correlation between the rise time of a cycle and its amplitude.

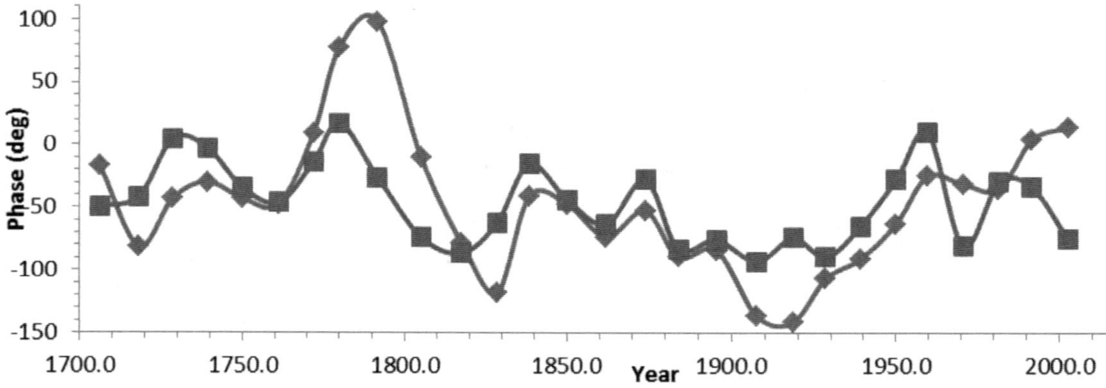

Figure 2. Observed (blue diamonds) and predicted phases (red squares) of the 11-year period calculated from a regression fit of the amplitude variations to the phases.

3. Discussion

This study has found evidence for long-term modulation of the solar cycle over periods of 28 to 450 years. At this stage there is no indication if any of the modulation periodicities are real or if they just represent a Fourier fit to a random variation. It can be noted though that a period of around 100 years, which matches the well-known Gleissberg cycle [9], does seem to be persistent in the data.

This study confirms the structure of the sunspot time series demonstrated in paper 1: a stable 11-year periodicity (the Schwabe cycle) that is amplitude and phase modulated by the long-term periodicities discussed above. This clearly implies that a clock mechanism must exist within the Sun for the 11-year periodicity to persist in the solar data as was first suggested by Dicke [10].

Modern theories provide a possible clock mechanism in the conveyor-belt-like meridional circulation between high polar latitudes and the equator. In the Babcock-Leighton models of Charbonneau and Dikpati [11] and [12] remnant magnetic flux from decaying sunspots is transported away from the equator by meridional circulation towards the poles generating the poloidal field of the following cycle. This field is transported to the base of the convection zone where shearing by differential rotation leads to a new toroidal field at low latitudes. Buoyant flux tubes rise to the surface as sunspots. In this way the meridional circulation provides the clock regulating the 11-year cycle and maintaining its continuity.

The models predict fluctuations in the meridional circulation and these fluctuations have been observed with the Michelson Doppler Imager on board the SOHO spacecraft [13]. However, according to Charbonneau and Dikpati [11] the meridional circulation can still act as a clock for most of their stimulations exhibit 'good phase locking, in the sense that their cycle periods rarely depart for more than a few consecutive cycles from their average value'. Moreover, the models also reproduce the amplitude-duration anticorrelation that is related to the amplitude-phase relationship discussed in this paper.

The Svalgaard modification of the sunspot number series would have to be widely accepted and further statistical tests would need to be done to establish the validity of the finding that the longer periods in the modified yearly sunspot data from 1700 to 2011 are all subharmonics of the main 11-year periodicity. It is worth noting though that subharmonics in the frequency spectrum would suggest a non-linear mechanism, possibly with chaotic instability [14].

4. References

[1] Lomb N R and Andersen A P 1980 *Mon. Not. R. astr. Soc.* **190**, 723–32
[2] Waldmeier M 1961 *The Sunspot Activity in the Years 1610-1960* (Zürich: Verlag Schulthess u. Co. AG)
[3] Lomb N R 1976 *Astrophysics and Space Science* **39** 447–62
[4] Scargle J D 1982 *Astrophys. J.* **263** 835–53
[5] Svalgaard L 2012 *Comparative Magnetic Minima: characterizing quiet times in the Sun and stars: Proc. IAU Symp. 286 (Mendoza, Argentina, 3–7 October 2011)* ed C H Mandrini and D Webb (Cambridge: Cambridge) 27–33
[6] Schwarzenberg-Czerny A 1991, *Mon. Not. R. astr. Soc.* **253** 198-206
[7] Kolláth Z and Oláh K 2009 *Astronomy & Astrophysics* **501** 695–702
[8] Waldmeier M 1935 *Astron. Mitt. Zurich* **14** 105–30
[9] Gleissberg W 1939 *The Observatory* **62** 158–9
[10] Dicke R H 1978 *Nature* **276** 676–80
[11] Charbonneau P and Dikpati M 2000 *Astrophys. J.* **543** 1027–43
[12] Charbonneau P 2010 *Living Rev. Solar Phys.* **7** 3
[13] Basu S and Antia H M 2010 *Astrophys. J.* **717** 488-95
[14] Adler L, Yost W T and Cantrell J H *International Congress on Ultrasonics: AIP Conf. Proc. 1433 (Gdańsk, Poland, 5–8 September 2011)* ed B B J Linde, J Pączkowski and N Ponikwicki (Melville: AIP) 527–30

Acknowledgments
An earlier poster paper version of this paper was co-authored with Martin Anderson of Sydney Observatory, whose assistance is gratefully acknowledged. Zoran Mikić of Predictive Science Inc suggested examining the sunspot data with the Svalgaard modification and helpful comments on the first draft of this paper were received from the referee, Robertus von Fay-Siebenburgen.

Journal of Physics: Conference Series **440** (2013) 012043 doi:10.1088/1742-6596/440/1/012043

On gravity waves in the Sun

Allan Sacha Brun[1], Lucie Alvan[1], Antoine Strugarek[1], Stéphane Mathis[1] & Rafael A. García[1]

Laboratoire AIM Paris-Saclay, CEA/Irfu Université Paris-Diderot CNRS/INSU, 91191 Gif-sur-Yvette, France

E-mail: sacha.brun@cea.fr

Abstract. We briefly present our recent progress to model in 3-D the excitation and propagation of internal waves in the deep solar radiative interior. By modeling a rotating spherical convection zone on top of a radiative interior with a realistic seismically calibrated stable stratification (i.e solar-like Brunt-Väisälä frequency), we are able to generate a large spectrum of internal waves and modes thanks to the continuous pummeling of convective plumes. When comparing with an adiabatic oscillation code we find a good overall agreement and confirm that those waves are gravity waves.

1. Waves in the Sun

Various waves are present in the Sun, such as acoustic, gravity, Alfvèn, and inertial waves to cite only a few. They can inform us on the internal structure and dynamics of our star as a function of depth and latitude with an accuracy that depends on their propagation properties and the time they spent traveling in any given region. Internal gravity waves (IGW) are essential for helio- and asteroseismology to probe the deep stably stratified interior of solar-like stars (e.g. Press 1981, Appourchaux 2010). Moreover, they are of great interest in the study of angular momentum transport over large distances and constitute an important ingredient in understanding the evolution of stellar rotation (Zahn et al. 1997; Charbonnel & Talon, 2005; Mathis 2009).

IGW propagate in the radiative zone under the condition $\omega < N$ where ω is the frequency of the waves and N the Brunt-Väisälä frequency, given by

$$N^2 = -g\left(\frac{1}{\bar{\rho}}\frac{\partial\bar{\rho}}{\partial r} - \frac{1}{\Gamma_1\bar{p}}\frac{\partial\bar{p}}{\partial r}\right), \quad (1)$$

with g the gravity, Γ_1 the adiabatic exponent, $\bar{\rho}$ and $\bar{p}$ the mean density, and pressure. The region where IGW propagate is visible in Figure 1 (left panel) where we have represented the path of propagation of two g-modes, using the linear dispersion relation (Christensen-Dalsgaard, 2003):

$$k_r^2 = \frac{\ell(\ell+1)}{r^2}\left(\frac{N^2}{\omega^2} - 1\right), \quad (2)$$

where ℓ is the spherical harmonic degree. Oscillations are trapped in the radiative zone between concentric spheres of radius r_0 such as $\omega = N(r_0)$. They are thus difficult to observe at the surface of solar-like stars as they are evanescent in convective envelope. However, Garcia et al.

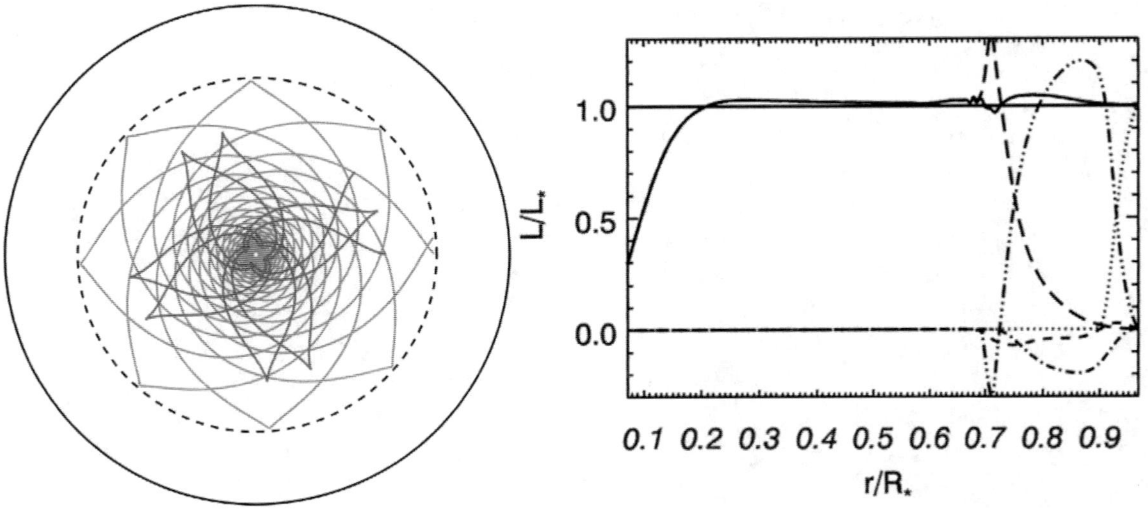

Figure 1. Left: Raytracing of 2 gravity modes with frequency of at 0.1 mHz (blue line) and 0.3 mHz (red line) in a stratified solar-like radiative interior. Right: Flux balance realized in a 3-D ASH simulation of most of the interior dynamics of the Sun (Brun et al. 2011). Radial fluxes have been converted to luminosity and normalized to the solar luminosity. We plot respectively the total (solid line), enthalpy (dash triple dots), radiative (long dash), unresolved (dot), viscous (dash), and kinetic (dash dot) luminosities.

(2007, 2008) have reported the detection at a 8.6 sigma level of the asymptotic period spacing of $\ell = 1$ solar gravity modes in Golf/SoHO data. By doing the power spectrum of the periodogram of 11.5 yr of Golf/SoHO data, they have been able to show an excess of power at 24-25 min characteristic of IGW. One is thus lead to ask: how is a turbulent convective envelope filtering the IGW signal likely to be present in the solar radiative interior? Along with theoretical work (Gough 1985, Kumar et al. 1996, Belkacem et al. 2009), a direct comparison with numerical simulation could help us further characterizing such waves. This was recently done in 2-D by Rogers & Glatzmaier in 2005 and later work. We here propose to expand to 3-D their pioneering work.

2. 3-D model of the Sun

In order to study the generation of internal (possibly gravity) waves by convective motions in solar-like star, we present new results obtained with the anelastic spherical harmonic (ASH) code (Brun et al, 2004). Our model nonlinearly couples a rotating convective envelope to a stable radiative interior (Brun et al, 2011), assuming a realistic solar stratification from $r = 0.07R_\odot$ up to $0.97R_\odot$ (Brun et al. 2002). On Figure 1 (right panel) we show the radial energy balance, with the radiative luminosity (respectively the convective luminosity) carrying most of the energy for $r < 0.7R = r_{bcz}$ (resp. $r > 0.7R$). We note that the convective (enthalpy) flux is negative at the base of the convection zone $r = r_{bcz}$. This is due to the penetration of convective plumes into the top of the stable radiative zone. Such continuous pummeling generates internal waves that carry energy and angular momentum inward. A 3-D rendering of such waves is shown on Figure 2 (left panel), where we plot the radial velocity normalized to the local (at each radii) rms velocity. We note the strikingly different patterns observed in the convective envelope with respect to that realized in the deep stable interior. The convective flows, possess a large irregular structure, with down flow lanes delineating broad up flows. By contrast in the radiative interior we see almost concentric wave fronts of alternative signs.

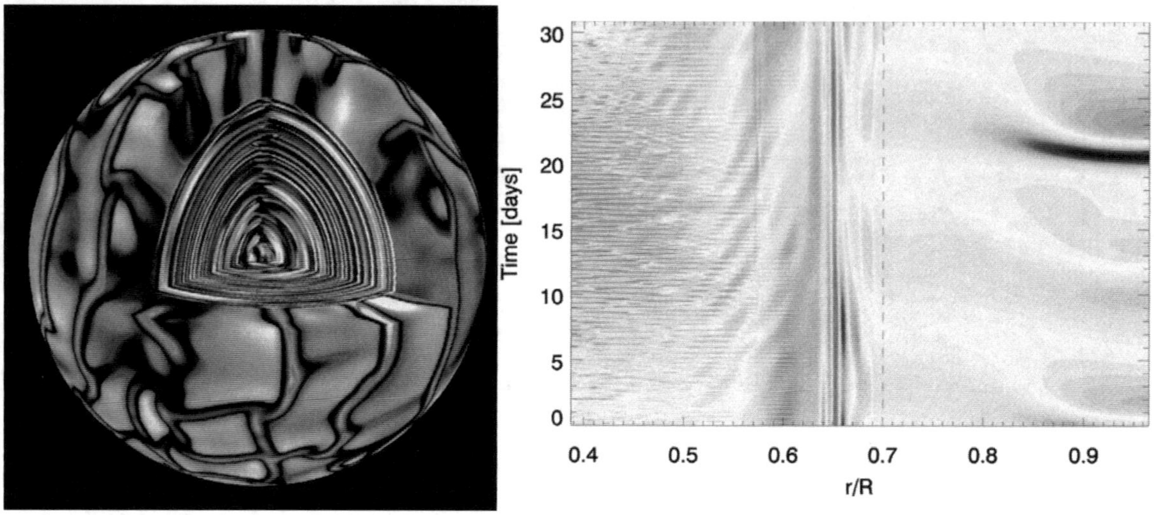

Figure 2. Left: 3-D rendering of the model, showing both the convective envelope and the wave patterns realized in the deep radiative interior of our simulations (Brun et al. 2011). Right: Time-radius plot showing a cut at fixed longitude in the equatorial plane of the normalized velocity with greenish tones indicating negative radial velocity and yellowish ones positive radial velocity.

In order to confirm that such wave-like patterns correspond to IGW, we plot on Figure 2 (right panel) the normalized radial velocity as a function of radius and time for a fixed longitude in the equatorial plane. We see that near the base of the convection zone, three to four almost vertical modulations are present, likely due to the dynamical response of the top of the radiative interior to the continuous pummeling of downflow lanes. These are apparent in the convective envelope as tilted blue lanes that appear at the surface, propagate downward and hit the base of the convection zone. Their numbers (almost 4) over the 30 days covered by the figure, is a consequence of the Reynolds number and rotation rate used in the simulation. Indeed only about 12 of those downflow lanes coexist at any given time over the 360 degree of longitude of the model (6 of these downflow sheets aligned with the rotation axis can be seen on the left panel of Figure 2). A more turbulent model would posses a larger number and more complex downflow lanes (Miesch et al. 2008). The forward tilt of the plumes is due to the presence of a large scale differential rotation which near the equator is faster near the base of the convection zone than at the surface. Hence as the plumes sink at low latitudes, they are advected forward by the radial shear. At larger latitude, the tilt is in the opposite direction. Deeper in the radiative interior we note inclined wave fronts going from the lower left to the top right of the figure. This is an indication that the wavefront is moving outward. IGW are known to have an opposite sign between their vertical group and phase velocities. Since the energy is carried downward, such fronts correspond to phase velocity going outward, and hence are compatible with IGW.

To further analyze the properties of the waves excited in the radiative interior of our model, we have computed the energy power spectra at various depths and show on Figure 3 the spectra realized at $r = 0.4R$. A series of ridges is clearly apparent in the $[0.1, 0.5]\mu Hz$ range. Their characteristic dispersion relation is clearly compatible with gravity modes, with all the ridges converging to a maximum corresponding to the Brunt-Väisälä (BV) frequency of that specific depth (we recall that BV frequency varies as a function of radius in the model). The analysis of the period spacing of the modes as in Alvan et al. (2012) and Figure 4 (middle panel) below is already a good indication that such waves are gravity waves since this spacing is constant.

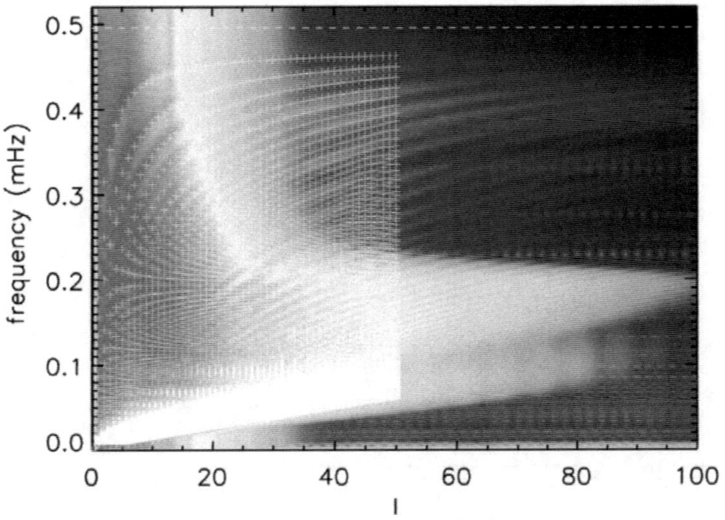

Figure 3. Color contours: Power spectrum as a function of the spherical harmonic degree ℓ of gravity mode frequencies at a depth of $r = 0.4R$ in the 3-D nonlinear ASH model of the Sun. Superimposed with white crosses the gravity mode frequencies as a function of ℓ computed with the *Adipls* code (Christensen-Dalsgaard 2011). The Brunt-Väsälä frequency at that depth is also shown (horizontal white dash line). We clearly see the ridges characteristic of IGW.

Indeed it is well known that the period spacing for gravity modes varies for a given ℓ as (e.g. Provost & Berthomieu 1986):

$$\Delta P_\ell \simeq \frac{\pi}{\sqrt{\ell(\ell+1)}\int_{r_1}^{r_2} N/r dr}. \tag{3}$$

To go beyond the straightforward analysis of the wave spectra done in Alvan et al. (2012), we are superimposing on the spectrum obtained with ASH in Figure 3, the frequencies computed with the *Adipls* code (Christensen-Dalsgaard 2011) using the background reference state structure used in the 3-D model. Each white cross corresponds to individual gravity modes, and they clearly match very well the ridge-pattern seen in the power spectra. We are thus confident that we have excited successfully for the first time in a 3-D spherical geometry IGW by the direct nonlinear action of convective plumes.

Zooming on one of the low degree ℓ mode, we plot on Figure 4 (left panels) the power spectra as a function of frequency and period. To further improve the analysis done in Alvan et al. (2012) and to be able to compare with observations as reported in Garcia et al. (2007) we have taken Fourier transforms of the periodogram (middle panel of Figure 4). By doing so we find the following values for respectively $\Delta P_{1,2,3} \sim 44, 27, 19$ min, as is evident for the latter on the bottom panel of Figure 4. Garcia et al. (2007) reported a value of about 24-25 min for ΔP_1 with a complex double peak structure of the signal and anticipated a $\Delta P_3 \sim 11$ min, hence about 40% smaller than our value of 19 min.

The difference between the observed value and the computed one is certainly due to our truncated model, with the inner domain stoping at $r = 0.07R$. On Figure 4 (right panel), we plot the ratio N/r as a function of radius. The solid line corresponds to the part we model whereas the dash-dot line emphasizes the missing part. Whereas we just omit 10% of the resonant cavity for the IGW, the integrant used in formula 3 is actually modified by 37%, explaining the larger period spacing found in our model compared to observations. While it

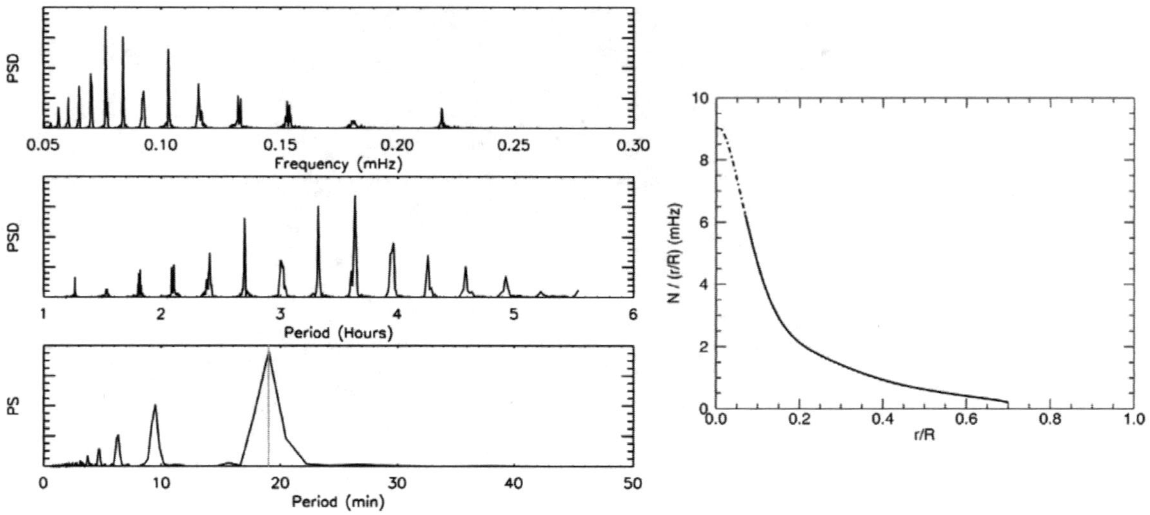

Figure 4. Left: $\ell = 3$ mode power at $r = 0.4R$ shown as a function of frequency (top) or period (middle); Power spectrum of the power spectrum for $\ell = 3$ modes (bottom), with a clear peak around 19 min. Right: truncated gravity-wave cavity (Brunt-Väisälä frequency) due to the inner sphere of radius $0.07R$. This truncation is at the origin of the observed frequency/period shift.

is very encouraging to see that we can apply similar techniques to our simulations and to real observations, we have yet to obtain such signal in the convective envelope as was reported in Garcia et al. (2007). Indeed Fig 4 is computed in the radiative zone of our 3-D simulation and so the wave's amplitude is higher than at the top of the domain as they do not suffer the damping due to their evanescent character in convective envelope. This clearly demonstrates the interest of doing these computations that model internal waves self-consistently and allow us to track their signal at all depths simulated. Looking at the eigenfunctions of the gravity modes excited in our simulations, we note that the high frequency, low degree ℓ modes are the most likely to be visible at the surface (Kumar et al. 1996). We thus intend to generate longer 3-D time series than the 100 days currently available to be able to boost the signature of IGW near the surface by improving the signal to noise ratio which will help us disentangling them from convection signal by integrating over many overturning times. We will also look for the most likely modes and study their stochastic variability and life time.

3. Conclusions

We have demonstrated that we can now model in 3-D the excitation and propagation of internal gravity waves by turbulent convection and study their properties and nonlinear interactions. When comparing the power spectra obtained in our 3-D model with adiabatic computations performed with the *Adipls* code (Christensen-Dalsgaard 2011) we find an excellent overall agreement, provide we take into account the truncated radius of our reference background state. In order to make direct comparison with observation and guide the detection of IGW in the Sun, we must build a 3-D dynamical solar model going all the way down to the center and introduce a radial dependence of the specific heat at constant pressure which is assumed constant for the time being. Further Brown et al. (2012) have demonstrated that all anelastic formulations of the MHD equations are not equivalent when modeling highly stratified radiative interior (such as the solar one) and IGW. Recent developments with the ASH code will allow us to improve our models by using a better anelastic description and by setting the inner boundary to $r = 0$.

Understanding the influence of an inner magnetic field such as the one stored in the tachocline on IGW is also of great interest (Strugarek et al. 2011; Mathis & de Brye 2012). Finally having long time series integrated over many convection overturning times will allow us to study the temporal variability of the mode excitation, to boost their surface signature by applying data analysis developed for the real Sun and to guide our knowledge of the wave dynamics by knowing their characteristic in the inner radiative interior.

Acknowledgments We are thankful to J. Christensen-Dalsgaard for letting us use the latest version of the *Adipls* code. We acknowledge useful discussions with B. Brown regarding the description of gravity waves using various anelastic prescriptions, with B. Hindman on helios-esimology subtleties and with J. Toomre, J.-P. Zahn and M.S. Miesch for long term scientific exchanges on solar dynamics. We acknowledge funding by ERC STARS2 grant (www.stars2.eu) and support by CNES for the Golf/SoHO instrument and Solar Orbiter related activities at CEA.

References

[1] Alvan, L., Brun, A.S., Mathis, S. 2012, proceedings of SF2A 2012, eds. Boissier et al., 289
[2] Appourchaux et al. 2010, A&ARv, 18, 197
[3] Belkacem, K., Samadi, R., Goupil, M. J., et al. 2009, A&A, 494, 191
[4] Brown, B.P., Vasil, G.M. & Zweibel, E.G. 2012, ApJ, 756, 109
[5] Brun, A. S., Antia, H. M., Chitre, S. M., & Zahn, J.-P. 2002, A&A, 391, 725
[6] Brun, A.S., Miesch, M.S., Toomre, J. 2004, ApJ, 614, 1043
[7] Brun, A.S., Miesch, M.S., Toomre, J. 2011, ApJ, 742, 79
[8] Charbonnel, C. & Talon, S. 2005, Science, 309, 2189
[9] Christensen-Dalsgaard, J. 2003, *Lecture Notes on Stellar Oscillations*
[10] Christensen-Dalsgaard, J. 2011, *ADIPLS: Aarhus Adiabatic Oscillation Package (ADIPACK)*
[11] Garcia, R.A. et al. 2007, Science, 316, 1591
[12] Garcia, R.A. et al. 2008, AN, 239, 476
[13] Gough, D. 1985, In: future missions on solar, heliospheric and space plasma physics, ESA SP-235, 183
[14] Kumar, P. Quartaert, E. & Bahcall, J.N. 1997, ApJ, 458, L83
[15] Mathis, S. 2009, A&A, 506, 811
[16] Mathis, S. & de Brye, N. 2012, A&A, 540, 37
[17] Miesch, M.S., Brun, A.S., DeRosa, M.L., & Toomre, J. 2008, ApJ, 673, 557
[18] Press, W. H. 1981, ApJ, 245, 286
[19] Provost, J. & Berthomieu, G. 1986, A&A, 165, 218
[20] Rogers, T.M., & Glatzmaier, G.A. 2005, MNRAS, 364, 1135
[21] Strugarek, A., Brun, A.S., Zahn, J.P. 2011, A&A, 532, 34
[22] Zahn, J.P., Talon, S. & Matias, J. 1997, A&A, 322, 320

Journal of Physics: Conference Series **440** (2013) 012044 doi:10.1088/1742-6596/440/1/012044

Statistics of Local Seismic Emission from the Solar Granulation

Charles Lindsey

NorthWest Research Associates, 3380 Mitchell Lane, Boulder, Colorado 80301, USA

E-mail: clindsey@cora.nwra.com

Alina-Catalina Donea

Center for Astrophysics, School of Mathematical Science, Monash University, Victoria 3800, Australia

Abstract. We apply computational seismic holography to high-frequency helioseismic observations of the quiet Sun from *SDO*/HMI to locate predominant sources of seismic emission with respect to the structure of the solar granulation. The regions of greatest seismic emission are the edges of photospheric granules. Seismic emission from regions whose continuum brightnesses are 95-100% of the mean, as resolved by HMI, are about 2.5 times as emissive as regions whose brightnesses are 100-104% of the mean. The greater seismic emissivity from regions whose brightnesses are somewhat less than the mean is roughly in line with expectations from an understanding that attributes most seismic emission to cool plumes plummeting from the edges of granules. However, seismic emission from regions whose continuum brightnesses significantly exceed 104% of the mean is also remarkably high. This unexpected feature of high-frequency seismic emission from the solar granulation begs to be understood.

1. Introduction

The discovery of the 300-second oscillations by Leighton, Noyes & Simon [8] and by Evans & Michard [7], and the recognition [11] that these are a manifestation of acoustic waves traveling through the solar interior opened a new era in solar research in the 20th century. Since then, our understanding of how these waves are generated has undergone a long evolution. Today, it is generally accepted that the solar oscillations, now popularly expressed as "p-modes", are driven mostly by convection in the Sun's outer convective envelope, hence, closely associated with the solar granulation [15]. Observational evidence of this includes the recognition of conspicuous episodes of outgoing ripples emanating from apparent source regions that tend to be located in intergranular lanes [16].

Since the late 1990s we have developed computational seismic holography for diagnostics of phenomena similar to seismic emission from the quiet Sun but more anomalous, most particularly to transient seismic emission from flares [3, 4] and to acoustic-emission halos surrounding active regions [5]. Recent applications of this by Alvarado Goméz et al. [1] to helioseismic observations by *SDO*/HMI of transient seismic emission from flares suggest that seismic holography in the high-frequency p-mode spectrum can resolve the source distributions of seismic emission

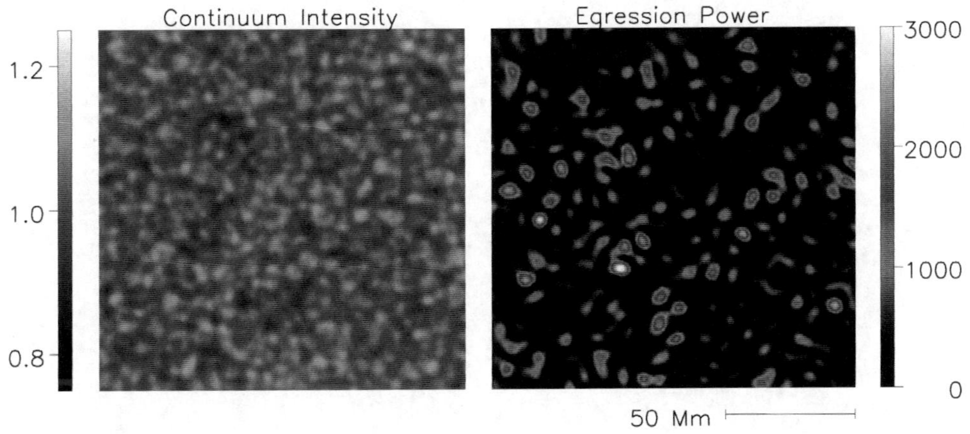

Figure 1. Cospatial, cotemporal samples of continuum intensity in the neighborhood of FeI λ6701 Å, as resolved by *SDO*/HMI (left), and egression power in the 5.5–6.5-mHz seismic spectrum (right).

sufficiently to locate it statistically with respect to the structure of the solar granulation.[1] This is the object of this study.

2. Acoustic-Power Holography of the Solar Granulation

2.1. Cospatial, Cotemporal Maps of Seismic Source-Power and Continuum Brigtness

This study shows preliminary results of a comparison between helioseismic source-power maps determined by seismic holography in the 6-mHz spectrum and the structure of the granulation as discriminated in terms of brightness. We apply "subjacent-vantage helioseismic holography" [10] to Doppler observations of the quiet Sun in the 5.5–6.5 mHz spectrum over a square region, $\mathcal{R}$, centered upon the Sun whose area, A, is 0.11 $R_\odot^2$ over a period, T, of 24 hr. For details on the general technique of seismic holography, we refer to [10]. For details on the application of the technique to episodic phenomena, we refer to [3].

Figure 1 shows cospatial maps of continuum intensity (left) and "egression power" (right), the holographic representation of seismic source-power density, over a small sub-region of $\mathcal{R}$. The egression-power (right) has a spatial resolution of 2.5 Mm, a diffraction limit attained by applying a computational "pupil" (see Fig 4, p. 269 of [1]) whose radial range was 7–28 Mm. This is greater than the $\sim$1.4-Mm cell size of a single granule [2]. However, as the analysis that follows will confirm, the availability of $\sim 5 \times 10^6$ granules appearing in $\mathcal{R}$ over T gives us the statistical leaverage us to discriminate mean relative locations of helioseismic signatures to within a small fraction of the diffraction limit.

Subjective statistics obtained by flickering between cospatial pairs of comparative maps such as those shown in Figure 1 show that kernels representing the largest egression powers tend to be centered near the edges of granules, i.e., in regions of intermediate brightness between the centers of granules and the centers of intergranular lanes. These subjective impressions are reinforced by cross-correlation statistics we will now describe.

[1] In fact, the solar photosphere is an efficient reflector of low-frequency p-modes [6, 9]. So, apparent sources of low-frequency ($\sim$3 mHz) emission from the neighborhood of any focus are mostly local reflections of p-mode energy generated elsewhere. At high frequencies (e.g., 6 mHz), the photosphere appears to be an efficient absorber of p-modes [6, 9]. So, seismic source signatures at high frequencies are a much stronger representation of acoustic power generated at or near the focus of the diagnostic.

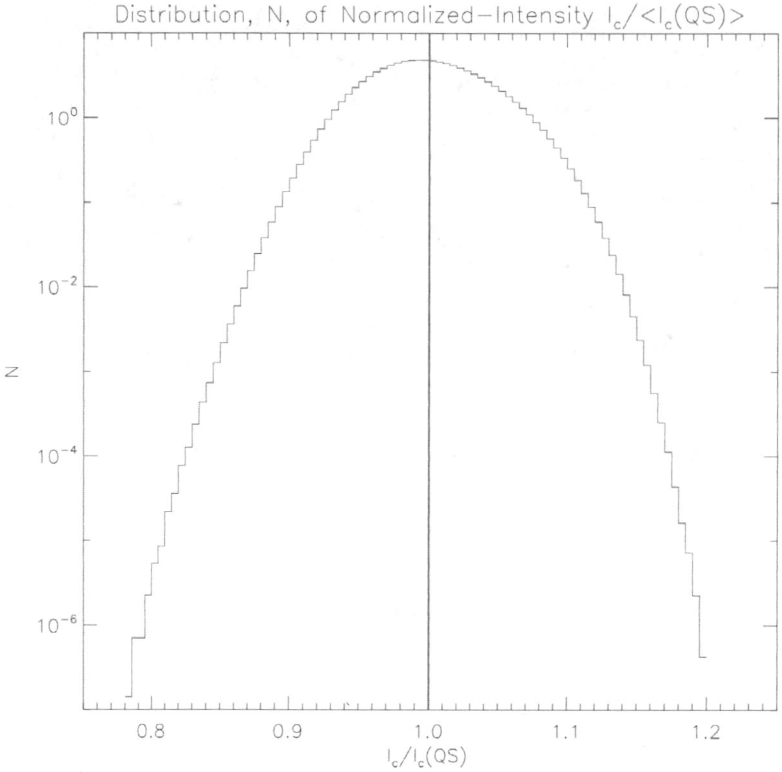

Figure 2. Histogram of the distribution of continuum intensity of the quiet Sun, as resolved by SDO/HMI, in the neighbohood of FeI λ6701 Å, normalized to unity for the mean quiet Sun.

2.2. Brightness and Seismic Source-Power Statistics

Figure 2 shows a histogram of the continuum brightness of the quiet Sun normalized to unity over $\mathcal{R}$ for the mean quiet Sun, over the period $T = 24$ hr. This histogram is partitioned into intervals of 0.005 units in quiet-Sun normalized intensity. Figure 3 plots the egression power averaged over the same intervals, over the range 0.85–1.15. This profile shows a strong statistical dependence of the mean egression power upon the local continuum brightness of the solar granulation. To discriminate regions of different seismic emissivity, we recognize three "continuum-brightness regions", I, II and II, delimited by vertical red lines in Figure 3. The strongest emitters appear to be the edges of granules, at which the continuum intensity is a few percent less than the mean, i.e., brightness region I. The weakest emission comes from regions a few percent brighter than the mean, i.e., brightness region II.

In quantifying these variations in terms of realistic seismic emmissivity, it has to be borne in mind that a significant part of the egression-power signature represents noise emitted in the pupil of the helioseismic computations, not just from the focus. For this analysis, we suppose that the contribution of the noise from the pupil is equal to the contribution from the focus. This can be roughly expected if the acoustic power emitted downward from a local source, into the solar interior to re-emerge into a surface pupil some distance away, is matched by that emitted upward, however unwelcomely, directly into a pupil whose focus is some distant location. Under this supposition, the acoustic power emanating from brigtness region I in Figure 3 is about 2.5 times that emanating in brightness region II.

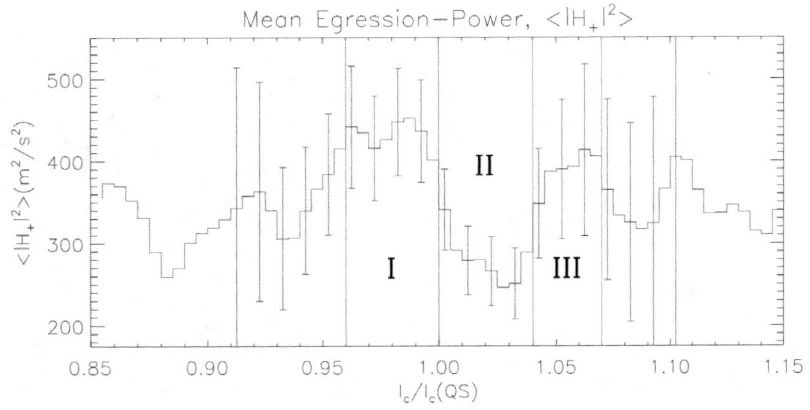

Figure 3. Egression power in the 5.5–6.5-mHz spectrum averaged over 0.005-unit intervals in normalized continuum brightness, as resolved by HMI, of regions from which seismic emission thereby represented emanates is plotted over the 0.85–1.15 range of normalized continuum brightness. Error bars indicate the standard statistical deviation of the egression power in respective 0.005-unit intervals of normalized continuum brightness.

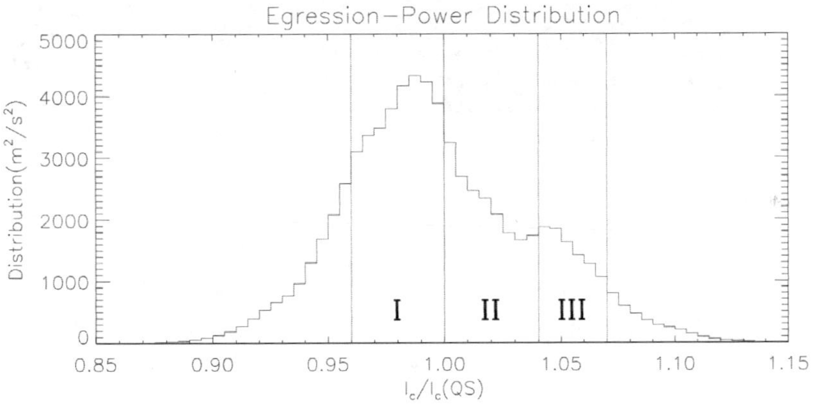

Figure 4. Statistical distribution of 5.5–6.5-mHz seismic source power is plotted over the brightness range 0.85–1.15 in normalized continuum intensity as resolved by HMI. This profile is scaled so that its integral over the unitless abscissa is the mean egression power.

The foregoing dependence may very well be expected based upon models that attribute seismic emission largely to down-flowing plumes [12] at the edges of granules. This might lead us to expect a maximum mean egression power at brightnesses that are intermediate between the bright granular interiors and the dark integranular lanes [14]. The mean egression power would then supposedly decrease from this relative maximum to reduced levels for brightnesses significantly greater, or less, than that of the maximum. This appears to be the case proceeding from left to right in Figure 3 through regions I and II, the maximum occuring in region I and dropping precipitously in region II. However, there appears to be a brightness region III, 4–7% brighter than the mean, of relatively strong seismic emission. The existence and placement of this feature brightward of the least emissive region, then, is not anticipated in the context of

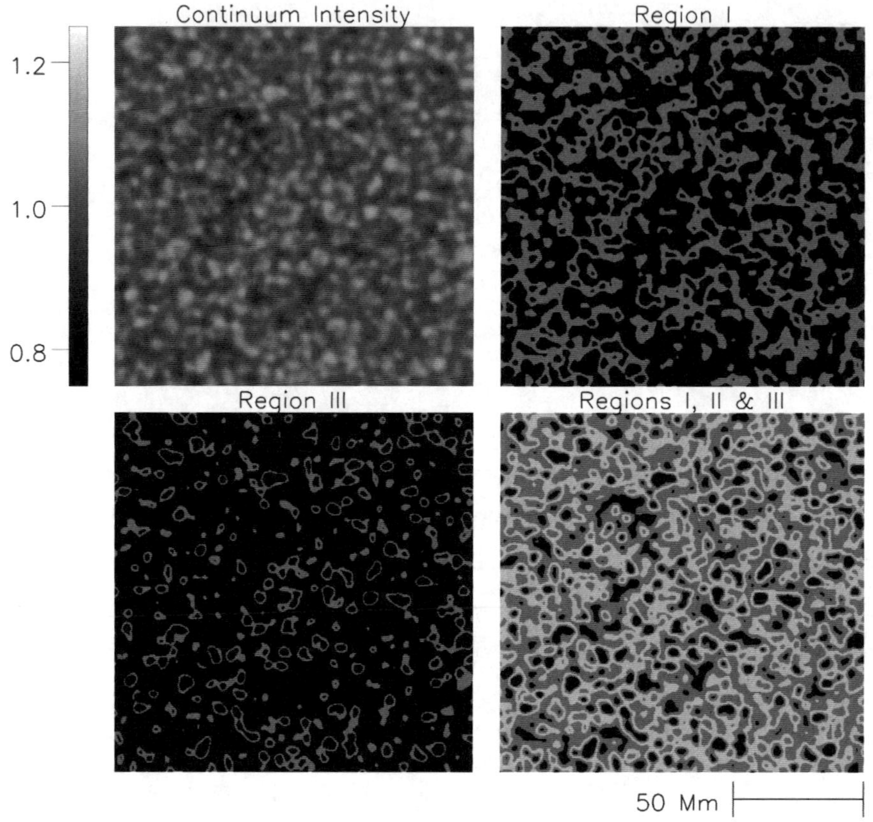

Figure 5. Regions whose continuum brightnesses (upper left) are in ranges in which 6-mHz egression power is statistically enhanced are mapped in the upper right (region I, 0–5% dimmer than the mean) and in the lower left (region III, 4–7% brighter than the mean) frames. Lower-right frame shows a composite map that renders in green the intervening range (0–4% brighter than the mean), in which acoustic emissivity is relatively suppressed.

the rough understanding formulated above.

Noting rapidly inflating error in Figure 3 for intensities more than 8% greater or less than unity, we forebear to comment on other interesting features of the profile plotted in Figure 3, except perhaps to note that, because of a rapidly decreasing distribution in the brightness histogram at brightnesses outside of 1.0 ± 0.08, the amount of seismic source power represented by these is not very large, while the statistical weight of the mean egression power diminishes, as indicated by the inflating error bars in Figure 3. The distribution in source power is shown by Figure 4. This profile is the product (see caption) of the profiles plotted in Figures 2 and 3.

3. Representation of Regions of Enhanced Seismic Source Power

Figure 4 identifies the brightness regions of anomalously high seismic emissivity in the upper right and lower left frames. The upper-left frame reproduces the continuum brightness rendered in the left frame of Figure 1. The upper-right frame renders region I as identified in Figures 3 and 4 in red, i.e., the brightness region of maximum mean egression power in Figure 3, whose continuum brightness is less than the mean. The lower left frame renders region III in light brown, i.e., the also-highly-acoustically-emissive region that is brighter than the mean. The

Journal of Physics: Conference Series **440** (2013) 012044 doi:10.1088/1742-6596/440/1/012044

lower-right frame shows a composite map that renders in green the intervening region, i.e., region II, whose brightness is greater than the mean but whose acoustic emissivity is relatively suppressed.

4. Discussion and Conclusions

The strong dependence of seismic source-power density upon continuum brightness as related to the structure of the solar granulation strongly supports the solar granulation as a major source of the solar oscillations. This dependence can be broadly summarized as follows:

(i) There is a significant correlation between the locations of the strongest episodes of high-frequency seismic source power and the edges of solar granules.

(ii) Statistically, the most seismically emissive regions are the edges of granules, where the continuum brightness is 0.95–0.99 times the quiet-Sun mean.

(iii) The least seismically emissive regions statistically are those whose continuum brightness is in the range 1.00–1.05 times the quiet-Sun mean.

(iv) However, regions in which the continuum brightness is 1.05–1.10 times the quiet-Sun mean, while less in area than the other two, are conspicuously emissive.

(v) The foregoing points are consistent with enhanced seismic emission in the neighborhoods of edges of photospheric granules. However, the significance of strong seismic emission from regions brighter than those from which the weakest seismic emission emanates, is somewhat of a mystery at this moment.

The last point should be moderated by the admission that the expectations challenged by this finding are based upon a relatively crude and highly qualitative understanding of how seismic emission from the solar granulation works. However, this result appears to invite some attention to aspects of p-mode generation that would lead to such a relationship. Some insight into this behavior might by acquired by applying the diagnostics developed in this study to simulations of the solar granulation such as those described by Skartlien, Stein & Nordlund [13].

4.1. Acknowledgements

We have have benefited deeply from the keen insights of Bob Stein and Mark Rast. This work has been supported by a contract from the Solar Heliospheric Physics program of the National Aeronautics and Space Administration.

5. References

[1] Alvarado Goméz J D, Buitrago Casas J C, Martíinez Oliveros J C, Lindsey C, Hudson H, Calvo-Mozo B 2012 *Solar Phys.* **280** 335
[2] Bray R J, Loughhead R E 1977 *Solar Phys.* **54** 319
[3] Donea A-C, Braun D C, Lindsey, C 1999 *Ap. J. Lett.* **513** 143
[4] Donea A-C, Lindsey C 2005 *Ap. J.* **630** 1168
[5] Donea, A-C Lindsey C, Braun D C 2000 *Solar Phys.* **192** 321
[6] Duvall T L Jr, Harvey J W, Pomerantz M A 1993 *Nature* **362** 430
[7] Evans J W, Michard R, Servajean R 1962 *Ap. J.* **136** 682
[8] Leighton R B, Noyes R W, Simon G W 1962 *Ap. J.* **135** 474
[9] Lindsey C, Braun D C 1999 *Ap. J.* **510** 494
[10] Lindsey C, Braun D C 2000 *Solar Phys.* **192** 261
[11] Noyes R W, Leighton R B 1963 *Ap. J.* **138** 631
[12] Rast, M. P. 1999 *Ap. J.* **524** 462
[13] Skartlien R, Stein, R F, Nordlund Å 2000 *Ap. J.* **541** 468
[14] Stein R F 2012 private communication
[15] Stein R F, Nordlund Å 1988 *Ap. J.* **499** 914
[16] Strous L H, Goode P R, and Rimmele T R 2000 *Ap. J.* **535** 1000

Journal of Physics: Conference Series **440** (2013) 012045 doi:10.1088/1742-6596/440/1/012045

Acoustic emission from magnetic flux tubes in the solar network

G Vigeesh[1] and S S Hasan[2]

[1] Department of Astronomy, New Mexico State University, Las Cruces, NM, U.S.A.
[2] Indian Institute of Astrophysics, Koramangala, Bangalore, India

E-mail: vigeesh@nmsu.edu, hasan@iiap.res.in

Abstract. We present the results of three-dimensional numerical simulations to investigate the excitation of waves in the magnetic network of the Sun due to footpoint motions of a magnetic flux tube. We consider motions that typically mimic granular buffeting and vortex flows and implement them as driving motions at the base of the flux tube. The driving motions generates various MHD modes within the flux tube and acoustic waves in the ambient medium. The response of the upper atmosphere to the underlying photospheric motion and the role of the flux tube in channeling the waves is investigated. We compute the acoustic energy flux in the various wave modes across different boundary layers defined by the plasma and magnetic field parameters and examine the observational implications for chromospheric and coronal heating.

1. Introduction

Observations of the solar surface in various filtergrams show a distribution of bright points, which are generally associated with an underlying concentrated magnetic field located in intergranular lanes [1]. Magnetic flux accumulates here after being swept in by granular flow forming strong vertical flux tubes at the junction of granules. The continually changing solar photosphere perturbs these magnetic structures resulting in the generation of magnetohydrodynamic (MHD) waves that propagate within the flux tube and also into the ambient plasma. The waves are excited by impulses imparted by granules. Eventually, the flux tubes are advected towards regions of stronger downdrafts present at the intersections of two or more granules where convectively driven vortex flows occur. The footpoints of the magnetic field structures are dragged in the vortex resulting in, presumably, a different source of wave excitation in the flux tube [2]. The conditions that prevail in the lower atmosphere impacts the upper layers, primarily through magnetic fields. Estimating the energy supplied by various photospheric disturbance will help us to determine the fraction of the overall heating of the upper atmosphere by wave sources. In order to examine energy transport to the outer layers of the quiet solar atmosphere mediated by magnetic fields [3–6], we need to clearly understand how the magnetic flux tube reacts to different types of perturbations. Despite being idealized representations of the actual processes, studies have shown that various wave-driven energy transport mechanisms are possible in a magnetized atmosphere [7–12]. Further investigations could reveal how these processes contribute to the overall energy budget and shed light on the significance of wave related sources. The purpose of this paper is to study two such scenarios with the aim of evaluating them in terms of their contribution to the heating of chromosphere and corona.

In this paper, we investigate the excitation of waves in magnetic flux tubes extending vertically through the solar atmosphere. This work is an extension of our previous study [12] that dealt with 3D modeling of MHD wave propagation in a magnetic flux tube embedded in a stratified atmosphere. We showed that there are possibly more than one mechanism of wave production effective which can lead to temporal variations of the emission that occur at chromospheric heights above these elements. In this paper our focus is to investigate the generations of acoustic waves excited by these perturbations and estimate the acoustic energy flux at different levels in the flux-tube. We calculate and compare the acoustic emission from various boundaries in the flux-tube and also from different β (the ratio of gas to magnetic pressure) surfaces as a result of the excitation.

2. Equilibrium Model and Boundary Conditions

Details of the equilibrium model that we consider in this study are given in [12]. Briefly, we consider an intense (kG field strength at the base) axially symmetric magnetic flux tube in a stratified solar model atmosphere that is in magnetostatic equilibrium. The temperature in the model increases with height to include the chromospheric temperature rise. The footpoint of the flux tube is located in the solar photosphere. The mathematical construction of the flux tube is described in the Appendix of [12] and the equilibrium properties of the flux tube are given in Table 1. The plasma-$\beta = 1$ surface essentially outlines the flux tube boundary in the lower part of the tube. In these layers, $\beta < 1$ inside the flux tube as shown in Fig 1.

Table 1. The equilibrium model parameters on the axis of the flux tube and ambient medium (values shown within brackets).

Height	T (K)	ρ (kg m^{-3})	P (N m^{-2})	c_S (km s^{-1})	v_A (km s^{-1})	B (G)	β
z=1 Mm	7263 (7195)	3.4×10^{-8} (1.0×10^{-7})	1.6 (4.8)	8.8 (8.7)	77.7 (39.2)	161 (141)	64 (16)
z=0 Mm	4768 (4766)	1.3×10^{-4} (4.0×10^{-4})	4.2×10^{3} (1.2×10^{4})	7.1 (7.1)	10.9 (0.003)	1435 (0.77)	1.9 (1.9×10^{-7})

Wave excitation is carried out by implementing velocity drivers at the bottom boundary. We use a horizontal driver to mimic the granular buffeting motion and a torsional driver to mimic the effect of vortex-like motion. The driving motions at the bottom boundary are specified using the following velocity drivers.

$$V_x(x,y,0,t) = \begin{cases} V_0 \sin(2\pi t/P) & \text{for} \quad 0 \le t \le P/2\,, \\ 0 & \text{for} \quad 0 > t > P/2\,. \end{cases} \quad \text{(Horizontal)} \tag{1}$$

$$V_\phi(x,y,0,t) = \begin{cases} -V_0 \tanh\left(\frac{2\pi r}{\delta r}\right) \sin\left(\frac{2\pi t}{P}\right) & 0 \le t \le P/2, \\ 0 & t > P/2. \end{cases} \quad \text{(Torsional)} \tag{2}$$

These driving motions generate slow (predominantly acoustic) and fast (predominantly magnetic) waves in the model along with the intermediate Alfvén wave which we do not consider in this study.

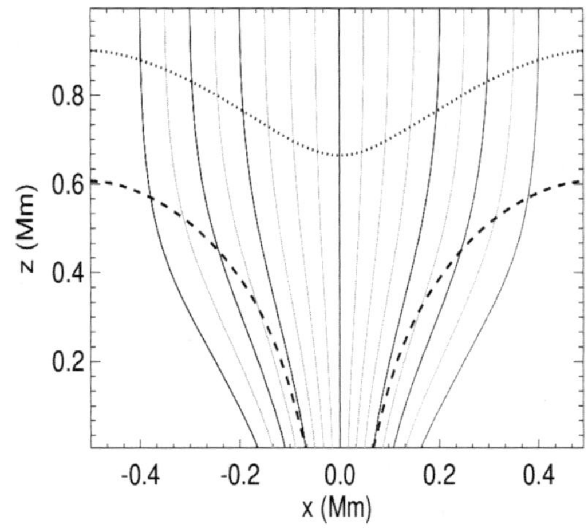

Figure 1. Two-dimensional representation of the flux-tube model. The solid lines mark the magnetic field lines on a x-z plane at $y = 0$ Mm. The $\beta = 1$ (- - - -) and $\beta = 0.1$ ($\cdots\cdots$) contours are also shown.

3. Numerical Simulation

The three-dimensional numerical simulations were carried out using the Sheffield Advanced Code [13]. The code uses a modified version of the set of MHD equations to deal with a strongly stratified magnetized medium. In this study, we solve the full set of ideal magnetohydrodynamic equations in three dimensions to study the propagation of waves in the computational domain. The domain is a 1 Mm $\times$ 1 Mm $\times$ 1 Mm cube discretized on 100 $\times$ 100 $\times$ 100 grid points. We use transmitting boundary conditions at the top and side boundaries to allow the waves to propagate out of the simulation box.

The simulation starts with a localized, time dependent perturbation at the bottom boundary resulting in the excitation of various kinds of MHD modes with different strength. The gas pressure perturbations drive slow magneto-acoustic wave (SMAW) within the β <1 region and fast magneto-acoustic wave (FMAW) in the ambient medium where β >1. The magneto-acoustic wave propagation can be seen in the velocity and temperature perturbation in the medium as shown in Fig 2. We notice that the horizontal excitation generates strong SMAW within the flux tube as can be seen in the temperature fluctuations. Since the SMAW propagates along field lines, we also see strong velocity parallel to the field lines. The torsional excitation on the other hand generates SMAW that are weaker than those generated by the horizontal driver. In the previous study [12], we estimated the energy transport by MHD waves for the two driving cases. The energy fluxes were calculated on a representative field line and we showed that there is a strong acoustic flux associated with SMAW in the case of the horizontal excitation, which is two orders of magnitude more than that for the case of torsional excitation. However, these studies were restricted to a single field line and hence it was not possible to assess the response of the whole flux tube to the different perturbations that were considered. In this paper, we extend the previous study by looking at the flux tube as a whole and estimating the total acoustic emission from various physical surfaces relevant to the model. Due to the nature of the drivers, the analysis on a single field line depends wholly on the location of the footpoint of the chosen field line with respect to the driving motion. Especially, in the case of a horizontal motion, where the strongest perturbations is localized to a small region on either side of the $y = 0$ plane. Also, since the velocity driver acts in region where $\beta < 1$ (close to the axis) as well as $\beta > 1$ (outer regions), a field line located in either of these region sees a different mode of magneto-acoustic wave. Considering an ensemble of field lines will give a better idea about the reaction of the flux tube as a whole to the excitation at the base.

Journal of Physics: Conference Series **440** (2013) 012045 doi:10.1088/1742-6596/440/1/012045

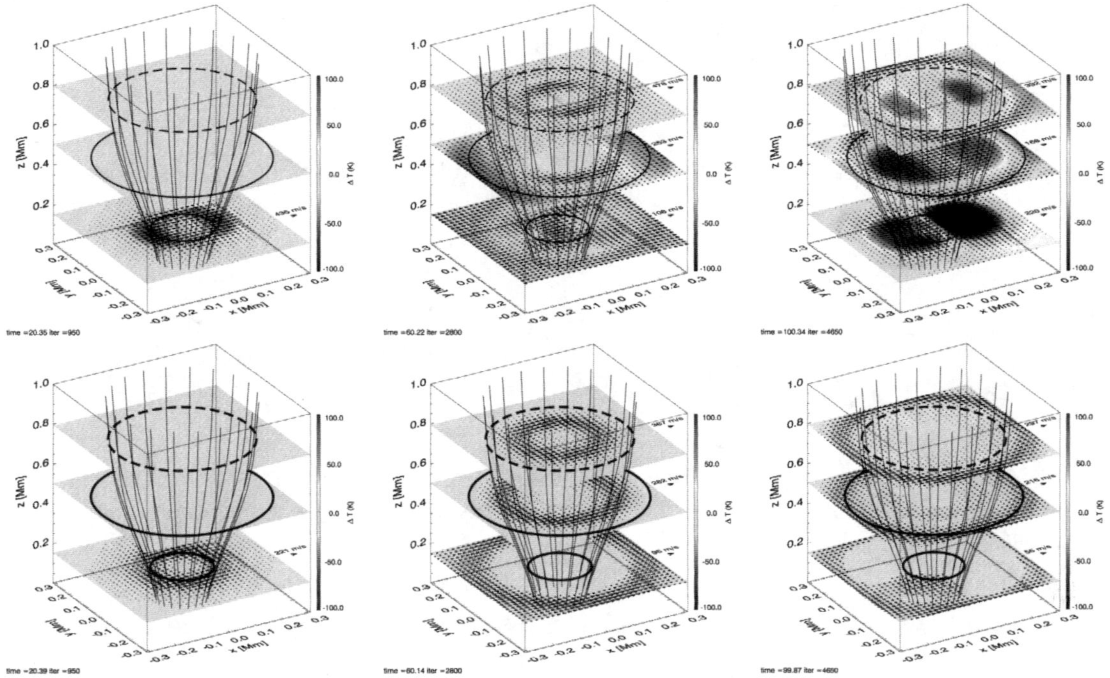

Figure 2. *Top:* Temperature perturbations at t=20 s,60 s, and 100 s at different heights as a result of a transversal uni-directional excitation. *Bottom*: Temperature fluctuations as a result of torsional excitation. The projected velocity vectors on the x-y plane are shown at each height. The thick blue curve depicts the β=1 region and the blue dashed curve shows the β=0.1 region.

4. Results

4.1. Longitudinal & Lateral emission

Apart from acting as a conduit for MHD waves to the overlying layers, the perturbed flux tube also transfers energy to the ambient medium in the form of acoustic waves. It is interesting to look at acoustic energy flux leaking out of the flux tube boundary and how it depends on the nature of the excitation that the flux tube undergoes. Since we consider here a thick flux tube, a strict definition of the flux tube boundary is ambiguous. However, for the purpose of this study, we choose the surfaces of equal magnetic potential or the magnetic isosurface as representative levels in the flux tube for purposes of calculating the fluxes. The cross-section of the magnetic isosurface at any given height is a circle centered at the axis, due to the cylindrical symmetry of the initial model. This allows us to define a circle at a radial distance r of the flux tube on any horizontal plane. We calculate the acoustic flux on three magnetic isosurfaces crossing radial distance of $r = 0.2$ Mm, 0.3 Mm and 0.4 Mm from the axis of the flux tube at $z = 1$ Mm. The magnetic field lines that define this isosurface can be identified from the equilibrium model and they remain the same throughout the simulation, allowing us to calculate physical quantities on this surface as it evolves. The velocity vector at any point on this surface can be decomposed into three orthogonal components, *viz.* parallel (v_s), normal (v_n) and azimuthal (v_ϕ) component. The calculation of these component for a single field line is described in [12]. Using these velocity components for a collection of field lines defining various isosurfaces, we calculate the parallel and normal acoustic fluxes according to,

$$F_{\rm s,n} = \Delta p v_{\rm s,n} \tag{3}$$

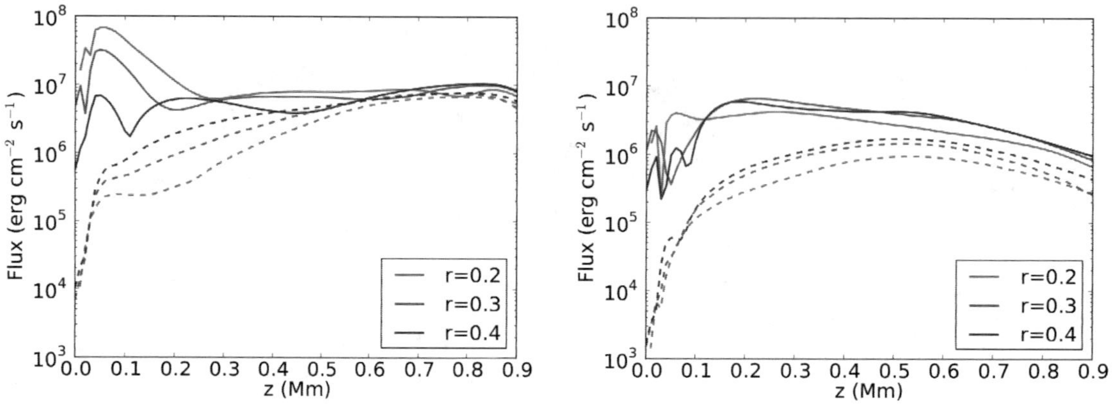

Figure 3. *Left*: Time-averaged longitudinal acoustic flux on the three isosurfaces for the horizontal (Solid lines) and torsional(dashed lines) excitation cases. *Right* Normal acoustic flux on the three isosurfaces.

where Δp is the gas pressure perturbation from the equilibrium. Acoustic fluxes are calculated on field lines corresponding to three magnetic isosurfaces. In Figure 3, the left panel shows the time averaged parallel acoustic fluxes (F_s) on field lines associated with different isosurfaces as a function of height. The horizontal excitation results in relatively stronger acoustic emission compared to torsional excitation in the lower part of the flux tube. But, the averaged fluxes tend to be similar as we go higher up in the atmosphere, since in the horizontal excitation case, only a small fraction of field lines on either side of the tube partake in transporting the longitudinal fluxes. The right panel of Fig. 3 shows the time averaged acoustic flux directed normal to the flux-tube boundaries, which is significantly lower than the fluxes directed along the field lines in the tube.

4.2. β-surface emission

The surface of equipartition (β=1 layer) between thermal and magnetic energy density influences the MHD modes by acting as region of wave transmission and conversion. As a preliminary step towards understanding this region more closely, we look at the acoustic energy flux crossing normal to the surfaces of constant plasma β surface (β isosurfaces). In this study, we mainly focus on the $\beta = 1$ isosurface, since this is where the different MHD modes undergo strong coupling which eventually leads to the generation of Alfvén waves in the upper chromosphere. The whole process proceeds in two stages. Initially, the magneto-acoustic modes encounter the equipartition zone as they propagate to the upper layers of the atmosphere. The energy in the modes get redistributed partially by transmission and partially by conversion to other modes. The transmission and conversion depends on the properties of the wave and the background magnetic field across this zone of influence [14]. In the second stage, the mode converted magnetic wave (FMAW) propagates up to a certain height after which it gets partially reflected down and partially gets converted to Alfvén waves due to steep gradients in the Alfvén speed [15]. A better understanding of the energy distribution during the initial stage where mode coupling occurs across the $\beta = 1$ surface is nessasary to further evaluate the energy available for Alfvén wave. To this end, we look at the acoustic flux directed normal to the surface of $\beta = 1$ and compare it with that from a $\beta = 0.1$ isosorface which lies well above in the atmosphere. It should be noted that in the lower part of the flux tube, the $\beta = 1$ surface normal points towards the axis of the flux-tube.

Journal of Physics: Conference Series **440** (2013) 012045 doi:10.1088/1742-6596/440/1/012045

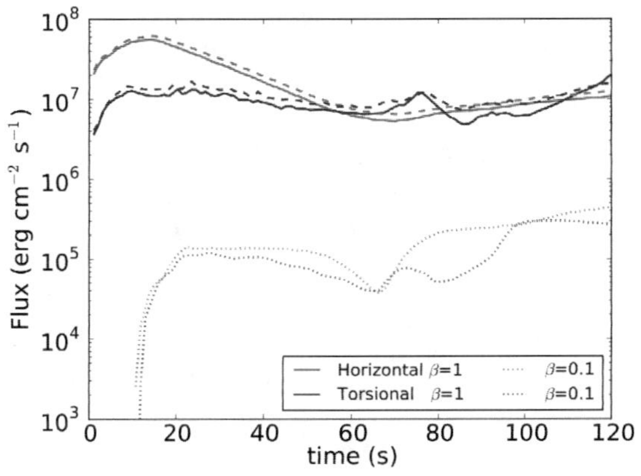

Figure 4. Time evolution of the normal acoustic flux on β=1 (solid line) and β=0.1 (dotted) surface for horizontal (red line) and torsional (blue) excitation of a magnetic flux tube

Figure 4 shows the spatially averaged acoustic flux directed normal to the surfaces of constant β for β=1 and 0.1 as a function of time. We clearly see that the acoustic emission from the β=1 surface is stronger compared to the emission on the β=0.1 surface for both horizontal and torsional excitations. The β=1 surface responds more efficiently to a horizontal uni-directional motion at the foot-point by emitting more acoustic flux normal to the surface. Our analysis is limited by the fact that we consider a strong flux tube where the $\beta = 1$ surface dips below the bottom boundary near the axis of the flux tube. To have a better understanding about this layer, we need to look at flux tubes models with $\beta = 1$ surfaces at different levels in the atmosphere.

5. Summary & Conclusions

Three dimensional numerical simulation of wave propagation in a magnetic flux tube embedded in a solar atmosphere were carried out. We investigated two types of excitation mechanism, viz. transversal and torsional, that are characteristic processes by which waves can be generated in the real solar atmosphere. We calculated the acoustic energy components at various levels in the flux tube as well as on the surfaces defined by constant plasma β of the flux tube due to the driving footpoint motions. We observe that the acoustic power is predominantly directed vertically upward along the flux tube in both cases. The lateral acoustic emission from the boundary of the flux tube in both cases of horizontal and torsional excitations is much lower. The magnetic flux tube acts as an efficient conduit for acoustic waves, with negligible acoustic leakage from the tube boundary. Most of the acoustic energy produced due to photospheric disturbances are efficiently transported and are made available to higher layers regardless of the source of the disturbance. As far as the comparison between the two excitation scenarios in terms of acoustic energy transport to upper layers is concerned, we have not been able to conclusively point out which of the two mechanisms is more efficient. This is partly due to the axisymmetric equilibrium model that we have used and the idealistic driving mechanisms that we consider in the study. Nevertheless, our analysis supports that the magnetic field mediates the coupling between various photospheric disturbances and the upper layers. However meagre their contribution to the overall energy output is, these scenarios must be considered when evaluating the available energy sources for the heating of chromosphere and corona. The surface of equipartition between magnetic and thermal energy density ($\beta = 1$) is a strong source of acoustic emission in both excitation scenarios and would be the focus of future investigations.

References

[1] de Wijn A G, Stenflo J O, Solanki S K and Tsuneta S 2009 *Space Sci. Rev.* **144** 275–315
[2] Jess D B, Mathioudakis M, Erdélyi R, Crockett P J, Keenan F P and Christian D J 2009 *Science* **323** 1582–1585
[3] Hasan S S and van Ballegooijen A A 2008 *Astrophys. J.* **680** 1542–1552
[4] van Ballegooijen A A, Asgari-Targhi M, Cranmer S R and DeLuca E E 2011 *Astrophys. J.* **736**
[5] Kato Y, Steiner O, Steffen M and Suematsu Y 2011 *Astrophys. J. Lett.* **730** L24
[6] Wedemeyer-Böhm S, Scullion E, Steiner O, Rouppe van der Voort L, de La Cruz Rodriguez J, Fedun V and Erdélyi R 2012 *Nature* **486** 505–508
[7] Hasan S S, van Ballegooijen A A, Kalkofen W and Steiner O 2005 *Astrophys. J.* **631** 1270–1280
[8] Khomenko E, Collados M and Felipe T 2008 *Sol. Phys.* **251** 589–611
[9] Vigeesh G, Hasan S S and Steiner O 2009 *Astron. Astrophys.* **508** 951–962
[10] Kitiashvili I N, Kosovichev A G, Mansour N N and Wray A A 2011 *Astrophys. J. Lett.* **727**
[11] Fedun V, Verth G, Jess D B and Erdélyi R 2011 *Astrophys. J. Lett.* **740** L46
[12] Vigeesh G, Fedun V, Hasan S S and Erdélyi R 2012 *Astrophys. J.* **755** 18
[13] Shelyag S, Fedun V and Erdélyi R 2008 *Astron. Astrophys.* **486** 655–662
[14] Cally P S 2007 *Astron. Nachr.* **328** 286
[15] Khomenko E and Cally P S 2012 *Astrophys. J.* **746** 68

On the Seismicity of September 7, 2011 X1.8-class Flare

S. Zharkov[1,2], **L.M. Green**[2], **S.A. Matthews**[2], **V.V. Zharkova**[3]
[1] Department of Physics and Mathematics, University of Hull, Cottingham Road, Kingston-upon-Hull, HU6 7RX, UK
[2] Mullard Space Science Laboratory, University College London, Holmbury St. Mary, Dorking, RH5 6NT, UK
[3] Department of Mathematics, University of Bradford, Bradford, BD7 1DP, UK

E-mail: s.zharkob@hull.ac.uk

Abstract. We present results of our preliminary analysis of acoustically active X-class flare of September 7, 2011. We report two acoustic sources detected via acoustic holography and verified by finding a ridge in time-distance diagrams. We compare the directional information extracted from time-distance and acoustic holography, showing a good agreement in this case. We report that the direction where amplitude of the wave-front is the largest lies through the strong magnetic field and sunspot, suggesting that absorption of the acoustic wave power by magnetic field can be ruled out as a wave anisotropy mechanism in this case.

1. Introduction

Although theoretically predicted [1] flare driven sun-quakes have only been discovered relatively recently [2] using SOHO/MDI data. Detection and analysis of sunquakes in the previous solar cycle 23 [3, 4, 5, 6, and others] has been considerably restricted by the availability of suitable observations, see recent review [7] for more details. Recent launch of SDO/HMI [8] combined with the approaching maximum of the current Solar Cycle 24 now ensures that we can observe this phenomenon in new unprecedented detail due to better data coverage as well as improved spatial and temporal resolution.

Sunquakes, seen as near circular ripples, propagating outward from impulsive solar flares along the solar surface, appear 20 - 60 minutes after the flare onsets. Strong anisotropy of the surface wave-field has often been reported [7, 4, 9]. This has normally been attributed to the suppression and absorption of acoustic waves by magnetic field, which has been extensively studied theoretically and modelled [10]. It has also been observed that, in sunquakes, the largest amplitudes of the generated surface wavefront are in the direction of the apparent moving source.

2. Data and Method

GOES Soft X-ray flare list reports the X1.8 class flare taking place in NOAA Active Region 11283 on September 7, 2011, starting at 22:32 UT and peaking at 22:38 UT. The flare was accompanied by a CME as reported by SOHO/LASCO and CaCTUS CME event catalogues.

We use full disk SDO/HMI intensity, dopplergram and line-of-sight magnetic field data at 45-second cadence to produce three hour-long datacubes, which are extracted by remapping

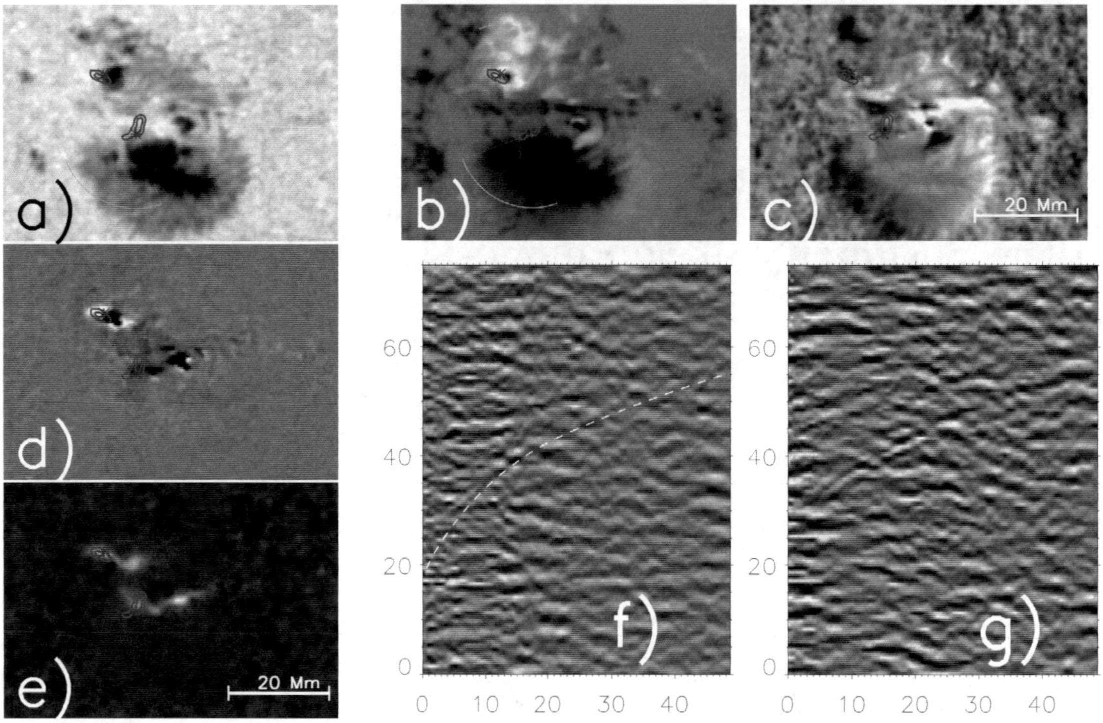

Figure 1. Context images taken at 22:37:22 UT: a) HMI continuum intensity saturated (black) at 30% of quiet sun intensity, b) line-of-sight magnetogram, scaled between ±1 kG and c) line-of-sight doppler velocity on the scale between $\pm 1.5\,\mathrm{km\,s^{-1}}$. The running difference images taken around 22:36-22:37 UT are shown in d) for HMI magnetogram (scaled between ±200 Gauss) and e) continuum showing variation between −4% and 20% of quiet sun intensity. The red contours represent the peak 6 mHz egression power at 2 and 3 factor the quiet Sun emission ($\approx 288\,\mathrm{m^2\,s^{-2}}$). Blue stars indicate the locations of found time-distance ridges, while orange arcs on images at the top row indicate the direction of the detected time-distance ridges. The FOV and scale are the same across a)-e). The time-distance diagrams computed for the two sources are presented in panels f) for the northern Source 1 with theoretical time-distance curve overplotted in white and g) for the southern Source 2. In time-distance plots, distance in Mm is plotted along x-axis, time in minutes since 22:20:07 UT is plotted along y-axis.

and derotating the region of interest using Postel projection and the Snodgrass differential rotation rate. The spatial resolution of the remapped data is 0.04 degrees per pixel. We apply acoustic holography to calculate the egression power maps from observations [11, 12, 13] along with computing time-distance diagram [2, 4, 9, 14]. The egression power is computed for each integral frequency from 3 to 10 mHz, by applying 1-mHz frequency bandwidth filters to the data and using Green functions built for surface monochromatic point source of corresponding frequency using geometrical optics approach. The choice of 1 mHz, $\Delta\nu$, bandwidth implies timing uncertainty of $\Delta t = 1/\Delta\nu = 1/1000$ seconds [3]. The pupil size is set from 10 to 45 Mm. The basic processing follows [15, 16].

Time–distance diagrams are computed by selecting a source location, rewriting the observed surface velocity signal $[v]$ in polar coordinates relative to the source $[v(r, \theta, t)]$ and then using

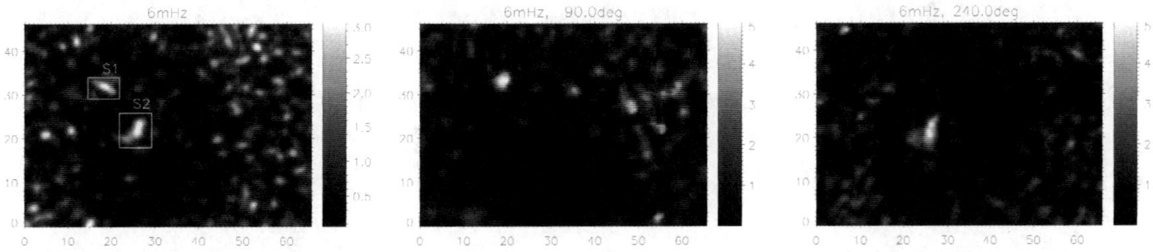

Figure 2. Egression power snapshot measured at 6mHz taken around 22:35:07 UT is presented on the left. The integration regions used in deriving statistics for Figure 3 are presented as white boxes encompassing each source. This is followed by two directional egression power snapshots with θ_0 measured from x-axis counterclockwise given in the title, and $\delta\theta = \pi/2$. See Section 2 for details. All axes are in Mm, field of view is the same as in context images presented in Figure 1.

the azimuthal transformation

$$V_m(r,t) = \int_{\alpha}^{\beta} v(r,\theta,t)\mathrm{e}^{-\mathrm{i}m\theta}\mathrm{d}\theta, \tag{1}$$

where integration is normally performed over the whole circle (i.e. $\alpha = 0$, $\beta = 2\pi$). Here, we use $m = 0$, and the integration limits are chosen to represent an arc where the signal is best observed. Acoustic wave-packets of a sufficiently strong amplitude relative to the noise are seen as a time–distance ridge that largely follows a theoretical time–distance curve.

The holography method works by using Green's function $[G_+(|\mathbf{r}-\mathbf{r}'|, t-t'),]$ which prescribes the acoustic wave propagation from a point source, to essentially "backtrack" the observed surface signal $[\psi(\mathbf{r},t)]$. This allows us to reconstruct egression images showing the subsurface acoustic sources and sinks:

$$H_+(\mathbf{r},z,t) = \int \mathrm{d}t' \int_{r'=a}^{r'=b} \int_{\theta=0}^{\theta=2\pi} G_+(r',t-t')\psi(r',\theta,t')\mathrm{d}^2\mathbf{r}', \tag{2}$$

where r' and θ are the polar coordinates describing the vector $|\mathbf{r}-\mathbf{r}'|$ at the solar surface. Thus, a, b define the dimensions of the holographic pupil. Integration over θ normally takes place over the whole range $[0, 2\pi]$, while here, in order to quantify and compare the helioseismic directional information, we use the interval of the form $[\theta_0 - \delta\theta, \theta_0 + \delta\theta,]$ varying θ_0 and taking $\delta\theta = \pi/2$.

The holographic method is generally more sensitive, but vulnerable to noise, variations from the assumed model and is susceptible to an increased possibility of false detections, which is normally handled by analysing the statistical significance of the detected signal [3, 17].

3. Results

The context images showing the NOAA 11283 sunspots and magnetic configuration are presented Figure 1. Egression power snapshot computed at 6mHz is presented in left panel of Figure 2. It shows two clear and compact acoustic sources, northern Source 1 (S1) and southern Source 2 (S2). We have found directional time-distance ridges for the two sources which are presented at the bottom right of Figure 1 with pointing information shown as orange arcs on top row HMI images. No ridges were found in other directions. To verify the observed acoustic signatures we carry out the rms analysis of the egression power, which is presented in Figure 3, showing statistical significance of the measured seismic signal, particularly at 5-6mHz range,

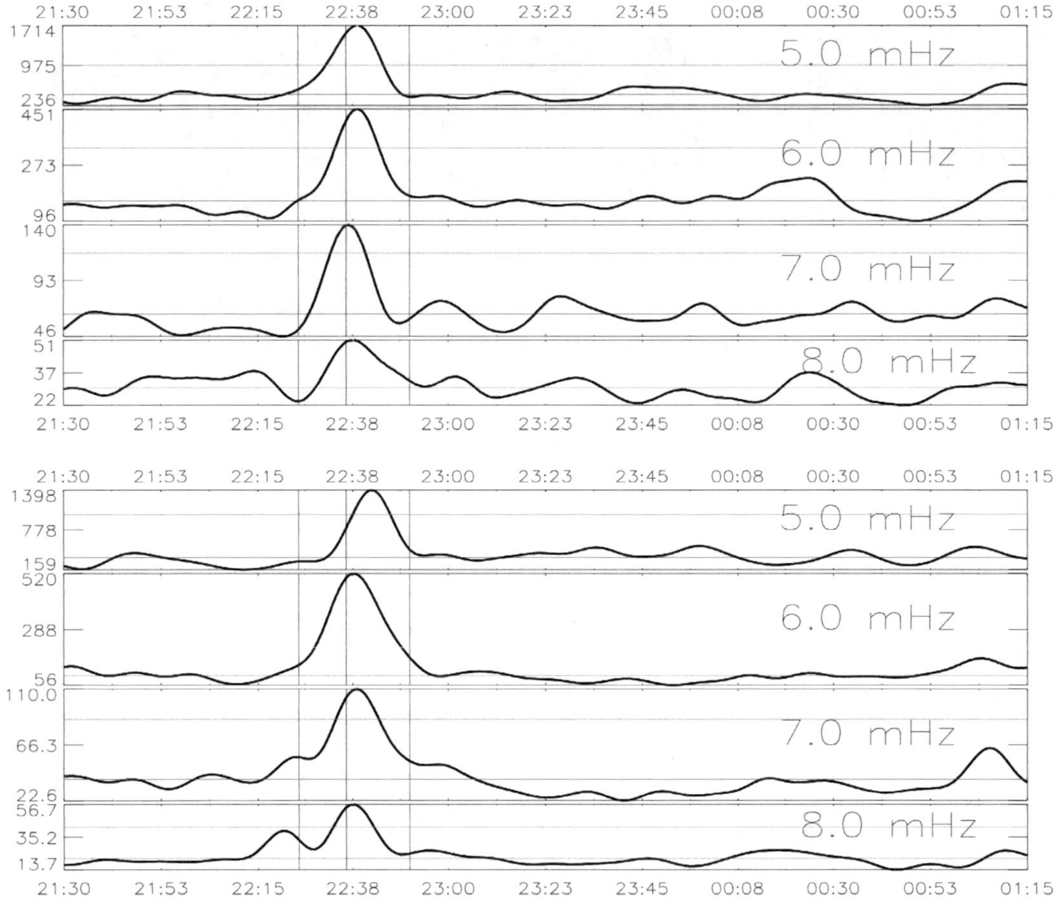

Figure 3. Egression rms measured at different frequencies for Source 1 *(top)* and Source 2 *(bottom)* as indicated in Figure 2. Time is along the x-axis, averaged egression power in $m^2\,s^{-2}$ is plotted along y-axis. Blue vertical lines represent GOES flare start and end times, red horizontal lines show the mean integrated rms measured away from the flare peak, with yellow horizontal line representing the mean plus 5σ variation significance level.

for both sources. Directional holography results are presented in two rightmost panels of Figure 2, showing a clear separation of the sources depending on the direction. The angles are in good general agreement with time-distance measurements, apparently supporting the derived preferred direction of surface wave-front propagation. Interestingly, as can be seen from Figure 1, for Source 2 this direction includes the strongly magnetised plasma of the main sunspot, suggesting that in terms of wave-front anisotropy a mechanism different from magnetic field suppression of acoustic wave is at play.

Further work is needed in understanding directional results, combined with analysis of RHESSI and SDO AIA observations to get better understanding of the event.

References

[1] Wolff C L 1972 *Astrophys. J.* **176** 833–842
[2] Kosovichev A G and Zharkova V V 1998 *Nature* **393** 317–318

[3] Donea A and Lindsey C 2005 *Astrophys. J.* **630** 1168–1183
[4] Kosovichev A G 2006 *Solar Phys.* **238** 1–11
[5] Moradi H, Donea A, Lindsey C, Besliu-Ionescu D and Cally P S 2007 *Mon. Not. Roy. Astron. Soc.* **374** 1155–1163
[6] Martínez-Oliveros J C, Moradi H and Donea A 2008 *Solar Phys.* **251** 613–626
[7] Donea A 2011 *Space Sci. Rev.* **158** 451–469
[8] Scherrer P H, Schou J, Bush R I, Kosovichev A G, Bogart R S, Hoeksema J T, Liu Y, Duvall T L, Zhao J, Title A M, Schrijver C J, Tarbell T D and Tomczyk S 2012 *Solar Phys.* **275** 207–227
[9] Kosovichev A G 2007 *Astrophys. J. Lett.* **670** L65–L68
[10] Moradi H, Baldner C, Birch A C, Braun D C, Cameron R H, Duvall T L, Gizon L, Haber D, Hanasoge S M, Hindman B W, Jackiewicz J, Khomenko E, Komm R, Rajaguru P, Rempel M, Roth M, Schlichenmaier R, Schunker H, Spruit H C, Strassmeier K G, Thompson M J and Zharkov S 2010 *Solar Phys.* **267** 1–62
[11] Braun D C and Lindsey C 1999 *Astrophys. J. Lett.* **513** L79–L82
[12] Donea A, Braun D C and Lindsey C 1999 *Astrophys. J. Lett.* **513** L143–L146
[13] Lindsey C and Braun D C 2000 *Solar Phys.* **192** 261–284
[14] Zharkova V V and Zharkov S I 2007 *Astrophys. J.* **664** 573–585
[15] Zharkov S, Green L M, Matthews S A and Zharkova V V 2011 *Astrophys. J. Lett.* **741** L35
[16] Zharkov S, Green L M, Matthews S A and Zharkova V V 2012 *Solar Phys.* 292
[17] Matthews S A, Zharkov S and Zharkova V V 2011 *Astrophys. J.* **739** 71

Journal of Physics: Conference Series **440** (2013) 012047 doi:10.1088/1742-6596/440/1/012047

Modelling Fast-Alfvén Mode Conversion Using SPARC

H Moradi, P S Cally

Monash Centre for Astrophysics, School of Mathematical Sciences, Monash University, Victoria 3800, Australia

E-mail: hamed.moradi@monash.edu

Abstract. We successfully utilise the SPARC code to model fast-Alfvén mode conversion in the region $c_A \gg c_S$ via 3-D MHD numerical simulations of helioseismic waves within constant inclined magnetic field configurations. This was achieved only after empirically modifying the background density and gravitational stratifications in the upper layers of our computational box, as opposed to imposing a traditional Lorentz Force limiter, to ensure a manageable timestep. We found that the latter approach inhibits the fast-Alfvén mode conversion process by severely damping the magnetic flux above the surface.

1. Introduction

A series of recent studies (e.g., [1–8]) have shown that an important feature not widely accounted for in sunspot seismology is fast-Alfvén wave mode conversion - a process which occurs at, and beyond, the fast wave reflection height (where $c_A \approx \omega/k_h$; c_A denotes the Alfvén wave speed, ω the wave frequency and k_h the horizontal wavenumber) - being spread over many scale heights for wavenumbers typical of local helioseismology. This process is most efficient for θ (field inclination from vertical) between $30° - 40°$, and ϕ (angle between the magnetic field and wave propagation planes) between $60° - 80°$, and appears to have the potential to modify the seismic wave-path through the solar atmosphere, thereby affecting the wave travel times that are the basis of our inferences about the subsurface (see [9–12] for recent reviews). Motivated by these studies, our aim is to use the Seismic Propagation through Active Regions and Convection (SPARC) code, a 3-D magnetohydrodynamic (MHD) wave-propagation code developed by [13] for computational heliosiesmology, to numerically simulate this process, and investigate the implications of fast-Alfvén wave mode conversion on the seismology of the photosphere. In this paper we describe our attempts at forward modelling this process with SPARC using a quiet-Sun background model permeated by homogenous inclined magnetic fields.

2. Numerical Setup

The SPARC code solves the 3-D linearized Euler and induction equations of magnetofluid motion in Cartesian geometries to investigate wave interactions with local perturbations (e.g., sound speed, pressure, density, flows, magnetic field etc.). Over the past few years, a number of various solar phenomena have been studied using this code (e.g., [14–18]). The computational box we employ for our simulations using SPARC spans 186.6 Mm in the horizontal direction (128 evenly-spaced grid points in the horizontal directions x and y; $\Delta x = \Delta y = 1.46$ Mm/pixel), and from

2.5 Mm above the surface ($z = 0$; where z denotes height in Mm) to 25 Mm below in the vertical direction (300 non-uniformly spaced grid points in z; Δz varies from several hundred kilometres at depth, to tens of kilometres in the near-surface layers). The vertical boundaries of the box are absorbent, with perfectly matched layer (PML) boundary layers spanning the top 10 and bottom 7 grid points in z, while periodic boundary conditions are imposed on the horizontal sides. In a similar manner to [5–7], we use a monochromatic ($\nu = \omega/2\pi = 5$ mHz) plane-parallel wave driver, which only excites waves propagating in the (x, z) planes. We do this by imposing a perturbation of the form:

$$\sin(\omega t)\ \mathrm{e}^{-(x-x_0)^2/2\delta_x^2}\ \mathrm{e}^{-(z-z_0)^2/2\delta_z^2} \tag{1}$$

in a few grid points near $z_0 = -5$ Mm and centred at around $x_0 = 0$ Mm, in pressure, density and velocity (x and z components only).

3. Model Atmosphere

The background model used is a convectively stabilised solar model (CSM_B) from [19]. On top of this background model we employ a constant, inclined magnetic field configuration using the prescription from [1]:

$$\mathbf{B}_0 = B_0(\sin\theta\cos\phi, \sin\theta\sin\phi, \cos\phi), \tag{2}$$

We choose $B_0 = 1500$ G and a number of different angle orientations in the range $0 < \theta < 90°$ and $0 < \phi < 90°$.

In 3-D MHD simulations, the timestep ($\Delta t \sim \Delta z/c_A$) is often highly constrained by the Courant-Friedrichs-Lewy (CFL) condition, due to the exponentially increasing value of $c_A = \mathbf{B}_0/(\mu_0\rho)^{1/2}$ (where ρ denotes density and μ_0 the magnetic permeability) above the surface. This causes the wavelengths of both the fast and Alfvén waves to become quite large, resulting in an extremely stiff numerical problem. The most common way of dealing with this issue, in computational helioseismology, has been to simply apply a c_A "limiter" to moderate the action of the Lorentz Force when the ratio c_A/c_S (where c_S denotes sound speed) becomes exceedingly large. Some choices for these limiters have been discussed before (e.g., [15; 20–23]). Generally, they tend to prefix the Lorentz-Force terms in the momentum equations, with c_A typically being capped at $\sim 20 - 60$ km/s. The physical implications of artificially limiting the Lorentz Force in such a manner (particularly on the seismology) have not been explored. Recently though [24] has shown that significant internal reflection of Alfvén waves can occur if the c_A profile is not adequately treated above the surface.

Our method of ensuring a reasonable Δt for our calculations involves empirically modifying the background density ($\rho_0(z)$) in the upper layers ($\sim 0.5 < z < 2.5$ Mm) of CSM_B, in conjunction with a commensurate modification of the gravitational acceleration ($g_0(z)$) profile over the same z range (in order to offset the change in the density scale height), to obtain a maximum global fast (c_F) and c_A of ≈ 80 km/s. All other background variables remain unaltered. The modified profiles of $\rho_0(z)$ and $g_0(z)$ are shown in Figure 1. The value of 80 km/s was chosen because it provides us with a reasonable time step ($\Delta t = 0.5$ s), and is safely higher than the largest horizontal phase-speeds (ω/k_h) typically sampled in sunspot seismology (e.g., [25]). This is important since $\omega/k_h \approx c_A$ also denotes the location of the fast mode reflection height in the solar atmosphere [6]. Figure 2 a) shows the resulting c_A and c_F profiles in our model as a function of height. For comparison purposes, the c_A and c_F profiles which would result from imposing a Lorentz Force limiter, instead of modifying $\rho_0(z)$ and $g_0(z)$ above the surface, are shown in Figure 2 b).

The obvious downside of empirically modifying $\rho_0(z)$ and $g_0(z)$ in order to satisfy the CFL condition is that background model will no longer be as 'solar-like' (i.e., in terms of eigenfrequencies, eigenfunctions and power spectrum) as CSM_B. The larger $\rho_0(z)$ profile which

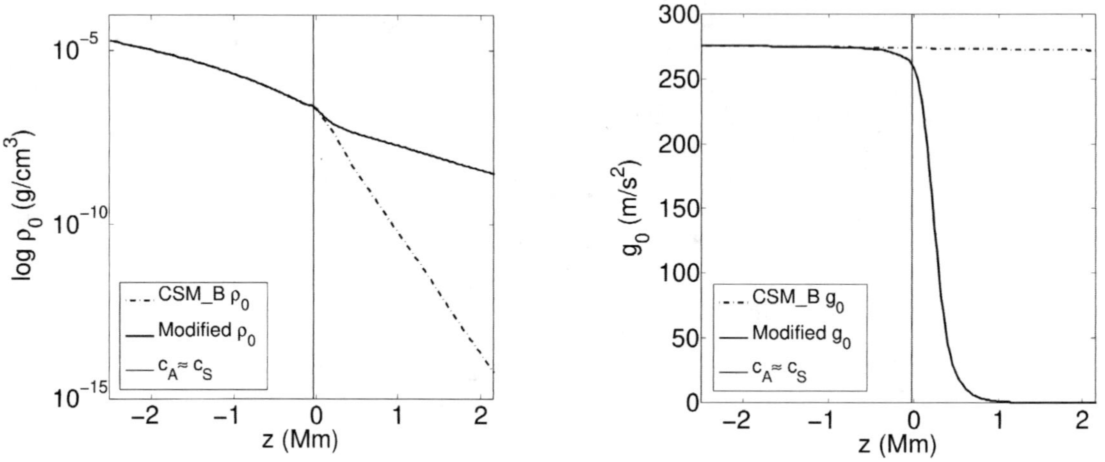

Figure 1. Plots of the original CSM_B ([19]; dot-dashed line) and modified (solid line) background density (left) and gravitational acceleration (right) profiles as a function of height. The solid vertical line represents the $c_A \approx c_S$ height.

now results above the surface also modifies acoustic cut-off frequency ($\omega_{ac} = c_S/2H_\rho$; where H_ρ denotes the density scale height), which is reduced from $\nu = 5.2$ mHz to 3.6 mHz. We also find that the modified atmosphere produces large-amplitude convective (g-) modes at $\nu \approx 1.5 - 1.7$ mHz (it is worth noting though that this is a frequency range which is typically associated with supergrannulation noise and is generally filtered out/ignored in sunspot seismology).

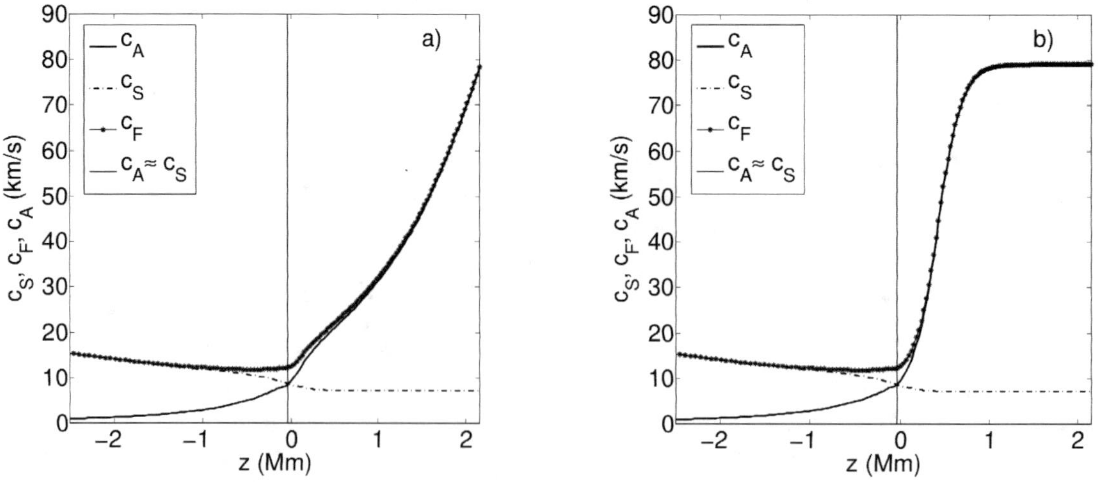

Figure 2. a): Plots of various wave speeds as a function of height resulting from the modified CSM_B background model. b): Same as a) but for an unmodified CSM_B background in conjunction with a Lorentz Force limiter (i.e, the Lorentz Force is multiplied by a factor of $200c_S^2/(200c_S^2 + c_A^2)$, leading to c_A and c_F being capped at ≈ 80 km/s). The solid vertical line represents the $c_A \approx c_S$ height.

However, more importantly for our concerns, this method ensures that we satisfy our CFL condition without any direct modification of the Lorentz Force via the introduction of an artificial term in Maxwell's equations, which as we shall show in the proceeding section, results in unphysical damping of the magnetic flux above the surface and inhibits the fast-Alfvén mode conversion process.

4. Results

Following [5–7], we use velocity projections onto three orthogonal directions ($\hat{\mathbf{e}}_{long}$, which selects the longitudinal component of wave propagation, i.e, the slow mode; $\hat{\mathbf{e}}_{trans}$, which selects the transversal component of wave propagation, i.e., the fast mode; $\hat{\mathbf{e}}_{perp}$, which selects the perpendicular component of wave motion, i.e., the Alfvén mode; see equations 1-3 in [5] for definitions) to separate the Alfvén mode from the fast and slow magneto-acoustic modes in the region $c_A \gg c_S$. We also calculate the temporally averaged acoustic ($\mathbf{F}_{ac} = \langle p_1 \mathbf{v}_1 \rangle$; where p represents pressure and $\mathbf{v}$ represents the 3-D velocity respectively, subscript "1" represents perturbations), magnetic ($\mathbf{F}_{mag} = \langle \mathbf{B}_1 \times (\mathbf{v}_1 \times \mathbf{B}_0)/\mu_0 \rangle$) and total ($\mathbf{F}_{tot} = \mathbf{F}_{ac} + \mathbf{F}_{mag}$) energy fluxes in order to measure the efficiency of conversion to Alfvén waves around the $c_A \approx c_S$ equipartition height.

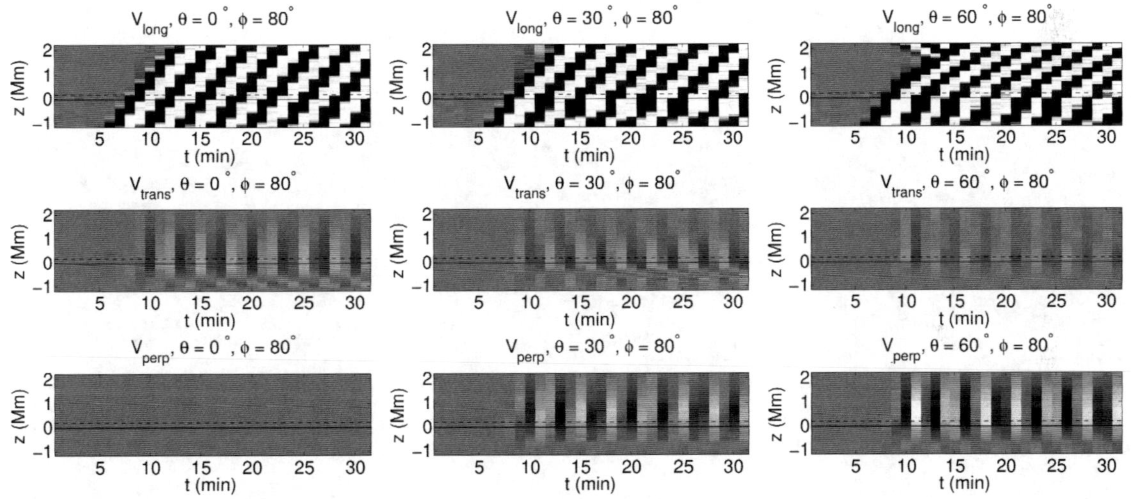

Figure 3. Figures above represent the velocity projections derived from simulations using constant inclined magnetic field configurations (left column: $\theta = 0°, \phi = 80°$; middle column: $\theta = 30°, \phi = 80°$; right column: $\theta = 60°, \phi = 80°$) as a function of height (Mm) and time (minutes). V_{perp} and V_{trans} amplitudes have been scaled by a factor of $(\rho_0 c_A)^{1/2}$, while V_{long} has been scaled by a factor of $(\rho_0 c_S)^{1/2}$. The grayscale is the same in all panels. The horizontal solid and dotted lines represent the $c_A \approx c_S$ and fast-mode reflection heights respectively.

The results of the projected velocities derived from simulations where ϕ is fixed at 80° and θ varies from 0° to 30° to 60° are shown in Figure 3. The inclination of the ridges in these projections indicates the wave propagation speed: the more inclined the ridges, the lower the propagation speed and vice versa. The presence of the (primarily acoustic) slow mode (V_{long}) is clearly visible above the $c_A \approx c_S$ level (solid line) in all three cases (with the reduced ω_{ac} resulting in a significant amount of acoustic waves propagating along the field both above and below the equipartition height), while the rapidly propagating Alfvén mode (V_{perp}) only appears when the magnetic field is sufficiently inclined and oriented out of the plane, i.e, $\theta = 30°, \phi = 80°$ and $\theta = 60°, \phi = 80°$, with the latter configuration appearing to be the more efficient in producing Alfvén waves. For these two cases, a faint presence of the magnetically dominated fast mode

Journal of Physics: Conference Series **440** (2013) 012047 doi:10.1088/1742-6596/440/1/012047

(V_{trans}) above the reflection height (dotted line) can still be made out. This is a result of our plane-parallel driver, which excites all wavenumbers, leading to a proportion of fast modes being transmitted through to the PML, rather than being reflected at $c_A \approx \omega/k_h$. In the $\theta = 60°, \phi = 80°$ projection, there appears to be a wavefront with an opposite inclination of ridges close to the top boundary of the domain. This could either be an artefact of the colour scheme/scaling, or a numerical artefact (i.e, reflection) from the upper boundary condition of the simulations. While we do not observe any reflection from the upper boundary in the corresponding acoustic flux (see Figure 4), given that the wavefornt appears to arrive at the upper boundary prior to the arrival of the slow waves, it could also be possible that the signature is a yet unmodelled product of the fast-Alfven conversion process. This is something which we hope investigate in a future work.

Figure 4 shows the vertical component of the averaged fluxes as a function of height. We observe that the flux variations are strongest near the conversion layer (solid vertical line), with the magnetic flux exceeding the acoustic flux when $\theta = 60°, \phi = 80°$ for $z > 1$ Mm. These results are in good agreement with previous numerical simulations of fast-Alfvén mode conversion using homogenous inclined magnetic fields (e.g., see Figure 3 from [5]).

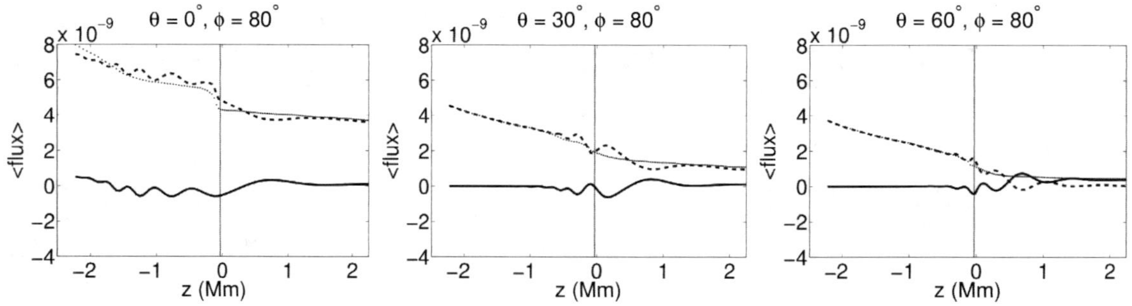

Figure 4. Time-averaged magnetic (bold sold line), acoustic (dashed line) and total (dotted line) fluxes (in non-dimensional units) calculated as a function of height. The solid vertical line represents the $c_A \approx c_S$ height.

For comparison purposes, we also conducted simulations where, instead of modifying $\rho_0(z)$ and $g_0(z)$ above the surface, we employ a Lorentz Force limiter (with c_A and c_F capped at ≈ 80 km/s, i.e, as shown in in Figure 2 b)). The resulting averaged fluxes are shown in Figure 5. While differences in the magnitude and height variations of the acoustic fluxes between these results, and those contained in Figure 4, can be explained by the change in ω_{ac}, the differences in the magnetic fluxes, particularly when considering the $\theta = 60°, \phi = 80°$ cases, are almost entirely due to the Lorentz Force limiter. With the Lorentz Force limiter in place, the magnetic flux above $z > 1$ Mm appears to just be able to creep above the acoustic flux for a couple of hundred kilometres, before being completely damped prior to reaching PML. We observed this phenomenon regardless of the value of the c_A cap that was used with the Lorentz Force limiter. As expected, the resulting velocity projections for these cases (figures not included) also confirmed the absence of any significant Alfvén modes above $c_A \approx c_S$.

5. Summary

Understanding the physics of propagating waves within regions of strong magnetic fields, of which fast-Alfvén wave mode conversion has recently been shown to be a critical component, is essential for helioseismic studies of sunspots and active regions. We used the 3-D linear MHD solver SPARC to simulate this process in a convectively stabilised solar model (CSM_B with

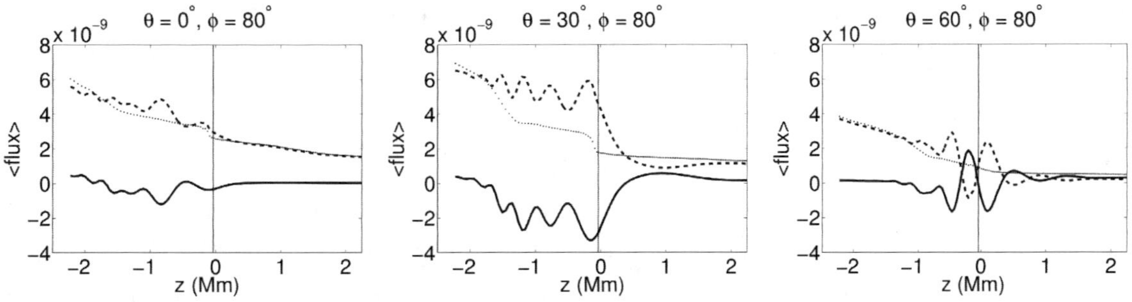

Figure 5. Time-averaged magnetic (bold sold line), acoustic (dashed line) and total (dotted line) fluxes (in non-dimensional units) calculated as a function of height for simulations where a Lorentz Force limiter is used with c_A and c_F capped as shown in Figure 2 b). The solid vertical line represents the $c_A \approx c_S$ height.

empirically modified $\rho_0(z)$ and $g_0(z)$ profiles in the upper layers to ensure a reasonable Δt) permeated by homogenous inclined magnetic fields. We found that employing a traditional Lorentz Force limiter to artificially cap c_A above the surface tends to inhibit the fast-Alfvén mode conversion process by significantly damping the magnetic flux above the surface.

The next steps in our forward modelling process will include the introduction of random stochastic sources and more realistic (i.e., sunspot-like) background atmospheres, in order to simulate artificial helioseismology data sets. With the aid of local helioseismic diagnostic tools, such as time-distance helioseismology and helioseismic holography, we will then be able to attempt to quantify the effects of fast-Alfvén mode conversion on the wave travel times.

Acknowledgments
This work was supported by an award under the Merit Allocation Scheme on the NCI National Facility at the ANU.

References
[1] Cally P S and Goossens M 2008 Sol. Phys. **251** 251–265
[2] Cally P S and Andries J 2010 Sol. Phys. **266** 17–38 (*Preprint* 1007.1808)
[3] Cally P S and Hansen S C 2011 ApJ **738** 119 (*Preprint* 1105.5754)
[4] Hanson C S and Cally P S 2011 Sol. Phys. **269** 105–110
[5] Khomenko E and Cally P S 2011 *Journal of Physics Conference Series* **271** 012042 (*Preprint* 1009.4575)
[6] Khomenko E and Cally P S 2012 ApJ **746** 68 (*Preprint* 1111.2851)
[7] Felipe T 2012 ApJ **758** 96 (*Preprint* 1208.5726)
[8] Hansen S C and Cally P S 2012 ApJ **751** 31 (*Preprint* 1203.3822)
[9] Gizon L, Schunker H, Baldner C S, Basu S, Birch A C, Bogart R S, Braun D C, Cameron R, Duvall T L, Hanasoge S M, Jackiewicz J, Roth M, Stahn T, Thompson M J and Zharkov S 2009 Space Sci. Rev. **144** 249–273 (*Preprint* 1002.2369)
[10] Moradi H, Baldner C, Birch A C, Braun D C, Cameron R H, Duvall T L, Gizon L, Haber D, Hanasoge S M, Hindman B W, Jackiewicz J, Khomenko E, Komm R, Rajaguru P, Rempel M, Roth M, Schlichenmaier R, Schunker H, Spruit H C, Strassmeier K G, Thompson M J and Zharkov S 2010 Sol. Phys. **267** 1–62 (*Preprint* 0912.4982)
[11] Gizon L, Birch A C and Spruit H C 2010 ARA&A **48** 289–338 (*Preprint* 1001.0930)

[12] Moradi H 2012 *Astronomische Nachrichten* **333** 1003 (*Preprint* 1210.5293)

[13] Hanasoge S M 2007 *Theoretical studies of wave interactions in the sun* Ph.D. thesis Stanford University

[14] Hanasoge S M, Duvall Jr T L and Couvidat S 2007 ApJ **664** 1234–1243

[15] Hanasoge S M 2008 ApJ **680** 1457–1466 (*Preprint* arXiv:0712.3578)

[16] Birch A C, Braun D C, Hanasoge S M and Cameron R 2009 Sol. Phys. **254** 17–27

[17] Moradi H, Hanasoge S M and Cally P S 2009 ApJ **690** L72–L75 (*Preprint* 0808.3628)

[18] Hanasoge S M, Duvall T L and DeRosa M L 2010 ApJ **712** L98–L102 (*Preprint* 1001.4508)

[19] Schunker H, Cameron R H, Gizon L and Moradi H 2011 Sol. Phys. **271** 1–26 (*Preprint* 1105.0219)

[20] Cameron R, Gizon L and Duvall Jr T L 2008 Sol. Phys. **251** 291–308 (*Preprint* 0802.1603)

[21] Cameron R H, Gizon L, Schunker H and Pietarila A 2011 Sol. Phys. **268** 293–308 (*Preprint* 1003.0528)

[22] Rempel M, Schüssler M and Knölker M 2009 ApJ **691** 640–649 (*Preprint* 0808.3294)

[23] Braun D C, Birch A C, Rempel M and Duvall T L 2012 ApJ **744** 77

[24] Cally P S 2012 Sol. Phys. **280** 33–50 (*Preprint* 1206.2114)

[25] Couvidat S, Gizon L, Birch A C, Larsen R M and Kosovichev A G 2005 ApJS **158** 217–229

Eclipse on the Coral Sea: Cycle 24 Ascending (GONG 2012, LWS/SDO-5, and SOHO 27) IOP Publishing
Journal of Physics: Conference Series **440** (2013) 012048 doi:10.1088/1742-6596/440/1/012048

Magnetohydrodynamic waves driven by p-modes

Elena Khomenko and Irantzu Calvo Santamaria

Instituto de Astrofísica de Canarias, 38205 La Laguna, Tenerife, Spain
Departamento de Astrofísica, Universidad de La Laguna, 38205, La Laguna, Tenerife, Spain

E-mail: khomenko@iac.es, irantzu@iac.es

Abstract. Waves are observed at all layers of the solar atmosphere and the magnetic field plays a key role in their propagation. While deep down in the atmosphere the p-modes are almost entirely of acoustic nature, in the upper layers magnetic forces are dominating, leading to a large variety of new wave modes. Significant advances have been made recently in our understanding of the physics of waves interaction with magnetic structures, with the help of analytical theories, numerical simulations, as well as high-resolution observations. In this contribution, we review recent observational findings and current theoretical ideas in the field, with an emphasis on the following questions: (i) Peculiarities of the observed wave propagation in network, plage and facular regions; (ii) Role of the mode transformation and observational evidences of this process; (iii) Coupling of the photosphere, chromosphere, and above by means of waves propagating in magnetic structures.

1. Introduction

Solar p-modes are generated by turbulent convection just beneath the photosphere and propagate through the solar interior [5; 53; 54; 94; 116; 120; 160; 161]. They also leak power to the atmospheric layers – photosphere, chromosphere and corona – and interact with magnetic structures present in these layers. Since the force balance changes with height from a gas pressure dominated region below the photosphere into a magnetically dominated region in the corona, the nature of waves changes as well and mode transformation occurs. These p-modes leaking to the upper atmosphere are a possible source of chromospheric/coronal heating. Besides, measuring and modeling the wave properties provides us with sensible indicators of the thermodynamic and magnetic structure of the atmosphere where they propagate [2; 30; 51; 115]. This article reviews the current advances in the field of atmospheric waves and oscillations driven by solar p-modes with an emphasis on their interaction with local magnetic fields.

Observationally, sunspot waves have been a usual and relatively easy target, due to the clear influence of the strong sunspot's magnetic field on them. Photospheric waves in the umbra and penumbra have been detected over decades, as well as their propagation to the chromosphere in the form of umbral flashes (see e.g., [1; 15; 16; 27; 32; 33; 45; 96–98; 100; 101; 103; 104; 144; 166; 167], and many others). Sunspot magnetic fields are easily detectable via polarimetric measurements by many instruments [10], and the measured parameters of oscillations and of the magnetic field can be linked. The general aspects of physics of these waves (such as the different types of waves, their physical properties, the conditions requires for transforming one into the other, etc.) are relatively well understood nowadays (for reviews on sunspot waves before 2008, see [9; 81]). Numerical modeling of sunspot waves has helped to explain many details [8; 18; 19; 25; 45–47; 55; 77; 78; 82; 83; 85; 86; 113; 114; 127–129; 142; 143; 152].

Journal of Physics: Conference Series **440** (2013) 012048 doi:10.1088/1742-6596/440/1/012048

Observations of waves in quiet regions have achieved important advances as well. Multi-line multi-layer observations have put in evidence how the propagation of waves in plage and network areas, and even in the quiet Sun, is influenced by the magnetic field (see, e.g., [29; 31; 48–50; 88–91; 106; 111; 131; 135; 158; 159; 169; 170], and references therein). Unlike in sunspots, magnetic field in the quiet regions is generally weaker, less organized, and, as a consequence, is more difficult to measure [151]. The question of the coupling between the different solar layers by means of waves, propagating in quiet Sun's structures, is particularly interesting. Waves with photospheric periodicities are systematically observed in the solar corona [112]. Nevertheless, it is still an open question how to get the p-mode energy efficiently up there because of a set of obstacles that waves encounter on their way up, such as: cut-off layer, wave speed gradients, transition region, etc (see, e.g. [57; 118]).

Yet another interesting topic is the practical application of the theoretical knowledge to infer information about the wave modes from observations. Diagnostics based on polarimetry or filter imaging have their advantages and shortcomings, and the information is often masked by radiative transfer effects and limited spatial and temporal resolution [4; 76; 121; 122; 148; 154]. Analytical/numerical modeling and Stokes diagnostics should be applied hand by hand to disentangle information about the wave modes and calculate wave energy fluxes, or to infer the parameters of magnetic structure, for example.

All these topics are discussed below with an emphasis on wave propagation in network, plage and facular regions and their interaction with magnetic elements present there, leaving aside sunspot waves. Both observational and theoretical/numerical aspects are highlighted.

2. Observations of waves in quiet Sun's magnetic structures

Magnetic elements in the quiet Sun are extremely complex and dynamic. Weak inter-network magnetic elements move, merge and cancel out, and small scale loops emerge on granular scale [3; 28; 69; 70; 105; 125; 126; 137; 150; 151]. Stronger magnetic elements in the network are generally more organized. Some recent works were able to resolve a network tube structure, including its canopy, from spectropolarimetric observations by Sunrise/IMaX [93; 107]. These real tubes are dynamic and asymmetric structures, unlike the "canonical" axisymmetric flux tubes used in many theoretical wave studies. Locally, acoustic p-modes and convective motions can drive waves in magnetic concentrations present over the quiet solar surface.

2.1. Network and inter-network regions

It has long been know that waves of 3-minute periodicity dominate quiet internetwork chromospheric regions [36; 67; 99; 102; 149]. These waves have been observed as bright intermittent Ca II H_{2V} and K_{2V} grains and are recognized to be acoustic shocks [26].

At the bright network cell borders, long period oscillations of 5–20 min are usually detected in the photosphere. These oscillations maintain their periodicity in the chromosphere and at larger heights [7; 91; 99; 131; 164; 169]. Shock occurrence is lower at the network cell borders compared to cell centers and the amplitudes remain below 2–4 km s^{-1} in the chromosphere.

Three-minute chromospheric oscillations inside supergranular network cells are well correlated with the underlying photospheric oscillations [36; 99]. It is not clear if the same happens to the 5-minute oscillations at the bright cell borders. Some earlier studies report that the correlation is difficult to trace [99], while others find a good correlation between the signals at two heights, suggesting nearly vertical wave propagation from the photosphere to the chromosphere in the proximity of the network magnetic elements [36; 95; 169]. Larger phase shifts between velocity signals at different heights, ϕ_{VV}, are found for waves at the network borders compared to cell centers [36; 95].

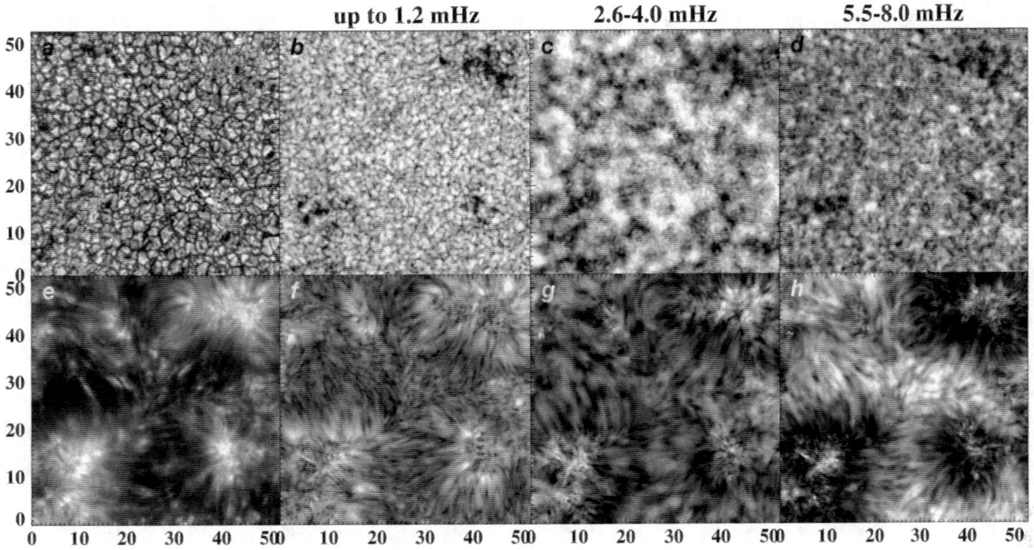

Figure 1. Example of IBIS observations of magnetic shadows from [169]. **a**: Speckle reconstructed broadband continuum at 710 nm for the FOV under analysis. **e**: Line core intensity of CaII 854.2 nm. **b–d** and **f–h**: Spatially resolved Fourier's velocity power maps, averaged over the range of frequencies indicated. The intensity scale is logarithmic, with bright indicating larger Fourier amplitudes. **b–d**: photospheric Fe I 709.0 nm line. **f–h**: chromospheric Ca II 854.2 nm line. *Courtesy of G. Cauzzi.*

Recent high-resolution observations from space and ground-based instruments revealed a much richer behavior of network oscillations related to the magnetic field topology. It was discovered that short-period waves (3 min range, $\nu = 5-8$ mHz) propagate from the photosphere to the chromosphere only in restricted areas of the network cell interiors, and that spatial distribution of 3-minute chromospheric shocks is highly dependent on the local magnetic topology [169; 170]. Network magnetic elements are surrounded by "magnetic shadows" where the power of short-period waves is reduced [48; 49; 74; 91; 136]. An example of this result borrowed from [169] is given in Figure 1. A prominent feature of these observations is the fine structuring of power maps at different frequencies, following the structure of chromospheric mottles.

Long-period waves (5 min range, $\nu = 1.2-4$ mHz) propagate efficiently to the chromosphere in the close proximity of the magnetic network elements, forming *enhancements* on the power maps ("power halos") [88]. These long-period network halos are most prominent in the photosphere, but are also present in the chromosphere; and are observed to be co-spatial with chromospheric "magnetic shadows" for 3 min waves. The halo/shadow areas show positive ϕ_{VV} phase shifts indicative of upward propagating waves at 3 min periodicity, but a mixture of positive and negative phase shifts for larger-period waves of 5-7 min, with a good coherence in all cases.

The peculiar power distribution in halos/shadows is apparently linked to the height of the magnetic canopy, defined as a layer where the plasma $\beta = 1$ [89; 90; 108; 109]. At the close proximity of network elements, where the canopies are low-lying (below 1600 km), the *chromospheric* power of short-period (3 min) waves was found to be suppressed, and the *photospheric* power of long-period waves (5-7 min) was found to be enhanced [89; 90]. It was suggested that the 5-7 min power enhancements in the photosphere may be due to reflection of waves on the overlying canopy. These enhancements are observed at all magnetic field inclinations, but are specially significant for more inclined canopy fields at the close proximity of magnetic elements, suggesting more efficient mode conversion there [89].

Table 1 summarizes the properties of oscillations in the quiet network areas, in terms of their periods.

2.2. Plage and facular regions

Plage and facular regions contain larger magnetic flux structures compared to the network regions discussed above. Early works (based on slit spectra) showed that the amplitude of 5 min oscillations in the photosphere decreases by about 25% in plage regions with an average flux above 80–100 G, compared to the quiet Sun, while shorter-period oscillations strengthen [6; 35; 68; 175]. More recent spectropolarimetric observations confirm that the amplitude of 5-min oscillations decreases by 20-40% in the facula photosphere [87]. These observations also reveal that the power of 5-min oscillations increases significantly in the chromosphere, dominating the power spectrum [29; 87]. The phase spectra of the 5-min facular waves are compatible with the vertical wave propagation from the photosphere to the chromosphere. One may conclude then, that the dominance of 5-min periodicities in the chromosphere is a common feature of both network and plage waves. However, the picture does not seem to be that simple.

The 2D filtergram observations show more pieces of the puzzle regarding the power distribution for long-period waves in relation to the magnetic field topology. It was argued that inclined magnetic field lines at the boundaries of large-scale enhanced network cells provide "portals" through which long-period (5 min) waves can propagate into the solar chromosphere, due to the lowering of the cut-off frequency in the low-β environment [58; 73; 110; 132; 133]. Larger propagation speeds of chromospheric waves were indeed found for strong fields inclined more than 40 degrees. At the center of plage magnetic elements (where the field is expected to be close to the vertical) preferably 3 min oscillations are observed to be transmitted to the

Table 1. Periods of network & inter-network oscillations

	Centers of mag. elements	Close surroundings	Inter-network beyond mag. elements
Photosphere	5 min	5 min	5 min
Chromosphere	Possibly 5 min	5 min enhanced 3 min shadowed	3 min shocks
Propagation	Up & down	Up & down	Up

Table 2. Periods of facula & plage oscillations

	Centers of mag. elements	Close surroundings	Halo areas
Photosphere	5 min damped 3 min enhanced	5 min damped 3 min enhanced	3 min enhanced
Chromosphere	Not clear	5 min for inclined B	3 min enhanced
Propagation	Possibly up	Possibly up	Up and down

chromosphere, while 5 min oscillations were transmitted at the edges of plage magnetic elements, where the field expands and becomes inclined [174].

The 2D filtergram observations rely on potential extrapolations from photospheric longitudinal magnetograms, unlike spectropolarimetric observations where the inversion of Stokes profiles gives information about the photospheric magnetic field vector (see e.g. [29]), suggesting it to be essentially vertical and thus giving strong arguments for the vertical propagation. On the other hand, 2D filtergrams have much higher resolution and field of view needed to detect the inclined propagation, if it takes place. The advantages of both observing techniques (imaging and spectropolarimetry) were exploited in [158; 159], by using simultaneous IBIS photospheric and chromospheric measurements together with spectropolarimetric inversion of the SOT/Hinode data. For a plage region containing a pore, maximum transmission of 3 min oscillations was found for field inclined by 15 degrees, while 5 min oscillations were transmitted for the field inclined by 25 degrees, providing support for the inclined propagation of long-period waves, at least for the strong field of a pore. It is not clear, however, if the same happens for the weaker magnetic elements, and more data as those by [158; 159] are clearly needed.

For shot-period waves (3 min range, $\nu = 5.5 - 7.5$ mHz), the power distribution over plage regions shows a suspicious enhancement, both in the photosphere [14] and in the chromosphere [11; 165]. These power enhancements are known as "halos" and have been much discussed in the literature on local helioseismic waves[1]. The acoustic power measured in halos is higher than in the quite Sun by about 40-60% [12; 38; 66; 72; 117]. The halos are observed at longitudinal magnetic fluxes $\langle B \rangle = 50 - 300$ G [66; 72; 162]. In the photosphere the halos are located at the edges of plage regions, while in the chromosphere they extend to a large portion of the nearby quiet Sun [11; 14; 162]. The characteristic period of maximum enhancement decreases with field strength and the wave spectrum in halos is shifted to higher wave numbers at constant frequency [52; 153]. The wave power in halos is observed to be modulated with the field inclination, being stronger for more inclined fields [153]. The height-dependent changes in halo properties seen in HMI/AIA SDO data provide signatures of magneto-acoustic wave refraction sensitive to magnetic field strength and topology [135] (the presence of up and down-going waves has also been reported previously by [13]).

The vast amount of information on the observed halo properties is mostly obtained from the observational datasets typical for local helioseismology (i.e. long time series over a large portion of the Sun, including a complete active region, but medium spatial resolution, as SOHO/MDI or SDO/HMI). It is not clear how to consistently compare these large-scale data with higher resolution (but smaller area) IBIS, SST and Hinode data [58; 131; 158; 159; 174] or chromospheric/transition region TRACE data [73; 110; 132; 133], resolving propagation in individual magnetic structures. In addition, there is much confusion about the waves observed in plage, facular, or enhanced network regions, often discussed in the same context. It is not evident beforehand if the observed wave behavior should be the same, since the field strength of the magnetic structure definitely plays a crucial role.

Table 2 makes an attempt to summarize the properties of oscillations in plage areas, though definite conclusions are hard to derive in some cases.

2.3. Magnetic field oscillations

Up to here, we have discussed oscillations of velocity and intensity. Magnetic field measurements, when performed, were mostly done to obtain the magnetic topology, but not oscillations of the magnetic flux (or magnetic field vector) itself.

Even in strong sunspot regions it has been hard to detect magnetic field oscillations due to

[1] Note that halos in plage are observed at short-periods, unlike long-period power halos surrounding network magnetic elements, discussed in Section 2.1.

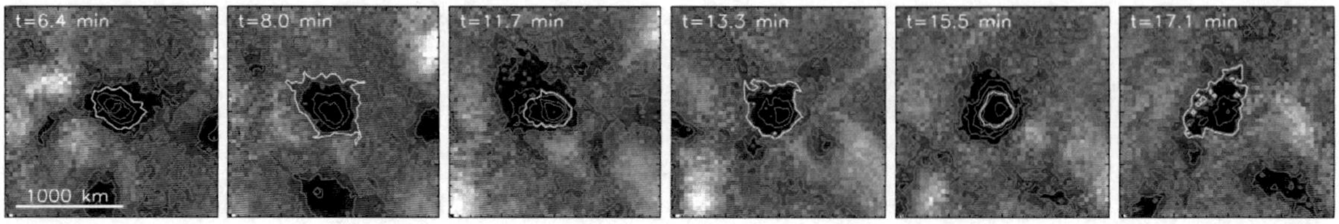

Figure 2. Example of time evolution of a weak circular polarization patch in the quiet Sun from [106]. The black and white background represents the magnetic flux density computed in the weak-field approximation, saturated to ± 20 Mx cm^{-2}. Yellow line is the iso-magnetic flux density contour containing a time-constant magnetic flux of -5$\times 10^{16}$ Mx. *Courtesy of M. Martínez González.*

the smallness of the expected amplitudes and due to radiative transfer and instrumental effects, masking the real oscillations [76; 87; 122; 147; 148; 154].

The detection of magnetic flux/field oscillations due to magneto-hydrodynamic waves in the quiet regions has been elusive. The relations between the magnetic flux, velocity and intensity oscillations in the photosphere of pores and intergranular magnetic elements, obtained from homogeneous, seeing-free SOT/Hinode data, reveal root-mean-square amplitudes of magnetic flux fluctuations of 4$-$17 G (0.3%$-$1.2% from the background value), corresponding to velocity oscillations of $\sim$100 m s^{-1} and intensity oscillations of 0.1%$-$1% [50]. Velocity was leading the magnetic flux oscillations by a quarter of cycle, and magnetic flux oscillations were 180 degree out of phase with the intensity oscillations. These phase lags were interpreted as due to superposition of the ascending and the descending sausage and kink mode waves reflected at the transition region [50].

Another recent detection of oscillations of magnetic flux density in quiet Sun magnetic elements was done from high-resolution Sunrise/IMaX data [106]. It was discovered that the area of patches with constant magnetic flux oscillates in response to granular forcing in a range of periods similar to granular life times (implying that magnetic field strength oscillates in antiphase). An example of such observation borrowed from this work is shown in Figure 2. Curiously, the amplitude of these area oscillations is highly non-linear with more than 100% variations from the mean value. No apparent correlation with velocity or intensity oscillations was found.

3. Models

This section gives an overview on recent, mostly numerical, models of the wave propagation in solar magnetic structures. We subdivide the simulations into several (overlapping) groups, one of them aiming to explain the power distribution in and around magnetic features, another group studying the energy propagation by waves from the photosphere to the chromosphere, transition region and corona, yet another one for the mode conversion.

3.1. Coupling of the photosphere, chromosphere and above

Small-scale magnetic field concentrations are ubiquitous in the quiet Sun and magnetic waves driven in these elements by *p*-mode or granular forcing are obvious candidates to transport energy to heat the upper layers. Solar magnetic elements are far more complex than canonical axisymmetric flux tubes. But despite the clear simplifications of the tube structure, theoretical works on propagation of waves in flux tubes are of great importance to understand the basic physics of the wave mode behavior in small scale magnetic elements. There have been numerous analytical works using the thin flux tube approximation, when the radius of the tube is considered

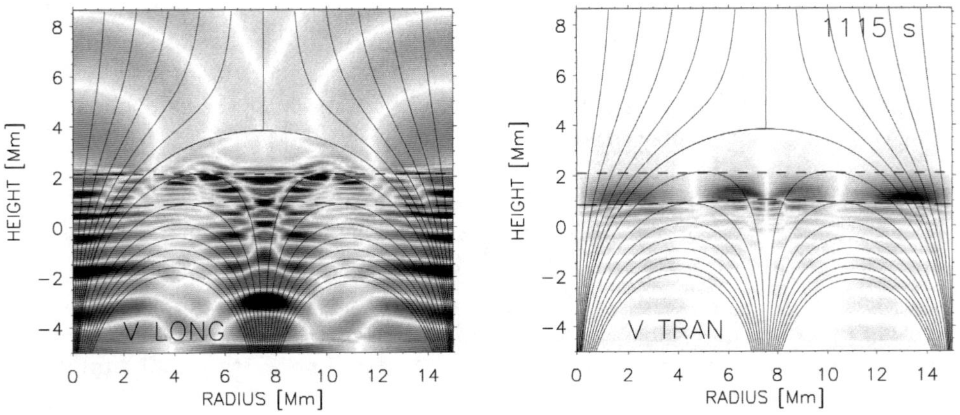

Figure 3. Snapshot of longitudinal (left) and transversal (right) velocities, scaled by a factor or $\sqrt{\rho c_s}$ and $\sqrt{\rho v_a}$, correspondingly, in the 2D simulations of fast (acoustic) mode propagation and conversion from sub-photosphere to the low corona from [24].

to be much smaller than the local vertical scale height, and the internal flux tube structure can be neglected [17; 37; 40; 138–141; 145; 146; 157] (see [155; 156] for a review). In fact, the comparison to simulations of magneto-convection suggests that a second-order thin flux tube approximation might do reasonably well for super-equipartition magnetic field concentrations like those existing in the simulations [176]. This argument can be used for wave studies if asymmetry and flows are not crucial for the considered problem.

The number of idealized simulations of flux tube waves propagating from the photosphere to the chromosphere and above is ever increasing during the last several years [41–43; 59–63; 75; 79; 123; 171–173]. Different ways of driving flux tube waves have been checked: horizontal motions producing transverse kink waves; pressure fluctuations producing longitudinal waves; twisting motions generating torsional Alfvén waves; observationally-driven or impulsive motions. The general idea behind all these works is to understand how much of the wave energy can reach the chromosphere and corona, and whether the wave modes reaching there can be easily dissipated to convert their energy into heat (as, e.g. compressible acoustic modes).

The transverse foot-point driving was extensively studied in a series of papers [60–62; 171; 173] by means of 1D and 2D simulations. These works mostly aim at explaining waves in bright network elements at cell borders observed in Ca II H (see Section 2.1) and their relation to G-band bright points. It was discovered an efficient mechanism of non-linear coupling between the kink waves (generated by transverse foot point driving) and longitudinal waves, forming shocks and supplying energy to the chromosphere. The efficiency of the coupling was found to be a function of the plasma β, being maximum for tubes with $\beta = 0.2$ when the kink and tube wave speeds are equal [61]. Adding a second dimension to the modeling (2D), the transverse foot point motions were seen to excite fast and slow magneto-acoustic waves inside the tube, and acoustic ones at the interface between the tube and the non-magnetic medium [62]. The acoustic waves at the interface steepen to shocks in the chromosphere and flux tubes with stronger magnetic field produce these shock waves more efficiently [171].

The above works were mostly focused on short-period (10-20 sec) waves or impulsive driving, and it is hard to make a direct comparison to the observations. Nevertheless, similar processes of magneto-acoustic mode generation and transformation were shown to take place when considering more realistic driving with 3 and 5 min periodicities [79]. Complex shock wave patterns are formed across the tube, with shock waves propagating alongside the tube borders. Sub-arcsec spatial resolution would be needed to observationally resolve this pattern [84; 172].

Apart from linear and non-linear mode transformation, slow chromospheric shocks in magnetic flux tubes were demonstrated to be produced by a mechanism of "turbulent pumping" [75], i.e. "massaging" of the magnetic tubes in intergranular lanes by external downflows. Possibly, a similar process is responsible for the periodic area variations of quiet Sun magnetic elements discovered in [106].

When the tube foot points are twisted instead of being transversally shaken, torsional Alfvén waves can be generated. There are indications that the energy input into the chromosphere is lower for the twisting driver, compared to the transverse shaking [173]. The torsional Alfvén waves, generated by twisting driver seem to be frequency-filtered by the tube's three-dimensional structure, with higher frequencies transmitted at the central part of the tube, and lower ones at the sides [42].

The next challenge for the wave numerical modeling in 2D and 3D is to extend the analysis into the transition region and corona, in order to explain how waves with photospheric periodicity leak their energy to the upper atmosphere. This is numerically challenging, since many pressure scale heights have to be modeled at the same time, and the wave speeds change considerably. The transition region represents a discontinuity in the wave speed and waves are expected to be efficiently reflected there. Thus, it is unclear how much energy can go through. Nevertheless, short-period (30 sec) magneto-acoustic waves were shown to be transmitted rather efficiently through the transition region, exciting horizontally propagating disturbances there [43]. Figure 3 shows an example of the modeling of wave propagation from the sub-photosphere to the chromosphere, transition region and corona for realistic periods in the 3-5 min range [24]. It demonstrates a complex wave pattern formed by multiple mode transformations at the $\beta = 1$ layer, wave refractions and reflections at the transition region and a 2D magnetic null point located above. More work will be needed in the future to quantify these complex interactions, and to evaluate the transmitted energy, depending on the magnetic field configuration and strength.

3.2. Mode transformation in 2D and 3D

Linear mode transformation is an important process that enables different wave modes to exchange their energy when their propagation speeds are close. Mode transformation between fast and slow magneto-acoustic modes in 2D was extensively studied analytically and numerically [8; 18–21; 34; 77; 142; 152; 178]. A review on these works before 2008 can be found in [81]. In brief, the direction and the effectiveness of the mode transformation depends on the wave frequency and the attacking angle between the wave vector $\vec{k}$ and the magnetic field $\vec{B}$ at the layer of plasma $\beta = 1$. The fast-to-slow mode transformation is complete for waves with $\vec{k} \parallel \vec{B}$. The fast-mode high-$\beta$ waves launched from their sub-photospheric lower turning points typically reach the transformation layer with an angle close to 20–30 degrees, so the transformation is particularly strong for magnetic fields inclined by this amount [21; 34; 152].

Recently, mode transformation in 3D has attracted a lot of attention. It enables to produce Alfvén waves from fast magneto-acoustic waves above their reflection height in the low chromosphere, a mechanism that may allow to transport wave energy through the transition region to the corona. Alfvén waves are claimed to be detected by numerous observations, but the discussion on the nature of the detected waves is still ongoing [112; 130; 134; 163; 168].

The pioneering study of 3D mode conversion from fast to Alfvén waves for homogeneous inclined fields discovered it to be most efficient for preferred magnetic field inclinations between 30 and 40 degrees, and azimuth angles between 60 and 80 degrees [22]. This orientation allowed the best alignment with $\vec{B}$ of the fast mode wave propagating from its sub-photospheric lower

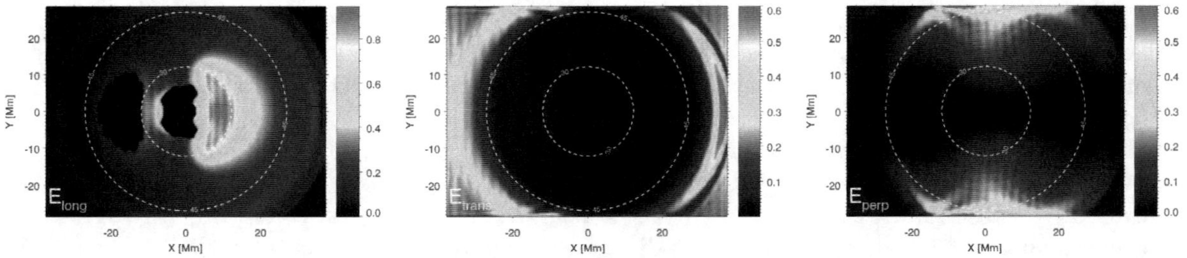

Figure 4. Example of wave energy fluxes in the chromosphere of a sunspot model from [47] showing the redistribution of the energy of different modes due to three-dimensional mode conversion. Left panel: acoustic flux due to slow longitudinal waves; middle panel: magnetic flux due to fast waves; right panel: magnetic flux due to Alfvén waves. The units of the color coding are 10^6 erg cm^{-2} s^{-1}. *Courtesy of T. Felipe.*

turning point. Alfvênic fluxes transmitted to the upper atmosphere are found to be similar, or larger, than acoustic fluxes at some orientations. Low-frequency gravity waves can also be converted efficiently into Alfvén waves at large magnetic field inclinations [118].

While the fast-to-slow mode conversion occurs at the layer where plasma $\beta = 1$, the fast-to-Alfvén conversion takes place around and above the fast wave reflection height $z_{\rm refl}$ in 3D (see Fig. 1 in [86]), localized more closely to $z_{\rm refl}$ as the frequency increases [23]. For 3-5 min waves, the conversion process is typically spread over much of the chromosphere. Both up-going or down-going Alfvén waves can be produced depending on the relative orientation of the wave vector and magnetic field at heights where the fast mode undergoes the refraction, going up and down [23; 86].

Numerical simulations of conversion to Alfvén waves require 3D or, at least, 2.5D geometries [44; 47; 85; 86]. Recent 2.5D simulations show that the global picture of conversion to Alfvén waves remains valid when considering complex field configurations, appropriate for large-scale sunspot magnetic structure [86]. The conversion to Alfvén waves is particularly important for strongly inclined fields like those existing in sunspot penumbrae. A fully 3D simulation, covering a large portion of a sunspot model with a wide range of inclinations, nicely demonstrated that a significant energy flux due to Alfvén waves can indeed be transmitted to the chromosphere at large field inclinations [47]. The upward-propagating flux due to Alfvén waves at the peripheric parts of the sunspot model was found to be as large as the acoustic flux due to slow longitudinal waves propagating close to the axis. An example of this result is shown in Figure 4.

Inclusion of radiative losses into the model does not change significantly the wave paths neither the conversion picture [119].

3.3. Tunneling of 5 min oscillations to the chromosphere

A specific question for the models of wave propagation in small scale structures is the explanation of the power spectrum of waves at chromospheric heights (see Section 2). Observations in network regions mostly agree that long-period 5 min oscillations are transmitted to the chromosphere in the close proximity of network magnetic elements (Section 2.1). In a plage, it seems not clear yet whether these long-period oscillations are transmitted at the plage borders with predominantly inclined field or vertically from the photosphere (Section 2.2).

There have been a number of works arguing that inclined magnetic field at the boundaries of enhanced network and plage regions allows long-period slow acoustic waves to propagate into the solar chromosphere due to a reduction of the cut-off frequency [58; 73; 110; 132; 133]. Numerical 1D modeling confirms the efficiency of this mechanism [64]: long-period 5 min waves can only propagate upwards where the magnetic field is inclined, whereas waves with periods around 3 min dominate when the field is vertical. This modeling includes a complex treatment of radiative transfer, but simplifications of a one-dimensional propagation in a constant inclined field.

Alternatively, some authors proposed that radiative losses in thin flux tubes can help propagating long-period oscillations vertically upward [29; 80]. The presence of radiative losses can lead to a significant reduction of the cutoff frequency, formally adding a propagating component to the oscillations [140]. Two-dimensional simulations of flux tube waves with a 5 min periodic photospheric driver, and including radiative losses by means of a simplified Newton's law with fixed radiative relaxation time, demonstrate the effectiveness of this mechanism in leaking 5 min oscillations into the chromosphere inside vertical flux tubes [80]. The power and phase spectrum of waves in these simulations reproduce observations in facular regions [29].

The treatment of radiative losses by Newton's law [80] is very rough. Recently, realistic radiative magneto-convection simulations were performed in a model extending from the convection zone to the corona, where oscillations were produced self-consistently by the turbulent motion in the convection zone without any imposed external driver [65]. A sophisticated state-of-the-art treatment of radiative losses was included. The simulations aimed at quantifying the contribution of both effects - field inclination and radiative losses - into the resulting spectrum of waves at chromospheric heights. It was found that 5 min oscillations are able to propagate in regions where the field is inclined and strong, including the edges of otherwise vertical flux tubes, whereas in regions where the field is vertical (center of flux tubes) or weak, oscillations with periods around 3 min propagate [65]. Radiative losses were found to play only a minor role in determining the propagation spectrum (though no reference simulation with radiative transfer switched off was performed).

There are also alternative mechanisms for transmission of 5 min oscillations into the chromosphere, by means of non-linear photospheric pulses that create nonlinear wakes leading to a trail of consecutive shocks with 5 min periodicity [177].

3.4. Power distribution at different frequencies

Many recent studies agree that the peculiar power shadowing and enhancements observed in plage and network areas are related to processes of mode conversion and refraction at the inclined magnetic canopy in the proximity of the $\beta = 1$ layer (see Sections 2.1 and 2.2).

There have been numerous theories to explain high-frequency acoustic halos in plage regions [11; 14; 55; 66; 71; 72; 82; 92]. Numerical simulations of interaction of waves with sunspot/plage flux tubes suggest that the power enhancement in halos is produced by mode mixing induced by the magnetic field, resulting in preferential scattering from low to high wave numbers [56]. Alternatively, an explanation has been offered based on the enhanced high-frequency wave excitation at the regions of intermediate magnetic field strength in halos [71] .

Single-source numerical simulations of magneto-acoustic wave propagation in a magneto-static sunspot model demonstrate that the halo effect happens in a natural way due to the additional energy input from the high-frequency fast mode waves, after their refraction above the $\beta = 1$ layer [82]. The frequency and location of the halo is extremely sensitive to the relative location of the $\beta = 1$ layer, the cut-off layer and the formation height of a given observed spectral line. The halo is produced in those photospheric regions where the field is intermediate, implying that the Alfvén speed is lower than the sound speed. The halo is not observed at low frequencies because these waves are already reflected by cut-off below the transformation layer, and is not observed in the umbra of the sunspot because the refraction happens below the layer visible to spectral line observations. This model [82], as well as the model of mode mixing [56] suggest that the halo power enhancements might be larger in the horizontal velocity component.

To clarify the nature of magnetic shadows, numerical simulations of high-frequency magneto-acoustic wave propagation in a snapshot of a three-dimensional model of magneto-convection were carried out [124]. On their way up, acoustic waves were converted into different mode types and were refracted. The Fourier analysis of this simulation clearly shown chromospheric shadows, similar to the observed ones. It was concluded that these shadows are linked to the mode conversion process and that power maps at these heights show the signature of the different magneto-acoustic wave modes. Yet another mechanism to explain power enhancements in halos was proposed based on acoustic waves trapped in field-free atmospheres lying below small-scale magnetic canopies of network cores and active regions [92].

Alternative pictures of wave transformation and refraction [82; 92; 124] are, in fact, similar, suggestive that both, halos and shadows, are different manifestation of the same physical mechanism. Recent observational results seem to favor the model based on wave transformation and refraction as well [135]. One can not exclude, though, that other effects may also play a role [56; 71]. For example, acoustic halos are found to be co-spatial with acoustic glories, locations with enhanced seismic emission surrounding active regions [39]. The nature of the glories is different to halos, since they tell us about an increase of seismic emission, i.e. power of waves emanating directly from the wave sources, whereas the acoustic halos are just measures of increased wave amplitudes, whatever is the reason. Thus, the last word on the nature of halos is not yet said.

4. Conclusions

Significant advances have been made in our understanding of the physics of atmospheric waves and their interaction with local magnetic structures in the photosphere, chromosphere and above. Summarizing the vast amount of theoretical and observational works reviewed here, we give below some brief conclusions and suggestions for perspective future work directions:

- The observed distributions of photospheric and chromospheric power of long and short period waves seems to be distinct over network and plage/facular regions. In the network, long-period 5 min waves seem to be transmitted in the close proximity of magnetic elements, while short-period 3 min waves are "shadowed" due to the interaction of magneto-acoustic waves with the more horizontal fields of the canopies. In the stronger-field plage regions, short-period (3 min, $\nu = 5 - 7$ mHz) halos dominate both in the photosphere and in the chromosphere, and the propagation of long-period waves is enhanced for inclined fields, but vertical propagation has also been reported.
  The reason for that different behavior may be the variation of the height of the magnetic canopy $\beta = 1$ layer, being lower for stronger plage fields, thus modifying the spectrum of waves reaching the heights sampled by photospheric and chromospheric observations.
  More high-resolution studies are needed in this direction, comparing regions with different magnetic fluxes, both at the disc center and closer to the limb to get information about horizontal velocities. Simultaneous measurements of the magnetic field vector are also very important.
- On the theoretical side, there seems to be still no agreement on the explanation of the presence of ubiquitous long-period waves in the chromosphere above network bright points and plage. Different observations give evidences both for inclined and vertical propagation (both upward and downward), at least for the network elements, while theoretical models seem to point toward the inclined propagation as a dominant mechanism. Inclined propagation receives more observational support especially for the strong field plage regions where it is expected that the slow acoustic waves propagate field-aligned in the low-β regime. Future studies should address the propagation in weaker structures, accompanied by observations of more network and inter-network elements.

- There have been few detections of magnetic field oscillations in quiet solar regions reported in the recent literature. Magnetic field and flux oscillations are very hard to detect, but an effort should be made to increase the number of such studies, since the amplitude and phase relations between the magnetic field variations and other parameters give valuable information on the observed wave types.
- Idealized simulations of wave propagation from the photosphere to the chromosphere in small scale flux tubes are rather well developed. These models have suggested a number of mechanisms to transport the energy of the wave to the upper layers. Three-dimensional simulations are being developed. A remaining challenge is to couple the sub-photospheric layers, photosphere, chromosphere, transition region and corona by means of waves. This is needed to understand how (and what amount of) the wave energy can reach the corona, depending on the driver properties, and taking into account all the wave physics such as 3D mode transformation, refraction, reflection, non-linear interaction, as well as dissipation mechanisms. It will be interesting in the future to get more insights on the efficiency of conversion to Alfvén waves on inclined canopy fields of the quiet Sun small scale flux tubes, by means of 3D numerical simulations.
- Oscillations in the surroundings of quiet Sun magnetic structures in network, plage and facular regions are being observed at extremely high spatial resolution, and their relation to the magnetic topology is being clarified. Theoretical models for acoustic halos and shadows indicate that both may have a similar origin, tracing mode transformation process, and fast mode refraction and reflection at the inclined field of the magnetic canopy. It remains to clarify the relation between the acoustic halos and glories, present on the same locations. Some theoretical models for halos suggest that the power enhancement in horizontal velocity component should be significantly stronger than in the vertical one. In the future it will be interesting to study halos and shadows at locations off the disc center to shed more light on their nature. Models of transformation to Alfvén waves suggest they will be reinforced at the periphery of active regions. Checking this in future observations is important as well.

Acknowledgements: This research has been supported by the Spanish Ministry of Economy and Competitiveness (MINECO) under the grants AYA2010-18029 and AYA2011-24808.

References

[1] Abdelatif T E, Lites B W and Thomas J H 1986 *ApJ* **311** 1015
[2] Basu S and Antia H M 2008 *Physics Reports* **457** 217
[3] Bello González N, Okunev O and Kneer F 2008 *A&A* **490** L23
[4] Bellot Rubio L R, Collados M, Ruiz Cobo B and Rodríguez Hidalgo I 2000 *ApJ* **534** 989
[5] Bi S and Li R 1998 *A&A* **335** 673
[6] Blondel M 1971 *A&A* **10** 342
[7] Bloomfield D S, McAteer R T J, Mathioudakis M and Keenan F P 2006 *ApJ* **652** 812
[8] Bogdan T J, Carlsson M, Hansteen V, et al. 2003 *ApJ* **599** 626
[9] Bogdan T J and Judge P G 2006 Observational aspects of sunspot oscillations *MHD wave and oscillations in the Solar Plasma* vol 364, (Phil. Trans. Royal. Soc.) p 313
[10] Borrero J M and Ichimoto K 2011 *Living Reviews in Solar Physics* **8** 4
[11] Braun D C, Lindsey C, Fan Y and Jefferies S M 1992 *ApJ* **392** 739
[12] Braun D C and Lindsey C 1999 *ApJ* **513** L79
[13] Braun D C and Lindsey C 2000 *Solar Phys.* **192** 307
[14] Brown T M, Bogdan T J, Lites B W and Thomas J H 1992 *ApJ* **394** L65

[15] Brynildsen N, Maltby P, Leifsen T, Kjeldseth-Moe O and Wilhelm K 2000 *Solar Phys.* **191** 129
[16] Brynildsen N, Maltby P, Fredvik T and Kjeldseth-Moe O 2002 *Solar Phys.* **207** 259
[17] Cally P S 1986 *Solar Phys.* **103** 277
[18] Cally P S, Bogdan T J and Zweibel E G 1994 *ApJ* **437** 505
[19] Cally P S and Bogdan T J 1997 *ApJ* **486** L67
[20] Cally P 2005 *MNRAS* **358** 353
[21] Cally P S 2006 *Phil. Trans. R. Soc. A* **364** 333
[22] Cally P S and Goossens M 2008 *Solar Phys.* **251** 251
[23] Cally P S and Hansen S C 2011 *ApJ* **738** 119
[24] Calvo Santamaria I, Khomenko E and Collados M 2013 *in preparation*
[25] Cameron R, Gizon L and Duvall Jr T L 2008 *Solar Phys.* **251** 291
[26] Carlsson M and Stein R F 1997 *ApJ* **481** 500
[27] Centeno R, Collados M and Trujillo Bueno J 2006 *ApJ* **640** 1153
[28] Centeno R, Socas-Navarro H, Lites B, et al. S 2007 *ApJ* **666** L137
[29] Centeno R, Collados M and Trujillo Bueno J 2009 *ApJ* **692** 1211
[30] Chaplin W J and Basu S 2008 *Solar Phys.* **251** 53
[31] Chitta L P, Jain R, Kariyappa R and Jefferies S M 2012 *ApJ* **744** 98
[32] Christopoulou E B, Georgakilas A A and Koutchmy S 2000 *A&A* **354** 305
[33] Christopoulou E B, Georgakilas A A and Koutchmy S 2001 *A&A* **375** 617
[34] Crouch A D and Cally P S 2003 *Solar Phys.* **214** 201
[35] Deubner F L 1967 *Solar Phys.* **2** 133
[36] Deubner F L and Fleck B 1990 *A&A* **228** 506
[37] Díaz A J and Roberts B 2006 *A&A* **458** 975
[38] Donea A C, Lindsey C and Braun D C 2000 *Solar Phys.* **192** 321
[39] Donea A and Newington M 2011 *Journal of Physics Conference Series* **271** 012004
[40] Dymova M V and Ruderman M S 2005 *Solar Phys.* **229** 79
[41] Erdélyi R, Pintér B and Malins C 2007 *Astronomische Nachrichten* **328** 305
[42] Fedun V, Verth G, Jess D B and Erdélyi R 2011 *ApJ* **740** L46
[43] Fedun V, Shelyag S and Erdélyi R 2011 *ApJ* **727** 17
[44] Felipe T, Khomenko E and Collados M 2010 *ApJ* **719** 357
[45] Felipe T, Khomenko E, Collados M and Beck C 2010b *ApJ* **722** 131
[46] Felipe T, Khomenko E and Collados M 2011 *ApJ* **735** 65
[47] Felipe T 2012 *ApJ* **758** 96
[48] Finsterle W, Jefferies S M, Cacciani A, Rapex P, Giebink C, Knox A, Dimartino V 2004 *Solar Phys.* **220** 317
[49] Finsterle W, Jefferies S M, Cacciani A, Rapex P, McIntosh S W 2004 *ApJ* **613** L185
[50] Fujimura D and Tsuneta S 2009 *ApJ* **702** 1443
[51] Gizon L and Birch A C 2005 *Living Reviews in Solar Physics* **2** 6
[52] Gizon L, Schunker H, Baldner C S, et al. 2009 *Space Science Reviews* **144** 249
[53] Goldreich P and Kumar P 1990 *ApJ* **363** 694
[54] Goldreich P, Murray N and Kumar P 1994 *ApJ* **424** 466

[55] Hanasoge S M 2008 *ApJ* **680** 1457

[56] Hanasoge S M 2009 *A&A* **503** 595

[57] Hansen S C and Cally P S 2012 *ApJ* **751** 31

[58] Hansteen V H, De Pontieu B, Rouppe van der Voort L, van Noort M and Carlsson M 2006 *ApJ* **647** L73

[59] Hasan S S, Kalkofen W and van Ballegooijen A A 2000 *ApJ* **535** L67

[60] Hasan S S, Kalkofen W, van Ballegooijen A A and Ulmschneider P 2003 *ApJ* **585** 1138

[61] Hasan S S and Ulmschneider P 2004 *A&A* **422** 1085

[62] Hasan S S, van Ballegooijen A A, Kalkofen W and Steiner O 2005 *ApJ* **631** 1270

[63] Hasan S S and van Ballegooijen A A 2008 *A&A* **680** 1542

[64] Heggland L, De Pontieu B and Hansteen V H 2007 *ApJ* **666** 1277

[65] Heggland L, Hansteen V H, De Pontieu B and Carlsson M 2011 *ApJ* **743** 142

[66] Hindman B W and Brown T M 1998 *ApJ* **504** 1029

[67] Hoekzema N M and Rutten R J 1998 *A&A* **329** 725

[68] Howard R 1967 *Solar Phys.* **2** 3

[69] Ishikawa R, Tsuneta S, Ichimoto K, et al. 2008 *A&A* **481** L25

[70] Ishikawa R and Tsuneta S 2009 *A&A* **495** 607

[71] Jacoutot L, Kosovichev A G, Wray A and Mansour N N 2008 *ApJ* **684** L51

[72] Jain R and Haber D 2002 *A&A* **387** 1092

[73] Jefferies S M, McIntosh S W, Armstrong J D, Bogdan T, Thomas J, Cacciani A and Fleck B 2006 *ApJ* **648** L151

[74] Judge P G, Tarbell T D and Wilhelm K 2001 *ApJ* **554** 424

[75] Kato Y, Steiner O, Steffen M and Suematsu Y 2011 *ApJ* **730** L24

[76] Khomenko E V, Collados M and Bellot Rubio L R 2003 *ApJ* **588** 606

[77] Khomenko E and Collados M 2006 *ApJ* **653** 739

[78] Khomenko E and Collados M 2008 *ApJ* **689** 1379

[79] Khomenko E, Collados M and Feliipe T 2008 *Solar Phys.* **251** 589

[80] Khomenko E, Centeno R, Collados M and Trujillo Bueno J 2008 *ApJ* **676** L85

[81] Khomenko E 2009 Simulations of Waves in Sunspots *Solar-Stellar Dynamos as Revealed by Helio- and Asteroseismology: GONG 2008/SOHO 21* (*Astronomical Society of the Pacific Conference Series* vol 416) ed Dikpati M, Arentoft T, González Hernández I, Lindsey C and Hill F p 31

[82] Khomenko E and Collados M 2009 *A&A* **506** L5

[83] Khomenko E, Kosovichev A, Collados M, Parchevsky K and Olshevsky V 2009 *ApJ* **694** 411

[84] Khomenko E, Collados M and Feliipe T 2009 Observational Signatures of Numerically Simulated MHD Waves in Small-scale Flux Sheets *Solar Polarization 5: In Honor of Jan Stenflo* (*Astronomical Society of the Pacific Conference Series* vol 405) ed Berdyugina S, Nagendra K N and Ramelli, R p 183

[85] Khomenko E and Cally P S 2011 *Journal of Physics Conference Series* **271** 012042

[86] Khomenko E and Cally P S 2012 *ApJ* **746** 68

[87] Kobanov N I and Pulyaev V A 2007 *Solar Phys.* **246** 273

[88] Kontogiannis I, Tsiropoula G and Tziotziou K 2010 *A&A* **510** A41

[89] Kontogiannis I, Tsiropoula G, Tziotziou K and Georgoulis M K 2010 *A&A* **524** A12

[90] Kontogiannis I, Tsiropoula G and Tziotziou K 2011 *A&A* **531** A66

[91] Krijger J M, Rutten R J, Lites B W, Straus T, Shine R A and Tarbell T D 2001 *A&A* **379** 1052

[92] Kuridze D, Zaqarashvili T V, Shergelashvili B M and Poedts S 2008 *Annales Geophysicae* **26** 2983

[93] Lagg A, Solanki S K, Riethmüller T L, et al. 2010 *ApJ* **723** L164

[94] Lighthill M J 1952 *Proc. R. Soc. London A.* **211** 564

[95] Lites B W, Chipman E G and White O R 1982 *ApJ* **253** 367

[96] Lites B W 1984 *ApJ* **277** 874

[97] Lites B W 1986 *ApJ* **301** 1005

[98] Lites B W 1988 *ApJ* **334** 1054

[99] Lites B W, Rutten R J and Kalkofen W 1993 *ApJ* **414** 345

[100] Lites B W, Thomas J H, Bogdan T J and Cally P S 1998 *ApJ* **497** 464

[101] López Ariste A, Socas-Navarro H and Molodij G 2001 *ApJ* **552** 871

[102] Lou Y Q 1995 *MNRAS* **274** L1

[103] Maltby P, Brynildsen N, Fredvik T, Kjeldseth-Moe O and Wilhelm K 1999 *Solar Phys.* **190** 437

[104] Maltby P, Brynildsen N, Kjeldseth-Moe O and Wilhelm K 2001 *A&A* **373** L1

[105] Martínez González M J and Bellot Rubio L R 2009 *ApJ* **700** 1391

[106] Martínez González M J, Asensio Ramos A, Manso Sainz R, et al. 2011 *ApJ* **730** L37

[107] Martínez González M J, Bellot Rubio L R, Solanki S K, et al. 2012 *ApJ* **758** L40

[108] McIntosh S W and Judge P G 2001 *ApJ* **561** 420

[109] McIntosh S W, Bogdan, T J, Cally, P S, et al. 2001 *ApJ* **548** L237

[110] McIntosh S W and Jefferies S M 2006 *ApJ* **647** L77

[111] De Moortel I, Ireland J, Hood A W and Walsh R W 2002 *A&A* **387** L13

[112] De Moortel I and Nakariakov V M 2012 *Royal Society of London Philosophical Transactions Series A* **370** 3193

[113] Moradi H and Cally P 2008 *Solar Phys.* **215** 309

[114] Moradi H, Hanasoge S M and Cally P S 2009 *ApJ* **690** L72

[115] Moradi H, Baldner C, Birch A and et al 2010 *Solar Phys.* **267** 1

[116] Musielak Z E, Rosner R, Stein R F and Ulmschneider P 1994 *ApJ* **423** 474

[117] Nagashima K, Sekii T, Kosovichev A G, et al. 2007 *PASP* **59** S631

[118] Newington M E and Cally P S 2010 *MNRAS* **402** 386

[119] Newington M E and Cally P S 2011 *MNRAS* **417** 1162

[120] Nordlund Å and Stein R F 2001 *ApJ* **546** 576

[121] Norton A A, Ulrich R K, Bush R I and Tarbell T D 1999 *ApJ* **518** L123

[122] Norton A A, Ulrich R K and Liu Y 2001 *ApJ* **561** 435

[123] Nutto C, Steiner O and Roth M 2010 *Astronomische Nachrichten* **331** 915

[124] Nutto C, Steiner O and Roth M 2012 *A&A* **542** L30

[125] Orozco Suárez D, Bellot Rubio L R, del Toro Iniesta J C and Tsuneta S 2008 *A&A* **481** L33

[126] Palacios J, Blanco Rodríguez J, Vargas Domínguez S,et al. 2012 *A&A* **537** A21

[127] Parchevsky K V and Kosovichev A G 2007 *ApJ* **666** L53

[128] Parchevsky K V and Kosovichev A G 2007b *ApJ* **666** 547

[129] Parchevsky K V and Kosovichev A G 2009 *ApJ* **694** 573

[130] Pascoe D J, Wright A N and De Moortel I 2010 *ApJ* **711** 990

[131] De Pontieu B, Erdelyi R and de Wijn A G 2003 *ApJ* **595** L63

[132] De Pontieu B, Erdelyi R J and Stewart P 2004 *Nat* 536

[133] De Pontieu B, Erdélyi R and De Moortel I 2005 *ApJ* **624** L61

[134] De Pontieu B, McIntosh S W, Carlsson M, et al. 2007 *Science* **318** 1574

[135] Rajaguru S P, Couvidat S, Sun X, Hayashi K and Schunker H 2012 *Solar Phys.* 303

[136] Reardon K P, Uitenbroek H and Cauzzi G 2009 *A&A* **500** 1239

[137] Riethmüller T L, Solanki S K, Martínez Pillet V, et al. 2010 *ApJ* **723** L169

[138] Roberts B and Webb A R 1978 *Solar Phys.* **56** 5

[139] Roberts B 1981 *Solar Phys.* **69** 27

[140] Roberts B 1983 *Solar Phys.* **87** 77

[141] Roberts B 2006 *Royal Society of London Transactions Series A* **364** 447

[142] Rosenthal C S, Bogdan T J, Carlsson M, et al. 2002 *ApJ* **564** 508

[143] Rosenthal C S and Julien K A 2000 *ApJ* **532** 1230

[144] Rouppe van der Voort L H M, Rutten R J, Sütterlin P, Sloover P J and Krijger J M 2003 *A&A* **403** 277

[145] Ruderman M S and Roberts B 2002 *ApJ* **577** 475

[146] Ruderman M S and Erdélyi R 2009 *Space Sci. Rev.* **149** 199

[147] Rüedi I, Solanki S K, Stenflo J, Tarbell T and Scherrer P H 1998 *A&A* **335** L97

[148] Rüedi I, Solanki S K, Bogdan T and Cally P 1999 *Solar Polarization* ed Nagendra K N and Stenflo J O (Dordrecht: Kluwer) p 337

[149] Rutten R J and Uitenbroek H 1991 *Solar Phys.* **134** 15

[150] Sánchez Almeida J, Márquez I, Bonet J A, Domínguez Cerdeña I and Muller R 2004 *ApJ* **609** L91

[151] Sánchez Almeida J and Martínez González M 2011 The Magnetic Fields of the Quiet Sun *Solar Polarization 6* (*Astronomical Society of the Pacific Conference Series* vol 437) ed Kuhn J R, Harrington D M, Lin H, Berdyugina S V, Trujillo-Bueno J, Keil S L and Rimmele T p 451

[152] Schunker H and Cally P S 2006 *MNRAS* **372** 551

[153] Schunker H and Braun D C 2011 *Solar Phys.* **268** 349

[154] Settele A, Carroll T A, Nickelt I and Norton A A 2002 *A&A* **386** 1123

[155] Solanki S K 1993 *Space Sci. Rev.* **63** 1

[156] Solanki S K, Inhester B and Schüssler M 2006 *Reports on Progress in Physics* **69** 563

[157] Spruit H C 1981 *A&A* **102** 129

[158] Stangalini M, Del Moro D, Berrilli F and Jefferies S M 2011 *A&A* **534** A65

[159] Stangalini M, Giannattasio F, Del Moro D and Berrilli F 2012 *A&A* **539** L4

[160] Stein R 1967 *Solar Phys.* **2** 385

[161] Stein R F and Nordlund Å 2001 *ApJ* **546** 585

[162] Thomas J H and Stanchfield D C H 2000 *ApJ* **537** 1086

[163] Tomczyk S, McIntosh S W, Keil S L, Judge P G, Schad T, Seeley D H and Edmondson J 2007 *Science* **317** 1192

[164] Tritschler A, Schmidt W, Uitenbroek H and Wedemeyer-Böhm S 2007 *A&A* **462** 303

[165] Toner C G and Labonte B J 1993 *ApJ* **415** 847

[166] Tziotziou K, Tsiropoula G, Mein N and Mein P 2006 *A&A* **456** 689

[167] Tziotziou K, Tsiropoula G, Mein N and Mein P 2007 *A&A* **463** 1153

[168] Van Doorsselaere T, Nakariakov V M and Verwichte E 2008 *ApJ* **676** L73

[169] Vecchio A, Cauzzi G, Reardon K P, Janssen K and Rimmele T 2007 *A&A* **461** L1

[170] Vecchio A, Cauzzi G and Reardon K P 2009 *A&A* **494** 269

[171] Vigeesh G, Hasan S S and Steiner O 2009 *A&A* **508** 951

[172] Vigeesh G, Steiner O and Hasan S S 2011 *Solar Phys.* **273** 15

[173] Vigeesh G, Fedun V, Hasan S S and Erdélyi R 2012 *ApJ* **755** 18

[174] de Wijn A G, McIntosh S W and De Pontieu B 2009 *ApJ* **702** L168

[175] Woods D T and Cram L E 1981 *Solar Phys.* **69** 233

[176] Yelles Chaouche L, Solanki S K and Schüssler M 2009 *A&A* **504** 595

[177] Zaqarashvili T V, Murawski K, Khodachenko M L and Lee D 2011 *A&A* **529** A85

[178] Zhugzhda Y D and Dzhalilov N S 1982 *A&A* **112** 16

Journal of Physics: Conference Series **440** (2013) 012049 doi:10.1088/1742-6596/440/1/012049

Atmosphere above a large solar pore

M Sobotka[1], M Švanda[1,2], J Jurčák[1], P Heinzel[1] and D Del Moro[3]

[1] Astronomical Institute, Academy of Sciences of the Czech Republic (v.v.i.), Fričova 298, CZ-25165 Ondřejov, Czech Republic

[2] Charles University in Prague, Faculty of Mathematics and Physics, Astronomical Institute, V Holešovičkách 2, CZ-18000 Prague 8, Czech Republic

[3] Department of Physics, University of Roma Tor Vergata, Via della Ricerca Scientifica 1, I-00133 Roma, Italy

E-mail: msobotka@asu.cas.cz

Abstract. A large solar pore with a granular light bridge was observed on October 15, 2008 with the IBIS spectrometer at the Dunn Solar Telescope and a 69-min long time series of spectral scans in the lines Ca II 854.2 nm and Fe I 617.3 nm was obtained. The intensity and Doppler signals in the Ca II line were separated. This line samples the middle chromosphere in the core and the middle photosphere in the wings. Although no indication of a penumbra is seen in the photosphere, an extended filamentary structure, both in intensity and Doppler signals, is observed in the Ca II line core. An analysis of morphological and dynamical properties of the structure shows a close similarity to a superpenumbra of a sunspot with developed penumbra. A special attention is paid to the light bridge, which is the brightest feature in the pore seen in the Ca II line centre and shows an enhanced power of chromospheric oscillations at 3–5 mHz. Although the acoustic power flux in the light bridge is five times higher than in the "quiet" chromosphere, it cannot explain the observed brightness.

1. Introduction

Pores are small sunspots without penumbra. The absence of a filamentary penumbra in the photosphere has been interpreted as the indication of a simple magnetic structure with mostly vertical field, e.g. [1, 2]. Magnetic field lines are observed to be nearly vertical in centres of pores and inclined by about 40° to 60° at their edges [3, 4]. Pores contain a large variety of fine bright features, such as umbral dots and light bridges that may be signs of a convective energy transportation mechanism.

Light bridges (hereafter LBs) are bright structures in sunspots and pores that separate umbral cores or are embedded in the umbra. Their structure depends on the inclination of local magnetic field and can be granular, filamentary, or a combination of both [5]. Many observations confirm that magnetic field in LBs is generally weaker and more inclined with respect to the local vertical. It was shown in [6] that the field strength increases and the inclination decreases with increasing height. This indicates the presence of a magnetic canopy above a deeply located field-free region that intrudes into the umbra and forms the LB. Above a LB, [7] found a persistent brightening in the TRACE 160 nm bandpass formed in the chromosphere. It was interpreted as a steady-state heating possibly due to constant small-scale reconnections in the inclined magnetic field.

In the chromosphere, large isolated sunspots are often surrounded by a pattern of dark, nearly radial fibrils. This pattern, called superpenumbra, is visually similar to the white-light penumbra but extends to a much larger distance from the sunspot. Dark superpenumbral fibrils are the locations of the strongest inverse Evershed flow – an inflow and downflow in the chromosphere toward the sunspot. Time-averaged Doppler measurements indicate the maximum speed of this flow equal to 2–3 km s^{-1} near the outer penumbral border [8].

2. Observations and data processing

A large solar pore NOAA 11005 was observed with the Interferometric Bidimensional Spectrometer IBIS [9] attached to the Dunn Solar Telescope (DST) on 15 October 2008 from 16:34 to 17:43 UT, using the DST adaptive optics system. The slowly decaying pore was located at 25.2 N and 10.0 W (heliocentric angle $\theta = 23°$) during our observation. According to [10], the maximum photospheric field strength was 2000 G, the inclination of magnetic field at the edge of the pore was 40° and the whole field was inclined by 10° to the west.

The IBIS dataset consists of 80 sequences, each containing a full Stokes (I, Q, U, V) 21-point spectral scan of the Fe I 617.33 nm line (see [10]) and a 21-point I-scan of the Ca II 854.2 nm line. The wavelength distance between the spectral points of the Ca II line is 6.0 pm and the time needed to scan the 0.126 nm wide central part of the line profile is 6.4 s. The exposure time for each image was set to 80 ms and each sequence took 52 seconds to complete, thus setting the time resolution. The pixel scale of these images was $0''.167$. Due to the spectropolarimetric setup of IBIS, the working field of view (FOV) was 228 × 428 pixels, i.e., $38'' \times 71''.5$. The detailed description of the observations and calibration procedures can be found in [10,11].

Complementary observations were obtained with the HINODE/SOT Spectropolarimeter [12,13]. The satellite observed the pore on 15 October 2008 at 13:20 UT, i.e., about 3 hours before the start of our observations. From one spatial scan in the full-Stokes profiles of the lines Fe I 630.15 and 630.25 nm we used a part covering the pore umbra with a granular LB.

According to [14], the inner wings (±60 pm) of the infrared Ca II 854.2 nm line sample the middle photosphere at the typical height $h \simeq 250$ km above the $\tau_{500} = 1$ level, while the centre of this line is formed in the middle chromosphere at $h \simeq$ 1200–1400 km. This provides a good tool to study the pore and its surroundings at different heights in the atmosphere and to look for relations between the photospheric and chromospheric structures.

The observations in the Ca II line are strongly influenced by oscillations and waves present in the chromosphere and upper photosphere. The observed intensity fluctuations in time are caused by real changes of intensity as well as by Doppler shifts of the line profile. To separate the two effects, Doppler shifts of the line profile were measured using the double-slit method [15], consisting in the minimisation of difference between intensities of light passing through two fixed slits in the opposite wings of the line. An algorithm based on this principle was applied to the time sequence of Ca II profiles. The distance of the two wavelength points ("slits") was 36 pm, so that the "slits" were located in the inner wings near the line core, where the intensity gradient of the profile is at maximum and the effective formation height in the atmosphere is approximately 1000 km. The wavelength sampling was increased by the factor of 40 using linear interpolation, thus obtaining the sensitivity of the Doppler velocity measurement 53 m s^{-1}. The reference zero of Doppler velocity was defined as a time- and space-average of all measurements. This way we obtained a series of 80 Doppler velocity maps. This method does not take into account the asymmetry of the line profile, e.g., the changes of Doppler velocity with height in the atmosphere. Using the information about the Doppler shifts, all Ca II profiles were shifted to a uniform position with a subpixel accuracy. This way we obtained an intensity data cube $(x, y, \lambda, \text{scan})$, where, if we neglect line asymmetries and inaccuracies of the method, the observed intensity fluctuations correspond to the real intensity changes. The oscillations and waves were separated from the slowly evolving intensity and Doppler structures by means of the 3D $k - \omega$ subsonic filter with the phase-velocity cutoff at 6 km s^{-1}.

Journal of Physics: Conference Series **440** (2013) 012049 doi:10.1088/1742-6596/440/1/012049

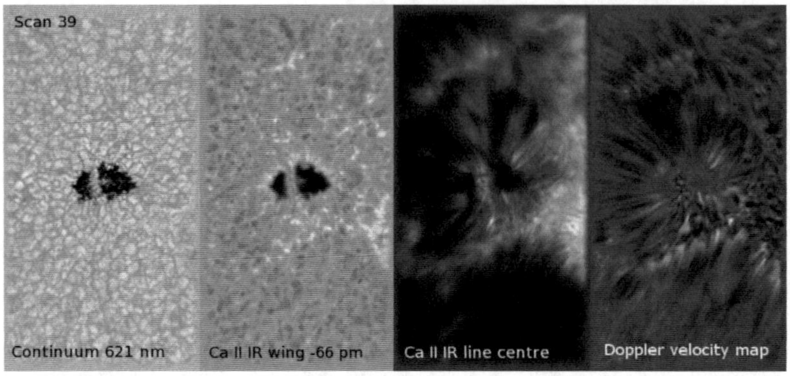

Figure 1. Pore NOAA 11005, full FOV 38″ × 71″.5. North is on the top, west on the left. From left to right: continuum 621 nm, Ca II 854.2 nm blue wing intensity, line centre intensity, Doppler map with velocity range from -2.3 km s^{-1} (black, toward the observer) to 4.7 km s^{-1} (white, away from the observer).

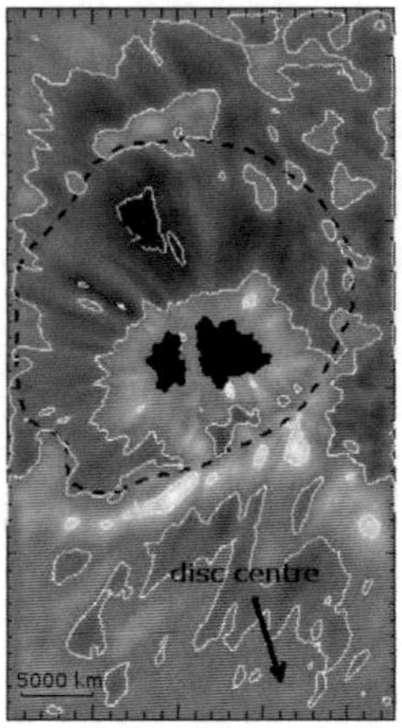

Figure 2. Time-averaged Doppler map with contours of zero, -1 and 1 km s^{-1}.

3. Results

Examples of intensity maps in the continuum 621 nm, blue wing and centre of the line Ca II 854.2 nm and a Doppler map are shown in Fig. 1. Oscillations and waves are filtered out from these images. A filamentary structure around the pore, composed of radially oriented bright and dark fibrils, is clearly seen in the line-centre intensity and Doppler maps. The area and shape of the filamentary structure are identical in all pairs of the line-centre intensity and Doppler images in the time series. The fibrils begin immediately at the umbral border and in many cases continue till the border of the filamentary structure. Their lengths are identical in the Doppler and intensity maps. However, the fibrils seen in the line centre are spatially uncorrelated with those in the Doppler maps.

Concentric running waves originating in the centre of the pore are observed in the unfiltered time series of the line-centre intensity and Doppler images. They propagate through the filamentary structure to the distance 3000–5000 km from the visible border of the pore with a typical speed of 10 km s^{-1}. These waves are very similar to running penumbral waves observed in the penumbral chromosphere and in superpenumbrae of developed sunspots [16].

The subsonic-filtered time series of Doppler maps was averaged in time to obtain a spatial distribution of mean line-of-sight (LOS) velocities around the pore. The result is shown in Fig. 2 together with contours of zero, -1 and 1 km s^{-1}. We can see from the figure that the inner part of the filamentary structure contains a positive LOS velocity (away from us, a downflow), while the outer part, located mostly on the limb side, shows a negative LOS velocity (toward us, an upflow). Taking into account the heliocentric angle $\theta = 23°$ and the fact that the plasma moves along magnetic field lines forming a funnel, the negative LOS velocity is only partly caused by real upflows but mostly by horizontal inflows into the pore. The picture of plasma moving toward the pore and flowing down in its vicinity is consistent with the inverse Evershed effect observed in the sunspots' superpenumbra.

All these facts are leading to the conclusion that the filamentary structure observed in the chromosphere above the pore is equivalent to a superpenumbra of a developed sunspot. The missing correlation between the fibrils seen in the line centre and those in the Doppler maps further supports this conclusion, because, according to [17], flow channels of the inverse Evershed effect are not identical with superpenumbral filaments.

A strong granular LB separates the two umbral cores of the pore. It is the brightest feature inside the pore in the photosphere as well as in the chromosphere, where it is by factor of 1.3 brighter than the average brightness in the FOV. The granular structure of the LB is preserved at all heights, from the photospheric continuum level to the formation height of the Ca II line centre. It is interesting that while a typical pattern of reverse granulation, observed in the Ca II wings, appears outside the pore in the middle photosphere ($h \simeq 250$ km), the LB is always composed of small bright granules separated by dark intergranular lanes (Fig. 1). A feature-tracking technique was applied to correlate the LB granules in position and time at different heights in the atmosphere. A correlation was found between the photospheric LB granules in the continuum and Ca II wings (correlation coefficient 0.46). On the other hand, there is no correlation between the chromospheric LB "granules" observed in the Ca II line centre and the photospheric ones in the wings and continuum.

The feature-tracking technique has shown that the mean size of the LB granules increases with height from $0''.45$ in the continuum to $0''.50$ in the Ca II wings and $0''.54$ in the Ca II line centre. Similarly, the average width of the LB increases with height from $2''$ (continuum) to $2''.5$ (Ca II wings) and $3''$ (Ca II line centre). On the other hand, it can be expected that the width of the LB magnetic structure decreases with height due to the presence of magnetic canopy. The complementary HINODE observations in two spectral lines Fe I 630.15 and 630.25 nm made it possible to obtain vertical stratification of temperature and magnetic field strength in the LB photosphere, using the inversion code SIR [18]. The results, summarised in Table 1, show that the width of LB in temperature maps really increases with height, while the width of the LB magnetic structure decreases and the magnetic field strength increases, confirming the magnetic canopy configuration.

Table 1. Widths of the light bridge measured in temperature and magnetic field at different geometrical heights in the photosphere. $B_{\rm min}$ is the minimum magnetic field strength in the light bridge.

Height (km)	Width in T	Width in B	$B_{\rm min}$ (G)
0	$2''.7$	$1''.7$	0
90	$2''.7$	$1''.5$	100
180	$2''.7$	$1''.3$	300
270	$2''.9$	$1''.1$	500
350	$3''.3$	$0''.8$	700

Power spectra of chromospheric oscillations were calculated using the unfiltered time series of the Ca II line-centre intensity and Doppler images. Power maps derived from the Doppler velocities at frequencies 3–6 mHz are shown in Fig. 3. At the LB position, we can see a strongly enhanced power around 3–5 mHz, comparable with that in a plage near the eastern border of the pore A similar power enhancement was reported in [19]. Usually, low-frequency oscillations do not propagate through the temperature minimum from the photosphere to the chromosphere due to the acoustic cut-off at 5.3 mHz [20]. To explain our observations, we assume that the low-frequency oscillations leak into the chromosphere along an inclined magnetic field [19,21], which is present in the LB and in the plage. The inclined magnetic field in the LB is verified by inversions of the full-Stokes Fe I 617.33 nm profiles [10]. The inclination angle, extrapolated to the height of the temperature minimum, is 40°–50° to the west thanks to inclined magnetic field lines at the periphery of the larger (eastern) umbral core. These field lines pass above the magnetic canopy of the LB.

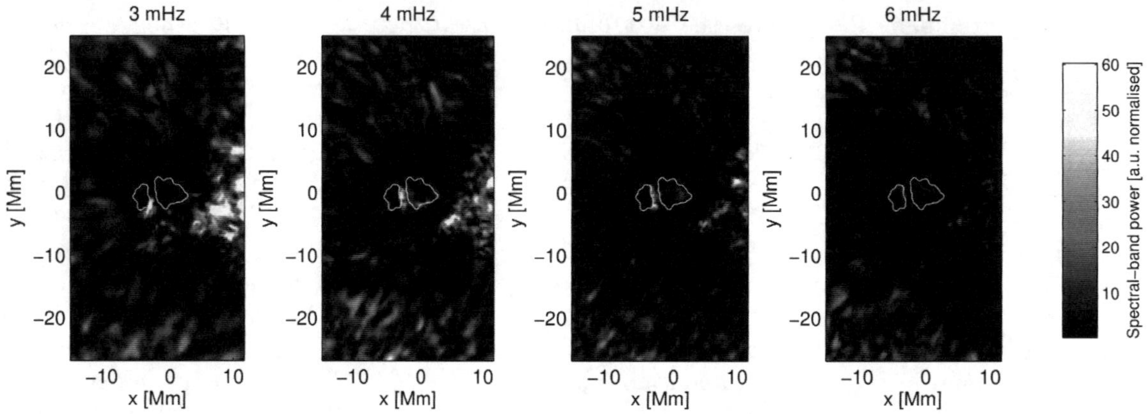

Figure 3. Power maps of Doppler velocities at frequencies 3–6 mHz. The contours outline the boundaries of the pore and light bridge, observed in the continuum.

Following [22], we estimate the acoustic energy flux in the LB chromosphere and compare it with the flux in the "quiet" region. The observed Doppler velocities are used for this purpose. With the estimated magnetic field inclination of 50°, the effective acoustic cut-off frequency decreases to the value of 3 mHz in the LB. The total calculated acoustic power flux is 550 W m^{-2} in the LB, while only 110 W m^{-2} in the "quiet" chromosphere. These results seem to be lower than 1840 W m^{-2} presented in [22] but one has to bear in mind that this value was obtained for the photosphere, where the power of the acoustic oscillations must be higher than in the chromosphere.

4. Discussion and conclusions

We studied the photosphere and chromosphere above a large solar pore with a granular LB using spectroscopic observations with spatial resolution of $0''.3$–$0''.4$ in the line Ca II 854.2 nm and photospheric Fe I lines. We have shown that in the chromospheric filamentary structure around the pore, observed in the Ca II line core and Doppler maps, the inverse Evershed effect and running waves are present. Chromospheric fibrils seen in the intensity and Doppler maps are spatially uncorrelated. From these characteristics and from the morphological similarity of the filamentary structure to superpenumbrae of developed sunspots we conclude that the observed pore has a kind of a superpenumbra, in spite of a missing penumbra in the photosphere.

A special attention was paid to the granular LB that separated the pore into two umbral cores. The magnetic canopy structure [6] is confirmed in this LB. In the middle photosphere ($h \simeq 250$ km, Ca II wings), the reverse granulation is seen around the pore but not in the LB. The reverse granulation is explained by adiabatic cooling of expanding gas in granules, which is only partially cancelled by radiative heating [23]. In the LB, hot (magneto)convective plumes at the bottom of the photosphere cannot expand adiabatically in higher photospheric layers due to the presence of magnetic field and the radiative heating dominates, forming small bright granules separated by dark lanes. The positive correlation between the LB structures in the continuum and Ca II line wings indicates that the middle-photosphere structures are heated by radiation from the low photosphere. Since the mean free photon path in the photosphere is larger than $1''$ for $h > 120$ km, the LB observed in the line wings is broader and its granules are larger than in the continuum due to the diffusion of radiation.

In the middle chromosphere ($h \simeq 1300$ km, Ca II centre), the LB is the brightest feature in the pore and it is brighter by factor of 1.3 than the average intensity in the FOV. Since the height in the atmosphere is well above the temperature minimum, the radiative heating cannot be expected. The heating by acoustic waves seems to be a candidate, because the acoustic power flux in the LB is five times higher than in the "quiet" chromosphere. To check this possibility, we have to compare the acoustic power flux with the total radiative cooling. An average profile of the Ca II line in the LB was used to derive a simple semi-empirical model based on the VAL3C chromosphere [24], with the temperature increased by 3000 K in the upper chromospheric layers ($h > 900$ km). The net radiative cooling rates were calculated using this model. The resulting height-integrated radiative cooling is approximately 6700 W m^{-2} in the LB chromosphere and 3000 W m^{-2} in the "quiet" chromosphere. The acoustic power fluxes (550 W m^{-2} and 110 W m^{-2}, respectively) are by an order of magnitude lower than the estimated total radiative cooling, so that the acoustic power flux does not seem to provide enough energy to reach the observed LB brightness.

Acknowledgments

This work was supported by the Czech Science Foundation under grants P209/12/0287 and P209/12/P568 and by the project RVO:67985815 of the Academy of Sciences of the Czech Republic. The Dunn Solar Telescope is located at the National Solar Observatory, operated by AURA for the National Sience Foundation. HINODE is a Japanese mission developed and launched by ISAS/JAXA, with NAOJ as domestic partner and NASA and STFC (UK) as international partners. It is operated by these agencies in cooperation with ESA and NSC (Norway).

References

[1] Simon G W and Weiss N O 1970 *Solar Phys.* **13** 85
[2] Rucklidge A M, Schmidt H U and Weiss N O 1995 *MNRAS* **273**, 491
[3] Keppens R and Martínez Pillet V 1996 *Astron. Astrophys.* **316** 229
[4] Sütterlin P 1998 *Astron. Astrophys.* **333** 305
[5] Sobotka M 1997 *1st Advances in Solar Physics Euroconference, Advances in the Physics of Sunspots*, ed B Schmieder, J C del Toro Iniesta and M Vázquez, ASP Conference series Vol 118 p 155
[6] Jurčák J, Martínez Pillet V and Sobotka M 2006 *Astron. Astrophys.* **453** 1079
[7] Berger T E and Berdyugina S V 2003 *Astrophys. J.* **589** L117
[8] Alissandrakis C E, Dialetis D, Mein P et al. 1988 *Astron. Astrophys.* **201** 339
[9] Cavallini F 2006 *Solar Phys.* **236** 415
[10] Sobotka M, Del Moro D, Jurčák J and Berrilli F 2012 *Astron. Astrophys.* **537** A85
[11] Viticchié B, Del Moro D, Criscuoli S and Berrilli F 2010 *Astrophys. J.* **723** 787
[12] Kosugi T, Matsuzaki K, Sakao T et al. 2007 *Solar Phys.* **243** 3
[13] Tsuneta S, Ichimoto K, Katsukawa Y et al. 2008 *Solar Phys.* **249** 167
[14] Cauzzi G, Reardon K P, Uitenbroek H et al. 2008 *Astron. Astrophys.* **480** 515
[15] Garcia A, Klvaňa M and Sobotka M 2010 *Cent. Eur. Astrophys. Bull.* **34** 47
[16] Christopoulou E B, Georgakilas A A and Koutchmy S 2000 *Astron. Astrophys.* **354** 305
[17] Tsiropoula G, Alissandrakis C E, Dialetis D and Mein P 1996 *Solar Phys.* **167** 79
[18] Ruiz Cobo B and del Toro Iniesta J C 1992 *Astrophys. J.* **398** 375
[19] Stangalini M, Del Moro D, Berrilli F amd Jefferies S M 2011 *Astron. Astrophys.* **534** A65
[20] Fossat E, Regulo C, Roca Cortes T et al. 1992 *Astron. Astrophys.* **266** 532
[21] Jefferies S M, McIntosh S W, Armstrong J D et al. 2006 *Astrophys. J.* **648**, L151
[22] Bello González N, Flores Soriano M, Kneer F and Okunev O 2009 *Astron. Astrophys.* **508** 941
[23] Cheung M C M, Schüssler M and Moreno Insertis F 2007 *Astron. Astrophys.* **461** 1163
[24] Vernazza J E, Avrett E H and Loeser R 1981 *Astrophys. J. Suppl. Ser.* **45** 635

Subsurface flows associated with non-Joy oriented active regions: a case study

Irene González Hernández[1], Rudolf Komm[1], Lidia van Driel-Gesztelyi[2,3,4], Deborah Baker[2], Louise Harra[2], Rachel Howe[5]

[1]National Solar Observatory, Tucson, Arizona, USA
[2]Mullard Space Science Laboratory, University College London, UK
[3]Observatoire de Paris, LESIA, FR
[4]Konkoly Observatory, Budapest, Hungary
[5]School of Physics and Astronomy, University of Birmingham, UK

E-mail: `iregh@nso.edu`

Abstract.
Non-Joy oriented active regions (ARs) are a challenge for solar magnetic field modelers. Although significant deviations from Joy's law are relatively rare for simple bipolar ARs, understanding the causes of their particularity could be critical for the big picture of the solar dynamo. We explore the possibility of the sub-surface local dynamics being responsible for the significant rotation of these ARs. We apply the ring-diagram technique, a local helioseismology method, to infer the flows under and surrounding a non-Joy oriented AR and present the results of a case study in this paper.

1. Introduction

The solar magnetic cycle is generally explained using magneto-hydrodynamic dynamo theory. Hence, understanding the mechanisms of the solar dynamo, in particular the formation an emergence of active regions, is key to understanding the big picture of the Sun's behavior in order to ultimately forecast its activity. One of the characteristics of emerging active regions is the tilt of their orientation (Joy's law, [1]), caused by the Coriolis force, with the leading polarity appearing closer to the equator. Although statistically active regions follow this pattern [2], there are cases of clear deviation, where the leading polarity stays closer to the pole. Our purpose is to research the possibility of the local dynamics surrounding the non-Joy oriented active regions being responsible for their unusual tilt. The ultimate goal is to get enough statistics by obtaining the flows under the solar surface from local helioseismology methods of a large set of non-Joy oriented ARs. In this paper, we present the results of applying ring-diagram analysis to a single anti-Joy law active region (NOAA 11073), but expect to extend the work to a significant sample of them, as well as to a similarly large set of regularly-oriented active regions, using the same technique in the near future.

NOAA 11073 is a well isolated active region that emerged in the Northern hemisphere around May 28th 2010. Its leading negative polarity is farther away from the equator than the trailing negative one. This situation persists for the four days (May 30-June 2) that we follow the active region as it crosses the solar central meridian (see Fig 1). Its location and isolation make it a perfect case for analysis.

2. Analysis method

The ring-diagram technique [3], one of the local helioseismology techniques, uses medium/high-degree waves propagating in localized areas of the Sun to obtain an averaged horizontal velocity vector for that particular region. The input data for our analysis are high-resolution Dopplergrams from the Helioseismic and Magnetic Imager (HMI, [4]) on board the Solar Dynamics Observatory. In the typical analysis, a particular area (patch) of the Sun is tracked and remapped into a data cube, and the analysis of the corresponding 3-D power spectrum of solar oscillations renders an average horizontal velocity vector for the area at different depths. By analyzing a mosaic of these patches, it is possible to infer 3-D velocity map in the depth range where the waves propagate.

We applied the ring-diagram analysis technique to 1664 minute series of HMI data, using a resolution of about 1.5 Mm (0.75 Mm/pixel) at the center of the disk. From each reprojected data cube, a 3-D power spectrum was obtained. The power spectrum is then fitted to a profile that includes the perturbation in the spatial frequencies due to the horizontal velocity fields [5]. Finally, the mode-dependent velocities from the fitting go through an inversion process to get the depth dependence of the flows.

To study the flows around active region NOAA 11073 we divide the surrounding area into 25 patches. Each area of $8^\circ \times 8^\circ$ is apodized into a 7.5° circular area before calculating the 3-D power spectrum. A horizontal velocity vector at several depths is obtained from the analysis of each single patch. The combination of the results from the 25 areas give a low resolution map of flows surrounding the active region from the surface to approximately 6-7 Mm depth for each analyzed day (see Fig 1).

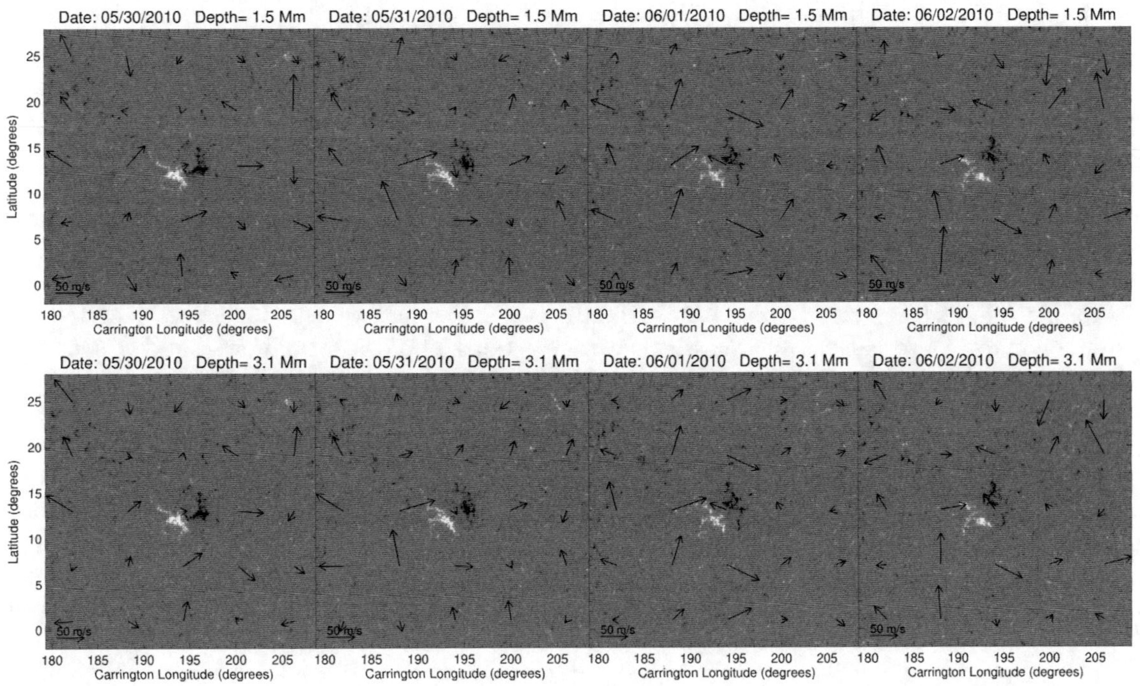

Figure 1. Subsurface flows surrounding AR NOAA 11073 at 1.5 Mm (top) and 3.1 Mm under the solar surface for four consecutive days as it crosses the solar central meridian.

We applied ring-diagram analysis as described to the 25 patches surrounding active region NOAA 11073 and obtained the horizontal subsurface flows for four consecutive days. In order

Journal of Physics: Conference Series **440** (2013) 012050 doi:10.1088/1742-6596/440/1/012050

to remove large scale flows, such as meridional circulation and rotation residuals, as well as contributions to the flows from projection effects, we subtract the average of three quiet solar areas located at exactly the same position as each of the individual analyzed patches on May 15-17. Figure 1 shows the results for the four days for two particular depths, 1.5 Mm and 3.1 Mm from the solar surface.

3. Results

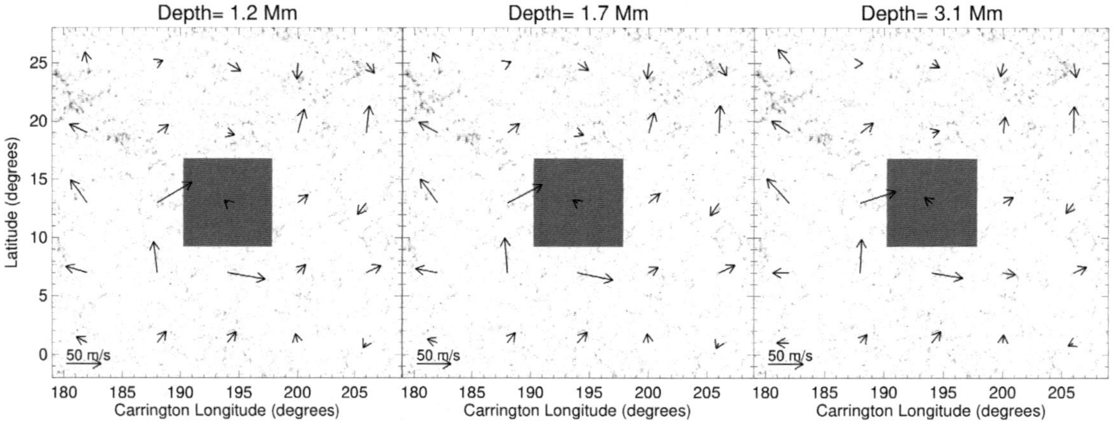

Figure 2. Temporal average of the subsurface flows surrounding active region NOAA 11073. The magnetic field background is the absolute value of the average for the four consecutive days. The central patch has been set to zero to show how quiet the contiguous analyzed regions are.

Temporal averages of the flows surrounding active region NOAA 11073 are presented in Fig. 2 for three different depths. Since we expect the subsurface macrodynamics to be relatively stable, we averaged the results from the four consecutive days of the analysis. The flows are superimposed on the absolute value of magnetic field averaged over the four continuous days from HMI magnetograms. The magnetic field from the central patch has been removed to emphasize how quiet the surrounding studied areas are.

From the horizontal velocity field an estimate of the vertical velocity can be calculated by assuming mass conservation [6]. Vertical vorticity (curl of horizontal flows) and associated helicity gives a sense of the curl of the material motions. The vertical vorticity is presented in Figure 3 for the 8° patches surrounding the active region. Figure 4 shows the vorticity of the synoptic map including the area of active region NOAA 11073 (approximate 13° latitude) from flows calculated using a lower resolution ring-diagram analysis of the full solar surface from NISP/Global Oscillations Network Group data. The vorticity is counter-clockwise (cyclonic) in both cases where the active region is located, which is what one might expect for an anti-Joy region, that is, flows that will drive the leading polarity away from the equator. However, previous work using large areas show a slight positive vorticity seems to be prevalent close to active regions in the Northern hemisphere [7]; whether AR NOAA 11073 have an excessive amount of positive vorticity compared to other regions of its size and strength needs to be determined by using this more targeted method to study a large data set of active regions.

4. Summary

With this single case analysis we have showed the potential of the ring-diagram analysis applied to HMI Dopplergrams for investigating the local dynamics around and below non-Joy orientated active regions. The effects of the magnetic field on helioseismology inferences are still a subject

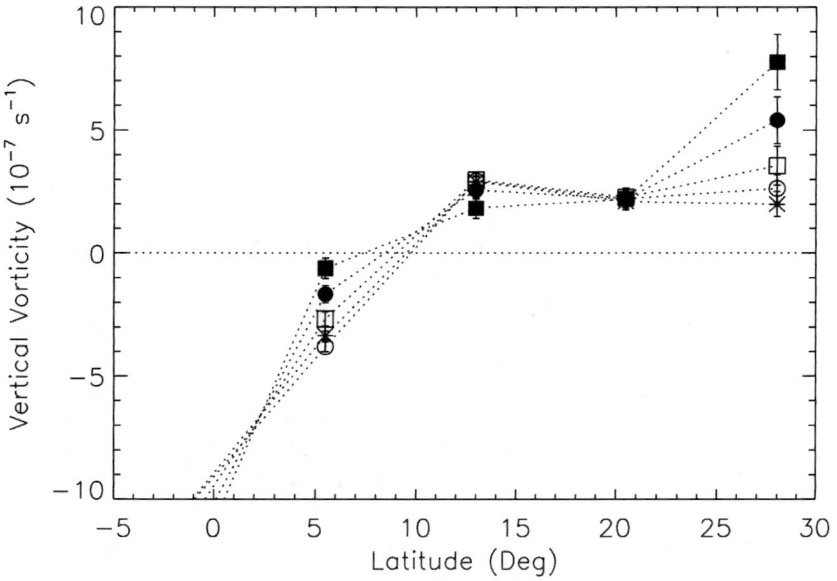

Figure 3. Calculated vertical vorticity for the areas surrounding AR NOAA 11073 using the 8° patches. There is positive (cyclonic) component close to the 13° latitude where AR NOAA 11073 is located.

of research. The power spectrum of solar oscillations is expected to be affected by the presence of magnetic field, making the interpretation of the fitted velocity parameters difficult. To avoid this, we have focus our analysis on the macrodynamics observed in the quieter areas surrounding the active region.

This particular case analysis shows counter-clockwise flows in the subsurface layers surrounding the active region, which may be responsible for driving the leading polarity away from the solar equator. However, statistics from the analysis of many such active regions are required to conclude whether the subsurface flows are responsible for the unusual orientation of these active regions. Customization of the method, such as a thorough determination of the trade-off parameter of the flows inversion (depth localization vs. error) need to be reviewed when more cases are added to the study. The subsurface flows associated with our case analysis (NOAA 11073) show the characteristic curl that would be expected to explain the leading polarity to be driven away from the equator. However we need to demonstrate that this is different for normally orientated active regions by analyzing with the same method a significant number of Joy oriented AR's as a control set.

Acknowledgments
We acknowledge the Leverhulme trust for the funding of *Probing the Sun: inside and out* project upon which this research is based. Usage of the SDO/HMI courtesy of NASA/SDO and the HMI science team. This work utilizes data obtained by the Global Oscillation Network Group (GONG) part of the NISP Program. NISP is managed by the National Solar Observatory, which is operated by AURA, Inc. under a cooperative agreement with the National Science Foundation. This work has been supported by the NASA Living with a Star *Targeted Research and Technology* program under grant NNH11AQ24I. LvDG acknowledges support by the Hungarian Research Grant OTKA K-081421.

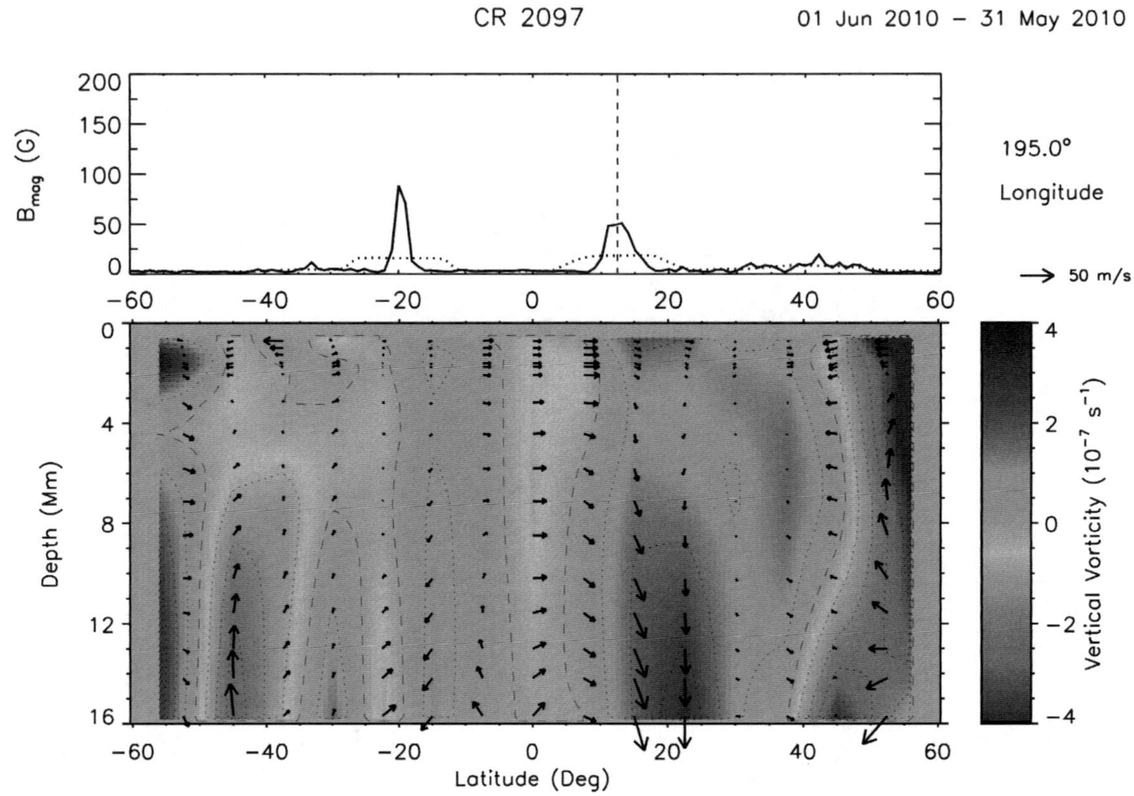

Figure 4. Calculated vertical vorticity for the areas surrounding AR NOAA 11073 using a low resolution ring-diagram analysis. A large area of the Sun at different depths is presented here, but there is clear positive (cyclonic) component close to the 13° latitude where AR NOAA 11073 is located.

References

[1] Hale G E, Ellerman F, Nicholson S B and Joy A H 1919 *Astrophys. J.* **49** 153
[2] Li J and Ulrich R K 2012 *Astrophys. J.* **758** 115
[3] Hill F 1988 *Astrophys. J.* **333** 996
[4] Scherrer P H, Schou J, Bush R I, , Kosovichev A G, Bogart R S, Hoeksema J T, Liu Y, Duvall T L, Zhao J, Title A M, Schrijver C J, Tarbell T D and Tomczyk S 2012 *Solar Phys.* **275** 207
[5] Haber D A, Hindman B W, J T, Bogart R S, Larsen R M and Hill F 2002 *Astrophys. J.* **570** 885
[6] Komm R, Corbard T, Durney B R, González Hernández I, Hill F, Howe R and Toner C 2004 *Astrophys. J.* **605** 554
[7] Komm R, Howe R, Hill F, Miesch M, Haber D and Hindman B 2007 *Astrophys. J.* **667** 571

AUTHOR INDEX

AUTHOR INDEX